DISCOVERY OF THE
HIGGS BOSON

DISCOVERY OF THE
HIGGS BOSON

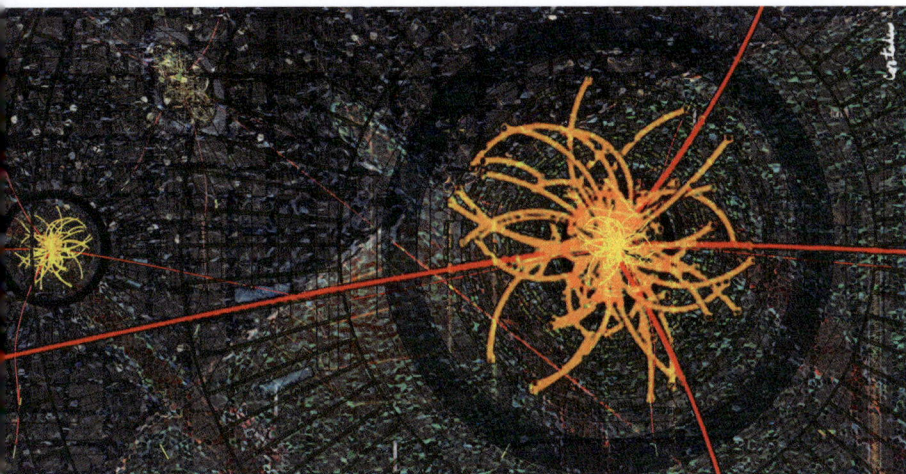

Editors

Aleandro Nisati
Istituto Nazionale di Fisica Nucleare — Sezione di Roma, Italy

Vivek Sharma
University of California, San Diego, USA

World Scientific

NEW JERSEY · LONDON · SINGAPORE · BEIJING · SHANGHAI · HONG KONG · TAIPEI · CHENNAI · TOKYO

Published by

World Scientific Publishing Co. Pte. Ltd.

5 Toh Tuck Link, Singapore 596224

USA office: 27 Warren Street, Suite 401-402, Hackensack, NJ 07601

UK office: 57 Shelton Street, Covent Garden, London WC2H 9HE

Library of Congress Cataloging-in-Publication Data

Names: Nisati, Aleandro, editor. | Sharma, Vivek, 1962–   editor.

Title: Discovery of the Higgs boson / edited by Aleandro Nisati (Istituto Nazionale di Fisica
    Nucleare - Sezione di Roma, Italy), Vivek Sharma (UC San Diego).

Description: Singapore ; Hackensack, NJ : World Scientific Publishing Co. Pte. Ltd., [2016] | 2016 |
    Includes bibliographical references and index.

Identifiers: LCCN 2016000699| ISBN 9789814425445 (hardcover ; alk. paper) |
    ISBN 9814425443 (hardcover ; alk. paper) | ISBN 9789814425872 (softcover ; alk. paper) |
    ISBN 9814425877 (softcover ; alk. paper)

Subjects: LCSH: Higgs bosons. | Particles (Nuclear physics)

Classification: LCC QC793.5.B62 D57 2016 | DDC 539.7/21--dc23

LC record available at http://lccn.loc.gov/2016000699

British Library Cataloguing-in-Publication Data

A catalogue record for this book is available from the British Library.

Cover image credit: Xavier Cortada (with the participation of physicist Pete Markowitz), "In search of the Higgs boson: $H \to ZZ$," digital art, 2013. (www.cortada.com)

For photocopying of material in this volume, please pay a copying fee through the Copyright Clearance Center, Inc., 222 Rosewood Drive, Danvers, MA 01923, USA. In this case permission to photocopy is not required from the publisher.

Desk Editor: Ng Kah Fee

Typeset by Stallion Press

Email: enquiries@stallionpress.com

Printed in Singapore

To our families, for their loving and enthusiastic support during the hunt for the Higgs boson.

Contents

Preface

The discovery in 2012 by the ATLAS and CMS experiments at CERN of a new boson with a mass near 125 GeV was an extraordinary milestone in the four-decade long experimental search for the mechanism that breaks the electroweak symmetry and generates the masses of all known fundamental particles.

The science that we describe in this book began about fifty years ago when three groups of theoreticians, Englert & Brout, Higgs, and Guralnik, Hagen & Kibble, conjectured a mechanism for the origin of vector boson masses consisting of a complex scalar quantum field with the quantum numbers of the vacuum, which permeates the entirety of our universe. The Higgs boson, a spin-zero excitation of this field, is the experimental footprint of this mechanism that hides the electroweak gauge symmetry. While the interactions of the Higgs boson with the known gauge bosons, quarks and leptons are completely predicted, its mass is a free parameter of the model. Therefore, the mass must be determined experimentally following a rigorous proof of its very existence.

The worldwide experimental search for the Higgs boson began in earnest soon after the consolidation of the electroweak theory into what is now known as the Standard Model (SM) of particle physics. While several model-dependent limits were placed on the existence of Higgs bosons with a mass below 3.6 GeV, it was the model-independent searches by the four experiments, ALEPH, DELPHI, OPAL and L3, at the Large Electron Positron (LEP) collider that ruled out the presence of a SM Higgs boson with a mass below 114.4 GeV at 95% confidence level (CL). Following the shutdown of the LEP collider in 2000, the direct searches for the Higgs boson shifted to the CDF and D0 experiments at the Tevatron $p\bar{p}$ collider at Fermilab. By the summer of 2010, the combined results from up to 6.7 fb^{-1} of data recorded by CDF and D0 had excluded, at 95% CL, a SM Higgs boson with a mass between 158 GeV and 175 GeV. Their search for a low mass Higgs boson decaying into $b\bar{b}$ continued with more data leading to a broad 3 standard deviations excess observed by the summer of 2012.

The Large Hadron Collider (LHC), the world's highest energy particle collider, was built primarily to collide protons at $\sqrt{s} = 14$ TeV, but on September 10, 2008, an electrical fault between two of the 1232 dipole magnets led to an explosive release of helium coolant from the magnets, causing mechanical damage to about 50 dipoles. Following an extraordinary effort to repair the magnets, on March 30, 2010, the first pp collisions by the LHC were obtained at a center-of-mass energy of 7 TeV, reduced below the design energy as a precautionary measure. While many had thought that, at this energy, the search for the Higgs boson would be a marathon spanning several years of data taking, due to the excellent performance of the LHC machine and the ATLAS and CMS detectors and their analysis tools, the hunt for the Higgs boson was reduced to a sprint! With just $5\,\text{fb}^{-1}$ of 7 TeV data collected in 2011, the LHC experiments ruled out the presence of the Standard Model Higgs boson in a large range of high masses. Interestingly, in the same data, both experiments saw a tantalizing excess near a mass of 125 GeV that, with another $5\text{-}6\,\text{fb}^{-1}$ data collected at 8 TeV in early 2012, grew to a significance of about 5 standard deviations in each experiment. This led to the historic announcement on the 4th of July 2012 at a jam-packed meeting at CERN, which was followed by an online audience exceeding a hundred thousand. When the full 8 TeV dataset of about 20 fb^{-1} were analyzed, a portrait of this new resonance consistent with the characteristics of the Standard Model Higgs boson came into view.

This book describes in scientific detail this epic search for the Standard Model Higgs boson. It is written by physicists who played leading roles in the search and discovery of the new boson. Most chapters are written by teams of two physicists from rival experiments. This book is targeted towards 3rd year graduate students (who have already taken courses in quantum field theory as well as experimental techniques in particle physics), young post-doctoral researchers as well as practicing particle physicists not directly involved in Higgs searches but curious about the scientific details of how we know what we know of this new particle discovered at the LHC. This book can also be used as a supplementary text in a graduate level course on electroweak physics.

The emphasis of this book is on pedagogy and on the big picture in the Higgs boson hunt. By describing not just *what* was done in the analyses but also explaining *why* it was done in that particular way, this book complements the dozens of ATLAS and CMS publications which form the basis of this book.

This book is organized as follows: in Chapter 1 we start with an introduction to the phenomenology of the Standard Model Higgs boson. This is followed in Chapter 2 by a description of the Higgs boson search at the LEP collider. Chapter 3 describes the pioneering searches carried out at the Tevatron. The narrative then shifts to the LHC starting with Chapter 4 which describes the general features of the LHC and the capabilities of the two major Higgs-hunting experiments there: ATLAS and CMS. The observation of a narrow resonance near 125 GeV (H_{125}) in the $H \to ZZ \to 4\ell$ channel and the measurements of its properties are described in Chapter 5. This is followed in Chapter 6 by the description of the properties of the H_{125} resonance in the $H \to \gamma\gamma$ channel. These two "golden" channels provide a precise measure of the mass of the new particle and reveal its bosonic nature. Studies in the $H \to WW^*$ final state described in Chapter 7 complete the profile of the coupling of H_{125} to dibosons. The discussion then turns to the search for the decays of H_{125} to fermions. Chapter 8 describes the evidence for the $H \to \tau\tau$ decay. Chapter 9 describes the search for H_{125} decay into $b\bar{b}$ final state, probably the most complicated channel of investigation at the LHC. Before the LHC experiments converged on the discovery of the H_{125} resonance, a variety of searches were conducted for a high mass Standard Model Higgs boson decaying to vector bosons. They are described in Chapters 10 and 11. Chapter 12 combines the results from all channels investigated by ATLAS and CMS and presents the experimentally measured properties of the H_{125} boson. Chapter 13 presents a short summary of the experimental measurements and concludes that the H_{125} resonance discovered at the LHC fits, within measurement uncertainties, the profile of the Standard Model Higgs boson. Lastly, the Higgs boson searches and its discovery benefited greatly from several sophisticated statistical and multivariate analysis techniques. The technical aspects of these procedures are reviewed in Appendices A and B, respectively.

We hope that the gentle reader will find the book pedagogically insightful and scientifically inspiring. We wish you happy reading!

A. Nisati and V. Sharma

Chapter 1

The Higgs boson in the Standard Model

Abdelhak Djouadi* and Massimiliano Grazzini†

Laboratoire Physique Théorique
U. Paris Sud, 91405V Orsay, France
†Physik-Institut, Universität Zürich
CH-8057 Zürich, Switzerland

The major goal of the Large Hadron Collider is to probe the electroweak symmetry breaking mechanism and the generation of the elementary particle masses. In the Standard Model this mechanism leads to the existence of a scalar Higgs boson with unique properties. We review the physics of the Standard Model Higgs boson, discuss its main search channels at hadron colliders and the corresponding theoretical predictions. We also summarize the strategies to study its basic properties.

1. Introduction

Establishing the precise mechanism of the spontaneous breaking of the electroweak gauge symmetry has been the central focus in high energy physics for many decades and is certainly the primary goal of the LHC. In the Standard Model (SM),[1-6] the theory that describes the electromagnetic, weak and strong interactions, explicit mass terms for the electroweak gauge bosons and the fermions are not allowed by the $SU(2)_L \times U(1)_Y$ gauge symmetry. Electroweak symmetry breaking (EWSB) is achieved via the Brout–Englert–Higgs mechanism,[7-9] in which a scalar field, a doublet under weak isospin, acquires a non-zero vacuum expectation value, thus providing masses for the electroweak gauge bosons and for the fermions. One of the four degrees of freedom of the original scalar field corresponds to a physical scalar particle: the *Higgs boson*.

Clearly, the discovery of this last missing piece of the SM has profound importance. In fact, despite of its phenomenal success in explaining the precision data, the SM could not be considered to be completely established until the Higgs boson was observed experimentally. The discovery in July 2012 of a scalar resonance compatible with the SM Higgs boson by the ATLAS and CMS collaborations[10,11] is, in this respect, historic as it crowns the SM as the correct theory of fundamental particles and interactions among them, at least up to the Fermi energy scale.

In addition, the fundamental properties of the Higgs particle such as its mass, spin and other quantum numbers, as well as its couplings to various matter and gauge particles and its self-couplings must be determined in the most precise way. These studies are important in order to achieve further clarity into the dynamics of the EWSB mechanism. The many important questions which one would like answered by probing the Higgs boson properties are: does the dynamics involve new strong interactions and is the Higgs a composite field? If elementary Higgs particles indeed exist in nature, how many fields are there and in which gauge representations do they appear? Does the EWSB sector involve sizable CP violation? Theoretical realizations span a wide range of scenarios and a complete discussion of Higgs physics thus touches upon almost all the issues under active investigation in theoretical and experimental particle physics. Nevertheless, in this chapter, only the phenomenology of the Higgs sector in the SM will be discussed (for a detailed review, see e.g. Ref. [12]). After summarizing the EWSB mechanism in the SM in Sec. 2 and the pre-LHC theoretical and experimental constraints on the Higgs boson mass in Sec. 3, its decay modes, production cross sections and detection channels at hadron colliders will be described in Secs. 4, 5 and 6, respectively. The theoretical predictions for differential distributions and the Monte Carlo generators used in the experimental analyses are briefly reviewed in Secs. 7 and 8. The strategies to measure the fundamental properties of the Higgs boson are finally summarized in Sec. 9.

2. Electroweak symmetry breaking and mass generation

The Standard Model is based on a very powerful principle: gauge symmetry. The fields corresponding to the particles, as well as the particle interactions, are invariant with respect to local transformations of an internal symmetry group. The model is a generalization of quantum electrodynamics (QED), the relativistic quantum theory of electromagnetism which describes the interaction of electrically charged particles through the exchange of photons. The QED Lagrangian is invariant under phase transformations on the charged fermionic fields collectively denoted by ψ,

$$\psi(x_\mu) \rightarrow e^{iQ\theta(x_\mu)}\psi(x_\mu), \qquad (1.1)$$

where $x_\mu = (t, \vec{x})$ is the spacetime four-vector and Q is the fermion electric charge. These transformations are called gauge or local transformations as the parameter θ depends on x_μ. The photon field mediating the interaction and described by the four-vector $A_\mu = (A_0, \vec{A})$, transforms as

$$A_\mu(x_\mu) \rightarrow A_\mu(x_\mu) + \partial_\mu\theta(x_\mu), \qquad (1.2)$$

where ∂_μ is the derivative with respect to x_μ. The gauge transformation group is noted $U(1)_Q$ for the group of unitary matrices of dimension one and conserves the quantum number that is the electric charge Q.

In fact, the interaction of charged fermions via the exchange of photons can be induced in a minimal way in the Lagrangian density of the free fermion and photon systems, by substituting the usual derivative ∂_μ by the so called *covariant derivative* $D_\mu \equiv \partial_\mu - iQA_\mu$.

In the SM, the symmetry group has a rich structure and is denoted by

$$SU(3)_C \times SU(2)_L \times U(1)_Y. \qquad (1.3)$$

For the strong interaction, based on the symmetry group $SU(3)_C$, the quarks appear in three different states differentiated by a quantum number called color that they exchange via eight intermediate massless gluons.

The electromagnetic and weak interactions are combined to form the electroweak interaction based on the symmetry group $SU(2)_L \times U(1)_Y$. The fermions of each family appear in two quantum configurations: the left-handed fermions f_L are assembled in doublets of weak isospin, while the right-handed fermions f_R are in iso-singlets. For instance, the left-handed electron and its associated neutrino always appear in a doublet $\binom{\nu_{e_L}}{e_L}$ of weak isospin $I_3 = +\frac{1}{2}$ and $I_3 = -\frac{1}{2}$ for ν_{e_L} and e_L respectively, while the right-handed electron e_R appears as a singlet. The same holds for quarks: the left-handed quarks form a doublet $\binom{u_L}{d_L}$ and the right-handed ones u_R, d_R are singlets. For a given fermion, the quantum number of hypercharge Y_f is given by the electric charge and the isospin, $Y_f = 2Q_f - 2I_3^f$. In this context, the neutrinos have only a left-handed chirality.

The electroweak interaction is mediated by the exchange of four gauge bosons: $W_{1,2,3}^\mu$ corresponding to the three generators of $SU(2)_L$ which can be identified with the three 2×2 Pauli matrices $\tau_{1,2,3}$ and B_μ corresponding to the generator $\frac{Y}{2}$ of $U(1)_Y$. The physical gauge bosons, $W^\pm$, Z and the photon γ, are linear combinations of the W_i^μ and B^μ:

$$W_\mu^\pm = \frac{1}{\sqrt{2}}(W_\mu^1 \mp iW_\mu^2), \quad Z_\mu = \frac{g_2 W_\mu^3 - g_1 B_\mu}{\sqrt{g_2^2 + g_1^2}}, \quad A_\mu = \frac{g_1 W_\mu^3 + g_2 B_\mu}{\sqrt{g_2^2 + g_1^2}},$$

(1.4)

with g_2 and g_1 the coupling constants of the $SU(2)_L$ and $U(1)_Y$ groups.

While the photon has zero mass, the $W^\pm$ and Z gauge bosons should be massive since they mediate the short-range weak force. However, the direct introduction of masses for the weak gauge bosons violates $SU(2)_L \times U(1)_Y$ gauge invariance as, in principle, gauge bosons should remain massless to preserve a local symmetry. For instance, a mass term for the photon and thus a term that is bilinear in the fields, $m_A^2 A_\mu A^\mu$, violates the invariance under the group $U(1)_Q$ of electromagnetism since:

$$m_A^2 A_\mu A^\mu \to m_A^2 \left(A_\mu - \frac{1}{Q}\partial_\mu\theta \right) \left(A^\mu - \frac{1}{Q}\partial^\mu\theta \right) \neq m_A^2 A_\mu A^\mu. \quad (1.5)$$

In addition, the fact that the fermions f_L and f_R do not have the same isospin quantum numbers, prevents them from acquiring a mass in an $SU(2)_L \times U(1)_Y$ gauge invariant way: in the case of the first family of leptons one cannot form a mass term for the electron, $m_e \bar{e}e = m_e(\bar{e}_L e_R + \bar{e}_R e_L)$, as this term violates the gauge symmetry.

In the SM, it is the Brout–Englert–Higgs mechanism,[7–9] commonly called the Higgs mechanism, which allows us to generate particle masses while preserving the gauge symmetry of electroweak interactions. The mechanism postulates the existence of a doublet (under isospin) of complex scalar fields,

$$\Phi = \begin{pmatrix} \mathrm{Re}\Phi^+ + i\mathrm{Im}\Phi^+ \\ \mathrm{Re}\Phi^0 + i\mathrm{Im}\Phi^0 \end{pmatrix}, \quad (1.6)$$

to which one associates a potential that is invariant under $SU(2)_L \times U(1)_Y$

$$V(\Phi) = \mu^2\Phi^\dagger\Phi + \lambda(\Phi^\dagger\Phi)^2, \quad (1.7)$$

where μ^2 is the mass term of the field Φ and λ for the (positive) coupling constant of its self-interaction. For $\mu^2 > 0$, the potential $V(\Phi)$ has the usual form of an inverted bell in which the minimum of the field Φ, corresponding to the state of vacuum which should be stable, has zero value. In this case we simply have four additional scalar fields corresponding

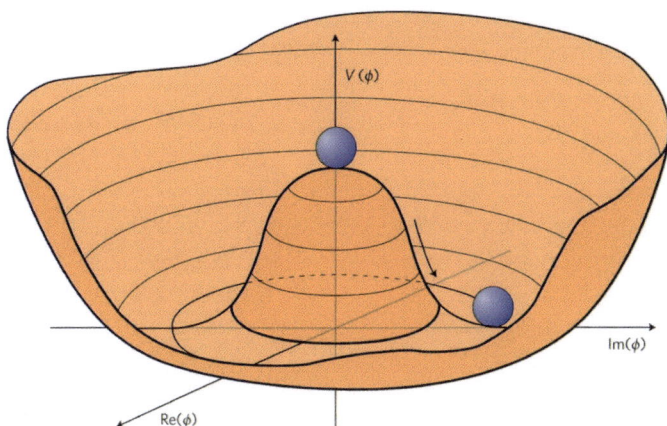

Fig. 1.1 The potential of the scalar field Φ with its minimum at the field value v.

to four new degrees of freedom or scalar particles, which does not help much towards the solution of the mass generation problem. The situation becomes much more interesting when $\mu^2 < 0$. In this case, the potential $V(\Phi)$ has the shape of a Mexican hat as shown in Fig. 1.1. The minimum of the potential is not reached for a zero value of the field Φ (or, rather, for its neutral component Φ^0) as usual, but at the non-zero value $v = \sqrt{-\mu^2/\lambda}$ that is called the non-zero vacuum expectation value (vev) of the field Φ.

When interpreting the field content of the theory starting from this non-symmetric but physical vacuum, one realizes that three among the four degrees of freedom of the field Φ have disappeared from the spectrum: they have been *absorbed* by three gauge bosons of the electroweak interaction. These spin-1 fields, initially massless and with only two transverse components, acquire an additional degree of freedom corresponding to a longitudinal component, a characteristic signature of massive spin-1 fields.

The Lagrangian density of the field Φ is given by:

$$\mathcal{L}_\Phi = (D^\mu \Phi)^\dagger (D_\mu \Phi) - V(\Phi), \quad V(\Phi) = \mu^2 \Phi^\dagger \Phi + \lambda (\Phi^\dagger \Phi)^2, \quad (1.8)$$

where the covariant derivative D_μ induces the interactions of Φ with the W_μ^a, B_μ gauge fields of $SU(2)_L \times U(1)_Y$:

$$D_\mu = \partial_\mu - ig_2 \frac{\tau_a}{2} W_\mu^a - ig_1 \frac{Y}{2} B_\mu. \quad (1.9)$$

To obtain the physical fields, Φ is first rewritten in terms of four real fields $\theta_{1,2,3}(x)$ and $H(x)$

$$\Phi(x) = \begin{pmatrix} \theta_2 + i\theta_1 \\ \frac{1}{\sqrt{2}}(v+H) - i\theta_3 \end{pmatrix} \simeq e^{i\theta_a(x)\tau^a(x)/v}\frac{1}{\sqrt{2}}\begin{pmatrix} 0 \\ v+H(x) \end{pmatrix}. \quad (1.10)$$

The freedom to perform a gauge transformation on Φ can be used to eliminate the three real fields $\theta_{1,2,3}$ called Goldstone bosons[13–15]

$$\Phi(x) \rightarrow e^{-i\theta_a(x)\tau^a(x)/v}\,\Phi(x) = \frac{1}{\sqrt{2}}\begin{pmatrix} 0 \\ v+H(x) \end{pmatrix}. \quad (1.11)$$

The kinetic term $|D_\mu\Phi|^2$ of the Lagrangian gives

$$\frac{1}{2}(\partial_\mu H)^2 + \frac{g_2^2}{8}(v+H)^2|W_\mu^1 + iW_\mu^2|^2 + \frac{1}{8}(v+H)^2|g_2W_\mu^3 - g_1B_\mu|^2. \quad (1.12)$$

After defining the new physical fields $W_\mu^\pm$, Z_μ and A_μ as in Eq. (1.4), we identify the terms that are bilinear in these fields with the particle masses

$$m_W = \frac{1}{2}vg_2, \quad m_Z = \frac{1}{2}v\sqrt{g_2^2 + g_1^2}, \quad m_A = 0, \quad (1.13)$$

with v given simply in terms of the Fermi constant of weak interaction,

$$v = 1/(\sqrt{2}G_F)^{1/2} = 246\,\text{GeV} \quad (1.14)$$

and g_2 and g_1 being measured, one recovers the correct masses of the $W^\pm$ and Z bosons. The photon, instead, remains massless, $m_A = 0$, as it should to maintain QED gauge invariance. The counting of degrees of freedom of the field Φ is respected since the three Goldstone bosons correspond to the longitudinal polarizations of the now massive $W^\pm$ and Z bosons.

The next major issue is the generation of the SM fermion masses. To do so, using the same scalar field Φ, one introduces a Lagrangian density that describes the fermion–Higgs interactions that is $SU(2)_L \times U(1)_Y$ gauge invariant. For the electron and the neutrino for instance, we would write[2]

$$\mathcal{L}_f = f_e(\bar{e},\bar{\nu})_L\Phi e_R \rightarrow f_e(\bar{\nu}_e,\bar{e}_L)\frac{1}{\sqrt{2}}\begin{pmatrix} 0 \\ v+H \end{pmatrix} e_R = \frac{f_e}{\sqrt{2}}(v+H)\bar{e}_L e_R \quad (1.15)$$

and identify the e and ν_e masses with the terms bilinear in the fields

$$m_e = f_e v/\sqrt{2} \quad \text{and} \quad m_\nu = 0. \quad (1.16)$$

In a similar way, mass terms for the down quarks can be obtained. To generate a mass term for up quarks it is enough to introduce the complex conjugate field $\tilde{\Phi} = i\tau_2\Phi$. Hence, with the same field Φ both gauge boson and fermion masses are generated, and $SU(2)_L \times U(1)_Y$ is still a symmetry of the Lagrangian but not a symmetry of the vacuum: it is said to be spontaneously broken to $U(1)_Q$.

Finally, among the four initial degrees of freedom of the field Φ, after three have been absorbed by the $W^\pm$ and Z gauge bosons to acquire their masses, one degree of freedom will be left over. This residual degree of freedom corresponds to a physical particle, the spin-zero Higgs boson.

The Higgs boson mass can be simply deduced from the scalar Higgs potential by isolating the terms that are bilinear in the H fields, $\frac{1}{2}m_H^2 H^\dagger H$:

$$m_H = \sqrt{2\lambda v^2}. \tag{1.17}$$

The Higgs boson has remarkable characteristics which gives it a unique status among elementary particles. First, in contrast to matter and gauge particles which have respectively spin 1/2 and spin 1, it has spin 0.

Another unique property of the Higgs boson is that it couples to particles proportionally to their masses. Indeed, in the Lagrangian density the bilinear terms involving the fields H and two fermionic or two gauge bosonic fields, are described by the same terms providing the masses since the field H always appears in the combination $H + v$. The interaction of the Higgs boson with the particles thus increases with their masses:

$$\mathcal{L}_{Hff} \propto m_f/v, \ \mathcal{L}_{HW^+W^-} \propto m_W^2/v, \ \mathcal{L}_{HZZ} \propto m_Z^2/v. \tag{1.18}$$

As a consequence, the Higgs particle couples more strongly to the $W^\pm$ and Z bosons, the masses of which are of the order of hundred GeV. It couples also more strongly to the top quark, the heaviest particle in the SM and, to a lesser extent, to the bottom quark and the τ lepton, than to the fermions of the first and second generations which have much smaller masses. It does not couple to the neutrinos which are considered as massless.

The Higgs boson does not couple directly to photons and gluons as they have no mass. However, couplings can be induced in an indirect way through quantum fluctuations. Higgs–photon–photon and Higgs–gluon–gluon couplings are then generated through triangular loops involving for instance the top quark.

Finally, the Higgs boson also has self-interactions, residual of those of the original field Φ appearing in the potential of Eq. (1.7); the magnitude of these triple and quartic self-interactions is also proportional to m_H^2:

$$\mathcal{L}_{HHH} \propto m_H^2/v, \quad \mathcal{L}_{HHHH} \propto m_H^2/v^2. \tag{1.19}$$

Hence, the only free parameter of the theory is the Higgs boson mass. Once this parameter is fixed, the Higgs profile is uniquely determined.

3. Pre-LHC constraints on the Higgs boson

In the SM, the mass of the Higgs boson is in principle completely undetermined. There are, however, both experimental and theoretical constraints on this fundamental parameter, which are summarized below.

One solid direct information on the Higgs boson mass was available from Higgs searches at the LEP2 collider with center-of-mass energies up to $\sqrt{s} = 209$ GeV. At LEP2, the dominant production process was Higgsstrahlung where the e^+e^- pair goes into an off-shell Z boson which then splits into a Higgs particle and a real Z boson, $e^+e^- \to Z^* \to HZ$. The cross section for the WW fusion process $e^+e^- \to H\bar{\nu}_e\nu_e$ is very small at these energies but is not completely negligible and is included to set the final Higgs mass limit. The searches by the LEP collaborations have been made in several topologies with the Higgs boson decaying into $b\bar{b}$ and $\tau^+\tau^-$ final states and the $Z \to e^+e^-, \mu^+\mu^-, \tau^+\tau^-$ and $Z \to b\bar{b}$ signatures of the associated Z state. Combining the results of the four LEP collaborations, no significant excess above the expected SM background has been seen, and the exclusion limit[16] $m_H > 114.4$ GeV has been established at the 95% confidence level (CL) with an expected exclusion at a mass $m_H > 115$ GeV; see Chapter 2.

More recently, the analysis of the full data set of the CDF and D0 collaborations at the Tevatron[17] allowed to exclude the SM Higgs boson in the mass range around $m_H \approx 160$ GeV and to observe a 3σ excess of events in the mass range between $m_H = 115-140$ GeV; see Chapter 3.

Furthermore, the high accuracy of the electroweak (EW) data[18] measured at LEP, SLC and the Tevatron provides an indirect constraint on m_H: the Higgs boson contributes logarithmically, $\propto \log(m_H/m_W)$, to the radiative corrections to the W/Z boson propagators and, thus, enters all electroweak observables at the quantum level. A recent analysis of the electroweak data,[19] which used $m_t = 173.18 \pm 0.94$ GeV for the top-quark mass,[20] yields a best-fit value $m_H = 94^{+25}_{-22}$ GeV.

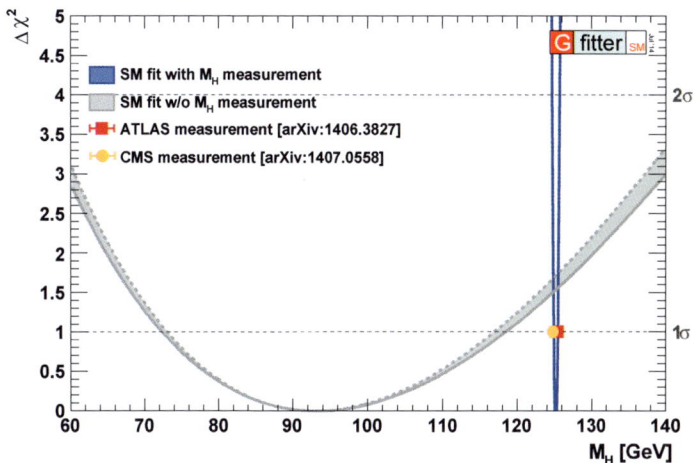

Fig. 1.2 $\Delta\chi^2$ as a function of the Higgs boson mass from the global fit to the precision electroweak data.[19] The solid and dashed lines give the results when including and ignoring theoretical errors, respectively. The blue curve represents the fit taking into account the ATLAS and CMS measurements of the Higgs boson mass.

Figure 1.2 shows the corresponding $\Delta\chi^2$ curve. The best-fit value for m_H is consistent to 1.3σ with the ATLAS and CMS mass measurement, $m_H \approx 125\,\text{GeV}$ to be discussed in various parts of this book.

From theoretical considerations, interesting constraints can be derived from assumptions on the energy range within which the SM is valid before perturbation theory breaks down and new phenomena would emerge. For instance, if the Higgs boson mass were larger than $\sim 1\,\text{TeV}$, the W and Z bosons would have to interact very strongly with each other so that their scattering at high energies respects unitarity. Imposing the unitarity requirement leads to the bound[21] $m_H \lesssim 700\,\text{GeV}$. If the Higgs boson were too heavy, unitarity would be violated in these processes at energies above $\sim 1\,\text{TeV}$ and new phenomena, observable at the LHC, should appear to restore it.

Another important theoretical constraint comes from the fact that the quartic Higgs self-coupling, which is proportional to the Higgs mass squared, $m_H^2 = 2\lambda v^2$, grows logarithmically with the energy scale. If m_H is small, the energy cut-off Λ at which the coupling grows beyond any bound and new phenomena should occur, is large; if m_H is large, the cut-off Λ is small. The condition $m_H \lesssim \Lambda$ sets an upper limit on the Higgs boson mass in the SM, the so-called triviality bound. A naive one-loop analysis assuming the validity of perturbation theory[22] as well as lattice simulations[23] leads to the estimate $m_H \lesssim 630\,\mathrm{GeV}$ for this limit. Requiring the SM to be extended to the Planck scale $\Lambda_P \sim 10^{18}\,\mathrm{GeV}$, the Higgs boson mass should be[24] $m_H \lesssim 180\,\mathrm{GeV}$. In fact, in any model beyond the SM in which the theory is required to be weakly interacting up to the grand unification or Planck scales, the Higgs boson should be lighter than $m_H \lesssim 200\,\mathrm{GeV}$.

Furthermore, top-quark quantum corrections tend to drive the quartic Higgs coupling λ to negative values which render the electroweak vacuum unstable. Recent analyses, including the state-of-the-art quantum corrections, give for the condition of absolute stability of the electroweak vacuum, $\lambda(M_P) \geq 0$, when the SM is extrapolated up to the Planck scale, $m_H \gtrsim 129\,\mathrm{GeV}$.[25,26] The bound critically depends on the top-quark mass and the strong coupling as well as on higher order contributions. The theoretical uncertainty on the maximal value of m_H is estimated to be between $1\,\mathrm{GeV}$ and $6\,\mathrm{GeV}$.[25,27] Recently, it has been observed[28] that new physics at the Plank scale could affect the stability bound.

It is interesting to observe that, as shown in Fig. 1.3, the Higgs boson mass measured by the ATLAS and CMS collaborations lies right at the limit between the stability and metastability regions of the scalar potential.

4. Higgs boson decay modes

In this section the Higgs boson decay modes and the corresponding branching ratios are discussed. The leading order (LO) Feynman diagrams

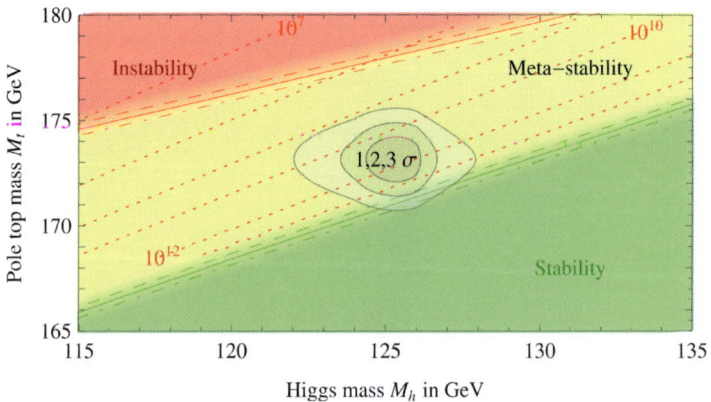

Fig. 1.3   Regions of absolute stability, metastability and instability of the SM vacuum (from Ref. [25]). The grey areas denote the allowed region at 1, 2 and 3σ.

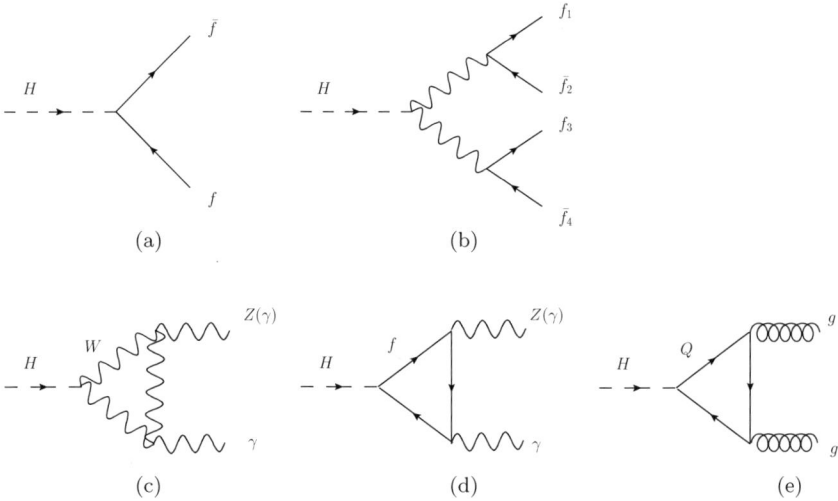

Fig. 1.4   Feynman diagrams illustrating the relevant Higgs boson decay modes.

for the decay channels are displayed in Fig. 1.4. Once its mass is fixed the profile of the Higgs particle is uniquely determined and its production rates and decay widths are fixed. As its couplings to different particles are proportional to their masses, the Higgs boson will tend to decay into the heaviest particles allowed by phase space.

As a consequence, in the *low-mass range*, $m_H \lesssim 130\,\text{GeV}$, the Higgs boson width is dominated by the $H \to b\bar{b}$ decay mode (see Fig. 1.5), with a branching ratio ranging between 50 and 80%. Other significant decay modes at low mass are $c\bar{c}$ and $\tau^+\tau^-$ with $\sim5\%$ and 10% branching ratios, respectively.

Higher order corrections play an important role. Next-to-leading order (NLO) QCD corrections to the $H \to q\bar{q}$ decay rate have been computed long ago[29–33] and, once the large logarithmic term of the form $\alpha_S \ln m_H/m_q$ is absorbed into the redefinition of the quark mass, the NLO corrections increase the LO rate by about 20%. Next-to-next-to-leading order (NNLO) corrections amount to a few percent.[34] The $\Gamma(H \to b\bar{b})$ decay width is now known to N^3LO, i.e. $\mathcal{O}(\alpha_S^4)$,[35] and fully differential calculations up to NNLO are available.[36,37] EW corrections to the $H \to f\bar{f}$ decay[38–41] are also known and their impact is at the few percent level.

Another set of important channels are the loop induced decays into $\gamma\gamma$, $Z\gamma$ and gg. Since gluons and photons are massless particles, they do not couple to the Higgs boson directly. Nevertheless, the corresponding

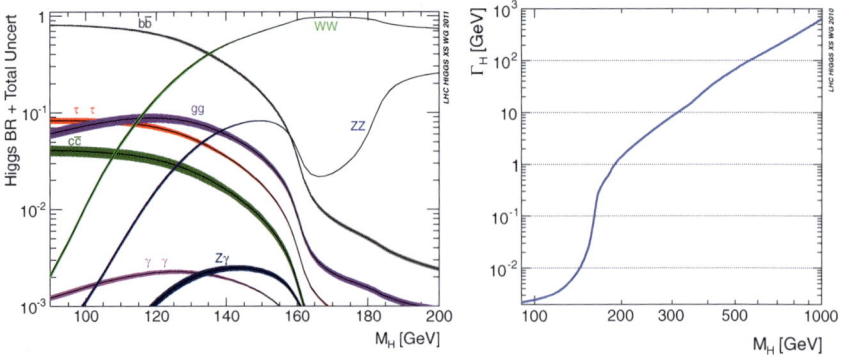

Fig. 1.5   The decays of the SM Higgs boson: the branching ratios (including uncertainties) (left) and the total decay width (right) as a function of its mass.

vertices can be generated at the quantum level through loops involving heavy particles. In particular, in the SM the $HZ\gamma$ and $H\gamma\gamma$ couplings are mediated by W and fermion loops (see Fig. 1.4 c,d), which interfere destructively. The Hgg coupling is mediated by a heavy-quark loop (see Fig. 1.4 d), where the dominant contribution is given by the top quark. The bottom-quark loop contributes mainly through its (negative) interference with the dominant top-quark contribution. These Higgs boson decays are extremely important, since, through the loops, they can probe scales well above the Higgs boson mass, e.g., particles coupling to the Higgs boson but too heavy to be produced directly. The $H \to \gamma\gamma$ and $H \to Z\gamma$ decays are both very rare, with branching ratios at the level of $\mathcal{O}(10^{-3})$, but they provide a very clear signature. The $H \to \gamma\gamma$ decay indeed offers the most important discovery channel for the SM Higgs boson in the region $110 \,\mathrm{GeV} \lesssim m_H \lesssim 140 \,\mathrm{GeV}$.

The $H \to \gamma\gamma$ decay rate has been computed to NLO in QCD, first in the large-m_t limit,[42–44] and then keeping the full top-mass dependence.[45–47] The effect of QCD corrections for $m_H \sim 125 \,\mathrm{GeV}$ is to increase the decay rate by about 2%. The two-loop EW corrections have also been computed[48–51] and their effect is opposite to the one of QCD corrections: for $m_H \sim 125$ GeV they decrease the $H \to \gamma\gamma$ decay rate by about 2%. Recently, three-loop QCD corrections to the $H \to \gamma\gamma$ rate have also been evaluated:[52] the combined impact of known QCD and EW radiative effects for $m_H = 125 \,\mathrm{GeV}$ turns out to be well below 1%.

The $H \to Z\gamma$ channel pays the additional suppression of the $Z \to l^+l^-$ decay and will only be important at high luminosity. NLO QCD corrections

Table 1.1   SM Higgs boson decay branching ratios (in %) for $m_H = 125$ GeV.

m_H (GeV)	$H \to b\bar{b}$	$H \to \tau\tau$	$H \to \gamma\gamma$	$H \to WW$	$H \to ZZ$
125	$57.7^{+3.2\%}_{-3.3\%}$	$6.32^{+5.7\%}_{-5.7\%}$	$0.22^{+5.0\%}_{-4.9\%}$	$21.5^{+4.3\%}_{-4.2\%}$	$2.64^{+4.3\%}_{-4.2\%}$

to the $Z\gamma$ decay rate have been computed first numerically[53] and then analytically,[54,55] and their effect is below 1%.

In the *high-mass range*, $m_H \gtrsim 140$ GeV, the Higgs boson decays predominantly into WW and ZZ pairs (see Fig. 1.4 b), and one of the two gauge bosons is necessarily virtual below the thresholds. In this region, a complete treatment of the $H \to WW/ZZ \to 4$ fermions decay is needed. Complete NLO QCD and EW corrections are now known, including off-shell effects.[56,57] Above the ZZ threshold, the branching ratios for the WW and ZZ decays are about 2/3 and 1/3, respectively and the opening of the $t\bar{t}$ channel for higher m_H does not alter this pattern significantly.

In the low-mass range, the Higgs boson is very narrow, with $\Gamma_H < 10$ MeV, but this width increases, reaching 1 GeV at the ZZ threshold. For very large masses, the Higgs boson becomes obese, since $\Gamma_H \sim m_H$.

The branching ratios and total decay widths are summarized in Fig. 1.5 which is taken from Ref. [58]. The corresponding numerical values for $m_H = 125$ GeV (from Ref. [59]) are reported in Table 1.1. These results are obtained with the numerical programs HDECAY[60,61] and Prophecy4f,[56,57] that include all the available QCD and EW corrections.

For a light Higgs boson, there is a significant contribution to the uncertainty coming from the parametric uncertainties on the bottom-quark mass and α_S.[58,62] Such uncertainty translates directly to an uncertainty of about 5% on the WW, ZZ and $\gamma\gamma$ branching ratios. In the case of the $b\bar{b}$ branching ratio, a cancellation reduces the uncertainty to about 3%.

It is interesting to note that the value $m_H \sim 125$ GeV, corresponding to the ATLAS and CMS measurement of the Higgs boson mass, is in a sense special as it allows at present and will allow in the future the study of many different and complementary decay modes.

5. Higgs boson production at hadron colliders

There are four production channels for a SM-like Higgs boson at hadron colliders: the relevant Feynman diagrams are shown in Fig. 1.6.

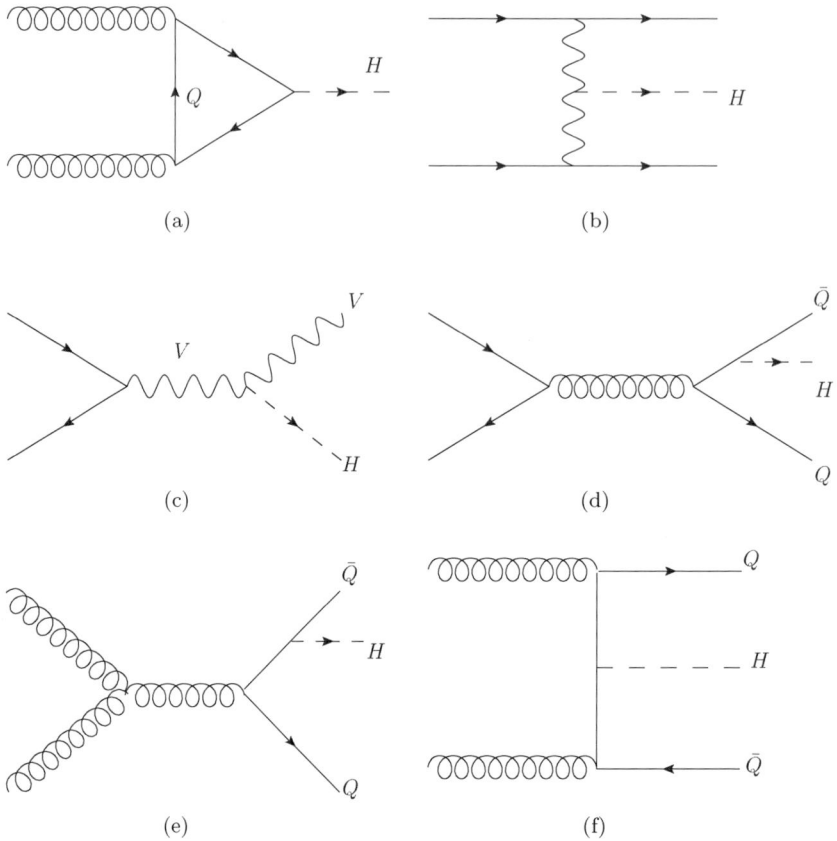

Fig. 1.6   Feynman diagrams for the main Higgs boson production channels at hadron colliders: gluon fusion (a); vector-boson fusion (b); associated production with a vector boson (c); associated production with a top-quark pair (d,e,f).

    The gluon–gluon fusion process[63] $gg \to H$ is by far the dominant production channel for SM-like Higgs particles at hadron colliders. The process, which proceeds through triangular heavy-quark loops (see Fig. 1.6 a), has been proposed in the late 1970s in Ref. [63] where the ggH vertex and the production cross section have been derived. In the SM, it is dominantly mediated by a top-quark loop, while the bottom-quark contribution does not exceed the 10% level at leading order. This process is known to be subject to extremely large QCD radiative corrections.

    The NLO corrections in QCD have been calculated in the 1990s. They are known both in the large-m_t limit[64,65] and by keeping the exact dependence on the masses of the top and bottom quarks.[66,67] It has been shown in Ref. [67] that working in an effective field theory approach[68,69] in which the top-quark mass is assumed to be infinite is a very good approximation for $m_H \lesssim 2m_t$, provided that the leading order cross section contains the full m_t and m_b dependence. The impact of NLO corrections is very large, of the order of 100% at the LHC, thus casting doubts on the reliability of the perturbative expansion.

The NNLO corrections have been computed in the large-m_t limit[70–72] and further increase the cross section at the LHC by about 25%. Since the completion of the NNLO calculation, the theoretical prediction has been improved in many ways. The accuracy of the large-m_t approximation at NNLO has been studied by computing subleading terms in the large-m_t limit.[73–76] Such works have shown that the approximation works remarkably well, to better than 1% for $m_H < 300$ GeV. The logarithmically enhanced contributions due to multiple soft emissions have been resummed up to next-to-next-to-leading logarithmic accuracy (NNLL) and the result has been consistently matched to the fixed order NNLO result.[77] Soft-gluon resummation leads to an increase of the cross section by about 9% at the LHC ($\sqrt{s} = 7$ TeV) and to a slight reduction of scale uncertainties. This result[77] has been used as reference theoretical prediction for many years. The quantitative impact of soft-gluon resummation is consistent with the evaluation of soft terms[78] at N^{3}LO. Very recently, the complete computation of the N^{3}LO corrections has been presented.[79] The N^{3}LO corrections are moderate, and they lead to a nice stabilization of the scale dependence.

Considerable work has been done also for the computation of EW corrections. Two-loop EW effects are known[47,80–83] and their effect strongly depends on the Higgs boson mass, ranging from +5% for $m_H = 120$ GeV to −2% for $m_H = 300$ GeV.[83] Mixed QCD-EW effects have been studied in Ref. [84] and EW corrections from real radiation have been studied in Refs. [85,86]. Both effects are at the 1% level or smaller. The QCD corrections to $gg \to H$ at the Tevatron are larger than at the LHC: at NLO they increase the cross section by more than 100%, while the NNLO effect is about 30%. Soft-gluon effects[77] further increase the cross section by about 13%.

Updates of the $gg \to H$ cross sections, including radiative corrections at various levels of theoretical accuracy (but not the recently computed N^{3}LO corrections) have been presented in Refs. [62,87–90].

The vector-boson fusion process[91] $qq \to Hqq$, in which the Higgs boson is produced in association with two hard jets and is denoted VBF, is an essential process for the Higgs boson searches and studies at the LHC. However, Higgs plus two-jet production receives two contributions at hadron colliders. The first one is from the genuine VBF process, in which the Higgs boson is radiated off a vector boson $V = W^\pm, Z$ that couples two quark lines (see Fig. 1.6 b). The hard jets have a strong tendency to be emitted in the forward and backward directions. The second contribution in Higgs plus two-jet production is through gluon fusion, which interferes with VBF starting from $\mathcal{O}(\alpha_S^2)$. As a consequence, the genuine VBF process is in principle not unambiguously defined (in fact, VBF interferes already at LO with the associated production with a vector boson, $pp \to HV \to Hjj$, but the effect is at the per mille level[92]). Nevertheless, it is possible to impose specific kinematical cuts which select mainly the VBF configuration. The interference contributions with gluon fusion[93-95] have been computed and found below the percent level.

The NLO QCD corrections to the total VBF rate were computed some time ago in the so-called structure function approach.[96] More recently, both QCD and EW NLO corrections have been computed.[92,97,98] QCD corrections turn out to be at the level of about 5–10%. The impact of EW corrections significantly depends on m_H and for a not-too-heavy Higgs boson is negative and tends to compensate the positive effect of QCD corrections. A refinement of the VBF cross section is the computation of the gluon induced terms[99] which however turns out to be well below 1%.

A good approximation of the NNLO QCD corrections to the total inclusive cross section (where some contributions have been neglected as they are expected to be both parametrically and kinematically suppressed) has been presented in Refs. [100,101]. The impact of these corrections is extremely small but they further reduce the scale uncertainty down to about $\pm 2\%$.

In summary, the VBF channel is under good theoretical control, since the theoretical predictions already have a precision comparable to the accuracy to which the process itself can be defined in perturbation theory. Thanks to this situation, and to the very clean signature, VBF offers the opportunity to study difficult decay channels like $H \to \tau\tau$.

<u>Higgs boson production in association with a vector boson[102] $q\bar{q} \to HV$</u> with $V = W^\pm, Z$ (see Fig. 1.6 c) is the third most important channel at the LHC, as far as the inclusive cross section is concerned and, as discussed below, is the most important in the low-mass region at the Tevatron, as the leptonic decay of the vector boson provides the necessary background rejection.

Up to NLO in QCD perturbation theory the process can be seen as Drell–Yan production of a vector boson that eventually radiates a Higgs boson. As such, the QCD corrections up to NLO are identical to those of Drell–Yan.[103] EW corrections are known and they typically decrease the cross section by about 5–10%.[104] In the case of WH production NNLO QCD corrections are still essentially given by those of Drell–Yan[105] and they increase the cross section by about 1−3% at the LHC, and by about 10% at the Tevatron.[106] In the case of ZH production, since the final state is electrically neutral, there are additional gluon-initiated diagrams that have to be evaluated at NNLO.[106] Their inclusion is particularly relevant at the LHC, where the effect ranges from 2% to 6%. (The NLO QCD corrections to the $gg \to ZH$ subprocess, which are formally N^3LO in α_S, have been recently estimated[107] within the large-m_{top} approximation.) Additional NNLO diagrams where the Higgs boson is produced through a heavy-quark loop have to be considered.[108] At the Tevatron, their effect is below (at) the % level in WH (ZH) production in the relevant Higgs mass range; at the LHC, the contribution of these terms is typically of the order of 1–3%.

Associated Higgs boson production with top quark pairs[109] $pp \to t\bar{t}H$ (see Fig. 1.6 d) is mostly relevant at the LHC. This channel is particularly important as it would allow a direct measurement of the $Ht\bar{t}$ coupling. The calculation of the NLO QCD corrections to this process was a real challenge which was met in Refs. [110–113]. The effect turned out to be rather small, being at most +20% at the LHC and −20% at the Tevatron, the precise impact depending on the PDFs used. The uncertainty from missing higher-order contributions is drastically reduced from a factor two at LO to the level of 10–20% at NLO.

The inclusive cross sections for the production channels discussed above, computed at the Tevatron and the LHC, with center-of-mass energies of $\sqrt{s} = 1.96$ TeV and $\sqrt{s} = 8$ TeV, respectively, are shown in Fig. 1.7.

Cross sections and uncertainties in the various channels for $m_H = 125$ GeV are reported in Table 1.2. At the LHC we consider $\sqrt{s} = 8$ and $\sqrt{s} = 13$ TeV. The former is the energy at which most of the data in Run 1 have been collected, while the latter is the starting energy of Run 2.

We conclude this section by adding a few comments on the use of the narrow-width approximation. In the present and previous sections, the production and decay stages of the Higgs boson have been considered as completely factorized, i.e., the narrow-width approximation has been assumed. However, a complete field theoretical description of the production of a resonance and its decay in a given channel requires not only the computation

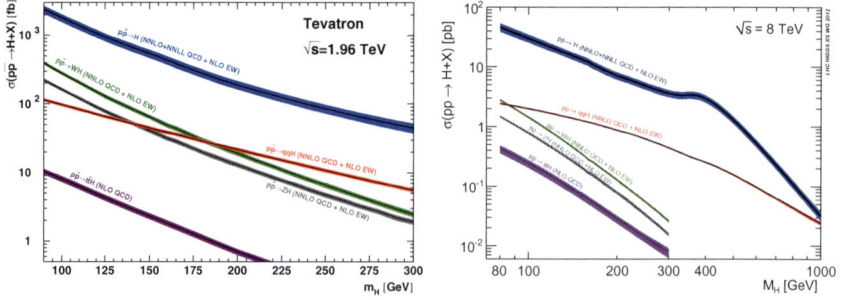

Fig. 1.7   Production cross sections for the SM Higgs boson in the main channels as a function of m_H: at the Tevatron (left, from Ref. [18]) and the LHC (right, from Ref. [59]).

Table 1.2   SM Higgs boson production cross sections (in pb) at the Tevatron (left), at the LHC with $\sqrt{s} = 8$ TeV (center) and $\sqrt{s} = 13$ TeV (right) for $m_H = 125$ GeV. The first uncertainty is from scale variations, the second from PDF+α_S.

	$\sqrt{s} = 1.96$ TeV	$\sqrt{s} = 8$ TeV	$\sqrt{s} = 13$ TeV
ggF	$0.949^{+7.1\%\ +10.1\%}_{-7.3\%\ -12.2\%}$	$19.27^{+7.2\%\ +7.5\%}_{-7.8\%\ -6.9\%}$	$43.92^{+7.4\%\ +7.1\%}_{-7.9\%\ -6.0\%}$
VBF	$0.065^{+3.3\%\ +4.7\%}_{-1.4\%\ -4.7\%}$	$1.578^{+0.2\%\ +2.6\%}_{-0.2\%\ -2.8\%}$	$3.748^{+0.7\%\ +3.2\%}_{-0.7\%\ -3.2\%}$
WH	$0.130^{+0.7\%\ +6.6\%}_{-1.0\%\ -6.7\%}$	$0.705^{+1.0\%\ +2.3\%}_{-1.0\%\ -2.3\%}$	$1.380^{+0.7\%\ +3.2\%}_{-1.5\%\ -3.2\%}$
ZH	$0.079^{+0.7\%\ +6.6\%}_{-1.0\%\ -6.7\%}$	$0.415^{+3.1\%\ +2.5\%}_{-3.1\%\ -2.5\%}$	$0.870^{+3.8\%\ +3.5\%}_{-3.8\%\ -3.5\%}$
ttH	—	$0.129^{+3.8\%\ +8.1\%}_{-9.3\%\ -8.1\%}$	$0.509^{+5.7\%\ +8.8\%}_{-9.3\%\ -8.8\%}$

of signal and background cross sections, but also of the signal–background interference. This is obviously important for searches of a heavy Higgs boson,[114–117] since (see Fig. 1.5) the Higgs boson width quickly increases with the mass. Unfortunately, a satisfactory theoretical treatment of this problem is missing and the recipe adopted by the ATLAS and CMS collaborations in the high-mass region is to assign a theoretical uncertainty of $150\% \, m_H$, where m_H is expressed in TeV.

Having discovered a Higgs boson at $m_H \sim 125$ GeV for which the narrow-width approximation is expected to work very well should not lead us to think that signal–background interferences are completely irrelevant. For example, signal–background interference effects are at the 5% level in the $H \to \gamma\gamma$ channel[118] and could lead to a sizable shift in the position of the mass peak.[119,120] In the case of $H \to WW \to l\nu l\nu$ and $H \to ZZ \to 4l$, interference effects can also be important, reaching the $\mathcal{O}(10\%)$

level, but can be reduced to the 1% level by using appropriate selection cuts.[121–124]

6. Detection channels at the LHC

From the discussion in the previous sections, producing the SM Higgs particle appears relatively easy; this is particularly the case at the LHC thanks to the high energy of the collider and its luminosity. However, detecting the particle in a very complex hadronic environment is another story. Indeed, in the main Higgs decay channels the backgrounds are simply gigantic. For instance, the rate for the production of light quarks and gluons is ten orders of magnitude larger than that of the Higgs boson. Even the cross sections for the production of W and Z bosons are three to four orders of magnitude larger. Detecting the Higgs particle in this hostile environment resembles finding a needle in a (million) haystack, the challenges to be met being simply enormous (see Chapter 4).

To be able to detect the Higgs particle, one should take advantage of the kinematical characteristics of the signal events which are, in general, quite different from that of the background events. In addition, one should focus on the decay modes of the Higgs particles (and those of the particles that are produced in association such as $W^{\pm}, Z$ bosons or top quarks) that are easier to extract from the background events. Pure hadronic modes such as Higgs decays into light quark or gluon jets have to be discarded although much more frequent in most cases.

The search strategy in the inclusive gluon fusion production channel at the LHC is discussed below. For Higgs boson masses below about $140\,\text{GeV}$ the dominant decay mode $H \to b\bar{b}$ is swamped by the huge QCD background and the Higgs boson can be observed in the clean but rare $H \to \gamma\gamma$ decay mode, complemented by the $H \to WW$ and $H \to ZZ$ channels. For $140\,\text{GeV} \leq m_H \leq 180\,\text{GeV}$ the $W^+W^- \to l^+l^-\nu\bar{\nu}$ mode is the most important. Despite the absence of a mass peak, there are strong angular correlations between the charged leptons that help to discriminate the signal over the irreducible WW background.[125] For $m_H \gtrsim 180\,\text{GeV}$ the $H \to ZZ \to 4l$ channel becomes the most important as it is much easier than $H \to WW$ because the invariant mass of the four charged leptons can be fully reconstructed. For higher m_H this channel can be supplemented by $H \to ZZ \to \ell\ell\nu\nu, \ell\ell jj$ and $H \to WW \to \ell\nu jj$ to increase the sensitivity.

The signal sensitivity, in particular in the $H \to WW$ channel, can be improved by considering jet categories, where the total $gg \to H$ cross section is broken into Higgs plus 0, 1 and ≥ 2 jet cross sections. By doing so the background can be significantly reduced, in particular in the zero-jet bin, which is mildly affected by the $t\bar{t}$ background.

Vector boson fusion is of particular interest since, as discussed previously, it has a large enough cross section which is affected only little by theoretical uncertainties. In addition, one can use specific cuts (forward-jet tagging, mini-jet veto for low luminosity as well as triggering on the central Higgs boson decay products), which render the backgrounds comparable to the signal, thereby allowing precision Higgs coupling measurements.

The associated production with gauge bosons, $q\bar{q} \to VH$ with $H \to b\bar{b}$ and possibly $H \to WW^* \to \ell\nu\ell\nu$ and $\ell\nu jj$, is the most relevant detection mode at the Tevatron for a light Higgs boson. The $H \to \gamma\gamma$ decay channel is unobservable due to the very small rate, while the inclusive $p\bar{p} \to H \to WW \to \ell\nu\ell\nu$ channel is instead important for Higgs boson masses around 160 GeV. At the LHC, $VH \to Vb\bar{b}$ production was considered not viable until recently, due to the large backgrounds but it was resurrected by the suggestion[126] to look at events where both the Higgs and the vector bosons have large transverse momenta.

Finally, associated $t\bar{t}H$ production can in principle be observed at the LHC and allows direct measurement of the top Yukawa coupling. The $H \to b\bar{b}$ channel would also allow us to access the bottom Yukawa coupling and, eventually, a determination of the Higgs CP properties (see Sec. 9). Nevertheless, isolating the $pp \to t\bar{t}H \to t\bar{t}b\bar{b}$ signal over the background is a real challenge.

7. Differential distributions and cross sections with cuts

In the previous sections, the theoretical predictions for Higgs boson decay and production at hadron colliders have been limited to fully *inclusive* quantities. However, experiments have always a finite acceptance and the analyses impose cuts on the final states in order to isolate a signal over a background. The computation of higher-order corrections to (multi)-differential distributions is typically much more involved than the corresponding calculation for fully inclusive quantities. Nonetheless, in the case of the main Higgs boson production channel, gluon fusion, numerous tools exist to obtain predictions which account for the experimental cuts, or to compute specific distributions. The fully differential NNLO cross section is available in two parton-level Monte Carlo simulation codes, FEHIPRO[127,128] and HNNLO.[129,130]

The Higgs transverse-momentum spectrum has been studied at NLO accuracy,[131–133] supplemented with next-to-next-to-leading logarithmic (NNLL) resummation of the logarithmically enhanced contributions at small transverse momenta.[134,135] Recently, this calculation has been extended to include the Higgs boson decay in a new program, HRES,[136] which includes the decay of the Higgs boson in $\gamma\gamma$, $WW \to l\nu l\nu$ and $ZZ \to 4$ leptons. The heavy-quark mass effects have been included up to NLL+NLO.[137] All of these results are actively used by the experimental community. In particular, the NNLL+NNLO calculation of the Higgs spectrum[135,136] is used to tune or to correct (reweight)[138] the p_T spectrum from Monte Carlo (MC) event generators, as a way to take into account higher-order perturbative effects. Very recently,[139,140] the calculation of the Higgs production cross section in association with a jet up to NNLO (i.e. $\mathcal{O}(\alpha_S^5)$ has been completed. This calculation allows us to further improve the perturbative predictions for the Higgs transverse-momentum spectrum, although still in the large-m_t approximation.

In associated Higgs boson production with a vector boson, fully differential NNLO computations exist in the case of WH[141] and ZH[142] production. Complete differential calculations of WH production and decay at NLO have also been performed.[143] These calculations allow to assess the impact of radiative effect when severe selection cuts are applied, as is the case in boosted object analyses.[126] A recent study[144] shows that QCD radiative corrections are well accounted for by Monte Carlo event generators.

In vector boson fusion, fully differential NLO QCD+EW calculations are available and implemented in the VBF@NLO[97] and HAWK[92,98] generators. In the case of the $t\bar{t}H$ process, the differential NLO predictions are available but still limited to the case of a stable top quark.[110–113] Results including the top-quark decay[145] have been obtained very recently by using the method of Ref. [146].

One final and important point to mention here is the issue of evaluating the uncertainties in the differential theoretical predictions. As mentioned in Sec. 6, the composition of the background to Higgs boson production in the dominant gluon fusion channel, especially in the case $H \to W^+W^-$, differs dramatically depending on how many jets are observed together with the Higgs boson in the final state. Many experimental analysis are thus optimized depending on the number of jets reconstructed in the event

(0, 1 or 2 and more jets). Such a split induces large logarithms associated
with the ratio of the Higgs boson mass over the defining p_T of the jet. The
most natural method of evaluating the theoretical uncertainty in the zero-
jet bin, that of performing scale variations, leads to an uncertainty which
is typically rather small. This indicates a possible cancellation between
logarithms of the p_T cut and the large corrections to the Higgs cross section,
and a potential underestimate of the theoretical uncertainty.

A prescription for estimating the perturbative uncertainty in this case
was proposed in Ref. [147]. The idea is to start from the *inclusive* cross
sections, $\sigma_{\geq N}$, for the production of N or more jets, and then define the
exclusive N-jet cross sections using $\sigma_N = \sigma_{\geq N} - \sigma_{\geq N+1}$. Assuming that the
uncertainties in the different $\sigma_{\geq N}$ are uncorrelated, the uncertainties and
correlations in σ_N follow from standard error propagation. This prescription
leads to a more conservative estimate of the uncertainty in the zero-jet bin,
but in some cases may overestimate the theoretical error.

Recently, the resummation of the logarithmically enhanced terms in the
jet vetoed cross section has been the subject of intense theoretical studies.
A NNLL prediction for the jet-veto efficiency matched to NNLO has been
presented[148] (see also Refs. [149,150]). The results show that, for the typical values of the p_T cut used by the ATLAS and CMS experiments, the
fixed order prediction is under good control. The new resummed result is
particularly useful to reduce the theoretical uncertainties.

8. Monte Carlo generators for Higgs physics

Monte Carlo (MC) event generators are indispensable tools for the analysis and interpretation of experimental data from high-energy collider
experiments. Through the complete simulation of all the stages of the
hadronic collision, from the initial state radiation to the hard scattering
process (including secondary scattering processes that may occur, the so-
called underlying event) to the final state radiation, the hadronization
and the detector resolution effects, they provide a realistic description
of the hadronic collisions. Historically, MC generators were based on a
set of Born level underlying processes, on top of which the initial and
final state QCD radiation was built up, by using the Monte Carlo *parton
shower* (PS): a cascade of subsequent multiparton emissions is generated
using soft and collinear approximations.[151] These generators were essentially LO and, as a consequence, could hardly describe an event with one

or more hard jets, since QCD radiation was obtained through the parton shower.

Recently, MC event generators have benefited from a number of theoretical achievements that have significantly improved their capability of making accurate predictions and simulations of events taking place at high-energy colliders. The most important of these achievements is the possibility to consistently include exact NLO corrections in the simulation, so as to achieve a NLO+PS generator.

The NLO+PS matching was pioneered in the work of Ref. [152]: it allows to combine the virtues and flexibility of the MC parton shower with the NLO accuracy of a fixed order calculation. For example, in the production of the Higgs boson through gluon fusion, the description of the recoiling jet is exact, through the corresponding tree-level $H + 1$ jet matrix element, and the virtual correction is included to achieve full NLO accuracy. NLO+PS matching is now implemented in three generators: MC@NLO (with its automated version aMC@NLO[153]), POWHEG-BOX[154] and SHERPA.[155] Thanks to the impressive progress in the automatization of the computation of NLO corrections, the main Higgs boson production channels at hadron colliders are now available in the context of NLO+PS,[153,156–160] together with some of the most important backgrounds.[161–164]

Recent studies consider the possibility to merge parton shower simulations with high-multiplicity tree-level matrix elements. For example, the computation of inclusive Higgs boson production through gluon fusion with a NLO+PS generator uses only $gg \to H + 0$ and $gg \to H + 1j$ matrix elements. A simulation of $H + 1$ jet which is accurate at NLO requires instead to have the exact $gg \to H + 1j$ and $gg \to H + 2j$ QCD matrix elements. The simultaneous description of observables which are exclusive in different jet multiplicities cannot simply be obtained by summing the corresponding results, but a suitable merging to avoid double counting is required. At LO, the solution to this problem exists with different recipes.[165,166] Work in the direction of NLO merging is being performed and solutions have been proposed.[167,168]

Another direction in which work is being carried out is the extension of the matching procedure to NNLO. The fully differential NNLO calculations mentioned in Sec. 7 can in fact be exploited in their full potential if matching to PS becomes possible. Work in this direction is being performed by different groups[169–171] and first NNLO+PS simulations for $gg \to H$ production have been presented in Refs. [172,173].

9. Measurements of the Higgs properties

After the discovery of a scalar resonance consistent with the Higgs boson has been established,[10,11] it becomes important to study its properties.

We recall that within the SM the Higgs boson mass is the only free parameter in the Higgs sector (see Sec. 2). The channels $H \to \gamma\gamma$ and $H \to ZZ \to 4\ell$, thanks to their excellent mass resolution, allow for a precise mass measurement, already at a sub-percent level with the LHC Run 1 data (see Chapter 12). Once the Higgs boson mass is fixed, its partial decay widths to the fermions and the gauge bosons follow, and hence, its total width is determined. The spin/CP quantum numbers of the SM Higgs boson are also fixed: $J^{CP} = 0^{++}$. Should measurements of Higgs boson properties reveal deviations from any of these predictions, this would be evidence for new physics. In this section we summarize the strategies to measure the fundamental properties of the Higgs boson.

We start our discussion from the Higgs couplings. Ratios of couplings squared can be determined by measuring ratios of production cross sections times decay branching ratios and accuracies at the 10–50% can be obtained in some cases.[174] However, it has been shown in Ref. [175] that with some theoretical assumptions, which are valid in general for multi-Higgs doublet models, the extraction of absolute values of the couplings, rather than just their ratios, becomes possible by performing a fit to the observed rates of Higgs boson production in different channels.

The current experimental studies (see Chapter 12) are based on the framework outlined in Ref. [145]. This framework relies on the following assumptions: i) The signals observed in the different search channels originate from a single resonance with a mass close to 125 GeV (the case of more resonances with similar mass is not considered). ii) The resonance width is very small, thus the use of the narrow width approximation is justified. iii) Only modifications of coupling strengths are considered, while the tensor structure of the Lagrangian is assumed to be the same as in the SM. This implies in particular that the observed state is assumed to be a CP-even scalar. Modifications of the coupling strength of the Higgs boson are introduced, by rescaling (some of) the couplings with appropriate scaling factors κ_i. Also the effective couplings to gluons and photons are modified by separate coupling scale factors. With this definition of the coupling scale factors κ_j, the cross sections σ_j and the partial decay widths Γ_j associated with the SM particle j scale with κ_j^2 compared to the SM prediction. Using this notation, and defining κ_H as the scale factor for the total Higgs boson width Γ_H, the cross section for e.g. the $gg \to H \to \gamma\gamma$ process can be expressed as:

$$\frac{\sigma \cdot \text{BR}\,(gg \to H \to \gamma\gamma)}{\sigma_{\text{SM}}(gg \to H) \cdot \text{B}_{\text{SM}}(H \to \gamma\gamma)} = \frac{\kappa_g^2 \cdot \kappa_\gamma^2}{\kappa_H^2}. \tag{1.20}$$

The SM decay rates and cross sections, including the relevant higher order corrections discussed in Secs. 4 and 5, are recovered in the limit $\kappa_i \to 1$. This procedure leads to different benchmark scenarios, according to which couplings are actually rescaled (see Chapter 12). It is important to point out that this strategy does not really correspond to a specific BSM scenario, but it is more an effective way to parametrize (small) deviations from the SM picture. With increasing precision of the measurements, this framework should eventually be replaced by a more rigorous approach that allows us to include radiative corrections consistently, by using effective-field theory methods, along the lines discussed in Ref. [145].

We now go on discussing the Higgs boson width. The width of a 125 GeV Higgs boson in the SM is $\Gamma_H \sim 4\,\text{MeV}$, and is too small to be directly measured. The direct upper bound that ATLAS and CMS can put on the width is about 1 GeV. The indirect method, which is used at lepton colliders, to measure the Higgs boson width from the missing mass distribution in ZH production cannot be used at the LHC because at hadron colliders the partonic centre-of-mass energy is not fixed. Recently, a new method to constrain the Higgs boson width has been proposed.[124,176,177] By using existing measurements of the ZZ and WW cross section off peak, an indirect determination of the width can be obtained (see Chapters 5 and 7). The following discussion is based on the ZZ final state. The production of a Higgs boson $gg \to H \to ZZ$ can be effectively described as

$$\frac{d\sigma}{dm_{ZZ}^2} \sim \kappa_g^2 \cdot \kappa_Z^2 \cdot g_{gg}^2(m_{ZZ}) \cdot g_{ZZ}^2 \frac{F(m_{ZZ})}{(m_{ZZ}^2 - m_H^2)^2 + m_H^2 \Gamma_H^2}, \tag{1.21}$$

where m_{ZZ} is the invariant mass of the ZZ final state, g_{gg} and g_{ZZ} are SM Higgs boson couplings to gluons and Z bosons, while κ_g and κ_Z are the coupling modifiers introduced above. The differential $d\sigma/dm_{ZZ}$ distribution for a 125-GeV Higgs boson remains fairly flat in the high-mass region due to the long tails of the Breit–Wigner function and the increasing phase space.[122] As mentioned at the end of Sec. 5, signal–background interference effects are very important to correctly model the tail of the distribution. In the SM the interference effect is negative and it is known only at LO.

From Eq. (1.21), one can see that the far off-shell rate ($m_{ZZ} - m_H \gg \Gamma_H$) is independent of the Higgs boson width: $\sigma_{\text{off-shell}} \sim \kappa_g^2 \kappa_Z^2$. On the other hand, the on-shell cross section ($|m_{ZZ} - m_H| \lesssim$ a few Γ_H) is inversely proportional to the width: $\sigma_{\text{on-shell}} \sim \kappa_g^2 \kappa_Z^2 / \Gamma_H$. Therefore, the ratio of off-shell to on-shell production is sensitive (proportional) to the Higgs boson width. However, this conclusion is model-dependent. Relatively light BSM particles could strongly affect off-shell production and decays of the 125-GeV Higgs boson as well as production of continuum background $gg \to ZZ$. The function $F(m_{ZZ})$ in Eq. (1.21) depends on spin/CP quantum numbers of the 125-GeV boson.[178] Also, the off-shell production would be very different if the dominant Higgs boson production mechanism were $q\bar{q} \to H$.

The total decay width can also be constrained by determining the invisible width of the Higgs boson. This can be done first directly by searching for missing transverse energy in Higgs boson production processes such as VBF[179] or the $gg \to H + j$ process.[180] Another possibility to constrain the invisible width would be through the Higgs boson signal strengths that are measured in the various channels. For instance, the signal strength for the $H \to ZZ \to 4\ell$ mode when one allows for arbitrary couplings of the Higgs boson leads to the constraint $\Gamma_H^{\text{inv}} / \Gamma_H^{\text{SM}} \lesssim 50\%$; see for instance Ref. [181].

We now move to the spin/CP properties of the new resonance. The observation in the decay modes $H \to \gamma\gamma$ (see Chapter 6), $H \to ZZ$ (see Chapters 5 and 11) and $H \to WW$ (see Chapter 10) allows multiple independent tests of these properties. Thanks to the Landau–Yang theorem,[182,183] the observation in the $H \to \gamma\gamma$ decay mode already rules out the possibility that the new resonance has spin 1. The argument goes as follows. Consider an initial state at rest decaying into two photons whose direction can be chosen to be the z axis and let $A(M \to h_1, h_2)$ be the decay amplitude for an initial state with $J_z = M$ along the $+z$ axis and final states having helicities $h_{1,2}$. Angular momentum conservation (i.e. rotational invariance) along the z axis implies that $M = h_1 - h_2$ so the only allowed amplitudes for a spin-0 or spin-1 particle to decay into $\gamma\gamma$ are $A_+ = A(0 \to 1, 1)$ and $A_- = A(0 \to -1, -1)$, i.e. the $J_z = 0$ state decays into two right-handed or two left-handed photons. A rotation $R_y(\pi)$ by an angle π around the y axis of the initial and final states leads for a spin J particle, $R_y(\pi)|J_z = 0\rangle = (-1)^J |J_z = 0\rangle$, using the usual angular momentum rotation matrices, so the initial state gets a phase $(-1)^J$. Since the photons are identical and have the same helicity, the final state is unchanged and rotational invariance implies that $A_\pm = (-1)^J A_\pm$. $A_\pm$ can be non-zero for spin-0 but for spin-1, $A_\pm = -A_\pm \Rightarrow A_\pm = 0$, and the decay amplitude vanishes. Note that the proof requires (a) rotational invariance; (b) identical gauge bosons; (c) massless gauge bosons which do not have longitudinal $h = 0$ polarizations, and no assumption is made about discrete symmetries such as P, C, etc. Note also that if one assumes that the Higgs sector is C conserving, the decay $H \rightarrow \gamma\gamma$ fixes $C = +1$ for the produced state.

The spin-2 possibility turns out to be extremely challenging from a theoretical view point. The naive coupling of a massive spin-2 particle with a U(1) field leads to well known pathologies[184,185] and the model has an intrinsic cut-off of the order of the mass itself.[186] These difficulties, however, should not prevent the experimental collaborations to pursue the spin-2 possibility.

Analogously to the pion in its decay[187] $\pi^0 \rightarrow e^+e^-e^+e^-$, the $H \rightarrow ZZ \rightarrow 4l^\pm$ decay mode is the ideal channel to study spin, parity and tensor structure of the coupling of the Higgs boson.[188–190] The invariant mass distribution of the off-shell gauge boson in the $H \rightarrow ZZ^*$ decay has a characteristic steep behavior just below the kinematical threshold. This behavior is related to the spin-zero nature of the SM Higgs boson and will rule out other spin assignments with the exception of the 1^+ and 2^+ cases, which can be ruled out through angular correlations. The pseudoscalar Higgs case 0^- can be instead discriminated against the 0^+ case by studying the distribution in the azimuthal angle between the two Z decay planes.

Besides specific kinematic variables, the best results are obtained by exploiting the full kinematic information on the event using the *matrix element method*. This method exploits the ratio of tree level event amplitudes calculated for alternative signal hypotheses in order to discriminate between them. The construction of the corresponding matrix element can be carried out by using two different strategies: the *effective Lagrangian* and *anomalous couplings* approaches. The former requires one to write the most general effective Lagrangian compatible with Lorentz and gauge invariance. The latter calls for writing out the most general *amplitude* compatible with Lorentz and gauge invariance, but does not assume a hierarchy in the scales, and thus the couplings become momentum dependent form factors. The effective Lagrangian approach has the advantage that it can be extended beyond LO. The anomalous coupling approach is restricted to LO, but somewhat more general, since it is valid also in the case in which new weakly coupled light degrees of freedom are present. This approach has been used to perform studies of the spin/CP properties of the Higgs boson.[191] The relevant matrix elements can be either extracted from LO event generators, where they are already explicitly implemented, such as JHUGEN,[192–194] or computed for any custom Lagrangian via FEYNRULES[195] interface[196,197] to such generators as MADGRAPH,[198] CALCHEP,[199] etc. The effective Lagrangian approach, with inclusion of NLO QCD corrections and parton shower effects, is pursued within the MADGRAPH framework.[197]

As far as a spin-0 boson is concerned, the general Lagrangian describing its on-shell decay $H \to VV$ ($V = W^\pm, Z$) can be written as follows:

$$\mathcal{L}_{HVV} \sim \kappa_V \frac{m_V^2}{v} H V^\mu V_\mu$$

$$+ \frac{\alpha_V}{v} H V^\mu \Box V_\mu + \frac{\beta_V}{v} H V^{\mu\nu} V_{\mu\nu} + \frac{\gamma_V}{v} H V^{\mu\nu} \tilde{V}_{\mu\nu} \qquad (1.22)$$

where v stands for the SM Higgs field vacuum expectation value and V_μ, $V_{\mu\nu} = \partial_\mu V_\nu - \partial_\nu V_\mu$, $\tilde{V}_{\mu\nu} = \frac{1}{2} \epsilon_{\mu\nu\rho\sigma} V^{\rho\sigma}$ represent the V boson field, field strength tensor, and dual field strength tensor, respectively. The first term in Eq. (1.22) is the minimal dim-3 operator and, with $\kappa_Z = 1$ and $\kappa_W = 2$, corresponds to the SM Higgs boson (0^+). The last three terms are dim-5 operators: the γ-term corresponds to a pseudo-scalar (0^-), while the α- and β-terms are even-parity scalars, which, however, are kinematically distinguishable from the SM Higgs boson. The β-term is referred to as 0_h^+. In fact, SM particles contribute to these three terms, but at a very small level.

The Higgs CP properties and the structure of the HVV vertex can also be studied in vector boson fusion (VBF), by looking at the azimuthal separation of the two tagging jets.[200] Recent studies on the determination of Higgs spin/CP properties in VBF are presented in Refs. [201–203].

One should bare in mind that if, as it happens in many BSM scenarios, the coupling of the CP-odd component to VV is suppressed compared to the HVV coupling, searches for the CP-odd component in $H \to ZZ$ and $H \to WW$ decays as well as in invariant mass distributions and azimuthal distributions in VH and VBF production may not be the best strategies. The couplings of the Higgs boson to fermions offer a complementary opportunity, since in this case the CP-even and odd components can have the same magnitude. In this respect, if the $t\bar{t}H$ production and $H \to \tau\tau$ decay modes can be exploited sufficiently well, they would offer a nice opportunity for studying the Higgs boson's CP properties.[204–207]

10. Outlook

It was expected since a long time that probing the EWSB mechanism would be at least a two-step process. The first one was the search for a Higgs-like particle to confirm (or refute) the hypothesis that the electroweak symmetry is spontaneously broken by a scalar field developing a non-zero vacuum expectation value. This test has been successfully passed with the historic discovery of the Higgs boson. A second step, that is as important as the first one, is to probe in all its facets the EWSB mechanism and to assess whether the Higgs sector of the theory is SM-like or involves new degrees of freedom. This step needs a much better determination of the basic Higgs properties compared to what has been achieved so far.

All the measurements performed in the channels that have been studied so far need to be significantly improved in order to better constrain the SM and to probe smaller effects of new physics. Although already rather precise, a better determination of m_H (and more importantly, a better and unambiguous determination of the top-quark mass) would reduce the uncertainties in the stability bound of the electroweak vacuum. The total decay width of the Higgs boson (that could involve a non-zero invisible component for instance) and its CP quantum numbers (in particular, the possibility that the Higgs boson is a CP-even state with a small admixture of a CP-odd component) need to be more precisely scrutinized not only in the $H \to ZZ \to 4\ell$ channel but in other channels as well.

Important production channels such as associated Higgs boson production with top quarks, $pp \to t\bar{t}H$, which allows for a direct probe of the top-quark Yukawa couplings and might provide an unambiguous determination of the Higgs CP properties, still need to be probed and rare decay modes such as $H \to Z\gamma$ (which could give complementary information to the $H \to \gamma\gamma$ decay) and $H \to \mu^+\mu^-$ (which would probe for the first time Higgs couplings to other fermions than those of the third generation) will only be accessible at high luminosity.

Besides that, a more precise measurement of the Higgs boson couplings to the fermions and the gauge bosons in the already probed channels should be performed. The goal would be to reach an accuracy at the few percent level, which is the size of the higher order electroweak corrections and hence of the potential new physics effects. From the theory side, recent years have witnessed a steady improvement in the theoretical predictions for the most important signal and background processes. For the main Higgs boson production channel at the LHC, gluon fusion, the recent computation of

N^3LO corrections,[79] together with a more precise extraction of PDFs using LHC data, should lead to a significant reduction of theoretical uncertainties in this channel. Nonetheless, other uncertainties, like those from the QCD coupling α_S, the effects of finite quark masses and missing EW corrections will make it difficult to push the overall accuracy in the gluon fusion channel to better than $\mathcal{O}(5\%)$ level. Ratios of couplings squared, which are free of many ambiguities such as the ones due to QCD and the total Higgs boson width, would play a very important role as they will be limited only by the statistical and systematical uncertainties.

Finally, of prime importance would be the measurement of the Higgs *self-coupling* λ which would allow us to fully determine the Higgs potential that is responsible of EWSB. The Higgs trilinear coupling can be studied by measuring *double* Higgs boson production. Like in the single Higgs case, Higgs boson pair production in the SM is dominated by the gluon fusion mechanism through a heavy (box or triangular) top-quark loop.[208–211] Unfortunately, the $gg \to HH$ cross section, which is known since some time at NLO within the large-m_t approximation[212] and which has been recently calculated at NNLO[213] QCD, is rather small. This is a fortiori also the case for all the other double Higgs boson production channels at the LHC.[211,214] This makes the observation of a double Higgs signal and a measurement of the self-coupling extremely difficult. A very high luminosity and eventually also a higher energy collider would be required in this context.

To conclude, the probing of the EWSB mechanism is still ongoing and the upgrade of the LHC with a center-of-mass energy up to 14 TeV and an integrated luminosity up to the level of several hundred inverse femtobarns and even a few atobarns, will be essential in this respect. The next decade is thus crucial for particle physics and it will hopefully provide us with many surprises. These surprises are highly expected for a reason that is deeply rooted in the Higgs sector: in the SM, there is no protection of the Higgs boson mass against very high scales, the so-called hierarchy problem. In order to have naturally $m_H = 125$ GeV, new degrees of freedom are expected at the TeV scale. Maybe nature was kind to us and made these new degrees of freedom accessible at the upcoming LHC upgrade.

References

1. S. Glashow, Partial symmetries of weak interactions, *Nucl.Phys.* **22**, 579–588 (1961). doi: 10.1016/0029-5582(61)90469-2.

2. S. Weinberg, A model of leptons, *Phys. Rev. Lett.* **19**, 1264–1266 (1967). doi: 10.1103/PhysRevLett.19.1264.

3. A. Salam, in *Elementary Particle Theory*, Stockholm (1969), p. 367.

4. D. J. Gross and F. Wilczek, Ultraviolet behavior of nonabelian gauge theories, *Phys. Rev. Lett.* **30**, 1343–1346 (1973). doi: 10.1103/PhysRevLett.30.1343.

5. H. D. Politzer, Reliable perturbative results for strong interactions?, *Phys. Rev. Lett.* **30**, 1346–1349 (1973). doi: 10.1103/PhysRevLett.30.1346.

6. H. Fritzsch, M. Gell-Mann, and H. Leutwyler, Advantages of the color octet gluon picture, *Phys. Lett.* **B47**, 365–368 (1973). doi: 10.1016/0370-2693(73)90625-4.

7. P. W. Higgs, Broken symmetries, massless particles and gauge fields, *Phys. Lett.* **12**, 132–133 (1964). doi: 10.1016/0031-9163(64)91136-9.

8. F. Englert and R. Brout, Broken symmetry and the mass of gauge vector mesons, *Phys. Rev. Lett.* **13**, 321–323 (1964). doi: 10.1103/PhysRevLett.13.321.

9. G. Guralnik, C. Hagen, and T. Kibble, Global conservation laws and massless particles, *Phys. Rev. Lett.* **13**, 585–587 (1964). doi: 10.1103/PhysRevLett.13.585.

10. ATLAS Collaboration, Observation of a new particle in the search for the Standard Model Higgs boson with the ATLAS detector at the LHC, *Phys. Lett.* **B716**, 1–29 (2012). doi: 10.1016/j.physletb.2012.08.020.

11. CMS Collaboration, Observation of a new boson at a mass of 125 GeV with the CMS experiment at the LHC, *Phys. Lett.* **B716**, 30–61 (2012). doi: 10.1016/j.physletb.2012.08.021.

12. A. Djouadi, The anatomy of electro-weak symmetry breaking. I: The Higgs boson in the Standard Model, *Phys. Rept.* **457**, 1–216 (2008). doi: 10.1016/j.physrep.2007.10.004.

13. Y. Nambu and G. Jona-Lasinio, Dynamical model of elementary particles based on an analogy with superconductivity. 1., *Phys. Rev.* **122**, 345–358 (1961). doi: 10.1103/PhysRev.122.345.

14. J. Goldstone, Field theories with superconductor solutions, *Nuovo Cim.* **19**, 154–164 (1961). doi: 10.1007/BF02812722.

15. J. Goldstone, A. Salam, and S. Weinberg, Broken symmetries, *Phys. Rev.* **127**, 965–970 (1962). doi: 10.1103/PhysRev.127.965.

16. LEP Working Group for Higgs boson searches, ALEPH Collaboration, DELPHI Collaboration, L3 Collaboration, OPAL Collaboration, Search for the Standard Model Higgs boson at LEP, *Phys. Lett.* **B565**, 61–75 (2003). doi: 10.1016/S0370-2693(03)00614-2.

17. CDF Collaboration, D0 Collaboration, Higgs boson studies at the Tevatron, *Phys. Rev.* **D88**, 052014 (2013). doi: 10.1103/PhysRevD.88.052014.

18. Particle Data Group, Review of particle physics (RPP), *Phys. Rev.* **D86**, 010001 (2012). doi: 10.1103/PhysRevD.86.010001.

19. M. Baak, M. Goebel, J. Haller, A. Hoecker, D. Kennedy, *et al.*, The electroweak fit of the Standard Model after the discovery of a new boson at the LHC, *Eur. Phys. J.* **C72**, 2205 (2012). doi: 10.1140/epjc/s10052-012-2205-9.

20. CDF Collaboration, D0 Collaboration, Combination of the top-quark mass measurements from the Tevatron collider, *Phys.Rev.* **D86**, 092003 (2012). doi: 10.1103/PhysRevD.86.092003.

21. B. W. Lee, C. Quigg, and H. Thacker, Weak interactions at very high-energies: The role of the higgs boson mass, *Phys.Rev.* **D16**, 1519 (1977). doi: 10.1103/PhysRevD.16.1519.

22. N. Cabibbo, L. Maiani, G. Parisi, and R. Petronzio, Bounds on the fermions and Higgs boson masses in grand unified theories, *Nucl.Phys.* **B158**, 295–305 (1979). doi: 10.1016/0550-3213(79)90167-6.

23. M. Luscher and P. Weisz, Is there a strong interaction sector in the standard lattice Higgs model?, *Phys.Lett.* **B212**, 472 (1988). doi: 10.1016/0370-2693(88)91799-6.

24. T. Hambye and K. Riesselmann, Matching conditions and Higgs mass upper bounds revisited, *Phys.Rev.* **D55**, 7255–7262 (1997). doi: 10.1103/PhysRevD.55.7255.

25. G. Degrassi, S. Di Vita, J. Elias-Miro, J. R. Espinosa, G. F. Giudice, *et al.*, Higgs mass and vacuum stability in the Standard Model at NNLO, *JHEP*. **1208**, 098 (2012). doi: 10.1007/JHEP08(2012)098.

26. F. Bezrukov, M. Y. Kalmykov, B. A. Kniehl, and M. Shaposhnikov, Higgs boson mass and new physics, *JHEP*. **1210**, 140 (2012). doi: 10.1007/JHEP10(2012)140.

27. S. Alekhin, A. Djouadi, and S. Moch, The top quark and Higgs boson masses and the stability of the electroweak vacuum, *Phys.Lett.* **B716**, 214–219 (2012). doi: 10.1016/j.physletb.2012.08.024.

28. V. Branchina and E. Messina, Stability, Higgs boson mass and new physics, *Phys.Rev.Lett.* **111**, 241801 (2013). doi: 10.1103/PhysRevLett.111.241801.

29. E. Braaten and J. Leveille, Higgs boson decay and the running mass, *Phys.Rev.* **D22**, 715 (1980). doi: 10.1103/PhysRevD.22.715.

30. N. Sakai, Perturbative QCD corrections to the hadronic decay width of the Higgs boson, *Phys.Rev.* **D22**, 2220 (1980). doi: 10.1103/PhysRevD.22.2220.

31. T. Inami and T. Kubota, Renormalization group estimate of the hadronic decay width of the Higgs boson, *Nucl.Phys.* **B179**, 171 (1981). doi: 10.1016/0550-3213(81)90253-4.

32. M. Drees and K.-I. Hikasa, Note on QCD Corrections to hadronic Higgs decay, *Phys.Lett.* **B240**, 455 (1990). doi: 10.1016/0370-2693(90)91130-4.

33. A. Djouadi, M. Spira, and P. Zerwas, QCD corrections to hadronic Higgs decays, *Z.Phys.* **C70**, 427–434 (1996). doi: 10.1007/s002880050120.

34. S. Gorishnii, A. Kataev, S. Larin, and L. Surguladze, Corrected three loop QCD correction to the correlator of the quark scalar currents and γ (Tot) ($H^0 \rightarrow$ Hadrons), *Mod.Phys.Lett.* **A5**, 2703–2712 (1990). doi: 10.1142/S0217732390003152.

35. P. Baikov, K. Chetyrkin, and J. H. Kuhn, Scalar correlator at O(alpha(s)**4), Higgs decay into b-quarks and bounds on the light quark masses, *Phys.Rev.Lett.* **96**, 012003 (2006). doi: 10.1103/PhysRevLett.96.012003.

36. C. Anastasiou, F. Herzog, and A. Lazopoulos, The fully differential decay rate of a Higgs boson to bottom-quarks at NNLO in QCD, *JHEP.* **1203**, 035 (2012). doi: 10.1007/JHEP03(2012)035.

37. V. Del Duca, C. Duhr, G. Somogyi, F. Tramontano, and Z. Trócsányi, Higgs boson decay into b-quarks at NNLO accuracy, *JHEP.* **1504**, 036 (2015). doi: 10.1007/JHEP04(2015)036.

38. J. Fleischer and F. Jegerlehner, Radiative corrections to Higgs decays in the extended Weinberg–Salam model, *Phys.Rev.* **D23**, 2001–2026 (1981). doi: 10.1103/PhysRevD.23.2001.

39. B. A. Kniehl, Radiative corrections for $H \to$ f anti-f (γ) in the Standard Model, *Nucl.Phys.* **B376**, 3–28 (1992). doi: 10.1016/0550-3213(92)90065-J.

40. D. Y. Bardin, B. Vilensky, and P. K. Khristova, Calculation of the Higgs boson decay width into fermion pairs, *Sov.J.Nucl.Phys.* **53**, 152–158 (1991).

41. A. Dabelstein and W. Hollik, Electroweak corrections to the fermionic decay width of the standard Higgs boson, *Z.Phys.* **C53**, 507–516 (1992). doi: 10.1007/BF01625912.

42. H.-Q. Zheng and D.-D. Wu, First order QCD corrections to the decay of the Higgs boson into two photons, *Phys.Rev.* **D42**, 3760–3763 (1990). doi: 10.1103/PhysRevD.42.3760.

43. A. Djouadi, M. Spira, J. van der Bij, and P. Zerwas, QCD corrections to gamma gamma decays of Higgs particles in the intermediate mass range, *Phys.Lett.* **B257**, 187–190 (1991). doi: 10.1016/0370-2693(91)90879-U.

44. S. Dawson and R. Kauffman, QCD corrections to $H \to \gamma\gamma$, *Phys.Rev.* **D47**, 1264–1267 (1993). doi: 10.1103/PhysRevD.47.1264.

45. J. Fleischer, O. Tarasov, and V. Tarasov, Analytical result for the two loop QCD correction to the decay $H \to 2\gamma$, *Phys.Lett.* **B584**, 294–297 (2004). doi: 10.1016/j.physletb.2004.01.063.

46. U. Aglietti, R. Bonciani, G. Degrassi, and A. Vicini, Analytic results for virtual QCD corrections to Higgs production and decay, *JHEP.* **0701**, 021 (2007). doi: 10.1088/1126-6708/2007/01/021.

47. S. Actis, G. Passarino, C. Sturm, and S. Uccirati, NNLO computational techniques: the cases $H \to \gamma\gamma$ and $H \to gg$, *Nucl.Phys.* **B811**, 182–273 (2009). doi: 10.1016/j.nuclphysb.2008.11.024.

48. A. Djouadi, P. Gambino, and B. A. Kniehl, Two loop electroweak heavy fermion corrections to Higgs boson production and decay, *Nucl.Phys.* **B523**, 17–39 (1998). doi: 10.1016/S0550-3213(98)00147-3.

49. F. Fugel, B. A. Kniehl, and M. Steinhauser, Two loop electroweak correction of $O(G(F)M(t)^{**}2)$ to the Higgs-boson decay into photons, *Nucl.Phys.* **B702**, 333–345 (2004). doi: 10.1016/j.nuclphysb.2004.09.018.

50. G. Degrassi and F. Maltoni, Two-loop electroweak corrections to the Higgs-boson decay $H \to \gamma\gamma$, *Nucl.Phys.* **B724**, 183–196 (2005). doi: 10.1016/j.nuclphysb.2005.06.027.

51. G. Passarino, C. Sturm, and S. Uccirati, Complete two-loop corrections to $H \to \gamma\gamma$, *Phys.Lett.* **B655**, 298–306 (2007). doi: 10.1016/j.physletb.2007.09.002.

52. P. Maierhöfer and P. Marquard, Complete three-loop QCD corrections to the decay $H \to \gamma\gamma$, *Phys.Lett.* **B721**, 131–135 (2013). doi: 10.1016/j.physletb.2013.02.040.

53. M. Spira, A. Djouadi, and P. Zerwas, QCD corrections to the H Z gamma coupling, *Phys.Lett.* **B276**, 350–353 (1992). doi: 10.1016/0370-2693(92)90331-W.

54. T. Gehrmann, S. Guns, and D. Kara, The rare decay $H \to Z\gamma$ in perturbative QCD (2015), *JHEP.* **1509**, 038 (2015). doi: 10.1007/JHEP09(2015)038.

55. R. Bonciani, V. Del Duca, H. Frellesvig, J. M. Henn, F. Moriello, *et al.*, Next-to-leading order QCD corrections to the decay width $H \to Z\gamma$ (2015), *JHEP.* **1508**, 108 (2015). doi: 10.1007/JHEP08(2015)108.

56. A. Bredenstein, A. Denner, S. Dittmaier, and M. Weber, Precise predictions for the Higgs-boson decay $H \to WW/ZZ \to 4$ leptons, *Phys.Rev.* **D74**, 013004 (2006). doi: 10.1103/PhysRevD.74.013004.

57. A. Bredenstein, A. Denner, S. Dittmaier, and M. Weber, Radiative corrections to the semileptonic and hadronic Higgs-boson decays $H \to WW/ZZ \to 4$ fermions, *JHEP.* **0702**, 080 (2007). doi: 10.1088/1126-6708/2007/02/080.

58. A. Denner, S. Heinemeyer, I. Puljak, D. Rebuzzi, and M. Spira, Standard Model Higgs-boson branching ratios with uncertainties, *Eur.Phys.J.* **C71**, 1753 (2011). doi: 10.1140/epjc/s10052-011-1753-8.

59. LHC Higgs Cross Sections Working Group webpage, URL http://twiki.cern.ch/twiki/bin/view/LHCPhysics/CrossSections.

60. A. Djouadi, J. Kalinowski, and M. Spira, HDECAY: A program for Higgs boson decays in the Standard Model and its supersymmetric extension, *Comput.Phys.Commun.* **108**, 56–74 (1998). doi: 10.1016/S0010-4655(97)00123-9.

61. A. Djouadi, M. Muhlleitner, and M. Spira, Decays of supersymmetric particles: The Program SUSY-HIT (SUspect-SdecaY-Hdecay-InTerface), *Acta Phys.Polon.* **B38**, 635–644 (2007).

62. J. Baglio and A. Djouadi, Higgs production at the lHC, *JHEP.* **1103**, 055 (2011). doi: 10.1007/JHEP03(2011)055.

63. H. Georgi, S. Glashow, M. Machacek, and D. V. Nanopoulos, Higgs bosons from two gluon annihilation in proton proton collisions, *Phys.Rev.Lett.* **40**, 692 (1978). doi: 10.1103/PhysRevLett.40.692.

64. S. Dawson, Radiative corrections to Higgs boson production, *Nucl.Phys.* **B359**, 283–300 (1991). doi: 10.1016/0550-3213(91)90061-2.

65. A. Djouadi, M. Spira, and P. Zerwas, Production of Higgs bosons in proton colliders: QCD corrections, *Phys.Lett.* **B264**, 440–446 (1991). doi: 10.1016/0370-2693(91)90375-Z.

66. D. Graudenz, M. Spira, and P. Zerwas, QCD corrections to Higgs boson production at proton proton colliders, *Phys.Rev.Lett.* **70**, 1372–1375 (1993). doi: 10.1103/PhysRevLett.70.1372.

67. M. Spira, A. Djouadi, D. Graudenz, and P. Zerwas, Higgs boson production at the LHC, *Nucl.Phys.* **B453**, 17–82 (1995). doi: 10.1016/0550-3213(95)00379-7.

68. J. R. Ellis, M. K. Gaillard, and D. V. Nanopoulos, A phenomenological profile of the Higgs boson, *Nucl.Phys.* **B106**, 292 (1976). doi: 10.1016/0550-3213(76)90382-5.
69. M. A. Shifman, A. Vainshtein, M. Voloshin, and V. I. Zakharov, Low-energy theorems for Higgs boson couplings to photons, *Sov.J.Nucl.Phys.* **30**, 711–716 (1979).
70. C. Anastasiou and K. Melnikov, Higgs boson production at hadron colliders in NNLO QCD, *Nucl.Phys.* **B646**, 220–256 (2002). doi: 10.1016/S0550-3213(02)00837-4.
71. R. V. Harlander and W. B. Kilgore, Next-to-next-to-leading order Higgs production at hadron colliders, *Phys.Rev.Lett.* **88**, 201801 (2002). doi: 10.1103/PhysRevLett.88.201801.
72. V. Ravindran, J. Smith, and W. L. van Neerven, NNLO corrections to the total cross-section for Higgs boson production in hadron hadron collisions, *Nucl.Phys.* **B665**, 325–366 (2003). doi: 10.1016/S0550-3213(03)00457-7.
73. S. Marzani, R. D. Ball, V. Del Duca, S. Forte, and A. Vicini, Higgs production via gluon-gluon fusion with finite top mass beyond next-to-leading order, *Nucl.Phys.B.* **800**, 127–145 (2008). doi: 10.1016/j.nuclphysb.2008.03.016.
74. R. V. Harlander and K. J. Ozeren, Finite top mass effects for hadronic Higgs production at next-to-next-to-leading order, *JHEP.* **0911**, 088 (2009). doi: 10.1088/1126-6708/2009/11/088.
75. R. V. Harlander, H. Mantler, S. Marzani, and K. J. Ozeren, Higgs production in gluon fusion at next-to-next-to-leading order QCD for finite top mass, *Eur.Phys.J.* **C66**, 359–372 (2010). doi: 10.1140/epjc/s10052-010-1258-x.
76. A. Pak, M. Rogal, and M. Steinhauser, Finite top quark mass effects in NNLO Higgs boson production at LHC, *JHEP.* **1002**, 025 (2010). doi: 10.1007/JHEP02(2010)025.
77. S. Catani, D. de Florian, M. Grazzini, and P. Nason, Soft gluon resummation for Higgs boson production at hadron colliders, *JHEP.* **0307**, 028 (2003).
78. S. Moch and A. Vogt, Higher-order soft corrections to lepton pair and Higgs boson production, *Phys.Lett.* **B631**, 48–57 (2005). doi: 10.1016/j.physletb.2005.09.061.
79. C. Anastasiou, C. Duhr, F. Dulat, F. Herzog, and B. Mistlberger, Higgs boson gluon-fusion production in QCD at three loops, *Phys.Rev.Lett.* **114**(21), 212001 (2015). doi: 10.1103/PhysRevLett.114.212001.
80. A. Djouadi and P. Gambino, Leading electroweak correction to Higgs boson production at proton colliders, *Phys.Rev.Lett.* **73**, 2528–2531 (1994).
81. U. Aglietti, R. Bonciani, G. Degrassi, and A. Vicini, Two loop light fermion contribution to Higgs production and decays, *Phys.Lett.* **B595**, 432–441 (2004). doi: 10.1016/j.physletb.2004.06.063.
82. G. Degrassi and F. Maltoni, Two-loop electroweak corrections to Higgs production at hadron colliders, *Phys.Lett.* **B600**, 255–260 (2004). doi: 10.1016/j.physletb.2004.09.008.

83. S. Actis, G. Passarino, C. Sturm, and S. Uccirati, NLO electroweak corrections to Higgs boson production at hadron colliders, *Phys.Lett.* **B670**, 12–17 (2008). doi: 10.1016/j.physletb.2008.10.018.

84. C. Anastasiou, R. Boughezal, and F. Petriello, Mixed QCD-electroweak corrections to Higgs boson production in gluon fusion, *JHEP.* **0904**, 003 (2009). doi: 10.1088/1126-6708/2009/04/003.

85. W.-Y. Keung and F. J. Petriello, Electroweak and finite quark-mass effects on the Higgs boson transverse momentum distribution, *Phys.Rev.* **D80**, 013007 (2009). doi: 10.1103/PhysRevD.80.013007.

86. O. Brein, Electroweak and bottom quark contributions to Higgs boson plus jet production, *Phys.Rev.* **D81**, 093006 (2010). doi: 10.1103/PhysRevD.81.093006.

87. D. de Florian and M. Grazzini, Higgs production through gluon fusion: Updated cross sections at the Tevatron and the LHC, *Phys.Lett.* **B674**, 291–294 (2009). doi: 10.1016/j.physletb.2009.03.033.

88. C. Anastasiou, S. Buehler, F. Herzog, and A. Lazopoulos, Inclusive Higgs boson cross-section for the LHC at 8 TeV, *JHEP.* **1204**, 004 (2012). doi: 10.1007/JHEP04(2012)004.

89. J. Baglio and A. Djouadi, Predictions for Higgs production at the Tevatron and the associated uncertainties, *JHEP.* **1010**, 064 (2010). doi: 10.1007/JHEP10(2010)064.

90. V. Ahrens, T. Becher, M. Neubert, and L. L. Yang, Updated predictions for Higgs production at the Tevatron and the LHC, *Phys.Lett.* **B698**, 271–274 (2011). doi: 10.1016/j.physletb.2010.12.072.

91. R. Cahn and S. Dawson, Production of very massive Higgs bosons, *Phys.Lett.* **B136**, 196 (1984). doi: 10.1016/0370-2693(84)91180-8.

92. M. Ciccolini, A. Denner, and S. Dittmaier, Electroweak and QCD corrections to Higgs production via vector-boson fusion at the LHC, *Phys.Rev.* **D77**, 013002 (2008). doi: 10.1103/PhysRevD.77.013002.

93. J. R. Andersen and J. M. Smillie, QCD and electroweak interference in Higgs production by gauge boson fusion, *Phys.Rev.* **D75**, 037301 (2007). doi: 10.1103/PhysRevD.75.037301.

94. J. Andersen, T. Binoth, G. Heinrich, and J. Smillie, Loop induced interference effects in Higgs Boson plus two jet production at the LHC, *JHEP.* **0802**, 057 (2008). doi: 10.1088/1126-6708/2008/02/057.

95. A. Bredenstein, K. Hagiwara, and B. Jager, Mixed QCD-electroweak contributions to Higgs-plus-dijet production at the LHC, *Phys.Rev.* **D77**, 073004 (2008). doi: 10.1103/PhysRevD.77.073004.

96. T. Han, G. Valencia, and S. Willenbrock, Structure function approach to vector boson scattering in p p collisions, *Phys.Rev.Lett.* **69**, 3274–3277 (1992). doi: 10.1103/PhysRevLett.69.3274.

97. T. Figy, C. Oleari, and D. Zeppenfeld, Next-to-leading order jet distributions for Higgs boson production via weak boson fusion, *Phys.Rev.* **D68**, 073005 (2003). doi: 10.1103/PhysRevD.68.073005.

98. M. Ciccolini, A. Denner, and S. Dittmaier, Strong and electroweak corrections to the production of Higgs + 2jets via weak interactions at

the LHC, *Phys. Rev. Lett.* **99**, 161803 (2007). doi: 10.1103/PhysRevLett.99. 161803.

99. R. V. Harlander, J. Vollinga, and M. M. Weber, Gluon-induced weak boson fusion, *Phys. Rev.* **D77**, 053010 (2008). doi: 10.1103/PhysRevD.77.053010.

100. P. Bolzoni, F. Maltoni, S.-O. Moch, and M. Zaro, Higgs production via vector-boson fusion at NNLO in QCD, *Phys. Rev. Lett.* **105**, 011801 (2010). doi: 10.1103/PhysRevLett.105.011801.

101. P. Bolzoni, F. Maltoni, S.-O. Moch, and M. Zaro, Vector boson fusion at NNLO in QCD: SM Higgs and beyond, *Phys. Rev.* **D85**, 035002 (2012). doi: 10.1103/PhysRevD.85.035002.

102. S. Glashow, D. V. Nanopoulos, and A. Yildiz, Associated production of Higgs bosons and Z particles, *Phys. Rev.* **D18**, 1724–1727 (1978). doi: 10.1103/PhysRevD.18.1724.

103. T. Han and S. Willenbrock, QCD correction to the $pp \to WH$ and ZH total cross-sections, *Phys. Lett.* **B273**, 167–172 (1991). doi: 10.1016/0370-2693(91)90572-8.

104. M. L. Ciccolini, S. Dittmaier, and M. Krämer, Electroweak radiative corrections to associated WH and ZH production at hadron colliders, *Phys. Rev.* **D68**, 073003 (2003). doi: 10.1103/PhysRevD.68.073003.

105. R. Hamberg, W. L. van Neerven, and T. Matsuura, A complete calculation of the order α_S^2 correction to the Drell–Yan K factor, *Nucl. Phys.* **B359**, 343–405 (1991). doi: 10.1016/0550-3213(91)90064-5.

106. O. Brein, A. Djouadi, and R. Harlander, NNLO QCD corrections to the Higgs-strahlung processes at hadron colliders, *Phys. Lett.* **B579**, 149–156 (2004). doi: 10.1016/j.physletb.2003.10.112.

107. L. Altenkamp, S. Dittmaier, R. V. Harlander, H. Rzehak, and T. J. Zirke, Gluon-induced Higgs-strahlung at next-to-leading order QCD, *JHEP*. **1302**, 078 (2013). doi: 10.1007/JHEP02(2013)078.

108. O. Brein, R. Harlander, M. Wiesemann, and T. Zirke, Top-quark mediated effects in hadronic Higgs-strahlung, *Eur. Phys. J.* **C72**, 1868 (2012). doi: 10.1140/epjc/s10052-012-1868-6.

109. Z. Kunszt, Associated production of heavy Higgs boson with top quarks, *Nucl. Phys.* **B247**, 339 (1984). doi: 10.1016/0550-3213(84)90553-4.

110. W. Beenakker, S. Dittmaier, M. Kramer, B. Plumper, M. Spira, *et al.*, Higgs radiation off top quarks at the Tevatron and the LHC, *Phys. Rev. Lett.* **87**, 201805 (2001). doi: 10.1103/PhysRevLett.87.201805.

111. W. Beenakker, S. Dittmaier, M. Kramer, B. Plumper, M. Spira, *et al.*, NLO QCD corrections to t anti-t H production in hadron collisions, *Nucl. Phys.* **B653**, 151–203 (2003). doi: 10.1016/S0550-3213(03)00044-0.

112. S. Dawson, L. Orr, L. Reina, and D. Wackeroth, Associated top quark Higgs boson production at the LHC, *Phys. Rev.* **D67**, 071503 (2003). doi: 10.1103/PhysRevD.67.071503.

113. S. Dawson, C. Jackson, L. Orr, L. Reina, and D. Wackeroth, Associated Higgs production with top quarks at the large hadron collider: NLO QCD corrections, *Phys. Rev.* **D68**, 034022 (2003). doi: 10.1103/PhysRevD.68. 034022.

114. E. N. Glover and J. van der Bij, Z boson pair production via gluon fusion, *Nucl.Phys.* **B321**, 561 (1989). doi: 10.1016/0550-3213(89)90262-9.

115. G. Valencia and S. Willenbrock, The heavy Higgs resonance, *Phys.Rev.* **D46**, 2247–2251 (1992). doi: 10.1103/PhysRevD.46.2247.

116. C. Anastasiou, S. Buehler, F. Herzog, and A. Lazopoulos, Total cross-section for Higgs boson hadroproduction with anomalous Standard Model interactions, *JHEP.* **1112**, 058 (2011). doi: 10.1007/JHEP12(2011)058.

117. S. Goria, G. Passarino, and D. Rosco, The Higgs boson lineshape, *Nucl.Phys.* **B864**, 530–579 (2012). doi: 10.1016/j.nuclphysb.2012.07.006.

118. L. J. Dixon and M. S. Siu, Resonance continuum interference in the diphoton Higgs signal at the LHC, *Phys.Rev.Lett.* **90**, 252001 (2003). doi: 10.1103/PhysRevLett.90.252001.

119. S. P. Martin, Shift in the LHC Higgs diphoton mass peak from interference with background, *Phys.Rev.* **D86**, 073016 (2012). doi: 10.1103/PhysRevD.86.073016.

120. D. de Florian, N. Fidanza, R. Hernández-Pinto, J. Mazzitelli, Y. Rotstein Habarnau, *et al.*, A complete $O(\alpha_S^2)$ calculation of the signal-background interference for the Higgs diphoton decay channel, *Eur.Phys.J.* **C73**, 2387 (2013). doi: 10.1140/epjc/s10052-013-2387-9.

121. J. M. Campbell, R. K. Ellis, and C. Williams, Gluon-gluon contributions to $W^+ W^-$ Production and Higgs interference effects, *JHEP.* **1110**, 005 (2011). doi: 10.1007/JHEP10(2011)005.

122. N. Kauer and G. Passarino, Inadequacy of zero-width approximation for a light Higgs boson signal, *JHEP.* **1208**, 116 (2012). doi: 10.1007/JHEP08(2012)116.

123. N. Kauer, Interference effects for H $\to$ WW/ZZ $\to$ $\ell\bar{\nu}_\ell\bar{\ell}\nu_\ell$ searches in gluon fusion at the LHC, *JHEP.* **1312**, 082 (2013). doi: 10.1007/JHEP12(2013)082.

124. F. Caola and K. Melnikov, Constraining the Higgs boson width with ZZ production at the LHC, *Phys.Rev.* **D88**, 054024 (2013). doi: 10.1103/PhysRevD.88.054024.

125. M. Dittmar and H. K. Dreiner, How to find a Higgs boson with a mass between 155-GeV–180-GeV at the LHC, *Phys.Rev.* **D55**, 167–172 (1997). doi: 10.1103/PhysRevD.55.167.

126. J. M. Butterworth, A. R. Davison, M. Rubin, and G. P. Salam, Jet substructure as a new Higgs search channel at the LHC, *Phys.Rev.Lett.* **100**, 242001 (2008). doi: 10.1103/PhysRevLett.100.242001.

127. C. Anastasiou, K. Melnikov, and F. Petriello, Fully differential Higgs boson production and the di-photon signal through next-to-next-to-leading order, *Nucl.Phys.* **B724**, 197–246 (2005). doi: 10.1016/j.nuclphysb.2005.06.036.

128. C. Anastasiou, S. Bucherer, and Z. Kunszt, HPro: A NLO Monte-Carlo for Higgs production via gluon fusion with finite heavy quark masses, *JHEP.* **0910**, 068 (2009). doi: 10.1088/1126-6708/2009/10/068.

129. S. Catani and M. Grazzini, An NNLO subtraction formalism in hadron collisions and its application to Higgs boson production at the LHC, *Phys.Rev.Lett.* **98**, 222002 (2007).

130. M. Grazzini, NNLO predictions for the higgs boson signal in the h→ww → *lνlν* and h→zz→4l decay channels, *JHEP.* **0802**, 043 (2008). doi: 10.1088/1126-6708/2008/02/043.

131. D. de Florian, M. Grazzini, and Z. Kunszt, Higgs production with large transverse momentum in hadronic collisions at next-to-leading order, *Phys.Rev.Lett.* **82**, 5209–5212 (1999). doi: 10.1103/PhysRevLett. 82.5209.

132. V. Ravindran, J. Smith, and W. L. Van Neerven, Next-to-leading order QCD corrections to differential distributions of Higgs boson production in hadron hadron collisions, *Nucl.Phys.* **B634**, 247–290 (2002).

133. C. J. Glosser and C. R. Schmidt, Next-to-leading corrections to the Higgs boson transverse momentum spectrum in gluon fusion, *JHEP.* **12**, 016 (2002).

134. G. Bozzi, S. Catani, D. de Florian, and M. Grazzini, Transverse-momentum resummation and the spectrum of the Higgs boson at the LHC, *Nucl.Phys.* **B737**, 73–120 (2006). doi: 10.1016/j.nuclphysb.2005.12.022.

135. D. de Florian, G. Ferrera, M. Grazzini, and D. Tommasini, Transverse-momentum resummation: Higgs boson production at the Tevatron and the LHC, *JHEP.* **1111**, 064 (2011). doi: 10.1007/JHEP11(2011)064.

136. D. de Florian, G. Ferrera, M. Grazzini, and D. Tommasini, Higgs boson production at the LHC: transverse momentum resummation effects in the $H \to 2\gamma$, $H \to WW \to l\nu l\nu$ and $H \to ZZ \to 4l$ decay modes, *JHEP.* **1206**, 132 (2012). doi: 10.1007/JHEP06(2012)132.

137. M. Grazzini and H. Sargsyan, Heavy-quark mass effects in Higgs boson production at the LHC, *JHEP.* **1309**, 129 (2013). doi: 10.1007/JHEP09 (2013)129.

138. G. Davatz, G. Dissertori, M. Dittmar, M. Grazzini, and F. Pauss, Effective K-factors for gg→H→WW→lνlν at the LHC, *JHEP.* **05**, 009 (2004).

139. R. Boughezal, F. Caola, K. Melnikov, F. Petriello, and M. Schulze, Higgs boson production in association with a jet at next-to-next-to-leading order in perturbative QCD, *JHEP.* **1306**, 072 (2013). doi: 10.1007/JHEP06(2013)072.

140. X. Chen, T. Gehrmann, E. Glover, and M. Jaquier, Precise QCD predictions for the production of Higgs + jet final states, *Phys.Lett.* **B740**, 147–150 (2015). doi: 10.1016/j.physletb.2014.11.021.

141. G. Ferrera, M. Grazzini, and F. Tramontano, Associated WH production at hadron colliders: a fully exclusive QCD calculation at NNLO, *Phys.Rev.Lett.* **107**, 152003 (2011). doi: 10.1103/PhysRevLett.107. 152003.

142. G. Ferrera, M. Grazzini, and F. Tramontano, Associated *ZH* production at hadron colliders: the fully differential NNLO QCD calculation, *Phys.Lett.* **B740**, 51–55 (2015). doi: 10.1016/j.physletb.2014.11.040.

143. A. Banfi and J. Cancino, Implications of QCD radiative corrections on high-pT Higgs searches, *Phys.Lett.* **B718**, 499–506 (2012). doi: 10.1016/j.physletb.2012.10.064.

144. G. Ferrera, M. Grazzini, and F. Tramontano, Higher-order QCD effects for associated WH production and decay at the LHC, *JHEP.* **1404**, 039 (2014). doi: 10.1007/JHEP04(2014)039.

145. LHC Higgs Cross Section Working Group, Handbook of LHC Higgs Cross Sections: 3. Higgs Properties (2013). doi: 10.5170/CERN-2013-004. URL http://arxiv.org/abs/1307.1347.

146. P. Artoisenet, R. Frederix, O. Mattelaer, and R. Rietkerk, Automatic spin-entangled decays of heavy resonances in Monte Carlo simulations, *JHEP.* **1303**, 015 (2013). doi: 10.1007/JHEP03(2013)015.

147. I. W. Stewart and F. J. Tackmann, Theory uncertainties for Higgs and other searches using jet bins, *Phys.Rev.* **D85**, 034011 (2012). doi: 10.1103/PhysRevD.85.034011.

148. A. Banfi, P. F. Monni, G. P. Salam, and G. Zanderighi, Higgs and Z-boson production with a jet veto, *Phys.Rev.Lett.* **109**, 202001 (2012). doi: 10.1103/PhysRevLett.109.202001.

149. T. Becher, M. Neubert, and L. Rothen, Factorization and $N^3 LL_p$+NNLO predictions for the Higgs cross section with a jet veto, *JHEP.* **1310**, 125 (2013). doi: 10.1007/JHEP10(2013)125.

150. I. W. Stewart, F. J. Tackmann, J. R. Walsh, and S. Zuberi, Jet p_T resummation in Higgs production at $NNLL' + NNLO$, *Phys.Rev.* **D89**(5), 054001 (2014). doi: 10.1103/PhysRevD.89.054001.

151. A. Buckley, J. Butterworth, S. Gieseke, D. Grellscheid, S. Hoche, *et al.*, General-purpose event generators for LHC physics, *Phys.Rept.* **504**, 145–233 (2011). doi: 10.1016/j.physrep.2011.03.005.

152. S. Frixione and B. R. Webber, Matching NLO QCD computations and parton shower simulations, *JHEP.* **0206**, 029 (2002). doi: 10.1088/1126-6708/2002/06/029.

153. R. Frederix, S. Frixione, V. Hirschi, F. Maltoni, P. Roberto, *et al.*, Scalar and pseudoscalar Higgs production in association with a top–antitop pair, *Phys.Lett.* **B701**, 427–433 (2011). doi: 10.1016/j.physletb.2011.06.012.

154. S. Frixione, P. Nason, and C. Oleari, Matching NLO QCD computations with parton shower simulations: the POWHEG method, *JHEP.* **0711**, 070 (2007). doi: 10.1088/1126-6708/2007/11/070.

155. T. Gleisberg, S. Hoeche, F. Krauss, M. Schonherr, S. Schumann, *et al.*, Event generation with SHERPA 1.1, *JHEP.* **0902**, 007 (2009). doi: 10.1088/1126-6708/2009/02/007.

156. K. Hamilton, P. Richardson, and J. Tully, A positive-weight next-to-leading order Monte Carlo simulation for Higgs boson production, *JHEP.* **0904**, 116 (2009). doi: 10.1088/1126-6708/2009/04/116.

157. S. Alioli, P. Nason, C. Oleari, and E. Re, NLO Higgs boson production via gluon fusion matched with shower in POWHEG, *JHEP.* **0904**, 002 (2009). doi: 10.1088/1126-6708/2009/04/002.

158. P. Nason and C. Oleari, NLO Higgs boson production via vector-boson fusion matched with shower in POWHEG, *JHEP.* **1002**, 037 (2010). doi: 10.1007/JHEP02(2010)037.

159. L. D'Errico and P. Richardson, A positive-weight next-to-leading-order Monte Carlo simulation of deep inelastic scattering and Higgs boson production via vector boson fusion in Herwig++, *Eur.Phys.J.* **C72**, 2042 (2012). doi: 10.1140/epjc/s10052-012-2042-x.

160. M. Garzelli, A. Kardos, C. Papadopoulos, and Z. Trocsanyi, Standard Model Higgs boson production in association with a top anti-top pair at NLO with parton showering, *Europhys.Lett.* **96**, 11001 (2011). doi: 10.1209/0295-5075/96/11001.

161. K. Hamilton, A positive-weight next-to-leading order simulation of weak boson pair production, *JHEP.* **1101**, 009 (2011). doi: 10.1007/JHEP01(2011)009.

162. R. Frederix, S. Frixione, V. Hirschi, F. Maltoni, R. Pittau, *et al.*, W and $Z/\gamma*$ boson production in association with a bottom–antibottom pair, *JHEP.* **1109**, 061 (2011). doi: 10.1007/JHEP09(2011)061.

163. T. Melia, P. Nason, R. Rontsch, and G. Zanderighi, W^+W^-, WZ and ZZ production in the POWHEG BOX, *JHEP.* **1111**, 078 (2011). doi: 10.1007/JHEP11(2011)078.

164. F. Cascioli, S. Hche, F. Krauss, P. Maierhfer, S. Pozzorini, *et al.*, Precise Higgs-background predictions: merging NLO QCD and squared quark-loop corrections to four-lepton + 0,1 jet production, *JHEP.* **1401**, 046 (2014). doi: 10.1007/JHEP01(2014)046.

165. S. Catani, F. Krauss, R. Kuhn, and B. Webber, QCD matrix elements + parton showers, *JHEP.* **0111**, 063 (2001). doi: 10.1088/1126-6708/2001/11/063.

166. J. Alwall, S. Hoche, F. Krauss, N. Lavesson, L. Lonnblad, *et al.*, Comparative study of various algorithms for the merging of parton showers and matrix elements in hadronic collisions, *Eur.Phys.J.* **C53**, 473–500 (2008). doi: 10.1140/epjc/s10052-007-0490-5.

167. R. Frederix and S. Frixione, Merging meets matching in MC@NLO, *JHEP.* **1212**, 061 (2012). doi: 10.1007/JHEP12(2012)061.

168. K. Hamilton, P. Nason, and G. Zanderighi, MINLO: Multi-scale improved NLO, *JHEP.* **1210**, 155 (2012). doi: 10.1007/JHEP10(2012)155.

169. K. Hamilton, P. Nason, C. Oleari, and G. Zanderighi, Merging H/W/Z + 0 and 1 jet at NLO with no merging scale: a path to parton shower + NNLO matching, *JHEP.* **1305**, 082 (2013). doi: 10.1007/JHEP05(2013)082.

170. S. Alioli, C. W. Bauer, C. Berggren, F. J. Tackmann, J. R. Walsh, *et al.*, Matching fully differential NNLO calculations and parton showers, *JHEP.* **1406**, 089 (2014). doi: 10.1007/JHEP06(2014)089.

171. S. Höche, Y. Li, and S. Prestel, Drell–Yan lepton pair production at NNLO QCD with parton showers, *Phys.Rev.* **D91**(7), 074015 (2015). doi: 10.1103/PhysRevD.91.074015.

172. K. Hamilton, P. Nason, E. Re, and G. Zanderighi, NNLOPS simulation of Higgs boson production, *JHEP.* **1310**, 222 (2013). doi: 10.1007/JHEP10(2013)222.

173. S. Höche, Y. Li, and S. Prestel, Higgs-boson production through gluon fusion at NNLO QCD with parton showers, *Phys.Rev.* **D90**(5), 054011 (2014). doi: 10.1103/PhysRevD.90.054011.

174. D. Zeppenfeld, R. Kinnunen, A. Nikitenko, and E. Richter-Was, Measuring Higgs boson couplings at the CERN LHC, *Phys.Rev.* **D62**, 013009 (2000). doi: 10.1103/PhysRevD.62.013009.

175. M. Duhrssen, S. Heinemeyer, H. Logan, D. Rainwater, G. Weiglein, *et al.*, Extracting Higgs boson couplings from CERN LHC data, *Phys.Rev.* **D70**, 113009 (2004). doi: 10.1103/PhysRevD.70.113009.

176. J. M. Campbell, R. K. Ellis, and C. Williams, Bounding the Higgs width at the LHC using full analytic results for $gg \to e^- e^+ \mu^- \mu^+$, *JHEP.* **1404**, 060 (2014). doi: 10.1007/JHEP04(2014)060.

177. J. M. Campbell, R. K. Ellis, and C. Williams, Bounding the Higgs width at the LHC: Complementary results from $H \to WW$, *Phys.Rev.* **D89**(5), 053011 (2014). doi: 10.1103/PhysRevD.89.053011.

178. J. S. Gainer, J. Lykken, K. T. Matchev, S. Mrenna, and M. Park, Beyond geolocating: Constraining higher dimensional operators in $H \to 4\ell$ with off-shell production and more, *Phys.Rev.* **D91**(3), 035011 (2015). doi: 10.1103/PhysRevD.91.035011.

179. O. J. Eboli and D. Zeppenfeld, Observing an invisible Higgs boson, *Phys.Lett.* **B495**, 147–154 (2000). doi: 10.1016/S0370-2693(00)01213-2.

180. A. Djouadi, A. Falkowski, Y. Mambrini, and J. Quevillon, Direct detection of Higgs-portal dark matter at the LHC, *Eur.Phys.J.* **C73**(6), 2455 (2013). doi: 10.1140/epjc/s10052-013-2455-1.

181. A. Djouadi and G. Moreau, The couplings of the Higgs boson and its CP properties from fits of the signal strengths and their ratios at the 7+8 TeV LHC, *Eur.Phys.J.* **C73**(9), 2512 (2013). doi: 10.1140/epjc/s10052-013-2512-9.

182. L. Landau, On the angular momentum of a two-photon system, *Dokl.Akad.Nauk Ser.Fiz.* **60**, 207–209 (1948).

183. C.-N. Yang, Selection rules for the dematerialization of a particle into two photons, *Phys.Rev.* **77**, 242–245 (1950). doi: 10.1103/PhysRev.77.242.

184. G. Velo and D. Zwanziger, Propagation and quantization of Rarita–Schwinger waves in an external electromagnetic potential, *Phys.Rev.* **186**, 1337–1341 (1969). doi: 10.1103/PhysRev.186.1337.

185. G. Velo and D. Zwanziger, Noncausality and other defects of interaction lagrangians for particles with spin one and higher, *Phys.Rev.* **188**, 2218–2222 (1969). doi: 10.1103/PhysRev.188.2218.

186. M. Porrati and R. Rahman, Intrinsic cutoff and acausality for massive spin 2 fields coupled to electromagnetism, *Nucl.Phys.* **B801**, 174–186 (2008). doi: 10.1016/j.nuclphysb.2008.05.013.

187. R. Plano, A. Prodell, N. Samios, M. Schwartz, and J. Steinberger, Parity of the neutral pion, *Phys.Rev.Lett.* **3**, 525–527 (1959). doi: 10.1103/PhysRevLett.3.525.

188. J. R. Dell'Aquila and C. A. Nelson, *P* or CP Determination by sequential decays: V1 V2 modes with decays into $\bar{\ell}$epton (A) $\ell(B)$ and/or $\bar{q}$ (A) $q(B)$, *Phys.Rev.* **D33**, 80 (1986). doi: 10.1103/PhysRevD.33.80.

189. V. D. Barger, K.-m. Cheung, A. Djouadi, B. A. Kniehl, and P. Zerwas, Higgs bosons: Intermediate mass range at e$^+$ e$^-$ colliders, *Phys.Rev.* **D49**, 79–90 (1994). doi: 10.1103/PhysRevD.49.79.

190. S. Choi, D. Miller, M. Mühlleitner, and P. Zerwas, Identifying the Higgs spin and parity in decays to Z pairs, *Phys.Lett.* **B553**, 61–71 (2003). doi: 10.1016/S0370-2693(02)03191-X.

191. A. De Rujula, J. Lykken, M. Pierini, C. Rogan, and M. Spiropulu, Higgs look-alikes at the LHC, *Phys.Rev.* **D82**, 013003 (2010). doi: 10.1103/PhysRevD.82.013003.

192. Y. Gao, A. V. Gritsan, Z. Guo, K. Melnikov, M. Schulze, *et al.*, Spin determination of single-produced resonances at hadron colliders, *Phys.Rev.* **D81**, 075022 (2010). doi: 10.1103/PhysRevD.81.075022.

193. S. Bolognesi, Y. Gao, A. V. Gritsan, K. Melnikov, M. Schulze, *et al.*, On the spin and parity of a single-produced resonance at the LHC, *Phys.Rev.* **D86**, 095031 (2012). doi: 10.1103/PhysRevD.86.095031.

194. I. Anderson, S. Bolognesi, F. Caola, Y. Gao, A. V. Gritsan, *et al.*, Constraining anomalous HVV interactions at proton and lepton colliders, *Phys.Rev.* **D89**(3), 035007 (2014). doi: 10.1103/PhysRevD.89.035007.

195. A. Alloul, N. D. Christensen, C. Degrande, C. Duhr, and B. Fuks, FeynRules 2.0 — A complete toolbox for tree-level phenomenology, *Comput.Phys.Commun.* **185**, 2250–2300 (2014). doi: 10.1016/j.cpc.2014.04.012.

196. P. Avery, D. Bourilkov, M. Chen, T. Cheng, A. Drozdetskiy, *et al.*, Precision studies of the Higgs boson decay channel $H \to ZZ \to 4\ell$ with MEKD, *Phys.Rev.* **D87**(5), 055006 (2013). doi: 10.1103/PhysRevD.87.055006.

197. P. Artoisenet, P. de Aquino, F. Demartin, R. Frederix, S. Frixione, *et al.*, A framework for Higgs characterisation, *JHEP.* **1311**, 043 (2013). doi: 10.1007/JHEP11(2013)043.

198. J. Alwall, M. Herquet, F. Maltoni, O. Mattelaer, and T. Stelzer, MadGraph 5: Going beyond, *JHEP.* **1106**, 128 (2011). doi: 10.1007/JHEP06(2011)128.

199. A. Belyaev, N. D. Christensen, and A. Pukhov, CalcHEP 3.4 for collider physics within and beyond the Standard Model, *Comput.Phys.Commun.* **184**, 1729–1769 (2013). doi: 10.1016/j.cpc.2013.01.014.

200. T. Plehn, D. L. Rainwater, and D. Zeppenfeld, Determining the structure of Higgs couplings at the LHC, *Phys.Rev.Lett.* **88**, 051801 (2002). doi: 10.1103/PhysRevLett.88.051801.

201. J. Frank, M. Rauch, and D. Zeppenfeld, Spin-2 resonances in vector-boson-fusion processes at NLO QCD, *Phys.Rev.* **D87**, 055020 (2013). doi: 10.1103/PhysRevD.87.055020.

202. C. Englert, D. Goncalves-Netto, K. Mawatari, and T. Plehn, Higgs quantum numbers in weak boson fusion, *JHEP.* **1301**, 148 (2013). doi: 10.1007/JHEP01(2013)148.

203. A. Djouadi, R. Godbole, B. Mellado, and K. Mohan, Probing the spin-parity of the Higgs boson via jet kinematics in vector boson fusion, *Phys.Lett.* **B723**, 307–313 (2013). doi: 10.1016/j.physletb.2013.04.060.
204. J. F. Gunion and X.-G. He, Determining the CP nature of a neutral Higgs boson at the LHC, *Phys.Rev.Lett.* **76**, 4468–4471 (1996). doi: 10.1103/Phys-RevLett.76.4468.
205. J. F. Gunion and J. Pliszka, Determining the relative size of the CP even and CP odd Higgs boson couplings to a fermion at the LHC, *Phys.Lett.* **B444**, 136–141 (1998). doi: 10.1016/S0370-2693(98)01364-1.
206. B. Field, Distinguishing scalar from pseudoscalar Higgs production at the CERN LHC, *Phys.Rev.* **D66**, 114007 (2002). doi: 10.1103/Phys-RevD.66.114007.
207. S. Berge, W. Bernreuther, and S. Kirchner, Determination of the Higgs CP-mixing angle in the tau decay channels (2014). arXiv: 1410.6362.
208. E. N. Glover and J. van der Bij, Higgs boson pair production via gluon fusion, *Nucl.Phys.* **B309**, 282 (1988). doi: 10.1016/0550-3213(88)90083-1.
209. O. J. Eboli, G. Marques, S. Novaes, and A. Natale, Twin Higgs boson production, *Phys.Lett.* **B197**, 269 (1987). doi: 10.1016/0370-2693(87)90381-9.
210. T. Plehn, M. Spira, and P. Zerwas, Pair production of neutral Higgs particles in gluon–gluon collisions, *Nucl.Phys.* **B479**, 46–64 (1996). doi: 10.1016/0550-3213(96)00418-X.
211. A. Djouadi, W. Kilian, M. Muhlleitner, and P. Zerwas, Production of neutral Higgs boson pairs at LHC, *Eur.Phys.J.* **C10**, 45–49 (1999). doi: 10.1007/s100529900083.
212. S. Dawson, S. Dittmaier, and M. Spira, Neutral Higgs boson pair production at hadron colliders: QCD corrections, *Phys.Rev.* **D58**, 115012 (1998). doi: 10.1103/PhysRevD.58.115012.
213. D. de Florian and J. Mazzitelli, Higgs boson pair production at next-to-next-to-leading order in QCD, *Phys.Rev.Lett.* **111**, 201801 (2013). doi: 10.1103/PhysRevLett.111.201801.
214. J. Baglio, A. Djouadi, R. Gröber, M. Mühlleitner, J. Quevillon, *et al.*, The measurement of the Higgs self-coupling at the LHC: theoretical status, *JHEP.* **1304**, 151 (2013). doi: 10.1007/JHEP04(2013)151.

Chapter 2

Searches for the Standard Model Higgs boson at the LEP collider

Peter Igo-Kemenes[*,†] and Alexander L. Read[‡]

*Gjøvik University College
P.O. 191, Teknologivn. 22, 2802 Gjøvik, Norway
†CERN, 1211 Geneva 23, Switzerland
‡Department of Physics, University of Oslo
Postbox 1048 Blindern, 0316 Oslo, Norway

The Large Electron Positron (LEP) collider installed at CERN provided unprecedented possibilities for studying the properties of elementary particles during the years 1989–2000. The four detectors associated to the collider, run by the ALEPH, DELPHI, L3, and OPAL Collaborations, were based on the latest available technologies. The conjunction of high collision energies, precise instrumentation and data analysis techniques allowed the Standard Model (SM) of elementary particles to be tested at the level of quantum corrections. The search for new particles, in particular the long-sought Higgs boson, was one of the primary research subjects. During the twelve years of LEP, data samples of the highest quality and statistical weight were analysed. Concerning the search for the SM Higgs boson, the domain extending from zero mass to the kinematic limit imposed by the collider energy was scrutinised. The spirit of scientific competition gradually gave way to a collaborative effort, allowing the final results of LEP to be optimised. The methodology of Higgs boson searches is summarised in this paper together with the statistical methods adopted to combine the data of the four collaborations.

1. The experimental environment

The Large Electron Positron (LEP) collider at CERN[1] was installed deep underground in a tunnel with a circumference of 26.7 km, located between the Jura mountains (France) and Lake Geneva (Switzerland). The counter-rotating beams of electrons and positrons crossed each other at four interaction points, where four large-scale, general purpose detector complexes, ALEPH,[2] DELPHI,[3] L3[4] and OPAL,[5] were installed.

The aim of the project was to test the predictions of the Standard Model (SM), notably through precision measurements of the properties of the Z, W^+, and W^- vector bosons. Differences between the observed and predicted properties would expose the limitations of the theory and possibly orient the research towards new physics beyond the SM.

The collisions of two elementary (or point-like) particles, such as electrons and positrons, are particularly suited for precision measurements since in such collisions the initial state is completely defined by the centre-of-mass (or collision) energy (E_{cm}).

The search for yet undetected particles, such as the SM Higgs boson or particles predicted by more speculative models, constituted another important part of the LEP programme. In such searches, the precise knowledge of the initial state can be used as a constraint in the interpretation of the observed final state.

Another great advantage of electron–positron colliders over hadron colliders is the relatively clean experimental environment, notably for the detection of rare processes such as Higgs boson production. The collisions of electrons and positrons are governed by the electroweak interaction, whereas collisions of protons are governed by the strong interaction and result in the production of hadron jets along with the rare processes of interest. In electron–positron collisions, the rare processes of interest are easily distinguished from the lower-energy background arising, e.g., from beam particles hitting the residual gas molecules in the evacuated beam pipe or scraping the collimator elements close to the detector zones.

The twelve years of LEP operation included two main phases. During the LEP 1 phase (1989–1995), the accelerator was tuned to energies close to the mass of the Z boson, around $\sqrt{s} = 91$ GeV. The collider was thus operated as a Z boson factory and the emphasis was put on precise measurements of the Z production and decay rates, its line-shape (production cross-section as a function of E_{cm}), partial and total decay widths, angular distributions of decay particles and various charge- and polarisation-asymmetries. The Higgs boson (H), as well as other new particles, were sought among the decay products of the Z boson.

During the LEP 2 phase (1995–2000), the collision energy was increased in steps until it reached the maximal value of 209 GeV, just before the final shutdown of LEP. The first big leap, from 90 GeV to 135 GeV, was followed by successive increases of about 10 GeV each. Soon the threshold of

160 GeV for the production of W^+W^- pairs was crossed, and the detailed study of the W boson could begin. At these energies, the Higgs boson was expected to be produced principally in association with a Z boson on the mass-shell, $e^+e^- \to HZ$. Hence, the search for the Higgs boson was limited by kinematics to masses (m_H) smaller than $E_{cm} - 91$ GeV, or 118 GeV at the maximal energy of 209 GeV. With each step in energy, the search entered into unknown territory and the hope for a rapid discovery of the Higgs boson was fully justified.

2. The LEP detectors

The main characteristics of the four LEP detectors, ALEPH, DELPHI, L3 and OPAL, were imposed by the physics programme laid out at the planning stage, which included precision measurements of the Z and W vector boson properties and the search for new particles. This broad programme made it necessary to conceive multi-purpose detectors, capable of responding simultaneously to the requirements of high precision and sensitivity to new phenomena.

The most judicious choice was that of a forward–backward symmetric geometry and a full (i.e. 4π) geometric coverage. To accommodate the solenoidal spectrometer magnets, which are typical for collider experiments with 4π coverage, each LEP detector was composed of a cylindrically symmetric barrel part enclosed between two planar end-caps, with two small holes that just allowed the beam-pipe to pass. Although this geometry was adopted by all four collaborations, the technical implementations of the detectors were quite different. For the purpose of illustration, a schematic representation of the OPAL detector is shown in Fig. 2.1.

It would be hard to do full justice in this short outline to all the effort that was invested during the ten years of planning and construction. The instruments used were at the cutting edge of technology and often specially developed for these experiments. Detailed technical descriptions are available in Refs. [2–5].

For a successful event reconstruction, all visible final-state particles created in the e^+e^- collision had to be identified (such that their invariant masses could be inferred); their momenta and energies had to be measured and their electric charges determined.

The momenta and electric charges were determined from the precise measurement of the particle trajectories in a large-volume tracking chamber

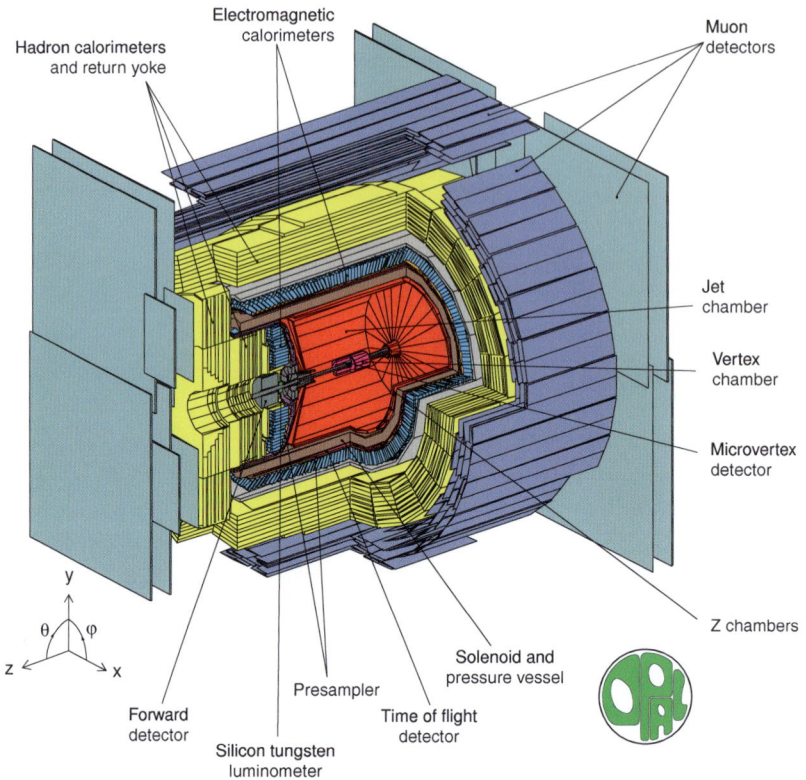

Fig. 2.1   A schematic representation of the OPAL detector. The tracking chamber (in red) consisted of three parts: the gas-filled jet and vertex chambers, and the silicon micro-strip vertex detector. The electromagnetic calorimeter was situated just outside the pressure vessel of the tracking chamber and the solenoid magnet. The iron return yoke (yellow) of the magnet was instrumented and used as hadron calorimeter. The muon chambers (blue) completed the setup.

filled with a special counting gas and immersed in the uniform axial magnetic field of the solenoid magnet surrounding the tracking chamber volume. The direction and bending radius of the helical trajectories determined, respectively, the charges and the momenta of the particles. The trajectories were made visible through the ionisation produced in the gas along the particle tracks. The ionisation charges were made to drift to a large number of parallel counting wires which spanned the whole chamber volume and acted as electrodes. The charges were collected, read out, and interpreted as coordinates along the particle tracks. In this way, typically about one hundred coordinates along the trajectories were measured.

The energies of the particles were measured by two types of calorimeters: one sensitive to the energy deposited by electrons, positrons and photons in the form of electromagnetic showers, and the other sensitive to the energy deposited by charged and neutral hadrons such as pions, kaons, protons, and neutrons, in the form of hadronic showers.

The electromagnetic calorimeters employed at LEP could be best described as insect-eyes seen from the inside. These consisted of a multitude of cells, all oriented towards the interaction point. The cells were made of scintillating material, lead-doped glass for OPAL, and BGO ($Bi_3Ge_4O_{12}$) crystals for L3. ALEPH used layers of lead and charge-sensitive wire-chamber pads to build up the cells while in the case of DELPHI a lead sampling calorimeter with high spatial resolution, based on the time-projection technique, was used. Since the calorimeter cells were sufficiently deep to fully absorb the electromagnetic showers, the signal size could be used directly as a measure of the energy of the initial electromagnetically interacting particle.

Passing hadronic particles only deposited a small fraction of their energy in the electromagnetic calorimeter; however, they produced showers of charged and neutral particles in the hadron calorimeter. The latter typically used the layers of iron of the magnet return yoke interleaved with layers of charge-sensitive detectors, scintillator plates or gaseous wire chambers. The shower shape (depth and width) and the total charge collected in the active detector layers were directly related to the energy of the initial hadronic particle.

Muons traversed both the electromagnetic and the massive hadron calorimeters producing only a minimum ionising signal; however, they created a detectable signal in the outer muon detector system that typically consisted of several layers of gaseous wire chambers.

The L3 detector differed from the other three detectors in that the tracking chamber, the electromagnetic and hadron calorimeters (the latter being made of depleted uranium), and the muon chambers, were all placed inside the magnet coil, with a diameter of about 12 meters. Hence, the bending of muon trajectories could be followed over about 6 meters, resulting in a highly accurate muon momentum measurement.

In summary, a typical LEP detector consisted of a large cylinder with two end-caps, about 10 to 12 meters in diameter and 10 meters long, with several layers of sophisticated detector elements, each reacting differently to the various types of particles emerging from the interaction point and traversing them. Particle identification was based on the full information that was collected from the various detector layers, fed into the online computer system and evaluated. A first online event selection was made on the basis of global criteria, such as the total visible energy or the overall event shape and orientation, that rejected most of the accelerator-related background. The storage of event data for later analysis was triggered only if the global criteria corresponded to those expected from energetic e^+e^- interactions including head-on collisions and peripheral interactions where a pair of intermediate vector bosons (photons, virtual Z, W^+, and W^- bosons) are produced and interact with each other. The detailed analysis and full reconstruction of the events were performed off line. The role of peripheral interactions was modest in Higgs boson production at LEP energies but was important for other aspects of LEP physics.

Neutrinos emerging from the decay of the Z or W bosons traversed the whole detector without leaving a signal in any of the layers. Such "invisible" particles appeared as missing energy and missing momentum once the whole event was reconstructed and the final state compared to the initial state. If the missing momentum vector pointed along the beam-pipe, then its origin was ambiguous since it could be due to particles escaping. If however it pointed into instrumented parts of the detector, then it was likely due to a genuine invisible particle. The analysis of missing energy and momentum was one of the powerful tools in the search for the Higgs boson.

The importance of detecting b-flavoured hadrons and their use for identifying Higgs boson decays were already mentioned in Chapter 1. While unstable hadrons of other flavours tend to fragment or decay immediately where they are produced, those carrying b-flavour travel typically a few millimetres before they decay, due to their comparatively long lifetime. One thus may observe secondary vertices displaced with respect to the interaction point, or at least some tracks that do not point back precisely to the interaction point. These are telltale signs of b-flavour. The backward extrapolation of particle trajectories was substantially improved by the usage of high-resolution tracking chambers and/or charge-sensitive microvertex detectors made of thin layers of silicon pixels or strips surrounding the beam-pipe as closely as possible. These layers provided a few high-precision coordinates at the very beginning of the trajectories.

Combining all the information available from the various detector elements, the invariant mass of a Higgs boson with hypothetical mass of 100 GeV decaying into $b\bar{b}$ could be reconstructed with a mass resolution of about 2 GeV and its decay vertex located with a precision of a few hundred microns.

3. Detection efficiencies, Monte Carlo simulation

A central problem in measurements using complex devices such as the LEP detectors is the determination of the detection efficiency for a given physics process, i.e. the probability that the process, if it took place, was indeed detected.

There are a number of potential sources of inefficiency, of which one has already been mentioned: particles may escape through the beam-pipe. However, even the instrumented parts usually add to the inefficiency due to inactive detector materials such as mechanical support frames, inlets for power cables and outlets for signal cables, supply lines for gas and cooling and small gaps between active detector elements. The detector regions most exposed to such inefficiencies were where the barrel and end-cap parts fitted together.

Other sources of inefficiency are due to the intrinsic properties of the detector materials (scintillators, counting gas, crystals, etc.) and the associated read-out electronics which pick up, amplify, transform and transmit the signals to the on-line computer system. The amount of light produced in a scintillating material as a result of the underlying random process may happen to be too low to be detected; similarly, the amount of ionisation produced locally in a multi-wire gas chamber may be just below the detection threshold of the associated read-out electronics. All these factors contribute to inefficiencies that are rather difficult to account for precisely.

The most common approach, widely adopted at LEP, is to use sophisticated Monte Carlo simulations that take into account, down to the finest detail, all known geometrical and intrinsic properties of the active and passive detector parts and related electronics. These properties were previously determined in a vast programme of test beam calibrations. An event generator is used to simulate the physics process according to previous knowledge (e.g. predictions of the SM). The four-momentum vectors of the simulated particles are fed into the detector simulation that makes them step through the different active and passive detector materials, mimicking interactions as if they were real particles. Some of these particles do not generate a signal in the active detector materials and go undetected, while others make it through the simulation chain, do create a detectable signal and are registered. The fraction of events registered at the end of the chain is an estimate of the true detection efficiency for the simulated physics process.

In the search for the Higgs boson, simulations were made separately for the Higgs production signal process and for all known SM processes which

might contribute to the physics background. The selection criteria which
were applied to the real data were also applied to the simulated events
and the signal and background detection efficiencies, including the effect
of selection criteria, were thus obtained. Typically, the simulated samples
contained many more events than the measured number of events, such
that the statistical error of the simulation did not contribute significantly
to the overall statistical error.

Such a complex simulation process introduces a number of uncertainties
which might systematically bias the outcome. Comparing measured and
simulated event distributions for well-known processes and repeating the
simulations while varying the relevant parameters within their uncertainty
ranges was one of many ways of evaluating the systematic uncertainties.
These are usually presented together with the statistical uncertainties as
part of the final result. Methods or event selections bearing large systematic
uncertainties were avoided in order to keep the total (quadrature sum of
statistical and systematic) uncertainty low.

4. Higgs boson searches during the LEP 1 phase

During the first five years of LEP operation, emphasis was put on the
detailed study of the Z boson. The accelerator was tuned to energies close to
the Z boson mass, 91 GeV. All accessible production and decay properties
of the Z resonance were evaluated, providing a fertile ground for testing
the predictions of the SM. The precise knowledge of the SM processes also
provided a firm ground for the detection of the rare Higgs boson production
process.

Searches for the Higgs boson, which is the focus of this book, were
already conducted well before the LEP era. These concentrated on the
decays of atomic nuclei, pions, kaons, b-flavoured hadrons and the upsilon $b\bar{b}$
mesonic state, and were therefore limited to masses less than about 5 GeV.[6]
These early searches all turned out negative, but no mass domain could
be unambiguously excluded due to a number of theoretical uncertainties.
Consequently, the searches at LEP had to start at zero mass.

After only one year of running, several LEP experiments presented
results on low-mass searches and excluded the range between zero and about
20 GeV.[7-15]

In the decay of the Z boson, the neutral spin-zero (scalar) Higgs particle
ought to be accompanied by a virtual (off-shell) Z^* boson carrying the same

quantum numbers as the real Z boson: $Z \to HZ^*$. The Z^* decays just like the real Z boson into fermion–anti-fermion pairs: e^+e^-, $\mu^+\mu^-$ and $\tau^+\tau^-$ (3.4% each), $\nu\bar{\nu}$ (20.1%) and $q\bar{q}$ (which may possibly be accompanied by gluons; 69.7%). Hence, in the final state, the decay products of the Higgs boson are accompanied by a fermion–anti-fermion pair.

If the Higgs boson is very light, the production is relatively abundant, representing about 1% of all Z boson decays.[16] However, the signal fraction decreases rapidly with increasing mass, reaching values as low as 5×10^{-5} for a Higgs boson mass of 50 GeV. The successive extensions of the searches towards higher masses were therefore limited throughout the LEP 1 phase by the available event statistics.

4.1. *Searches for a low-mass Higgs boson*

The searches conducted during the first year of LEP were mostly "cut-based". Selection criteria (cuts) were devised to suppress the background, at the cost of a moderate loss of the expected signal obtained from Monte Carlo simulations, as discussed earlier. The number of observed events remaining after selection was compared to that expected on the basis of SM predictions, using standard statistical methods, in order to conclude about the existence or absence of a signal.

In the low-mass domain, i.e. below about 10 GeV, the decay pattern of the Higgs boson varies rapidly as a function of the mass, as shown in Fig. 2.2. The threshold behaviour shown is typical for the Yukawa coupling (see Chapter 1): the decay into the most massive fermion–anti-fermion pair that is allowed by energy conservation is always dominant.

The ALEPH, L3 and OPAL Collaborations used a number of complementary searches to investigate the domain of lowest masses (see particularly Ref. [7]).

An important channel, covering a relatively wide mass range, was $Z \to HZ^* \to (H \to$ anything$)$ $(Z^* \to \nu\bar{\nu})$. The final state consists of a mono-jet formed by the decay products of the boosted Higgs boson and a large amount of missing energy from the undetected neutrinos. This search was complemented by a search for $Z \to HZ^* \to (H \to$ anything$)(Z^* \to e^+e^-,$ $\mu^+\mu^-$, $\tau^+\tau^-)$ where the mono-jet is accompanied by a pair of oppositely charged leptons in the hemisphere opposite to the mono-jet.

In general, for a Higgs boson with mass below 10–12 GeV, the decay into two oppositely charged particles only would be predominant; hence, the channel $Z \to HZ^* \to (H \to 2$ particles$)(Z^* \to q\bar{q}(g))$ could also be exploited.

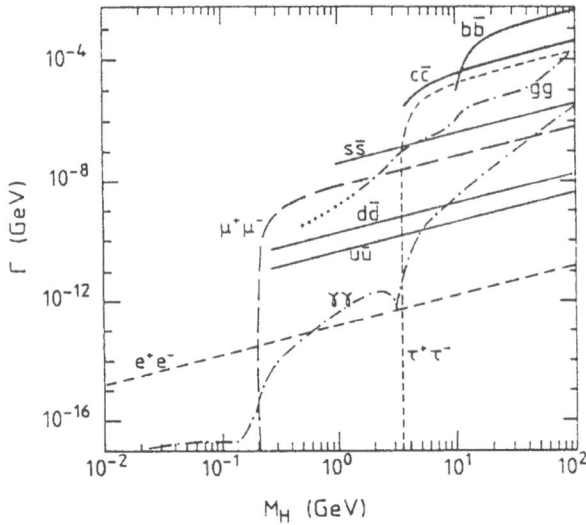

Fig. 2.2   Partial decay widths of the Higgs boson over a wide mass range, showing in detail the lowest mass domain, as predicted by the SM (from Ref. [6]).

For a Higgs boson lighter than a $\mu^+\mu^-$ pair (i.e. 212 MeV), the decay would be into an electron–positron pair. Moreover, such a light Higgs boson would have a small decay rate (see Fig. 2.2), corresponding to a long lifetime (see Chapter 1), and would therefore travel a measurable distance before decaying. The lower the mass, the larger the typical separation of the Higgs decay vertex from the interaction vertex. However, for masses less than about 50 MeV, the decay would occur beyond the tracking chamber limits and hence the Higgs boson would escape undetected. The effective size of the tracking detector therefore fixed a lower limit to the mass this search was sensitive to.

The complementary and overlapping searches just described typically retained a small number of candidate events which could be attributed either to a Higgs boson signal or to background. In statistical terms, a given Higgs boson mass can be regarded as excluded if the expected signal rate is higher than the 95% confidence level (CL) upper bound, given the observed event count. Taking the search described in Ref. [7], where only one event survived, the comparison of the upper bounds derived from the observed event count and expected rates excluded the mass range between 32 MeV and 15 GeV, at the 95% CL.

The small gap from zero to 32 MeV was closed by a search for the process $Z \rightarrow HZ^* \rightarrow (Z^* \rightarrow e^+e^-, \mu^+\mu^-)(H \rightarrow \text{invisible})$[10] which made use of the fact that a Higgs boson with such low mass would be essentially stable, i.e. it would travel hundreds of meters before decaying. Since the Higgs boson is a neutral, weakly-interacting particle, such a low-mass Higgs boson would leave no trace in the detector layers, thus appearing as a small invisible mass recoiling against the Z^*, the latter being unambiguously identified by its decay into a pair of oppositely charged electrons or muons.

4.2. *Extensions to higher masses*

For masses higher than about 10 GeV, the Higgs boson was expected to decay predominantly into quark–antiquark pairs of various flavours, $b\bar{b}$, $c\bar{c}$, $s\bar{s}$, ..., of which $b\bar{b}$ was expected to be the most abundant, with a small contribution from $H \rightarrow \tau^+\tau^-$ (see Fig. 2.2). The extension of the searches towards higher masses was increasingly difficult due to the decreasing signal cross-section and the high background from hadronic (QCD) processes ($Z \rightarrow q\bar{q}$, $q\bar{q}g$, $q\bar{q}\gamma$, ...). To cope with these problems, more sophisticated event selections and larger data samples were required.

Altogether, the four collaborations collected more than five million Z boson decays each. The sensitivity to Higgs bosons was continuously extended to higher masses and reached about 60 GeV towards the end of the LEP 1 phase.

The most promising search channels were those in which the decay of the virtual Z^* boson provided a distinctive signature that could be exploited to separate a signal from the QCD background. This was indeed the case for two channels: the missing energy channel $Z \rightarrow HZ^* \rightarrow (H \rightarrow q\bar{q})(Z^* \rightarrow \nu\bar{\nu})$ where the search was based on the large missing energy carried by the neutrinos, and the leptonic channel $Z \rightarrow HZ^* \rightarrow (H \rightarrow q\bar{q})(Z^* \rightarrow e^+e^-, \mu^+\mu^-)$ where the pair of oppositely charged electrons or muons provided a clear signature. A complete picture of these searches can be obtained by consulting the latest papers from LEP 1,[17–20] which point back to earlier publications.

In the missing energy channel, the quality of the search relied on the accurate determination of jet energies and directions, which implied the combination of momenta and energies of individual particles, avoiding false particle identifications or assignments to jets, and double-counting of energies and momenta. The missing energy and momentum vector was obtained after full reconstruction of the event, by comparing the visible final state to the precisely known initial state. The data were searched for events with a hadronic mono-jet or di-jet, and sizeable missing energy. The mono-jet topology would occur when the Higgs boson is light and the Z^* has a large invariant mass, while the di-jet topology characterises events with a heavy Higgs boson and light Z^*, typically less than 25 GeV. Events were classified as mono-jet or di-jet by dividing them into two hemispheres using the plane orthogonal to the thrust axis.[a] If there were neither charged tracks nor energy clusters in one of the hemispheres, the event was classified as mono-jet, otherwise it was classified as di-jet.

The signature in the leptonic channels, He^+e^- and $H\mu^+\mu^-$, required a pair of energetic electrons or muons of opposite charge, isolated from other decay products. Electron identification relied on the precise evaluation of particle momenta and electromagnetic calorimeter information where electrons produce clusters of adjacent activated cells. The two electron clusters were required to be well separated from each other and isolated from other nearby activity in the detector. Muons generated tracks which could be followed from the collision point through the tracking chamber and calorimeters, and provided signals in the muon detector planes. Isolation and momentum requirements imposed on the two muon tracks allowed virtually all background to be eliminated.

The rest of the event, excluding the pair of leptons, was assigned to the Higgs boson decay. These were assembled into jets using standard jet-forming algorithms. Appropriate cuts were applied on the jet directions, angular separations, invariant masses and recoil masses, which all helped reducing the hadronic (QCD) and four-fermion backgrounds.

Towards the end of the LEP 1 phase, the four LEP collaborations achieved rather similar search sensitivities for the SM Higgs boson. Unfortunately, all searches yielded negative results and hence only mass exclusion limits could be produced. When all LEP 1 data were analysed, the excluded domains started at zero mass and reached 58.4 GeV for ALEPH, 57 GeV for DELPHI, 60.2 GeV for L3, and 59.6 GeV for OPAL, all at 95% CL (see Refs. [17–20]).

5. The LEP 2 phase

In 1995, following a long technical shutdown, the LEP collider restarted operating with an increased collision energy. The shutdown was used to

---

[a]The thrust axis is the one that maximises the sum of the particle momentum projections onto that axis.

revisit the whole collider system in anticipation of several years of running at collision energies reaching 200 GeV or more. In particular, superconducting accelerating cavities were installed at the rate of their production and some LEP 1 copper cavities were put back to fill the remaining space; corrector magnets were used for extra bending and to make the dipole field more spread out and gentle. In order to maximise the energy, so-called "mini-ramps" were practiced towards the end of the fills when the beam currents dropped below a certain value. Also, the shielding against synchrotron radiation was reinforced.

The four collaborations used the opportunity to change faulty detector elements and to effectuate upgrades to meet the challenges of the new experimental environment. The capabilities of tagging b-flavour were improved for a better efficiency for the $H \to b\bar{b}$ decay channel.

According to the plans, the collision energy was increased in successive steps, crossed the threshold for W boson pair production of about 160 GeV, and continued to increase towards even higher energies. Besides measuring precisely the W mass and studying its production and decay properties, great emphasis was put on pursuing the search for the Higgs boson in the higher mass domains made available.

5.1. *A new experimental environment*

The principal mechanism for producing the SM Higgs particle at LEP 2 energies is the Higgs-strahlung process $e^+e^- \to Z^* \to HZ$, where the virtual Z^* boson becomes real while radiating a Higgs particle. Hence, the search for the Higgs boson is limited by kinematics to $m_H^{\max} = E_{\mathrm{cm}} - m_Z$ or about 118 GeV at $E_{\mathrm{cm}}^{\max} = 209$ GeV, the largest energy ever attained.

In order to visualise the experimental challenges of the LEP 2 search environment, let us inspect Fig. 2.3, which compiles the cross-sections of the most relevant physics processes over the entire range of LEP 2 energies, including Higgs production for masses of 60 GeV (the LEP 1 exclusion limit) and more.

One observes that in the LEP 2 domain the signal-to-background ratio becomes altogether more favourable than at LEP 1 energies (at about 90 GeV). However, important new sources of background are introduced from the vector-boson pair production processes $e^+e^- \to W^+W^-$ and ZZ, which come in addition to the familiar QCD processes $e^+e^- \to$ hadrons. These produce event topologies rather similar to those of $e^+e^- \to HZ$ signal process, especially if the Higgs boson mass is close to the W or Z masses (80 and 91 GeV respectively).

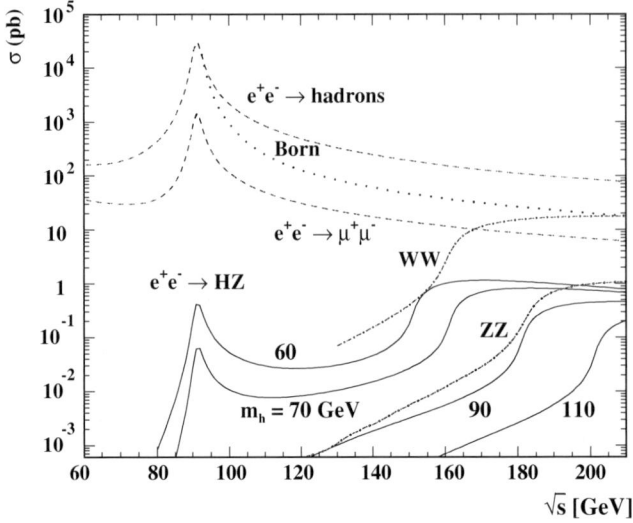

Fig. 2.3   Cross-sections for e^+e^- collision processes that are the most relevant for Higgs boson searches at LEP 2 energies (from Ref. [21]).

The most relevant decays of the Higgs boson are summarised in Fig. 1.5. For masses below about 120 GeV accessible at LEP 2, the decays into fermion pairs would dominate, with $H \to b\bar{b}$ having the largest branching ratio (74% for a mass of 115 GeV), while the decays to $\tau^+\tau^-$, $WW^{(*)}$, gg (approximately 7% each), $c\bar{c}$ (approximately 4%) and $ZZ^{(*)}$ (<1%) make up the rest of the decays.

The four final states that were expected to provide the best search sensitivity are the following: The four-jet topology $e^+e^- \to HZ \to (H \to b\bar{b})(Z \to q\bar{q})$, which is the most abundant process. The invariant mass of two jets is close to m_Z while the other two jets, assigned to the Higgs boson, contain b-flavour. The missing energy channel $e^+e^- \to HZ \to (H \to b\bar{b})(Z \to \nu\bar{\nu})$ produces events with two b-flavoured jets, substantial missing momentum and a missing mass compatible with m_Z. In the leptonic final states, $e^+e^- \to HZ \to (H \to b\bar{b})(Z \to e^+e^-, \mu^+\mu^-)$, the two leptons reconstruct to m_Z and the two jets assigned to the Higgs boson have b-flavour and an invariant mass close to m_H. Although the branching ratio is small, these channels are significant due to a very low background. In the final states with τ leptons produced in the processes $e^+e^- \to HZ \to (H \to \tau^+\tau^-)(Z \to q\bar{q})$ and $(H \to q\bar{q})(Z \to \tau^+\tau^-)$, the τ

leptons can be identified by their decays into electrons, muons, or low-multiplicity hadron jets, accompanied by missing energy.

Each of the four collaborations based their searches on these four signal topologies and profited immensely from the upgraded b-tagging capabilities during the long shutdown preceding the LEP 2 phase. Secondary vertices from the b-quark decays could thus be identified with increased sensitivity. Although the lifetime of the τ lepton is only one-fifth of the lifetime of the b-quark, it also produces secondary vertices. However, b-quarks produce a higher number of decay tracks on average than τ leptons do; also, b-quark decay products are less collimated than those from τ leptons, due to the larger b-quark mass. These features all helped to distinguish between b- and τ-jets and were useful complements to the tagging methods based on secondary vertices.

The stepping up in collision energy started with values of 135 GeV and 161 GeV, followed by successive steps of about 10 GeV. Each new energy was seen as a journey into unknown territory and generated great expectations. The statement so often heard "It is just around the corner" expresses well the general mood of excitement among the Higgs searchers, and the competition between the four collaborations was rather intense.

5.2. *Progress in analysis methods*

The technical upgrades were accompanied by improvements in the analysis. Besides the improved b-tagging methods, better jet finding algorithms and kinematic fits were developed in order to achieve the best possible invariant mass resolution of jet–jet systems. The latter was essential in separating the HZ signal from the otherwise irreducible W^+W^- and ZZ background.

The search results of the four experiments were published in a number of successive papers (see: Refs. [22–35]), which, besides reporting mass limits, also show the progress in analysis techniques developed over the years. For example, new methods were worked out to optimise simultaneously all selections to achieve optimal signal-to-background ratios and to minimise systematic errors. The purely cut-based selections, familiar from LEP 1, were replaced by multivariate techniques which attributed individual weights to selection criteria according to their discriminating power. Instead of cutting on discrete variables, global variables were invented, and the selection was done on the distributions of these. Finally, neural networks were introduced that were trained on simulated event samples to recognise small differences between signal and background events. With time, even more complex likelihood-based interpretations were adopted (see e.g. Refs. [36–38]) where events were not discarded any more but rather classified according to their statistical likelihoods of being signal or background.

6. Combination: The LEP Higgs Working Group (LHWG)

Already a few years before the planned shutdown of the LEP collider, several LEP-wide working groups were formed with the aim of optimally combining the data of the four experiments. Clearly, the combination would considerably improve the statistical power; furthermore, common sources of systematic uncertainty, e.g. affecting the collider energy or due to common theoretical assumptions, could be treated as correlated when presenting final results.

The LEP working group for Higgs boson searches (LHWG) concentrated on two issues.

First of all, a common platform had to be found where the data of the four experiments, with widely different contents and formats, could be combined. The exchange of low-level data, e.g. raw detector information, would have required the sharing of a large amount of software and often undocumented internal knowledge, and was found impracticable. Rather, a high level of abstraction was adopted where the event reconstruction and most of the interpretation remained in the hands of the individual collaborations, while the statistical combination and the evaluation of the common statistical and systematic errors were passed onto the working group.

The common platform for the combination of the data was finally defined as a two-dimensional space where each event occupied a well-defined position. The two axes along which the events were classified were the reconstructed Higgs boson mass m_H^{rec}, and a global discriminating variable G. The latter, while formally equivalent, was widely different from one experiment to another and from one channel to another regarding its information content. It summarised for each event all (detector-dependent) information on kinematics, b-flavour likelihoods, neural network outputs, etc., that could be used to distinguish between signal and background. The observed data events and the distributions (in the form of histograms) of the predicted backgrounds and the hypothetical Higgs boson signal (as a function of the hypothetical mass or "test-mass" m_H) in the (m_H^{rec}, G) plane were exchanged for the combined analysis. The collaborations also exchanged the systematic uncertainties on the signal and background predictions, identified by a "source name", so that correlations could be taken into account; however, the inclusion of correlations did not affect the combined search results significantly.

The second important issue was to identify the most suitable statistical procedure, i.e. the one providing the highest sensitivity to the potential Higgs boson signal while minimising biases in the event interpretation. Finally, a "frequentist" approach was adopted. Details about the method and the quantities used to quantify the interpretation of the data (likelihood ratio $Q = L(s + b)/L(b)$, where $s + b$ and b represent the signal-plus-background and background-only hypotheses, respectively, and p-values CL_b and CL_{s+b} that measure the compatibility of the observed data configuration with the two respective hypotheses), implemented in the LEP-wide combination of Higgs data, can be found in Appendix A and in the common publication, Ref. [39].

The searches carried out prior to the year 2000 did not reveal any evidence for the production of a SM Higgs boson. Both the individual searches and the successive combinations yielded negative results (see Ref. [40]). However, in the data gathered during the summer of 2000 at collision energies mostly beyond 205 GeV, ALEPH reported an intriguing excess of events, arising from a few four-jet candidates with clean b-tags (secondary vertices) and kinematic properties compatible with a SM Higgs boson with mass in the vicinity of 115 GeV.[41] The statistical significance of the reported effect, considered as preliminary, was 3.9 standard deviations (σ). One of these suggestive ALEPH candidates is shown in Fig. 2.4.

Due to this observation, the experiments requested that the LEP 2 run be extended. In response CERN asked the experiments to publish their results.[35,42–44] Based on these results, the shutdown was postponed by one month, during which the collision energy was pushed to its extreme, reaching 209 GeV. The new data did not contain any outstanding Higgs boson candidates. When the new data of the four experiments were added and the complete dataset was analysed, a significance of $2.9\,\sigma$ was found for the excess, just below the value of $3\,\sigma$ where evidence can customarily be claimed.[45]

The decision to shut down the LEP collider despite the ambiguity of this outcome was thus taken in a climate of tense discussions within the particle physics community worldwide.

Finally, it took two more years for the collaborations to fully analyse their data and to publish their final results[43,46–48] following a thorough

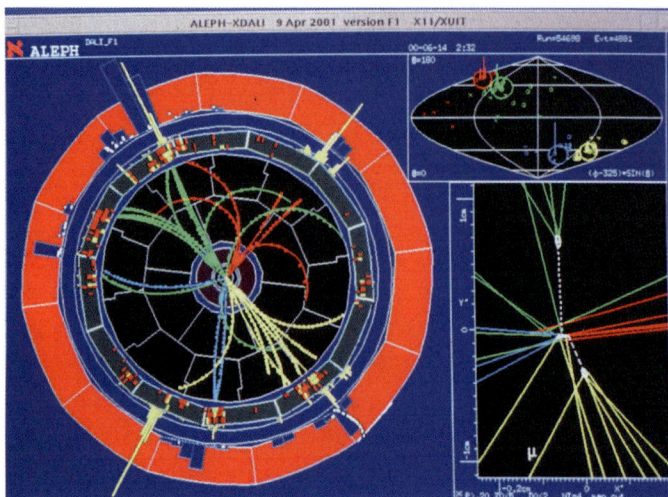

Fig. 2.4   Four-jet Higgs boson candidate $(H \to b\bar{b})(Z \to q\bar{q})$ from the ALEPH experiment, with a reconstructed Higgs boson mass of 114.3 GeV. Secondary vertices are clearly visible in two of the jets, strongly suggesting b-flavour.

revision of their detector calibrations, analysis methods, and systematic uncertainties. The revised final inputs provided to the LEP Higgs working group yielded a combined significance of $1.7\,\sigma$ for the excess at 115 GeV, which is not uncommon for statistical fluctuations.[39] The same combined analysis provided a lower bound of 114.4 GeV for the mass of the SM Higgs boson (95% CL).

Figure 2.5 shows the test statistic $-2\ln Q$ for the ALEPH data alone and for the final LEP data combined. For a hypothetical mass $m_H = 115$ GeV, p-values $1 - \mathrm{CL_b}$ of 0.09 for the background-only hypothesis and $\mathrm{CL_{s+b}} = 0.15$ for the signal-plus-background hypothesis were obtained.

Figure 2.6 shows the distribution of the reconstructed Higgs boson mass for events from all four experiments with a loose selection on G, compared to Monte Carlo simulations of the background and the expected signal for a hypothetical SM Higgs boson with 115 GeV mass. The data showed no significant excess in the vicinity of 115 GeV. The same conclusion was reached if the selection adopted for the purpose of this figure was tightened.

Recently, the ATLAS and CMS Collaborations at the LHC reported the observation of a new particle, with mass of 125 GeV and characteristics suggesting the SM Higgs boson[49,50] and, at the same time, exclude a signal at 115 GeV. In the light of this observation, one is bound to conclude that

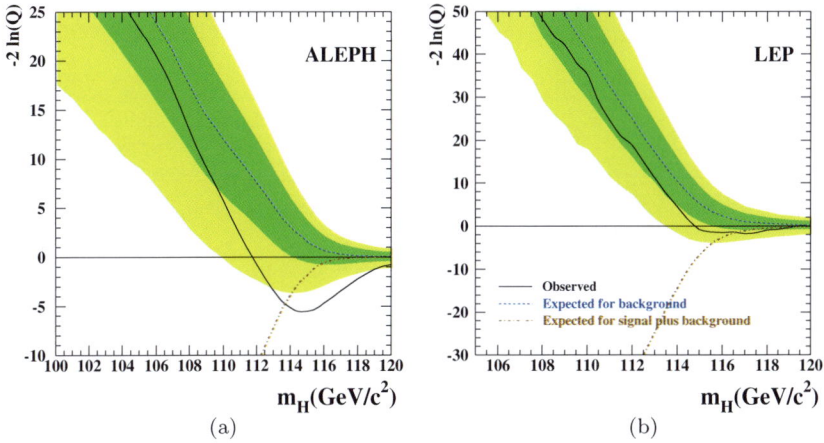

Fig. 2.5 Observed and expected behaviour of the test statistic $-2\ln Q$ as a function of the hypothetical Higgs boson mass m_H, obtained from using the ALEPH data alone (a) and the final data of the four LEP experiments (b). The full curve represents the observation; the dashed curve shows the median expectation for the background; the dark (green) and light (yellow) shaded bands represent the 68% and 95% probability bands about the median background expectation. The dash-dotted curve indicates the position of the minimum of the median expectation for the signal-plus-background hypothesis when the signal mass given on the abscissa is tested.

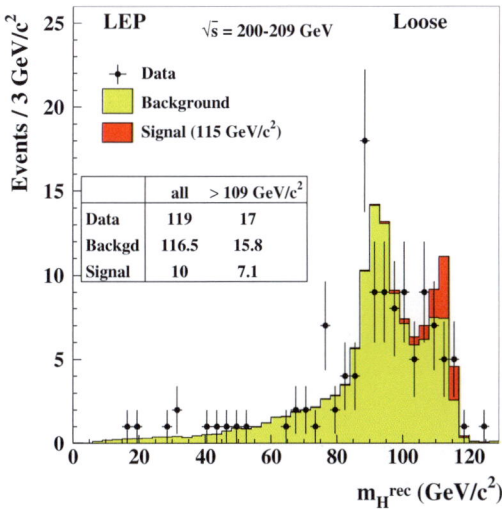

Fig. 2.6 Distribution of the reconstructed Higgs boson mass m_H^{rec} for the sum of the data of the four LEP experiments. The histogram shows the Monte Carlo predictions, lightly shaded (yellow) for the background, heavily shaded (red) for an assumed SM Higgs boson of mass 115 GeV, together with the data (from Ref. [39]).

the excess observed at LEP was indeed an artefact due to a statistical fluctuation.

In retrospect, the observation of a Higgs boson with 125 GeV mass (i.e. the LHC signal) would have required a collision energy of about 219 GeV, which was technically out of reach at LEP2.

7. The LEP "legacy"

Besides the mass limit of 114.4 GeV derived from direct searches for the SM Higgs boson, indirect experimental bounds were obtained from global fits to the precision measurements of electroweak observables that were measured at LEP and elsewhere. Within the SM, these observables are all inter-dependent and weakly sensitive to the Higgs boson mass through higher-order radiative corrections. A complete and up-to-date discussion can be found in Ref. [51]. Such fits were performed at regular intervals by the joint Electroweak (EW) Working Group and produced interesting restrictions on the mass of the SM Higgs boson. Already at the end of the LEP 1 phase, the fit suggested a mass value smaller than about 200 GeV (at 95% CL), with an optimum in the vicinity of 100 GeV. One of the recent fits, performed in March 2012, incorporating the recently obtained precise values of the top-quark and W masses[52,53] is reproduced in Fig. 2.7.

The best-fit value of the Higgs boson mass, indicated by the minimum of the χ^2 curve, lies between 95 and 105 GeV (depending slightly on theoretical assumptions), with an uncertainty of about $\pm$ 40 GeV. The yellow regions are excluded by direct searches: the exclusion from below is the one produced by the LEP Higgs Working Group, while the exclusion from above is the LHC exclusion of early 2012. The narrow band between the two exclusions is the range that was still open at the time when the fit was produced, just a few months before the announcement of the discovery in July 2012,[49,50] with a mass of $\sim$125 GeV.

In summary, the LEP experiments did not produce evidence for the SM Higgs boson. The legacy of LEP is thus limited to a lower bound for the mass of 114.4 GeV (at 95% CL) from direct searches and a large number of precision measurements contributing to an indirect constraint on the Higgs boson mass.

If the particle discovered at the LHC is indeed a Higgs boson, the question remains open whether or not it is really the particle predicted by the SM. More elaborate models, in particular the supersymmetric (SUSY)

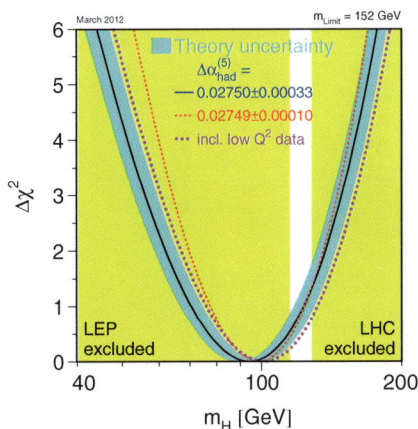

Fig. 2.7 χ^2 fit performed in March 2012 to a large number of measurements of electroweak observables, including the top-quark and W-boson masses, by the joint Electroweak Working Group.[52,53] The position of the minimum indicates the optimum Higgs
boson mass derived from the fit. The yellow bands represent exclusions by direct searches,
the LEP searches excluding low masses up to 114 GeV and the LHC measurements limiting the allowed mass range from above.

extensions of the SM, predict a Higgs sector with (at least) five Higgs
bosons, three neutral and two electrically charged. The question thus
becomes relevant if the particle discovered is the lightest of the Higgs
bosons or one of the heavier kinds. The LEP collaborations, having scanned
the low-mass domain where the LHC collider is virtually blind, have contributed to answering that question by placing stringent upper bounds on
the HZZ coupling for non-standard models with various assumptions concerning the decay of the Higgs boson by combining the LEP 1 and LEP
2 data of the four collaborations (see Fig. 2.8 and Ref. [39]). In the ratio
$\xi^2 = (g_{HZZ}/g_{HZZ}^{SM})^2$, the variable g_{HZZ} designates the non-standard coupling and g_{HZZ}^{SM} the same coupling in the SM.

Furthermore, the four collaborations searched explicitly for Higgs particles predicted by SUSY models and even more exotic scenarios, the discussion of which goes beyond the present text. The exclusion limits produced
by LEP strongly constrain the parameters of such models. In the Minimal
Supersymmetric Standard Model (MSSM), the LEP searches provided hard
lower bounds in the vicinity of 90 GeV for the masses of the lightest scalar
and pseudo-scalar Higgs bosons h and A[54] and a lower bound of about
80 GeV for the mass of the charged Higgs boson in models (which include
the MSSM) with two Higgs field doublets.[33,55–59]

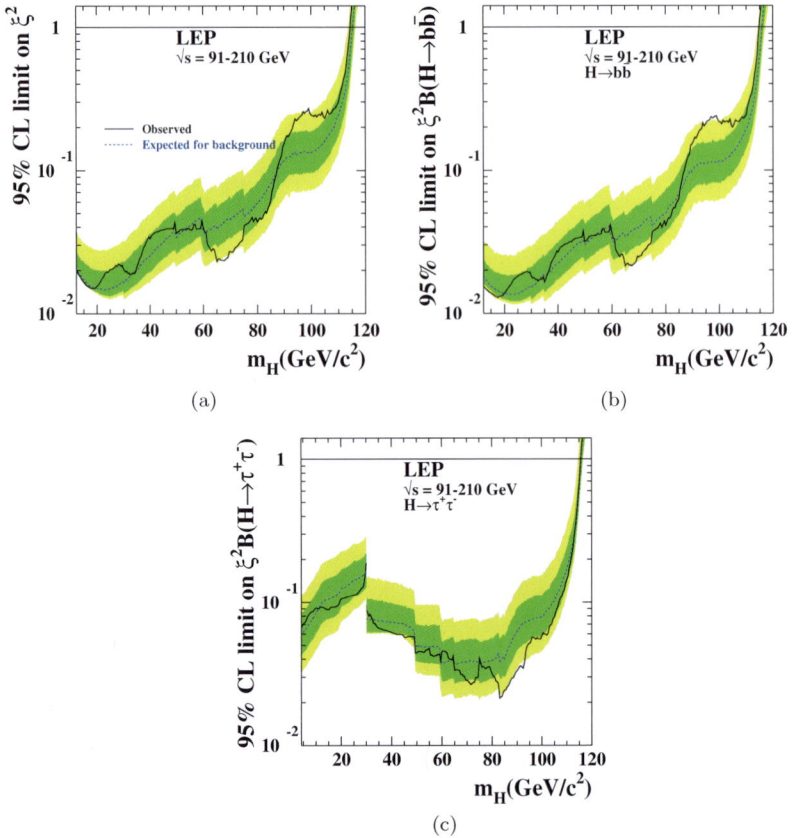

Fig. 2.8   The 95% confidence level upper bound (solid line) on the strength of the coupling of the Higgs boson to the Z boson $\xi^2 = (g_{HZZ}/g_{HZZ}^{\mathrm{SM}})^2$, as described in the text, versus the hypothetical Higgs boson mass m_H (from Ref. [39]). The green and yellow bands around the median expected (dashed) line correspond to the 68% and 95% probability bands. The horizontal lines correspond to the Standard Model coupling. (a): For all Higgs boson decays as predicted by the Standard Model; (b): for 100% $H \to b\bar{b}$ and (c): for 100% $H \to \tau^+\tau^-$.

In parallel to the effort at LEP, Higgs boson searches were also carried out at the Tevatron proton–antiproton collider at Fermilab (Chicago). These searches are addressed in Chapter 3.

References

1. The Large Electron–Positron Collider at CERN. http://home.web.cern.ch/about/accelerators/large-electron-positron-collider.

2. ALEPH Collaboration, ALEPH: A detector for electron–positron annnihilations at LEP, *Nucl. Instrum. Meth.* **A294**, 121–178 (1990). doi: 10.1016/0168-9002(90)91831-U.
3. DELPHI Collaboration, The DELPHI detector at LEP, *Nucl. Instrum. Meth.* **A303**, 233–276 (1991). doi: 10.1016/0168-9002(91)90793-P.
4. L3 Collaboration, The construction of the L3 experiment, *Nucl. Instrum. Meth.* **A289**, 35–102 (1990). doi: 10.1016/0168-9002(90)90250-A.
5. OPAL Collaboration, The OPAL detector at LEP, *Nucl. Instrum. Meth.* **A305**, 275–319 (1991). doi: 10.1016/0168-9002(91)90547-4.
6. G. Altarelli, R. Kleiss, and C. Verzegnassi (eds.), *Z Physics at LEP1: Higgs Search and New Physics* (1989). P.J. Franzini and P. Taxil (conveners), pp. 59–120.
7. ALEPH Collaboration, Search for the neutral Higgs boson from Z^0 decay, *Phys. Lett.* **B236**, 233–244 (1990). doi: 10.1016/0370-2693(90)90834-S.
8. ALEPH Collaboration, Search for neutral Higgs bosons from supersymmetry in Z decays, *Phys. Lett.* **B237**, 291–302 (1990). doi: 10.1016/0370-2693(90)91447-J.
9. ALEPH Collaboration, Search for the neutral Higgs boson from Z^0 decay in the Higgs mass range between 11 and 24 GeV, *Phys. Lett.* **B241**, 141–149 (1990). doi: 10.1016/0370-2693(90)91501-2.
10. ALEPH Collaboration, Search for a very light Higgs boson in Z decays, *Phys. Lett.* **B245**, 289–297 (1990). doi: 10.1016/0370-2693(90)90148-Y.
11. L3 Collaboration, Search for the neutral Higgs boson in Z^0 decay, *Phys. Lett.* **B248**, 203–210 (1990). doi: 10.1016/0370-2693(90)90040-D.
12. L3 Collaboration, Search for a low mass neutral Higgs boson in Z^0 decay, *Phys. Lett.* **B252**, 518–524 (1990). doi: 10.1016/0370-2693(90)90581-P.
13. OPAL Collaboration, Mass limits for a Standard Model Higgs boson in e^+e^- collisions at LEP, *Phys. Lett.* **B236**, 224–232 (1990). doi: 10.1016/0370-2693(90)90833-R.
14. OPAL Collaboration, Limits on a light Higgs boson in e^+e^- collisions at LEP, *Phys. Lett.* **B251**, 211–222 (1990). doi: 10.1016/0370-2693(90)90255-5.
15. OPAL Collaboration, Decay mode independent search for a light Higgs boson and new scalars, *Phys. Lett.* **B268**, 122–136 (1991). doi: 10.1016/0370-2693(91)90934-I.
16. A. Djouadi, The anatomy of electro-weak symmetry breaking. I: The Higgs boson in the Standard Model, *Phys. Rept.* **457**, 1–216 (2008). doi: 10.1016/j.physrep.2007.10.004.
17. ALEPH Collaboration, Search for the Standard Model Higgs boson, *Phys. Lett.* **B313**, 299–311 (1993). doi: 10.1016/0370-2693(93)91227-E.
18. DELPHI Collaboration, Search for the Standard Model Higgs boson in Z^0 decays, *Nucl. Phys.* **B421**, 3–37 (1994). doi: 10.1016/0550-3213(94)90222-4.
19. L3 Collaboration, Search for neutral Higgs boson production through the process $e^+e^- \to Z^*H^0$, *Phys. Lett.* **B385**, 454–470 (1996). doi: 10.1016/0370-2693(96)00987-2.
20. OPAL Collaboration, Search for neutral Higgs bosons in Z^0 decays using the OPAL detector at LEP, *Z. Phys.* **C73**, 189–199 (1997). doi: 10.1007/s002880050309.

21. P. Igo-Kemenes, Review on searches for Higgs bosons. In K. Hagiwara *et al.* (eds.), *Review of Particle Physics*. Particle Data Group, Vol. D66, p. 010001 (2002). doi: 10.1103/PhysRevD.66.010001.

22. ALEPH Collaboration, Search for the Standard Model Higgs boson in e^+e^- collisions at $\sqrt{s} = 161$, 170 and 172 GeV, *Phys. Lett.* **B412**, 155–172 (1997). doi: 10.1016/S0370-2693(97)01110-6.

23. ALEPH Collaboration, Searches for neutral Higgs bosons in e^+e^- collisions at center-of-mass energies from 192 to 202 GeV, *Phys. Lett.* **B499**, 53–66 (2001). doi: 10.1016/S0370-2693(00)01364-2.

24. DELPHI Collaboration, Search for neutral and charged Higgs bosons in e^+e^- collisions at $\sqrt{s} = 161$ GeV and 172 GeV, *Eur. Phys. J.* **C2**, 1–37 (1998). doi: 10.1007/PL00021558.

25. DELPHI Collaboration, Search for neutral Higgs bosons in e^+e^- collisions at $\sqrt{s} = 183$ GeV, *Eur. Phys. J.* **C10**, 563–604 (1999). doi: 10.1007/s100520050600.

26. DELPHI Collaboration, Searches for neutral Higgs bosons in e^+e^- collisions around $\sqrt{s} = 189$ GeV, *Eur. Phys. J.* **C17**, 187–205 (2000). doi: 10.1007/s100520000401.

27. L3 Collaboration, Search for the Standard Model Higgs boson in e^+e^- interactions at $161 \leq \sqrt{s} \leq 172$ GeV, *Phys. Lett.* **B411**, 373–386 (1997). doi: 10.1016/S0370-2693(97)01033-2.

28. L3 Collaboration, Search for the Standard Model Higgs boson in e^+e^- interactions at $\sqrt{s} = 183$ GeV, *Phys. Lett.* **B431**, 437–450 (1998). doi: 10.1016/S0370-2693(98)00568-1.

29. L3 Collaboration, Search for Standard Model Higgs boson in e^+e^- interactions at $\sqrt{s} = 189$ GeV, *Phys. Lett.* **B461**, 376–386 (1999). doi: 10.1016/S0370-2693(99)00853-9.

30. L3 Collaboration, Search for the Standard Model Higgs boson in e^+e^- collisions at $\sqrt{s}$ up to 202 GeV, *Phys. Lett.* **B508**, 225–236 (2001). doi: 10.1016/S0370-2693(01)00326-4.

31. OPAL Collaboration, Search for the Standard Model Higgs boson in e^+e^- collisions at $\sqrt{s} = 161$ GeV, *Phys. Lett.* **B393**, 231–244 (1997). doi: 10.1016/S0370-2693(96)01645-0.

32. OPAL Collaboration, Search for the Standard Model Higgs boson in e^+e^- collisions at $\sqrt{s} = 161\text{-}172$ GeV, *Eur. Phys. J.* **C1**, 425–438 (1998). doi: 10.1007/s100520050094.

33. OPAL Collaboration, Search for Higgs bosons in e^+e^- collisions at 183-GeV, *Eur. Phys. J.* **C7**, 407–435 (1999). doi: 10.1007/s100529901102.

34. OPAL Collaboration, Search for neutral Higgs bosons in e^+e^- collisions at $\sqrt{s} \approx 189$ GeV, *Eur. Phys. J.* **C12**, 567–586 (2000). doi: 10.1007/s100520000286.

35. OPAL Collaboration, Search for the Standard Model Higgs boson in e^+e^- collisions at $\sqrt{s} \approx 192\text{-}209$ GeV, *Phys. Lett.* **B499**, 38–52 (2001). doi: 10.1016/S0370-2693(01)00070-3.

36. A. L. Read, Optimal statistical analysis of search results based on the likelihood ratio and its application to the search for the MSM Higgs boson at

$\sqrt{s}$ = 161 and 172 GeV, DELPHI Note 97-158 PHYS 737 (1997). URL http://delphiwww.cern.ch/pubxx/delnote/public/97_158_phys_737.ps.gz.

37. A. Read, Modified frequentist analysis of search results (The CL_s method). In F. James, Y. Perrin, and L. Lyons (eds.), *Workshop on Confidence Limits*, CERN, Geneva, Switzerland, 17–18 Jan 2000. URL https://cds.cern.ch/record/451614.

38. A. L. Read, Presentation of search results: The CL_s technique, *J. Phys.* **G28**, 2693–2704 (2002). doi: 10.1088/0954-3899/28/10/313.

39. LEP Working Group for Higgs boson searches, ALEPH Collaboration, DELPHI Collaboration, L3 Collaboration, OPAL Collaboration, Search for the Standard Model Higgs boson at LEP, *Phys. Lett.* **B565**, 61–75 (2003). doi: 10.1016/S0370-2693(03)00614-2.

40. P. Igo-Kemenes, Searches for new particles in e^+e^- collisions. In C. Lim and Y. Yamanaka (eds.), *High Energy Physics. Proceedings, 30th International Conference, ICHEP 2000*, Osaka, Japan, July 27–August 2, 2000. Vol. 1, 2, World Scientific (2001).

41. D. Schlatter. ALEPH Status Report, Sep. 5, 2000. http://aleph.web.cern.ch/aleph/ALPUB/seminar/lepc_sep00/lepc_0509.pdf. Presentation to LEP Committee.

42. ALEPH Collaboration, Observation of an excess in the search for the Standard Model Higgs boson at ALEPH, *Phys. Lett.* **B495**, 1–17 (2000). doi: 10.1016/S0370-2693(00)01269-7.

43. DELPHI Collaboration, Search for the Standard Model Higgs boson at LEP in the year 2000, *Phys. Lett.* **B499**, 23–37 (2001). doi: 10.1016/S0370-2693(01)00069-7.

44. L3 Collaboration, Higgs candidates in e^+e^- interactions at $\sqrt{s}$ = 206.6-GeV, *Phys. Lett.* **B495**, 18–25 (2000). doi: 10.1016/S0370-2693(00)01281-8.

45. P. Igo-Kemenes. Status of the Higgs boson searches, Nov. 3, 2000. http://lephiggs.web.cern.ch/LEPHIGGS/talks/index.html. Presentation to LEP Committee.

46. ALEPH Collaboration, Final results of the searches for neutral Higgs bosons in e^+e^- collisions at $\sqrt{s}$ up to 209 GeV, *Phys. Lett.* **B526**, 191–205 (2002). doi: 10.1016/S0370-2693(01)01487-3.

47. L3 Collaboration, Standard Model Higgs boson with the L3 experiment at LEP, *Phys. Lett.* **B517**, 319–331 (2001). doi: 10.1016/S0370-2693(01)01010-3.

48. OPAL Collaboration, Search for the Standard Model Higgs boson with the OPAL detector at LEP, *Eur. Phys. J.* **C26**, 479–503 (2003). doi: 10.1140/epjc/s2002-01092-3.

49. ATLAS Collaboration, Observation of a new particle in the search for the Standard Model Higgs boson with the ATLAS detector at the LHC, *Phys. Lett.* **B716**, 1–29 (2012). doi: 10.1016/j.physletb.2012.08.020.

50. CMS Collaboration, Observation of a new boson at a mass of 125 GeV with the CMS experiment at the LHC, *Phys. Lett.* **B716**, 30–61 (2012). doi: 10.1016/j.physletb.2012.08.021.

51. J. Erler and P. Langacker, Electroweak model and constraints on new physics. In J. Beringer *et al.* (eds.), *Review of Particle Physics (RPP)*, Vol. D86, p. 010001 (2012). doi: 10.1103/PhysRevD.86.010001.
52. ALEPH, DELPHI, L3, OPAL, SLD Collaborations, LEP Electroweak Working Group, SLD Electroweak and Heavy Flavour Groups, Precision electroweak measurements on the Z resonance, *Phys. Rept.* **427**, 257–454 (2006). doi: 10.1016/j.physrep.2005.12.006.
53. ALEPH, DELPHI, L3, OPAL Collaborations, LEP Electroweak Working Group, Electroweak measurements in electron–positron collisions at W-boson-pair energies at LEP, *Phys. Rept.* **532**, 119–244 (2013). doi: 10.1016/j.physrep.2013.07.004.
54. ALEPH Collaboration, DELPHI Collaboration, L3 Collaboration, OPAL Collaboration, LEP Working Group for Higgs Boson Searches, Search for neutral MSSM Higgs bosons at LEP, *Eur. Phys. J.* **C47**, 547–587 (2006). doi: 10.1140/epjc/s2006-02569-7.
55. ALEPH Collaboration, Search for charged Higgs bosons in e^+e^- collisions at energies up to $\sqrt{s} = 209$ GeV, *Phys. Lett.* **B543**, 1–13 (2002). doi: 10.1016/S0370-2693(02)02380-8.
56. DELPHI Collaboration, Search for charged Higgs bosons in e^+e^- collisions at $\sqrt{s} = 189$-202 GeV, *Phys. Lett.* **B525**, 17–28 (2002). doi: 10.1016/S0370-2693(01)01282-5.
57. L3 Collaboration, Search for charged Higgs bosons at LEP, *Phys. Lett.* **B575**, 208–220 (2003). doi: 10.1016/j.physletb.2003.09.057.
58. OPAL Collaboration, Search for charged Higgs Bosons in e^+e^- collisions at $\sqrt{s} = 189$-209 GeV, *Eur. Phys. J.* **C72**, 2076 (2012). doi: 10.1140/epjc/s10052-012-2076-0.
59. ALEPH Collaboration, DELPHI Collaboration, L3 Collaboration, OPAL Collaboration, LEP working group for Higgs boson searches, Search for charged Higgs bosons: combined results using LEP data, *Eur. Phys. J.* **C73**, 2463–2471 (2013). doi: 10.1140/epjc/s10052-013-2463-1.

Chapter 3

Searches for the Standard Model Higgs boson at the Tevatron collider

Wade C. Fisher[*] and Thomas R. Junk[†]

[*] *Department of Physics and Astronomy*
Michigan State University
East Lansing, Michigan, 48824, USA
[†] *Fermi National Accelerator Laboratory*
Batavia, Illinois, 60510, USA

During Run II of the Tevatron collider, which took place from 2001 until 2011, the CDF and D0 detectors each collected approximately 10 fb^{-1} of $p\bar{p}$ collision data at a center-of-mass energy of $\sqrt{s} = 1.96$ TeV. This dataset allowed for tests for the presence of the SM Higgs boson in the mass range 90–200 GeV in the production modes $gg \to H$, W/ZH, vector-boson fusion, and $t\bar{t}H$, with H decay modes $H \to b\bar{b}$, $H \to W^+W^-$, $H \to \tau^+\tau^-$, $H \to \gamma\gamma$, and $H \to ZZ$. This chapter summarizes the search methods and the results of the Higgs boson search at the Tevatron. The increased sophistication of the analysis techniques as the collider run progressed is discussed, covering the strategies used over time to improve the sensitivity and breadth of the analyses. Using the full Tevatron data sample for both experiments, the combined Higgs search in all channels observes an excess consistent with the predicted SM Higgs boson signal with mass of 125 GeV, with a significance of 3.0 standard deviations above the background prediction.

1. Introduction

The Tevatron proton–antiproton ($p\bar{p}$) collider was part of a series of particle accelerators at the Fermi National Accelerator Laboratory (FNAL or Fermilab) which is located approximately 60 kilometers west of the city of Chicago. Fermilab hosts a large program of particle physics experiments and has a long history of particle discovery and innovation in high-energy particle accelerators. Following the shutdown of the LEP e^+e^- collider in 2000 (see Chapter 2) and before LHC operations, the Tevatron was the highest-energy particle collider in the world. A primary goal of the Tevatron physics program was the discovery of the top quark and subsequent measurement of its properties. After the discovery of the top quark at the Tevatron in 1995, the accelerator and its two experiments (CDF and D0) were upgraded for a second running period.

The Tevatron collider and its experiments were not optimized to search for Higgs bosons. However, the upgrades to the Tevatron and the experiments provided an opportunity to expand the Tevatron's physics program to include Higgs boson searches. In the early years of the current millennium there were several studies performed to determine the potential of a Tevatron Higgs boson search.[1,2] These studies suggested that with integrated luminosities of 10–20 fb^{-1} per experiment, a search for the Higgs boson could reach sensitivity thresholds of 3–5 standard deviations above background predictions over a Higgs boson mass range of 100–200 GeV. Sensitivity to beyond-Standard-Model (BSM) Higgs searches was also studied with similar conclusions. With the support of these Higgs boson search sensitivity studies, the CDF and D0 experiments began a Higgs physics program in Run II of Tevatron operations. This program began in 2001 and continued into 2013 beyond the end of Tevatron operations with final studies of Higgs boson properties being finalized at the time of this writing.

2. The Tevatron collider and experimental apparatus

The Tevatron collider operated in two distinct data collection epochs, referred to as Run I and Run II, encompassing the periods of 1992–1996 and 2001–2011, respectively. The Tevatron produced $p\bar{p}$ collisions that were recorded by two multipurpose particle detectors, CDF[3,4] and D0.[5–8] During Run I, the Tevatron operated at a center-of-mass energy of 1.8 TeV and delivered an integrated luminosity of approximately 145 pb^{-1}(110 pb^{-1} recorded by the detectors). For Run II, the detectors were upgraded and the Tevatron's center-of-mass energy was increased to 1.96 TeV, and delivered an integrated luminosity of 11.9 fb^{-1} to each of the two experiments (10.7 fb^{-1} recorded). The search for Higgs bosons at the Tevatron was performed using data from Run II and henceforth this article will solely refer to the period of Run II.

The Tevatron had a circumference of 6.82 kilometers and used 774 niobium-titanium (NbTi) superconducting dipole magnets that produced a field of 4.2 T to maintain the particle orbit. The beam focusing was performed using 240 NbTi quadrupole magnets. The Tevatron's luminous region was approximately 30 μm wide in the directions transverse to the beam axis and 120 cm long. The time between bunch crossings at the Tevatron was 396 ns, with 36 counter-rotating bunches each of protons and antiprotons. The remaining orbit spacing was taken by gaps in the bunch trains to allow time for abort kicker magnets to fire. The Tevatron reached a maximum instantaneous luminosity of 4.31×10^{32} cm^{-2}s^{-1}.

The CDF and D0 detectors were solenoidal spectrometers with precise silicon-strip tracking detectors surrounded by a drift chamber (CDF) or a scintillating fiber tracker (D0), inside a magnetic field of strength 1.4 T (CDF) and 2.0 T (D0). Time-of-flight, calorimetry, and muon detectors were located outside of the magnet solenoids, and provided energy measurements and particle identification. For illustration purposes, the organization of detector systems for the D0 detector is shown in schematic format in Fig. 3.1 (CDF's construction was highly similar).

For both experiments, the maximum data logging rate was approximately 200 Hz, and so both detectors employed three-level triggering systems to reduce the 2.5 MHz beam collision rate to acceptable levels. The

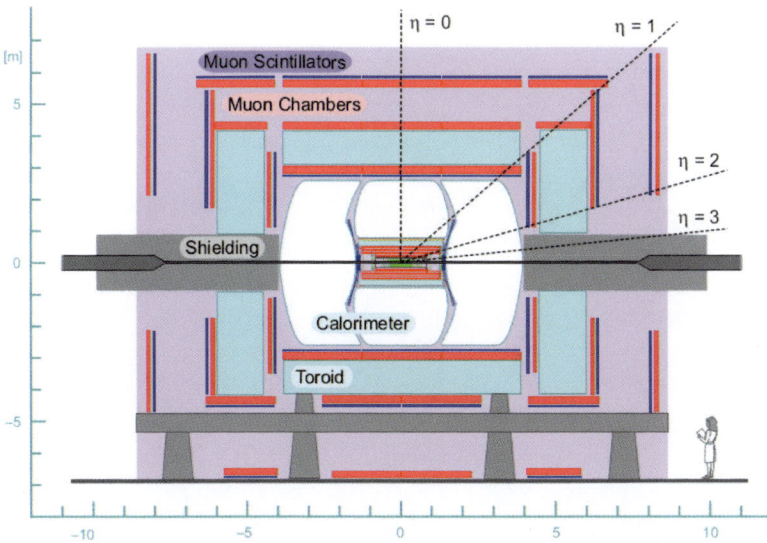

Fig. 3.1   A schematic representation of the Run-II D0 detector.[6] The inner tracking system resides at the center of the detector apparatus and within a 2 T solenoid magnet. The electromagnetic and hadronic calorimeters are located outside the solenoid. The muon system surrounds the calorimetry and incorporates a 1.8 T toroidal magnet.

first trigger level operated in dedicated hardware and summed calorimeter energies for fast jet/electron identification and $E_{\mathrm{T}}^{\mathrm{miss}}$ calculation, and, using separate hardware, identified charged-particle tracks in the trackers. The first-level trigger reduced the event rate to approximately 2–10 kHz. A second-level trigger considered a more complete readout of the detector and operated in specialized hardware and software. At this trigger level, tracks were associated with calorimeter clusters and track segments in the muon detectors in order to form trigger decisions sensitive to the presence of high-p_{T} leptons. Using tracking information, displaced vertexes could be found at this level of triggering. The trigger rate at this level was approximately 1–2 kHz. A third level of triggering consisted of processing the data with a speed-optimized version of the offline reconstruction software, at which point tighter requirements were made on the missing transverse energy, jet energies, and lepton identification criteria.

3. Data samples and Monte Carlo simulation

The data used to perform the Tevatron Higgs search correspond roughly to 10 fb^{-1} per experiment and were recorded over the Run II period of 2001–2011. This represents a subset of the total recorded data and reflects the removal of periods in which the sub-detectors were not operating optimally. These data were used to test for the presence of Higgs bosons by comparing them to predictions of the Higgs boson signal strength yields and those of all known background sources. These predictions require detailed Monte Carlo (MC) simulation of the differential distributions of particles in the final states as well as the detector responses and the trigger and analysis requirements. The MC simulation was typically performed using leading-order (LO) or next-to-leading-order (NLO) event generators. These MC samples were then normalized using the highest-order cross section calculations available for each production process, although for many of the dominant background processes, data-driven techniques were used instead. Predictions of the integral and differential rates require knowledge of the parton momenta inside the proton and the antiproton. The MC simulation used parton distribution functions (PDFs) from CTEQ:[9,10] CTEQ5L (CDF) and CTEQ6L1 (D0). The cross section calculations also used MSTW[11] PDFs in many cases. The parton shower and hadronization were either performed by the generator or via PYTHIA for generators without this functionality. More detail on how these processes were predicted can be found in the specific publications for each data analysis performed.[12]

All MC samples were processed through a GEANT[13] simulation of the detector, and reconstructed with the same algorithms used for the data. The effects of instrumental noise and additional $p\bar{p}$ interactions were modeled using MC in the CDF analyses, while recorded data from randomly selected beam crossings with the same instantaneous luminosity profile as data were overlaid on to the MC events in the D0 analyses. Over the entire Run II data sample, the average number of reconstructed primary vertices was approximately 3, including the hard scatter.

In several instances, data control samples were used to predict the rates and differential distributions of backgrounds, as well as to refine the predictions of MC simulations and cross section calculations. The normalization and shape of the instrumental and multijet backgrounds were predicted from subsidiary data samples orthogonal to those used for the Higgs boson search. For the CDF searches, the rate of vector bosons produced in association with heavy-flavor jets was constrained in data samples where the signal-to-background ratio is too small to bias the rate estimation. For the D0 searches, the kinematic distributions of V+jets samples were corrected empirically using data samples which are insensitive to the presence or absence of a Higgs boson signal.

4. Basic event selection and efficiencies

The data and simulated samples were reconstructed using algorithms designed to identify the signatures of particles in the detector. The input to these algorithms are the digitized, calibrated measurements of energy deposition made by particles as they interact with the detector systems. The energy depositions ("hits") were grouped into basic components such as tracks indicating the paths of charged particles in the tracking detectors, clusters of energy depositions in the calorimeter modules and localized track segments ("stubs") in the muon systems. These basic components were associated together to identify likely candidates for muons, electrons, taus and parton-initiated jets, based on their respective signatures. The presence of final-state neutrinos was inferred from these measurements by imposing the conservation of momentum in the plane transverse to the beam direction. This "missing transverse energy", or $E_{\mathrm{T}}^{\mathrm{miss}}$, is indicative of the presence of one or more neutrinos in an event. The algorithms used to reconstruct tracks, calorimeter clusters and muon system hits have an associated inefficiency due to the need to operate the detector digitization well above analog noise levels. There are also inefficiencies and mis-identifications associated with the algorithms themselves in cases where energy depositions gave rise to ambiguous topologies in the detector.

Electron candidates were identified via the association of charged-particle tracks with clusters of energy in the electromagnetic calorimeter. Muon candidates were identified by matching tracks with hits in the muon detectors. Tau leptons were identified based on the nature of their decay. Leptonic tau decays were typically detected as electrons or muons, provided the decay products satisfied the minimum lepton reconstruction requirements. Hadronic tau decays were identified using charged-particle tracks associated with energy clusters in the calorimeters. For all leptons, an associated quality of the particle identification was generated based on the nature of the contributing energy measurements, which may depend on the location of the candidate particle in the detector, the amount of energy deposited, the degree to which the energy deposition is isolated from other energy depositions, and the number of hits in a detector system. These quality criteria were used to control the relative rates of true or mis-identified leptons, and this property allowed the optimization of Higgs boson searches by analyzing high-quality leptons separately from lower-quality leptons, as described in detail in Sec. 6.

Highly collimated sprays of particles (jets) originating from gluon or quark fragmentation and hadronization were identified by grouping energy clusters in the electromagnetic and hadronic calorimeters using iterative algorithms that associate clusters based on their angular separation. These "cones" of energy can also be combined with charged-particle tracks to improve the identification efficiency, the purity, and the precision of the energy estimation. The transverse energy vector $\vec{E}_\mathrm{T}$ of a calorimeter energy deposit is $E \sin\theta \cdot \hat{n}$, where E is the measured energy, θ is the angle with respect to the proton beam axis of a line drawn from the collision point to the energy deposit, and $\hat{n}$ is a unit vector in the plane perpendicular to the beam pointing along that line. The missing transverse energy ($E_\mathrm{T}^\mathrm{miss}$) is the magnitude of the vector opposite to the sum of the $\vec{E}_\mathrm{T}$ vectors measured in the calorimeter, after propagation of all corrections to the calorimetric objects and for identified muons (which deposit only small amounts of energy in the calorimeters).

Jets that originate from bottom or charm quarks were identified by exploiting the relatively long lifetimes and large masses of the hadrons containing the bottom and charm quarks. These hadrons travel a few hundred microns before decaying into final states with multiple hadrons, often with charged leptons. By identifying tracks with a significant displacement from the primary $p\bar{p}$ interaction point, these heavy hadron decays were reconstructed. These *"b-tagging"* algorithms were typically designed to create a selection criterion (or operating point) that allowed a customizable relationship between the efficiency to identify true heavy quark decays and the efficiency to reject light quark jets. These operating points were optimized to allow non-overlapping selections of different Higgs boson signal purity to improve the significance of $H \to b\bar{b}$ searches. The b-tagging algorithms were also able to perform classification of bottom and charm decays based on the difference in their lifetimes, though with a lower efficiency than classification relative to light quarks.

As noted above, these particle identification algorithms were designed to provide flexibility in the design of data analyses. They thus gave rise to adjustable efficiencies in the selection process. The careful optimization of these efficiencies is a large consideration in analysis design. These efficiencies have a strong dependence on the momentum transverse to the beam line (p_T) as well as their angle relative to the beam line (measured as the Lorentz-invariant quantity pseudo-rapidity, or η). Particles with large p_T and large angles relative to the beam line tend to have the highest identification efficiencies. Characteristic selection requirements are $p_T > 20$ GeV and $|\eta| < 2$. Electron, muon and tau selections were categorized into *"tight"* and *"loose"* measurements with typical efficiencies to select real leptons of 82/94% (electrons), 72/89% (muons) and 65/82% (taus) for leptons in the tight/loose categories, respectively. Jets have typical identification efficiencies of 85–95% depending on the region of the detector in which they are detected and their transverse momentum. The efficiency for b-tagging algorithms can be described in generic tight and loose categories with bottom/charm/light quark identification of 50/12/0.1% and 70/17/3%, respectively.[14] Finally, there is an inefficiency associated with the hardware and software algorithms used to reduce the data rate in the triggering process. Triggers for single electrons or muons have typical efficiencies of 90% and 75%, respectively. Data analyses commonly use several dedicated triggers in parallel to achieve average triggering efficiencies above 95%.

Further details of the object reconstruction algorithms used in the Tevatron Higgs boson searches can be found in the references for the individual analyses (see Table 3.1).

5. Analysis channel summary

The Tevatron Higgs searches were performed in mutually exclusive final states motivated by the Higgs boson production and decay modes

Table 3.1   Luminosities, explored mass ranges, and references for the different processes and final states in the Tevatron Higgs search ($\ell = e$ or μ, τ_{had} denotes a hadronic tau decay and $V = W$ or Z).

Experiment	Channel	Luminosity (fb^{-1})	m_H range (GeV)	Reference
CDF	$WH \to \ell\nu b\bar{b}$; 2-jet	9.45	90–150	15
CDF	$WH \to \ell\nu b\bar{b}$; 3-jet	9.45	90–150	15
D0	$WH \to \ell\nu b\bar{b}$; 2-jet	9.7	90–150	16
D0	$WH \to \ell\nu b\bar{b}$; 3-jet	9.7	90–150	16
CDF	$ZH \to \nu\bar{\nu}b\bar{b}$	9.45	90–150	17,18
D0	$ZH \to \nu\bar{\nu}b\bar{b}$	9.5	100–150	19
CDF	$ZH \to \ell^+\ell^- b\bar{b}$; 2-jet	9.45	90–150	20
CDF	$ZH \to \ell^+\ell^- b\bar{b}$; 3-jet	9.45	90–150	20
D0	$ZH \to \ell^+\ell^- b\bar{b}$	9.7	90–150	21
CDF	$H \to W^+W^- \to \ell\nu\ell\nu$	9.7	110–200	22
CDF	$H \to W^+W^- \to \ell\nu\tau_{\text{had}}\nu$	9.7	130–200	22
D0	$H \to W^+W^- \to \ell\nu\ell\nu$	9.7	115–200	23
D0	$H \to W^+W^- \to \mu\nu\tau_{\text{had}}\nu$	7.3	115–200	24
D0	$H \to W^+W^- \to \ell\bar{\nu}jj$	9.7	100–200	16
CDF	$WH \to WW^+W^-$	9.7	110–200	22
CDF	$ZH \to ZW^+W^-$	9.7	110–200	22
D0	$VH \to e^\pm\mu^\pm + X$	9.7	100–200	25
D0	$VH \to \ell\ell\ell + X$	9.7	100–200	25
D0	$VH \to \tau_{\text{had}}\tau_{\text{had}}\mu + X$	8.6	100–150	25
CDF	$H \to \gamma\gamma$	10.0	100–150	26
D0	$H \to \gamma\gamma$	9.6	100–150	27
CDF	$H \to ZZ \to \ell\ell\ell\ell$	9.7	120–200	28
CDF	$H \to \tau^+\tau^-$	6.0	100–150	29
CDF	$WH + ZH \to jjb\bar{b}$	9.45	100–150	30
CDF	$t\bar{t}H \to WW b\bar{b}b\bar{b}$	9.45	100–150	31
D0	$H+X \to \ell^\pm \tau_{\text{had}}^\mp jj$	9.7	105–150	32
D0	$VH \to \ell\bar{\nu}jjjj$	9.7	100-200	16

kinematically accessible at the Tevatron, as discussed below in this section. Lists of the analyses performed at CDF and D0 can be found in Table 3.1. To maximize the sensitivity of the Higgs boson search, the data samples were separated into final state categories defined by the numbers of charged leptons and jets, the amount of missing transverse energy, the number of b-tagged jets, and b-tagging criteria. This subdivision allows, for example, the efficient use of poorly reconstructed leptons, the exploitation of the different signal and backgrounds in each sub-channel, and the reduction of the impact of systematic uncertainties.

The Tevatron Higgs boson searches were focused on the Higgs boson mass range of 90–200 GeV, which was primarily motivated by two considerations. First, the Higgs boson production cross sections above $m_H = 200$ GeV were too small to provide sufficient sensitivity to higher masses. And second, indirect constraints on the Higgs boson mass from top quark and W and Z boson measurements at SLC, LEP and the Tevatron disfavored the regions below $m_H = 90$ GeV and above $m_H = 200$ GeV.

5.1. *Searches for Higgs boson masses below 135 GeV*

For Higgs boson masses below 135 GeV, the dominant Higgs boson decay mode is $H \to b\bar{b}$. Though the largest production rate is via the fusion of gluons, the signature of two bottom quarks cannot be easily isolated from the continuum production of bottom quark pairs. Thus the associated production modes W/ZH with Higgs boson decays to $b\bar{b}$ provide the largest sensitivity to Higgs boson production for the Tevatron experiments. The leptonic decays of the vector bosons in $W \to \ell\nu$, $Z \to \ell^+\ell^-$ and $Z \to \nu\bar{\nu}$ provide the most distinct signatures relative to the dominant background of jet production. When coupled with a pair of jets with b-tags, these signatures provide a detector topology that can be isolated in a sample of events with an overall purity of 2–3%. This purity is improved to 50% or more for a small subset of events when using multivariate analysis methods (discussed later), typically corresponding to a phase space where the dijet invariant mass is close to the simulated Higgs boson mass. The channels investigated at the Tevatron are $WH \to \ell\nu b\bar{b}$, $ZH \to \ell^+\ell^- b\bar{b}$, and $(WH + ZH) \to E_T^{\mathrm{miss}} b\bar{b}$. The dominant background to these searches is the production of vector bosons in association with jets (W/Z+jets), where the jets may arise from gluons or from light, charm or bottom quarks. The requirement of one or more b-tags also enhances the $t\bar{t}$ background to this final state. Finally, an irreducible background arises from the production of two vector bosons (WW, WZ and ZZ) with semi-leptonic decays.

Additional sensitivity in this Higgs boson mass search range comes from a channel in which the W boson or the Z boson decays hadronically, resulting in a four-jet final state, which is also sensitive to VBF Higgs production. Higgs boson decays to $\tau^+\tau^-$ were sought in analyses that also require an additional one or two jets, which suppresses Drell–Yan background, while focusing on WH, ZH, and VBF production.[a] Searches for $H \to \gamma\gamma$ and $t\bar{t}H \to t\bar{t}b\bar{b}$ also are most sensitive for $m_H < 135$ GeV, but are less sensitive searches due to smaller decay rates and production rates.

5.2. Searches for Higgs boson masses above 135 GeV

For Higgs boson masses between 135 GeV and 200 GeV, the decay $H \to W^+W^-$ becomes dominant.[b] Leptonic decays of the W bosons provide striking signatures easily separated from hadronic multijet events by requiring the presence of two oppositely-charged leptons and large $E_{\mathrm{T}}^{\mathrm{miss}}$. This allows the $H \to W^+W^-$ decay to be paired with the gluon fusion production mode to maximize the rate of Higgs production in the search. The most significant backgrounds to these analyses are non-resonant W^+W^-, $t\bar{t}$, and Drell–Yan production. The production of W+jets events in which a jet is mis-identified as a lepton is also important and was estimated using data-based techniques. A large sample of W+jets events has been collected by each collaboration, and lepton fake rates measured as functions of the lepton candidate energy and η in inclusive samples of multijet events. These fake rates were applied to the jets in W+jets events in order to estimate the rates and distributions of the background due to mis-identified jets in each of these analyses without relying on Monte Carlo simulations.

The semi-leptonic decays for this mode were also considered ($H \to W^+W^- \to \ell\nu q\bar{q}$). While the signal rate is significantly larger, the W/Z+jets background also rises and limits the ultimate sensitivity of this search. The associated production of Higgs and vector bosons was also studied ($VH \to VVV$, $V = W$ or Z) in both fully leptonic and semi-leptonic final states through specific data analyses. At higher masses, $m_H > 170$ GeV, the $H \to ZZ \to \ell^+\ell^-\ell^+\ell^-$ search also contributes to the combined search sensitivity. A smaller contribution to the sensitivity comes in at $m_H = 130$ GeV due to the branching ratio dependence on m_H.

The relative search power of each contributing analysis can be visualized in Fig. 3.2 wherein the expected upper limit on the Higgs boson production

---

[a] In the case of $H \to \tau^+\tau^-$ searches, the small signal expected from $gg \to H$ was also included for completeness.
[b] For $H \to W^+W^-$ decays with $m_H < 2m_W$, at least one W boson must be off mass shell. For simplicity, we do not denote the off-shell W boson in formulas.

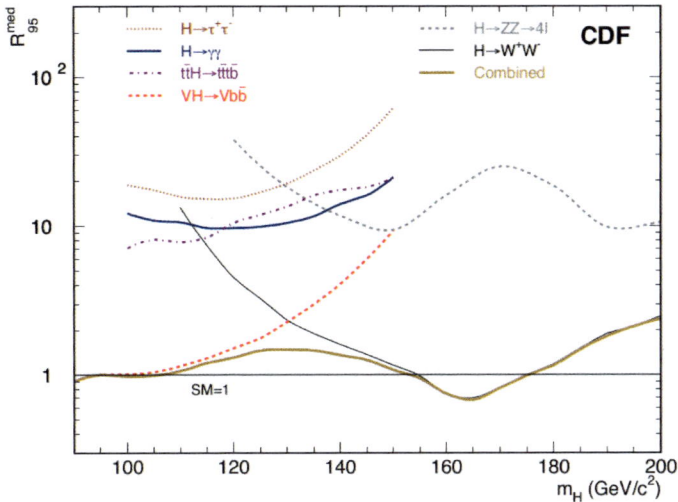

Fig. 3.2   Expected upper limits on the cross section times branching fraction for Higgs bosons as a function of Higgs boson mass in the major search channels at the CDF experiment.[33] The expected upper limits are expressed as a ratio to the SM prediction for the cross section times branching fraction.

cross section times branching fraction is shown for the major search channels at the CDF experiment.[c] In this case, the expected upper limits are expressed as a ratio to the SM prediction for the cross section times branching fraction. It is clear from the figure that the dominant search channels are $VH \to Vb\bar{b}$ for masses below 135 GeV and $H \to WW$ above.

6. Analysis optimization strategies

Over the years, the searches for the SM Higgs boson at the Tevatron steadily improved in their sensitivity and comprehensiveness. The initial versions of the analyses selected particles consistent with the dominant features of the signal processes, within the ranges the triggers selected. Jet reconstruction and lepton candidate requirements were the same as those used in $t\bar{t}$ cross section measurements, as were the b-tagging algorithms. Lepton identification requirements included tight cuts on the track/cluster or track/stub

---

[c]Details of limit setting and interpretation are discussed in Sec. 8.

matching, track χ^2, shower shape, hadronic contamination, and other variables, and were restricted to the most central portions of the detector. Lepton transverse momenta were typically required to exceed 20 GeV. These reconstruction and selection requirements were originally optimized to provide a purified sample of $t\bar{t}$ events in analyses with just one sample of selected events, and applied for the Higgs boson search.

The sensitivity to the SM Higgs boson in the low mass range was predicted to be very low with these initial analyses. Calculations indicated that improving the detection efficiency, even at the cost of increased backgrounds, would improve the search sensitivity. Lowering jet E_T cuts, loosening lepton ID requirements, and including events with just one b-tag instead of two (as well as looser b-tagging criteria) were strategies that improved the signal selection efficiency in the low-mass channels. The triggering also improved over time, especially the E_T^{miss} reconstruction in CDF[34] and the use of trigger suites with efficiencies near 95% at D0.

In early data collection periods, the size of the selected search samples was insufficient to constrain the systematic uncertainties in MC predictions of differential distributions. During this time, the Higgs boson search analyses were designed to isolate regions of kinematic phase space with large signal vs background purity. These analyses used only the total number of selected expected and observed events to test the presence or absence of Higgs boson signal, rather than use the additional information in the final state kinematic distributions. As the dataset size increased, the use of multivariate analysis (MVA) techniques became commonplace (see Appendix B for more details). The use of MVA methods led to a strategy of separating search selections into orthogonal search samples in which the MVA could be independently optimized. The statistical combination of non-overlapping channels provided the mechanism by which additional samples of events could be included in a way that improved the sensitivity. With a single set of selection requirements, the cuts can be optimized to maximize the predicted sensitivity. Loosening the cuts adds additional data, with both signal and background components. The additional data by itself has a nonzero sensitivity to the Higgs boson, but adding a sample of data with a low signal-to-background ratio (s/b) to a sample of higher s/b can dilute the sensitivity of the latter and reduce the total sensitivity. Placing events selected with a lower s/b selection strategy in a separate channel from those with a higher s/b and combining them with the statistical methods described in Sec. 8 allows for a broad search program including many additional samples. Existing searches were broken into sub-samples. For example, the $H \rightarrow W^+W^-$

searches were broken into separate jet categories: $\ell^+\ell^- + E_T^{\mathrm{miss}} + 0$ jets, $+1$ jet, and $+2$ or more jets.

The MVA techniques used in the Tevatron Higgs analyses were also studied in depth in the context of the measurements of top quark production, both in pairs and singly via electroweak interactions.[35,36] Both cases provided a validation of the methods through careful study of the agreement between data and simulation in kinematic regimes and final states shared with Higgs boson searches. Notably, the behavior of MVA techniques in a discovery scenario was scrutinized by both experiments in the Tevatron's observation of single-top production in 2009.[37–40] These tests continued in the Higgs boson search *in situ* by using diboson production (WW, WZ, ZZ) as a signal. This is discussed later in this chapter in Sec. 9.2.1.

7. Systematic uncertainties

The predictions for the expected signal and background rates in the Tevatron Higgs search depend on three factors used to normalize the MC simulations: selection efficiency, cross section times branching ratio ($\sigma\times$BR) and integrated luminosity. Each of these factors was estimated using methods with varying degrees of uncertainty. These uncertainties propagate to the expected signal and background rates, resulting in the potential for over- or under-estimation of rate in the simulation relative to the data. Furthermore, these uncertainties on efficiency and cross section depend on the kinematics of the process in question. These systematic uncertainties degrade the power of the statistical analysis of the data and their causes and effects must be carefully considered. Systematic uncertainties were evaluated independently for each final state, background, and signal process.

Uncertainties that impact the predicted signal and background rates arise from both experimental and theoretical sources. For example, the uncertainty on the identification of electrons was determined using large auxiliary samples of $Z \to e^+e^-$ decays. This estimation has both statistical uncertainty based on the number of Z decays used and systematic uncertainty associated with the efficiency estimation methodology and its extrapolation to other data samples. These sources of uncertainty were combined in quadrature to describe the total uncertainty on the rate for electron identification. Uncertainties also arose for the theoretical calculations used to determine $\sigma\times$BR. These uncertainties were estimated from the variations in the factorization and renormalization scales, uncertainties due to PDFs and the dependence on the strong coupling constant, α_s. In addition, in certain instances it is informative to make comparisons with alternative theory calculations. These variations in the parameters of theory calculations sometimes also result in changes in the predicted kinematic distributions for the relevant physics processes, which were considered as described below.

To study uncertainties on the shapes of the probability density functions of the final discriminants, the parameter associated with the uncertainty is varied within one standard deviation (s.d.) of its estimated uncertainty and the full analysis procedure is repeated using the modified distribution. For example, for the jet energy scale uncertainty, the parameters of the energy scale were varied within one s.d. of their associated uncertainties and the analysis is carried out using the resulting modified jets, which may change the features of each event and its selection efficiency. This process can change the distribution of the final variable and the sample composition. No retraining of MVAs was performed during the propagation of systematic uncertainties to the distributions of the discriminants, as this would introduce a false bias. Correlations between signal and background, across different channels within an experiment and between CDF and D0 were taken into account. Full details on the treatment of the systematic uncertainties in the individual channels can be found in the relevant references, available in Table 3.1.

The largest sources of uncertainty on the predicted signal and background rates arise from the integrated luminosities used to normalize the simulated event yields (6%) and $\sigma \times$BR. For the luminosity uncertainty, there is a component due to the total inelastic $p\bar{p}$ cross section of 4% which is considered to be 100% correlated between CDF and D0. The uncertainties on cross sections vary from the level of 6–7% for diboson and $t\bar{t}$ production to 20–40% for the $V+$ heavy flavor quark, instrumental and multijet backgrounds.

Sources of systematic uncertainty that affect both the normalization and the shape of the final discriminant distribution include jet energy scale (1–4%), jet energy resolution (1–3%), trigger efficiencies, and b-tagging. Many of these uncertainties were constrained using auxiliary data samples. For example, the b-tagging efficiencies and rates for falsely tagged light jets were similarly constrained using samples such as inclusive jet data or $t\bar{t}$ events. The uncertainty on the per-jet b-tag efficiency is approximately 4%, and the mistag uncertainties vary between 7% and 15%.

In total, 326 independent sources of systematic uncertainty were evaluated and included in the Tevatron Higgs boson searches. Most of these arose from the independent experimental estimations of selection efficiency and were considered to be uncorrelated between CDF and D0. Uncertainties considered to be correlated between the experiments come from the total inelastic $p\bar{p}$ cross section and the differential and inclusive theoretical predictions for the signal and background rates.

8. Statistical analysis methods

After categorizing the data into independent search channels, an estimation of the relative agreement between the data and the MC simulation was made to evaluate the presence or absence of Higgs boson signal. This is facilitated as a hypothesis test in which the data are compared to a hypothesis in which no Higgs boson signal exists (*i.e.*, background-only) and one that includes the known SM backgrounds in addition to the Higgs boson signal (signal-plus-background). The techniques used are described in detail in Appendix A. The Tevatron experiments interpreted their results using two different statistical techniques to provide a robust cross check of the results of each method. The results were calculated using both a Bayesian integration with a uniform prior on the number of signal events and also a modified Frequentist calculation. The results were calculated at each value of m_H studied, every 5 GeV over the range 90–200 GeV. The results of the two calculations typically agree very well to within 5% for individual m_H hypotheses and within 1% when averaged over all m_H values tested. These two calculations typically provide different information on the Higgs search and the technique used for each interpretative result was agreed upon *a priori*.

Both statistical techniques are based on the joint likelihood evaluated over each histogram bin in each analysis channel. Of particular importance in the comparison between the two techniques are the methods used to determine the most probable value of the likelihood function. The Bayesian calculation integrates over all values of the systematic uncertainties that result in a non-negative prediction for the signal and background rates in all channels and histogram bins. The modified Frequentist calculation estimates the likelihood maximum by performing a fit to the likelihood function by varying the systematic uncertainties within their priors. Upper limits and measured cross sections are evaluated in the Bayesian calculation by integrating over the likelihood posterior relative to the observed data. Confidence levels and p-values are evaluated in the modified Frequentist method by generating large numbers of pseudo-data produced by randomly varying the values of systematic uncertainties and by randomly sampling the Poisson rate of predicted events in each histogram bin. These pseudo-data are evaluated using the log-likelihood ratio (LLR) and the resulting LLR distributions are integrated relative to the observed data to calculate p-values.

9. Results of the Tevatron Higgs boson searches

9.1. *The Tevatron Higgs Working Group*

The CDF and D0 collaborations performed their Higgs boson searches independently, with very limited interactions between the two collaborations on their data analysis techniques. Over the many years that the data analyses were performed, there was a healthy competition between the collaborations to present data analyses with more stringent expected limits on Higgs boson production. This competition pushed each collaboration to innovate in data analysis techniques, to work quickly to incorporate the newest data into the searches and to ensure the highest level of understanding for the data itself to avoid mistakes that could slow down progress. Following an intense period from 2001–2006 during which the Higgs boson searches matured significantly, the two collaborations created a formal group to combine the results of the CDF and D0 searches. This provided an independent cross check of each experiment's results, as well as a validation of the combined result. The group's roles also included collecting theoretical predictions and related uncertainties and agreeing on prescriptions for their use within CDF and D0 to facilitate combination, agreeing on statistical procedures and treatment of shared systematic uncertainties, and preparation for conference presentations and publications. The combined results represent the Tevatron's most powerful statement on Higgs boson production and typically improved over the individual experiments by a factor of roughly $\sqrt{2}$ due to the doubling of search statistics. Though the Tevatron experiments each adopted similar analysis strategies, there were many instances in which one experiment could capitalize on detector-specific capabilities that allowed unique search channels to be studied. Furthermore, differences in detection efficiencies often caused common search channels to have different search sensitivities. Thus the full combination had a highly complex improvement relative to each individual experiment as a function of m_H. Though each experiment evaluated their own combination of search channels independently, we present only the Tevatron combined results here. The results from CDF and D0 alone can be found in Refs. 33 and 41, respectively.

9.2. *Higgs boson search results*

9.2.1. *Tests of SM diboson production*

As a validation procedure to assess the modeling of background processes and the data analysis methods, the experiments performed measurements of SM diboson production in the same final states used for the SM Higgs boson searches. The techniques used in the two largest contributions to the search ($H \to b\bar{b}$ and $H \to W^+W^-$) were also able to efficiently isolate diboson production from other processes. The $VH(\to b\bar{b})$ analysis were able to study $VZ(\to b\bar{b})$ production, while the $H \to W^+W^-$ analyses measured $VV' \to \ell\ell' + X$ cross sections. These studies used the same data sample, reconstruction, background modeling, systematic uncertainties and search methods as the corresponding SM Higgs boson searches. The only differences were in the final discriminants used to classify SM diboson production relative to the remaining background processes. As an example, we consider here the case of $VZ(\to b\bar{b})$ production. The rate of VZ production with $Z \to b\bar{b}$ is approximately 6–7 times larger than the analogous $VH(\to b\bar{b})$ processes for $m_H = 125$ GeV. This larger rate makes VZ production not only important for testing analysis methods, but also provides a crucial means to study the dijet invariant mass spectrum, which is the most important means of discriminating between $H \to b\bar{b}$ production and its associated backgrounds. By analyzing the distribution of the dijet invariant mass distribution in the data, a comparison can be made with the simulated WZ and ZZ processes. This distribution, for the combined Tevatron data, is shown in Fig. 3.3. In the figure, the predicted VZ signal and backgrounds are fit to the data by varying systematic uncertainties, then the fitted background is subtracted from the data. The resulting data distribution is then compared to the fitted VZ distribution, and the expectation for a SM Higgs boson signal is overlaid for illustration purposes. The measured cross section, using the dijet mass discriminants, for VZ is 4.3 ± 0.7 (stat) ± 0.7 (syst) pb whereas the SM prediction is 4.4 ± 0.3 pb.[42] The VV' boson cross sections measured by the $H \to W^+W^-$ analyses also agree well with SM predictions.[22,23] This result provides a critical validation

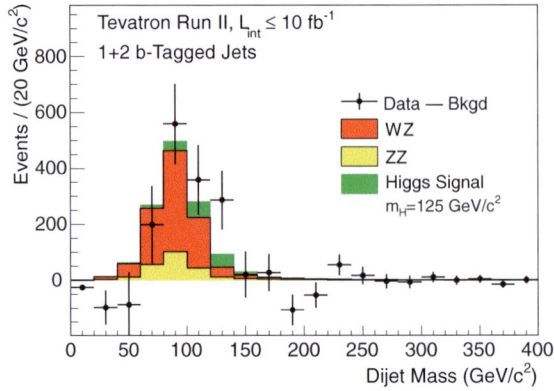

Fig. 3.3 Distribution of background-subtracted data for the reconstructed dijet mass, summed over CDF and D0 measurements of VZ production. The VZ signal and the backgrounds are fit to the data, and the resulting background is subtracted from the data. The fitted VZ and the expected SM Higgs ($m_H = 125$ GeV) contributions are shown as filled histograms. The error bars shown on the data points correspond in each bin to the square root of the sum of the expected signal and background yields. (From Ref. 12.)

of the ability to detect and measure a well-known decay of the Z boson, and hence proof of the robustness of the Higgs search method for objects of similar mass and decay mechanisms.

9.2.2. *Search for Higgs boson production*

The Tevatron Higgs boson search results were obtained by combining results from all contributing channels over the decay modes $H \to b\bar{b}$, $H \to W^+W^-$, $H \to \tau^+\tau^-$, $H \to \gamma\gamma$ and $H \to ZZ$, and in the production modes $gg \to H$, W/ZH, vector-boson fusion and $t\bar{t}H$ (see Table 3.1). These results were obtained on a grid of hypothesized Higgs boson masses in 5 GeV steps from 90 GeV to 200 GeV. We describe here the final, published results.

Historically, the $H \to b\bar{b}$ results were ready for combination first, and these were performed in the range 90 GeV $< m_H <$ 150 GeV. The combined $H \to b\bar{b}$ results were published in 2012 for CDF,[43] D0,[44] and the Tevatron.[45] These results were used to calculate an upper limit on the $WH + ZH$ production cross section times the branching ratio BR($H \to b\bar{b}$) as a function of m_H, and the result is shown in Fig. 3.4. Here the upper limit on $\sigma \times$BR at the 95% C.L. is expressed as a ratio to the values predicted by the SM. In the figure, the upper limit evaluated with the observed data is

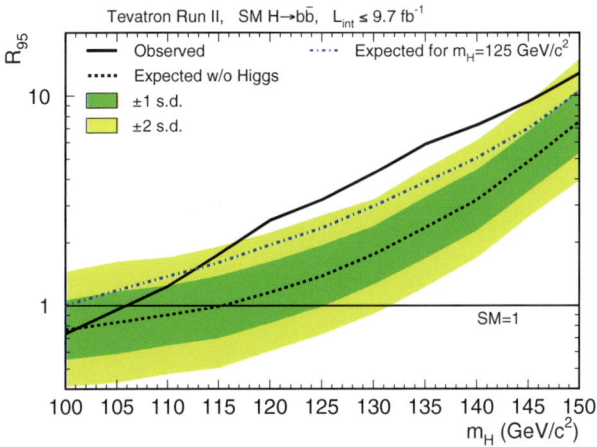

Fig. 3.4 Observed and median expected (for the background-only hypothesis) 95% C.L. Bayesian upper limits on Higgs boson production expressed as multiples of the SM cross section as a function of Higgs boson mass for the combined CDF and D0 searches in the $H \to b\bar{b}$ decay mode. The green- and yellow-shaded bands indicate, respectively, the one and two s.d. probability regions in which the limits are expected to fluctuate in the absence of signal. The blue dot-dashed line shows median expected limits assuming the SM Higgs boson is present at $m_H = 125$ GeV. (From Ref. 45.)

compared to the median upper limit expected if there were no Higgs boson production. A broad excess is seen in the range 120 GeV $< m_H <$ 135 GeV. The significance of this excess is quantified in terms of the p-value $1-\text{CL}_b$ for the background-only hypothesis.[d] This metric is the probability to find an excess as extreme as that observed in the data, assuming no Higgs boson signal is present. The resulting p-values are shown in Fig. 3.5. At the time of publication, the presence of the Higgs boson had not been established and its mass had not been measured, except for preliminary conference reports, and so the statistical tests were made on the Tevatron's search region of interest, 115 GeV $< m_H <$ 150 GeV, which is determined by the LEP lower bound and the upper end of sensitivity to $H \to b\bar{b}$. The most significant observed local p-value in this range corresponds to 3.3 standard deviations, and was observed at $m_H = 135$ GeV. Because the resolution in the reconstructed mass m_{jj} at CDF and D0 is poor ($\sim$25 GeV), Monte Carlo studies indicated that the Look-Elsewhere Effect (LEE) factor in the mass range 115 GeV $< m_H <$ 150 GeV is approximately 2.0. The resulting

---

[d]See Appendix B for a discussion of p-values.

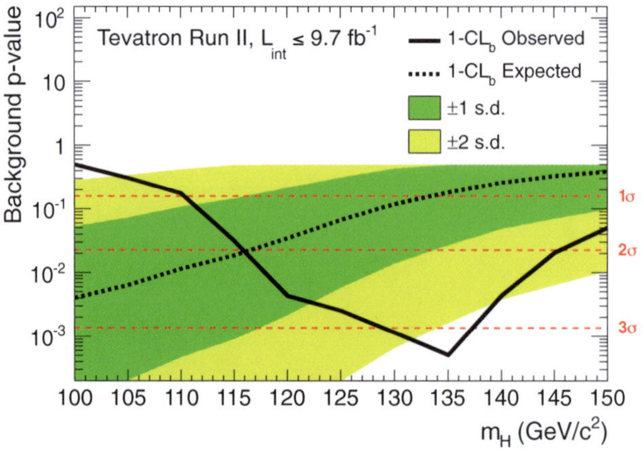

Fig. 3.5   The solid black line shows the background p-value as a function of m_H for the combined CDF and D0 SM Higgs boson searches in the $H \to b\bar{b}$ decay mode. The dotted black line shows the median expected values assuming a SM signal is present, evaluated separately at each m_H. The associated green- and yellow-shaded bands indicate the one and two s.d. fluctuations of possible experimental outcomes under this scenario. (From Ref. 45.)

LEE-corrected global significance corresponds to 3.1 standard deviations, and was used to claim evidence for the Higgs boson.[45]

The combinations of the results for all channels listed in Table 3.1 were published in 2013 for CDF,[33] D0,[41] and the Tevatron.[12] Between the 2012 publication of combined results and the 2013 round of final publications, the CDF $E_T^{\mathrm{miss}}\, b\bar{b}$ channel[17] was updated to include the latest neural-network b-tagging algorithm[14] used in the $WH \to \ell\nu b\bar{b}$ and $ZH \to \ell^+\ell^- b\bar{b}$ searches, and the result was published in Ref. 18. In the process of updating to the newer b-tagging algorithm, the previous result was repeated and verified, demonstrating consistency of the two results within statistical and systematic uncertainties. Nonetheless, with the more efficient b-tagger, fewer events in the highest-score regions of the neural-network outputs were observed, and the best-fit cross section for $H \to b\bar{b}$ decreased.

Figure 3.6 shows the observed upper limit on $\sigma\times$BR at the 95% C.L., expressed as a ratio to the rates predicted by the SM over the full range 90 GeV $< m_H <$ 200 GeV, which is now possible due to the inclusion of the $H \to W^+W^-$ and $H \to ZZ$ search channels. The excess of events in the $H \to b\bar{b}$ channels, noted above, causes the observed limits to be weaker than the expected limits. The shape of this excess is very similar to the excess expected if there were Higgs boson production, shown in dashed

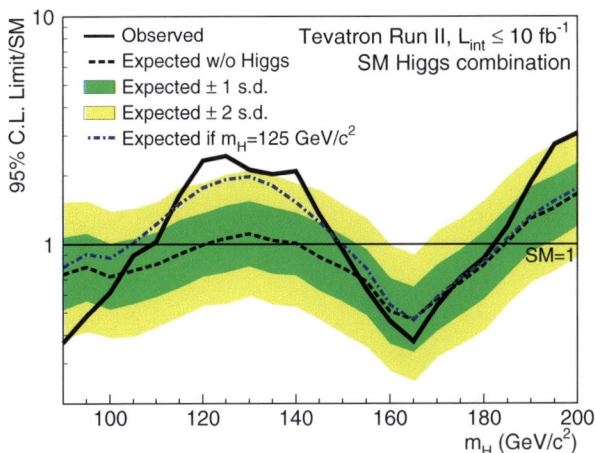

Fig. 3.6   Observed and median expected (for the background-only hypothesis) 95% C.L. Bayesian upper limits on Higgs boson production expressed as multiples of the SM cross section as a function of Higgs boson mass for the combined CDF and D0 searches. The green- and yellow-shaded bands indicate, respectively, the one and two s.d. probability regions in which the limits are expected to fluctuate in the absence of signal. The blue short-dashed line shows median expected limits assuming the SM Higgs boson is present at $m_H = 125$ GeV. (From Ref. 12.)

blue. To test the agreement of this excess in data with the SM Higgs predictions, the SM Higgs boson signal and background processes are fit to the data with an unconstrained signal rate. The resulting values for the signal rate are shown in Fig. 3.7 along with the expectation for a SM Higgs boson with $m_H = 125$ GeV. The observed signal rate agrees with the prediction for a SM Higgs boson within 1 s.d. with a peak in the rate near $m_H = 125$ GeV where the fit yields $\sigma(\text{Fit})/\sigma(\text{SM}) = 1.44^{+0.49}_{-0.47}$ (stat)$^{+0.33}_{-0.31}$ (syst) ± 0.10 (theory).

The significance of the excess in the data over the predicted background is quantified by the background-only p-value, which is shown in Fig. 3.8 as a function of m_H along with the expected p-value assuming a SM Higgs boson signal is present at each tested Higgs mass. The median expected p-values assuming the SM Higgs boson is present with $m_H = 125$ GeV for signal strengths of 1.0 and 1.5 times the SM prediction are also shown. The median expected excess at $m_H = 125$ GeV corresponds to 1.9 standard deviations assuming the SM Higgs boson is present at that mass. The observed local significance at $m_H = 125$ GeV corresponds to 3.0 standard deviations. The maximum observed local significance is at $m_H = 120$ GeV

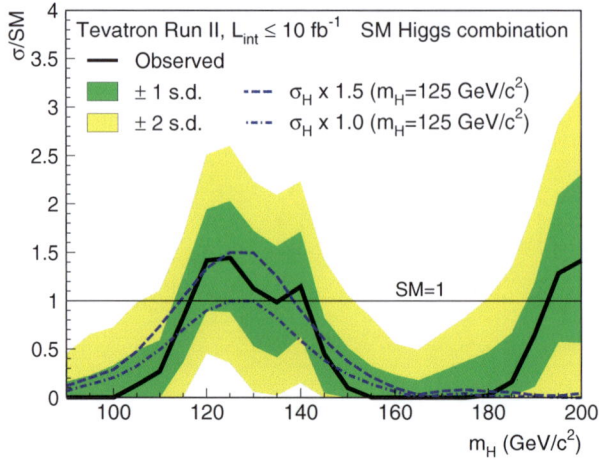

Fig. 3.7   The best-fit signal cross section expressed as a ratio to the SM cross section as a function of Higgs boson mass for all CDF and D0 SM Higgs boson searches. The green- and yellow-shaded bands show the one and two s.d. uncertainty ranges on the fitted signal, respectively. Also shown with blue lines are the median fitted cross sections expected for a SM Higgs boson with $m_H = 125$ GeV at signal strengths of 1.0 times (short-dashed) and 1.5 times (long-dashed) the SM prediction. (From Ref. 12.)

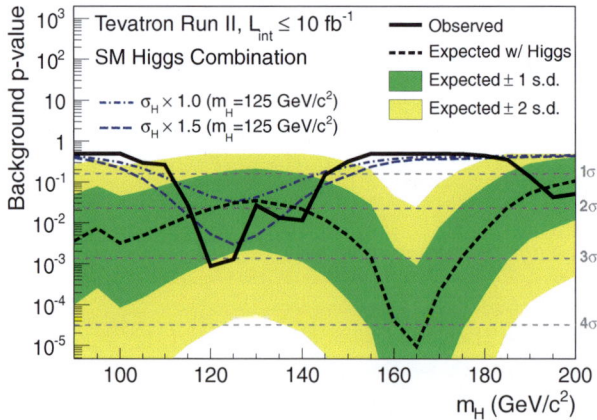

Fig. 3.8   The solid black line shows the background p-value as a function of m_H for all of CDF and D0's SM Higgs boson searches in all decay modes combined. The dotted black line shows the median expected values assuming a SM signal is present, evaluated separately at each m_H. The associated green- and yellow-shaded bands indicate the one and two s.d. fluctuations of possible experimental outcomes under this scenario. The blue lines show the median expected p-values assuming the SM Higgs boson is present with $m_H = 125$ GeV at signal strengths of 1.0 times (short-dashed) and 1.5 times (long-dashed) the SM prediction. (From Ref. 12.)

and corresponds to 3.1 standard deviations. The width of the dip in the observed p-values from 115 to 140 GeV/c^2 is consistent with the resolution of the combination of the $H \to b\bar{b}$ and $H \to W^+W^-$ channels, as illustrated by the injected signal curves. The effective resolution of this search comes from two independent sources of information. The reconstructed candidate masses help constrain m_H, but more importantly, the expected cross sections times the relevant branching ratios for the $H \to b\bar{b}$ and $H \to W^+W^-$ channels are strong functions of m_H in the SM. The observed excess in the $H \to b\bar{b}$ channels and the slight excess in the $H \to W^+W^-$ channels determine the shape of the observed p-value as a function of m_H.

9.2.3. *Studies of Higgs boson properties*

Higgs boson couplings: The Higgs boson is expected to couple more strongly to more massive particles than to less massive ones, and thus may provide sensitivity to non-SM particles whose interactions become more relevant at higher energies. Thus, the CDF and D0 searches are also grouped into combinations by the dominant decay modes: $H \to b\bar{b}$, $H \to W^+W^-$, $H \to \gamma\gamma$, and $H \to \tau^+\tau^-$. For each of these four subset combinations the best-fit rate of Higgs boson production is evaluated for $m_H = 125$ GeV and the results are shown in Fig. 3.9. The results for each decay mode are consistent with each other, with the full combination, and with the production of the SM Higgs boson at that mass.

These individual results provide useful insight into the data excess, but their interpretation cannot be directly extended to study potential Higgs boson couplings because the searches mix various couplings in their production and decay modes. Also, each search channel has acceptance to a range of production and decay modes, causing further ambiguity. However, most of the Tevatron's search channels are sensitive to the product of fermionic and bosonic coupling strengths. For example, in the $VH \to Vb\bar{b}$ searches the production depends on the coupling of the Higgs boson to the weak vector bosons, while the decay is to fermions. Some search channels, such as the trilepton searches for $VH \to VW^+W^-$ are sensitive only to Higgs boson couplings to vector bosons, while the $t\bar{t}H \to t\bar{t}b\bar{b}$ searches are sensitive only to the couplings of the Higgs boson to fermions. By introducing multiplicative scaling factors to describe a model with enhanced or suppressed Higgs boson couplings to fermions (κ_f) or to vector bosons (κ_V), the Tevatron data can be tested for deviations from the SM predictions of $\kappa_f = \kappa_V = 1$. To illustrate, the production and decay of Higgs bosons in $VH(H \to b\bar{b})$

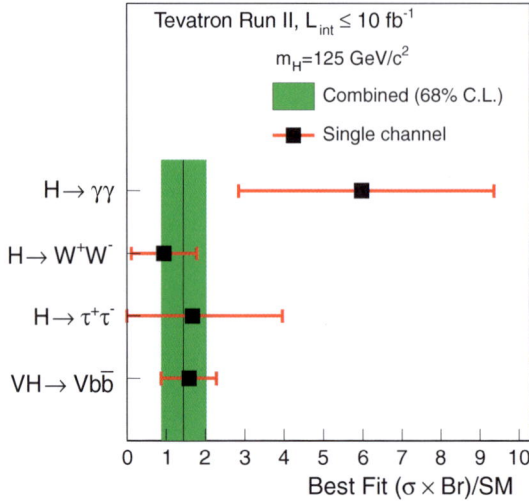

Fig. 3.9 Best-fit values of $R = (\sigma \times \mathcal{B})/\text{SM}$ using the Bayesian method for the combinations of CDF and D0 Higgs boson search channels in the $H \to W^+W^-$, $H \to b\bar{b}$, $H \to \gamma\gamma$, and $H \to \tau^+\tau^-$ decay modes for a Higgs boson mass of 125 GeV. The green-shaded band corresponds to the one s.d. uncertainty on the best-fit value of R for all SM Higgs boson decay modes combined. (From Ref. 12.)

could be expressed as a function of the scaling factors and the SM prediction as $\sigma(VH) \cdot BR(H \to b\bar{b}) = (\sigma(VH) \cdot BR(H \to b\bar{b}))^{SM} \times (\frac{\kappa_V^2 \kappa_f^2}{\kappa_H^2})$, where κ_H is introduced to normalize changes to the total rate if individual decays are modified. Assuming that custodial symmetry[46] holds (*i.e.*, $\kappa_W/\kappa_Z = 1$), the values of κ_V and κ_f can be varied. The results of this test are shown in Fig. 3.10 with the regions preferred at the 68% C.L. and the 95% C.L. in the two-dimensional plane (κ_V, κ_f). The results are bimodal with a local maximum in both the positive and negative κ_f planes. The asymmetry results from the interference in diagrams contributing to the $H \to \gamma\gamma$ decay mode. The maximum in the positive plane agrees with the SM prediction at the level of 1 s.d., providing no evidence for significant deviations from the Higgs couplings predicted by the SM Higgs mechanism.

Higgs boson spin and parity: The Tevatron data have been used to constrain models of the Higgs boson with exotic spin J and parity P, using the $VX \to Vb\bar{b}$ channels. The kinematics of VX production depend strongly on the spin and parity of the Higgs-like boson X.[47] Two models are considered: a pseudoscalar, Higgs-like boson with $J^P = 0^-$, and

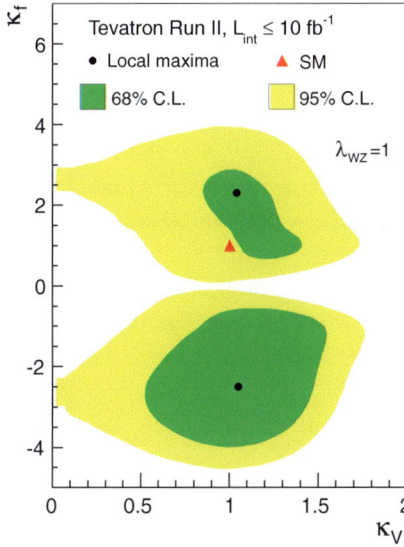

Fig. 3.10   Two-dimensional constraints in the (κ_V, κ_f) plane, for the combined Tevatron searches for a SM-like Higgs boson with mass 125 GeV assuming custodial symmetry ($\lambda_{WZ} = 1$). The points that maximize the local posterior probability densities are marked with dots, and the 68% and 95% C.L. intervals are indicated with the green- and yellow-shaded regions, respectively. The SM prediction for (κ_V, κ_f) is marked with a triangle. (From Ref. 12.)

a graviton-like boson with $J^P = 2^+$. For the SM Higgs boson, which has $J^P = 0^+$, the production rate near threshold is proportional to β, where $\beta = 2p/\sqrt{\hat{s}}$, p is the momentum of the X boson in the VX ($V = W$ or Z) reference frame, and $\sqrt{\hat{s}}$ is the total energy of the VX system in its rest frame. For the pseudoscalar model, the dependence is proportional to β^3. For the graviton-like model, the dependence is proportional to β^5; however, not all $J^P = 2^+$ models share this β^5 factor.[47] These powers of β alter the distributions of kinematic variables measured from the decay products of the vector boson and the X boson, most notably the invariant mass of the VX system, which has a higher average value in the $J^P = 0^-$ hypothesis than in the SM 0^+ case, and higher still in the $J^P = 2^+$ hypothesis. These models predict neither the production rates nor the decay branching fractions of the X particles.

Both D0[48] and CDF[49] have sought evidence for exotic Higgs-like bosons, taking advantage of the expected differences between the SM Higgs boson and the predictions of the exotic models. The three VH search channels,

$WH \to \ell\nu b\bar{b}$, $ZH \to \ell^+\ell^- b\bar{b}$, and $ZH \to \nu\bar{\nu}b\bar{b}$, are combined together using each experiment's results. Because the particle discovered by ATLAS and CMS resembles the SM Higgs boson in all measured ways including its spin and parity, it is natural that this boson is also present in the Tevatron data. Performing the search in the Tevatron data for exotic Higgs-like bosons however has significant value because the associated production mode with decays of $X \to b\bar{b}$ may sample an admixture of new particles, if more than one boson were present in the data. CDF and D0 thus sought admixtures of a SM-like Higgs boson and an exotic Higgs boson with the same mass but different spin and parity. The SM Higgs boson search channels were augmented to use the M_{VX} variable, in order to separate the X boson signals from the SM Higgs boson signal and from the much larger backgrounds. The CDF and D0 results are combined in Ref. 50 using statistical techniques similar to those use to combine the SM Higgs boson search results. Fits for the production rate times the decay branching fraction for the SM Higgs boson and the X boson are shown in Fig. 3.11 for the $J^P = 0^-$ and $J^P = 2^+$ models. Figure 3.12 shows upper bounds at the 95% C.L. on the fraction of exotic Higgs boson production that may be present alongside a SM-like Higgs boson, as a function of the total production rate. If the total production rate is assumed to be that predicted by the SM and the exotic Higgs boson is assumed to entirely replace the SM Higgs boson, then the exotic Higgs boson models are excluded with significances of 5.0 and 4.9

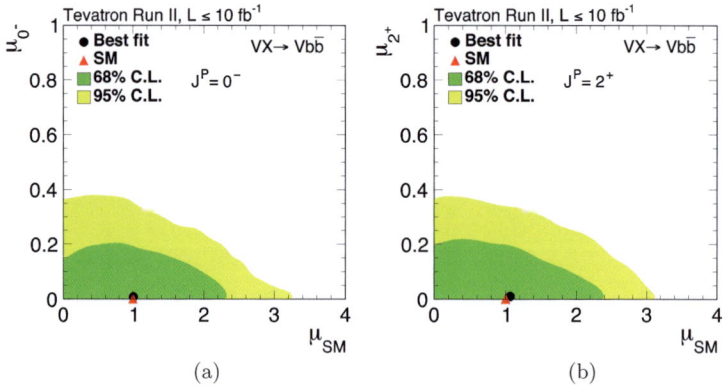

(a)   (b)

Fig. 3.11   Two-dimensional constraints in the (μ_{SM}, μ_{0^-}) (a) and (μ_{SM}, μ_{2^+}) (b) planes, in the combined Tevatron searches for Higgs bosons with exotic spin and parity. The points that maximize the local posterior probability densities are marked with dots, and the 68% and 95% C.L. intervals are indicated with the green- and yellow-shaded regions, respectively. The SM predictions are marked with red triangles. (From Ref. 50.)

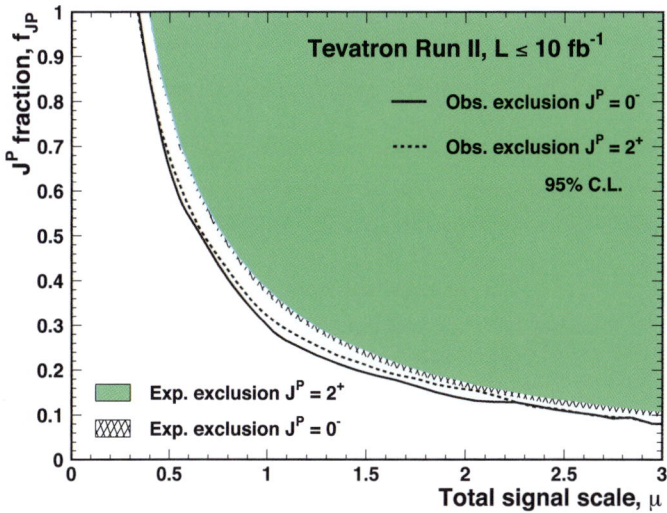

Fig. 3.12 Observed and expected 95% C.L. upper bounds on the fraction of exotic Higgs boson production as functions of the total assumed Higgs boson production rate, in units of the SM Higgs boson production rate. Curves are shown separately for the 0^- and the graviton-line 2^+ hypotheses. (From Ref. 50.)

standard deviations for the $J^P = 0^-$ and $J^P = 2^+$ models, respectively. While the Tevatron results constrain models with exotic Higgs-like bosons, at the time of this writing the LHC experiments have performed measurements of the spin and parity of the Higgs boson in bosonic decay modes. These measurements suggest the particle observed at the LHC does indeed have $J^P = 0^+$ (See Chapter 12, Sec. 7, Chapter 7, Sec. 6.3, Chapter 6, Sec. 6.4 and Chapter 5, Sec. 11).

10. Summary

The Tevatron Higgs boson search represented a broad program that probed the theoretically interesting mass range for the Higgs boson. In addition to providing the first Higgs mass exclusions beyond those set at LEP, the results of the Tevatron Higgs boson search ultimately indicated an excess of data over the expected background processes with no Higgs boson signal present. This excess has a significance of 3.0 standard deviations and can be associated with a Higgs boson production rate at a factor of $1.44^{+0.59}_{-0.56}$ times the SM prediction. This evidence for Higgs boson production is realized through the analysis of four primary final states: $H \rightarrow b\bar{b}$, $H \rightarrow \tau^+\tau^-$, $H \rightarrow W^+W^-$, and $H \rightarrow \gamma\gamma$. The analysis of each of these final states exhibits excesses in data beyond background predictions. While the Tevatron experiments did not discover the Higgs boson, the data collected and analyzed provide significant insight into the nature of the electroweak symmetry breaking mechanism.

The data excess observed at the Tevatron is consistent with the predictions for the Standard Model Higgs boson, and consistent with the boson discovered by the ATLAS and CMS collaborations.[51,52] The Tevatron search results are highly compatible with a Higgs boson mass of $m_H \simeq 125$ GeV (as observed at the LHC), and the apparent couplings of this new boson to the SM particles agree with the SM predictions within measurement uncertainties. The Tevatron Higgs boson search results provided early evidence that the Higgs boson does indeed provide mass for both the gauge bosons and the fermions. This is evidenced by the significant excess of data observed by the Tevatron experiments in the $H \rightarrow b\bar{b}$ search channels. Furthermore, the Tevatron Higgs search data did not show evidence for Higgs bosons with exotic spin and parity and the CDF and D0 collaborations set limits on their production rates.

While searches for the SM Higgs boson at the Tevatron were predicted to be very challenging, the Tevatron Higgs physics program was ultimately very successful in its execution. Because the Higgs boson mass is relatively light, the Tevatron experiments were kinematically capable of studying it, but were ultimately limited by statistics. With a very broad range of search topologies studied and through an extensive effort to optimize and improve the search techniques, the early exclusions and the later measurements of Higgs boson properties will remain major legacies of the Tevatron physics program.

Acknowledgments

We thank the members of the CDF and D0 collaborations, their theoretical colleagues, the Fermilab staff, and the technical staffs of the participating institutions for their vital contributions to the Tevatron program. Fermilab is operated by Fermi Research Alliance, LLC, under contract with the U.S. Department of Energy.

References

1. Tevatron Higgs Working Group, Report of the Tevatron Higgs working group (2000). URL http://arxiv.org/abs/hep-ph/0010338.
2. CDF and D0 Working Group Members, Results of the Tevatron Higgs sensitivity study (2003). URL http://lss.fnal.gov/archive/preprint/fermilab-pub-03-320-e.shtml.
3. CDF Collaboration, Measurement of the cross section for $t\bar{t}$ production in $p\bar{p}$ collisions using the kinematics of lepton + jets events, *Phys. Rev.* **D72**, 052003 (2005). doi: 10.1103/PhysRevD.72.052003.
4. CDF Collaboration, Measurements of inclusive W and Z cross sections in p anti-p collisions at $\sqrt{s} = 1.96$ TeV, *J. Phys.* **G34**, 2457–2544 (2007). doi: 10.1088/0954-3899/34/12/001.
5. D0 Collaboration, The D0 detector, *Nucl. Instrum. Meth.* **A338**, 185–253 (1994). doi: 10.1016/0168-9002(94)91312-9.
6. D0 Collaboration, The upgraded D0 detector, *Nucl. Instrum. Meth.* **A565**, 463–537 (2006). doi: 10.1016/j.nima.2006.05.248.
7. M. Abolins, M. Adams, T. Adams, E. Aguilo, J. Anderson, *et al.*, Design and implementation of the new D0 level-1 calorimeter trigger, *Nucl. Instrum. Meth.* **A584**, 75–97 (2008). doi: 10.1016/j.nima.2007.10.014.
8. D0 Collaboration, The layer 0 inner silicon detector of the D0 experiment, *Nucl. Instrum. Meth.* **A622**, 298–310 (2010). doi: 10.1016/j.nima.2010.04.148.
9. H. Lai, J. Huston, S. Kuhlmann, F. I. Olness, J. F. Owens, *et al.*, Improved parton distributions from global analysis of recent deep inelastic scattering and inclusive jet data, *Phys. Rev.* **D55**, 1280–1296 (1997). doi: 10.1103/PhysRevD.55.1280.
10. J. Pumplin, D. Stump, J. Huston, H. Lai, P. M. Nadolsky, *et al.*, New generation of parton distributions with uncertainties from global QCD analysis, *JHEP.* **0207**, 012 (2002).
11. A. Martin, W. Stirling, R. Thorne, and G. Watt, Parton distributions for the LHC, *Eur. Phys. J.* **C63**, 189–285 (2009). doi: 10.1140/epjc/s10052-009-1072-5.
12. CDF Collaboration, D0 Collaboration, Higgs boson studies at the Tevatron, *Phys. Rev.* **D88**, 052014 (2013). doi: 10.1103/PhysRevD.88.052014.
13. R. Brun, F. Bruyant, M. Maire, A. C. McPherson, and P. Zanarini, *GEANT 3: user's guide Geant 3.10, Geant 3.11; rev. version.* CERN, Geneva (1987).
14. J. Freeman, T. Junk, M. Kirby, Y. Oksuzian, T. Phillips, *et al.*, Introduction to HOBIT, a b-jet identification tagger at the CDF experiment optimized for light Higgs boson searches, *Nucl. Instrum. Meth.* **A697**, 64–76 (2013). doi: 10.1016/j.nima.2012.09.021.
15. CDF Collaboration, Search for the Standard Model Higgs boson decaying to a bb pair in events with one charged lepton and large missing transverse

energy using the full CDF data set, *Phys. Rev. Lett.* **109**, 111804 (2012). doi: 10.1103/PhysRevLett.109.111804.

16. D0 Collaboration, Search for the Standard Model Higgs boson in $\ell\nu$ + jets final states in 9.7 fb^{-1} of $p\bar{p}$ collisions with the D0 detector, *Phys. Rev.* **D88**, 052008 (2013). doi: 10.1103/PhysRevD.88.052008.

17. CDF Collaboration, Search for the Standard Model Higgs boson decaying to a $b\bar{b}$ pair in events with no charged leptons and large missing transverse energy using the full CDF data set, *Phys. Rev. Lett.* **109**, 111805 (2012). doi: 10.1103/PhysRevLett.109.111805.

18. CDF Collaboration, Updated search for the Standard Model Higgs boson in events with jets and missing transverse energy using the full CDF data set, *Phys. Rev.* **D87**, 052008 (2013). doi: 10.1103/PhysRevD.87.052008.

19. D0 Collaboration, Search for the Standard Model Higgs boson in the $ZH \to \nu\bar{\nu}b\bar{b}$ channel in 9.5 fb^{-1} of $p\bar{p}$ collisions at $\sqrt{s} = 1.96$ TeV, *Phys. Lett.* **B716**, 285–293 (2012). doi: 10.1016/j.physletb.2012.08.034.

20. CDF Collaboration, Search for the Standard Model Higgs boson decaying to a bb pair in events with two oppositely-charged leptons using the full CDF data set, *Phys. Rev. Lett.* **109**, 111803 (2012). doi: 10.1103/PhysRevLett.109.111803.

21. D0 Collaboration, Search for $ZH \to \ell^+\ell^-b\bar{b}$ production in 9.7 fb^{-1} of $p\bar{p}$ collisions with the D0 detector, *Phys. Rev.* **D88**, 052010 (2013). doi: 10.1103/PhysRevD.88.052010.

22. CDF Collaboration, Searches for the Higgs boson decaying to $W^+W^- \to \ell^+\nu\ell^-\bar{\nu}$ with the CDF II detector, *Phys. Rev.* **D88**, 052012 (2013). doi: 10.1103/PhysRevD.88.052012.

23. D0 Collaboration, Search for Higgs boson production in oppositely charged dilepton and missing energy final states in 9.7 fb^{-1} of $p\bar{p}$ collisions at $\sqrt{s} = 1.96$ TeV, *Phys. Rev.* **D88**, 052006 (2013). doi: 10.1103/PhysRevD.88.052006.

24. D0 Collaboration, Search for the Standard Model Higgs boson in tau lepton pair final states, *Phys. Lett.* **B714**, 237–245 (2012). doi: 10.1016/j.physletb.2012.07.012.

25. D0 Collaboration, Search for Higgs boson production in trilepton and like-charge electron-muon final states with the D0 detector, *Phys. Rev.* **D88**, 052009 (2013). doi: 10.1103/PhysRevD.88.052009.

26. CDF Collaboration, Search for a Higgs boson in the diphoton final state using the full CDF data set from proton–antiproton collisions at $\sqrt{s} = 1.96$ TeV, *Phys. Lett.* **B717**, 173–181 (2012). doi: 10.1016/j.physletb.2012.08.051.

27. D0 Collaboration, Search for a Higgs boson in diphoton final states with the D0 detector in 9.6 fb^{-1} of $p\bar{p}$ collisions at $\sqrt{s} = 1.96$ TeV, *Phys. Rev.* **D88**, 052007 (2013). doi: 10.1103/PhysRevD.88.052007.

28. CDF Collaboration, Novel inclusive search for the Higgs boson in the four-lepton final state at CDF, *Phys. Rev.* **D86**, 072012 (2012). doi: 10.1103/PhysRevD.86.072012,10.1103/PhysRevD.86.099902.

29. CDF Collaboration, Search for a low mass Standard Model Higgs boson in the $\tau-\tau$ decay channel in $p\bar{p}$ collisions at $\sqrt{s} = 1.96$ TeV, *Phys. Rev. Lett.* **108**, 181804 (2012). doi: 10.1103/PhysRevLett.108.181804.

30. CDF Collaboration, Search for the Higgs boson in the all-hadronic final state using the full CDF data set, *JHEP*. **1302**, 004 (2013). doi: 10.1007/JHEP02(2013)004.
31. CDF Collaboration, Search for the Standard Model Higgs boson produced in association with top quarks using the full CDF data set, *Phys. Rev. Lett.* **109**, 181802 (2012). doi: 10.1103/PhysRevLett.109.181802.
32. D0 Collaboration, Search for the Higgs boson in lepton, tau and jets final states, *Phys. Rev.* **D88**, 052005 (2013). doi: 10.1103/PhysRevD.88.052005.
33. CDF Collaboration, Combination of searches for the Higgs boson using the full CDF data set, *Phys. Rev.* **D88**, 052013 (2013). doi: 10.1103/PhysRevD.88.052013.
34. CDF Collaboration, The CDF level 2 calorimetric trigger upgrade, *Nucl. Instrum. Meth.* **A598**, 331–333 (2009). doi: 10.1016/j.nima.2008.08.035.
35. CDF Collaboration, D0 Collaboration, Combination of measurements of the top-quark pair production cross section from the Tevatron collider, *Phys. Rev.* **D89** (7), 072001 (2014). doi: 10.1103/PhysRevD.89.072001.
36. CDF Collaboration, D0 Collaboration, Tevatron combination of single-top-quark cross sections and determination of the magnitude of the Cabibbo–Kobayashi–Maskawa matrix element $\mathbf{V_{tb}}$ (2015).
37. D0 Collaboration, Evidence for production of single top quarks, *Phys. Rev.* **D78**, 012005 (2008). doi: 10.1103/PhysRevD.78.012005.
38. D0 Collaboration, Observation of single top quark production, *Phys. Rev. Lett.* **103**, 092001 (2009). doi: 10.1103/PhysRevLett.103.092001.
39. CDF Collaboration, First observation of electroweak single top quark production, *Phys. Rev. Lett.* **103**, 092002 (2009). doi: 10.1103/PhysRevLett.103.092002.
40. CDF Collaboration, Observation of single top quark production and measurement of —Vtb— with CDF, *Phys. Rev.* **D82**, 112005 (2010). doi: 10.1103/PhysRevD.82.112005.
41. D0 Collaboration, Combined search for the Higgs boson with the D0 experiment, *Phys. Rev.* **D88**, 052011 (2013). doi: 10.1103/PhysRevD.88.052011.
42. J. M. Campbell and R. K. Ellis, An Update on vector boson pair production at hadron colliders, *Phys. Rev.* **D60**, 113006 (1999). doi: 10.1103/PhysRevD.60.113006.
43. CDF Collaboration, Combined search for the Standard Model Higgs boson decaying to a bb pair using the full CDF data set, *Phys. Rev. Lett.* **109**, 111802 (2012). doi: 10.1103/PhysRevLett.109.111802.
44. D0 Collaboration, Combined search for the Standard Model Higgs boson decaying to $b\bar{b}$ using the D0 Run II data set, *Phys. Rev. Lett.* **109**, 121802 (2012). doi: 10.1103/PhysRevLett.109.121802.
45. CDF Collaboration, D0 Collaboration, Evidence for a particle produced in association with weak bosons and decaying to a bottom–antibottom quark pair in Higgs boson searches at the Tevatron, *Phys. Rev. Lett.* **109**, 071804 (2012). doi: 10.1103/PhysRevLett.109.071804.

46. P. Sikivie, L. Susskind, M. B. Voloshin, and V. I. Zakharov, Isospin breaking in technicolor models, *Nucl. Phys.* **B173**, 189 (1980). doi: 10.1016/0550-3213(80)90214-X.
47. J. Ellis, D. S. Hwang, V. Sanz, and T. You, A fast track towards the 'Higgs' spin and parity, *JHEP.* **1211**, 134 (2012). doi: 10.1007/JHEP11(2012)134.
48. D0 Collaboration, Constraints on spin and parity of the Higgs boson $VH \rightarrow Vb\bar{b}$ final states, *Phys. Rev. Lett.* **113**, 161802 (2014). doi: 10.1103/PhysRevLett.113.161802.
49. CDF Collaboration, Constraints on models of the Higgs boson with exotic spin and parity using decays to bottom–antibottom quarks in the full CDF data set, *Phys. Rev. Lett.* **114** (14), 141802 (2015). doi: 10.1103/PhysRevLett.114.141802.
50. CDF Collaboration, D0 Collaboration, Tevatron constraints on models of the Higgs boson with exotic spin and parity using decays to bottom–antibottom quark pairs, *Phys. Rev. Lett.* **114** (15), 151802 (2015). doi: 10.1103/PhysRevLett.114.151802.
51. ATLAS Collaboration, Observation of a new particle in the search for the Standard Model Higgs boson with the ATLAS detector at the LHC, *Phys. Lett.* **B716**, 1–29 (2012). doi: 10.1016/j.physletb.2012.08.020.
52. CMS Collaboration, Observation of a new boson at a mass of 125 GeV with the CMS experiment at the LHC, *Phys. Lett.* **B716**, 30–61 (2012). doi: 10.1016/j.physletb.2012.08.021.

Chapter 4

Overview of the Large Hadron Collider
and of the ATLAS and CMS experiments

Aleandro Nisati* and Vivek Sharma[†]

*INFN, Sezione di Roma
P.le A. Moro 2, Rome 00185, Italy
[†]University of California San Diego
La Jolla, California, 92093, USA

The Large Hadron Collider is the most powerful particle accelerator ever built. It has allowed the discovery of a Higgs boson with mass near 125 GeV in 2012 by the ATLAS and CMS experiments. This chapter provides first an overview of the main characteristics of this collider, as well as a short description of the two general purpose experiments, ATLAS and CMS, which discovered in 2012 a Higgs boson with mass close to 125 GeV. This is followed by a summary of the main aspects of particle identification and reconstruction by these two detectors, together with a short presentation of the main analysis tools used to extract the LHC results of the Higgs boson(s) searches and measurements.

1. The Large Hadron Collider: Machine and experiments

The *Large Hadron Collider* (LHC) is to date the largest and highest-energy particle accelerator in the world. Although the idea of a large hadron collider had been around since at least 1977, it is generally accepted that the birth of the LHC was in the Lausanne Workshop in 1984.[1] The proposal was to install this new machine in the LEP tunnel (see Chapter 2), once the scientific programme of this electron–positron collider was completed. Approved by the CERN Council in December 1996, the construction of this accelerator started in 1998 and lasted about ten years. A nice recollection is available in Ref. [2].

This machine has been designed primarily to collide protons at a center-of-mass energy $\sqrt{s} = 14\,\text{TeV}$ with a nominal instantaneous luminosity $\mathcal{L} = 1 \times 10^{34}\,\text{cm}^{-2}\text{s}^{-1}$ and a bunch-spacing of 25 ns. Details on the design can be found in Refs. [3–5]. Furthermore, the LHC has been designed to also collide heavy ions, in particular lead nuclei, to study the properties of the "quark–gluon plasma".

Four main particle physics experiments have been installed around the ring: two general purpose detectors, ATLAS[6] and CMS[7]; a detector dedicated to study heavy flavour physics, LHCb[8]; and a detector to study heavy ion collisions, ALICE.[9]

The LHC accelerator began to operate on September 10th, 2008. After 9 days, an electrical fault induced a mechanical damage on about 50 magnets. After the necessary repairs and consolidation of the accelerator, LHC operation started again in December 2009, with a test-run at the center-of-mass energy $\sqrt{s} = 0.9$ TeV. In 2010 and 2011, this accelerator collided protons at an energy of $\sqrt{s} = 7$ TeV. A total of 5.6 fb^{-1} integrated luminosity was delivered. In 2012 the energy was increased to 8 TeV, and a total of 23.3 fb^{-1} integrated luminosity was delivered. In March 2013, the LHC entered a shutdown period planned for machine maintenance and to perform the needed interventions to bring the collision energy close to the design value of $\sqrt{s} = 14$ TeV.

1.1. *The LHC*

The Large Hadron Collider is located near Geneva, Switzerland, about 100 meters underground in the 27 km LEP tunnel, between the Jura mountains and Lake Geneva, see Fig. 4.1.

The whole accelerator complex of CERN is used to inject the proton beams to the LHC with the initial energy of 450 GeV (see Fig. 4.2(a)).

A total of about 9600 superconducting magnets are distributed along the circumference of the LHC. Among these, 1232 dipoles are used to bend the proton trajectory to follow the design orbit. Each dipole is 15 meters long and produces a magnetic field of 8.3 Tesla (T) for a 7 TeV proton beam energy. The superconducting cable used for these dipoles (a titanium-niobium alloy) is cooled down to 1.9 K in a liquid helium bath, a temperature lower than in the intergalactic space of the Universe. Superconducting quadrupoles, sextupoles, octupoles and other types of magnets are also used to correct the beam orbit and to achieve the needed beam quality.

The LHC has been designed to circulate two counter-rotating beams of protons each grouped in about 2808 bunches. Each bunch is composed of 1.15×10^{11} protons, with longitudinal and transverse dimensions of 7.6 cm and 16 μm, respectively. The bunch-to-bunch design separation has been

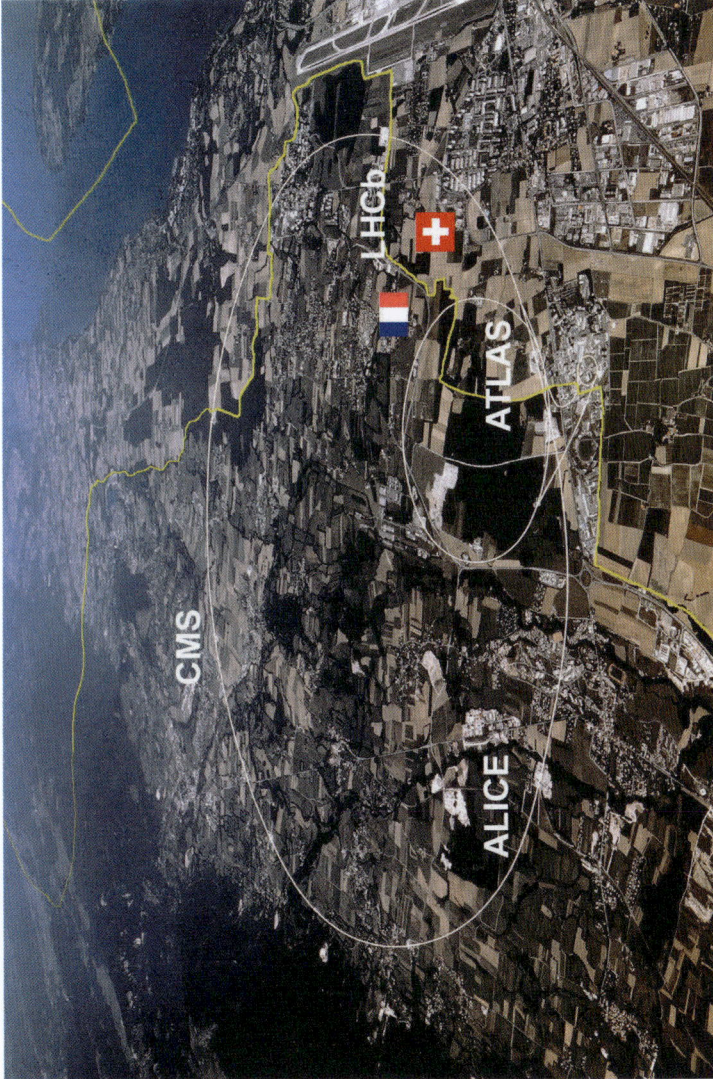

Fig. 4.1 Aerial view of the region between the Jura mountains (left hand side) and the Lake Geneva (right hand side). The Large Hadron Collider is represented by the big circle of 27 km circumference. The positions of the four experiments ATLAS, CMS, LHCb and ALICE are also shown. Also visible is the Super Proton Synchrotron accelerator (inner circle) used to inject the proton (or ion) beams into the LHC. Very close to the ATLAS location is the CERN site. (Photograph courtesy of CERN)

The LHC injection complex

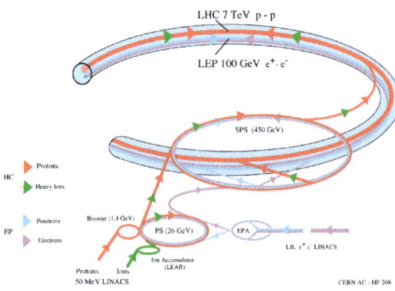

(a) Simplified scheme of the LHC injection complex.

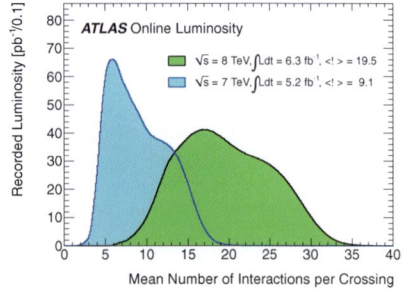

(b) Event pile-up levels in 2011 and 2012 run.

Fig. 4.2 (a) Protons accelerated by a linear accelerator (LINAC) to 50 MeV are transferred to the Proton Synchrotron (PS) Booster and then to the PS from where they get out with an energy of 26 GeV. Finally, they are accelerated by the Super Proton Synchrotron (SPS) to 450 GeV before being injected into the LHC, where they reach the final energy. (b) Distribution of the average number of proton–proton collisions per bunch-crossing.

Table 4.1 The main operating parameters of the Large Hadron Collider during 2011 and 2012, compared to the design values.[3–5] The expected average number of proton–proton collisions per bunch-crossing is also reported.

Parameter	2011	2012	Design
Beam energy [TeV]	3.5	4.0	7.0
Bunch-spacing [ns]	50	50	25
Number of bunches per beam	1380	1380	∼2808
β^* [m]	1.0	0.6	0.55
Protons/bunch	1.45×10^{11}	1.70×10^{11}	1.15×10^{11}
Norm. emittance [mm×mrad]	$\simeq 2.4$	$\simeq 2.5$	3.75
Peak luminosity [cm^{-2}s^{-1}]	3.7×10^{33}	7.7×10^{33}	1.0×10^{34}
Expected pp collisions per bunch-crossing	13	27	20

set to 25 ns. During the 2011 and 2012 runs, the bunches were separated by 50 ns, and up to 1380 bunches were collected in a single beam. Other machine parameters were below design values. However, the bunch population exceeded the design figure, reaching the value of 1.7×10^{11} protons. This allowed an instantaneous luminosity $\mathcal{L} = 0.77 \times 10^{34}$ cm^{-2}s^{-1} to be reached. The main parameters that characterise the LHC during the operations of 2011 and 2012, compared to the nominal ones, are collected in Table 4.1.

Running the machine at a bunch-spacing of twice the design value had the advantage of achieving better beam stability and large instantaneous luminosity, allowing large data samples to be collected in relatively short time. However, this has the disadvantage of leading to multiple proton–proton collisions in the same bunch-crossing. In fact the average number n of collisions occurring per unit of time is $\langle n \rangle = \mathcal{L} \times \sigma$, where σ is the inelastic proton–proton inclusive cross section evaluated at the collision energy. Hence, the number of collisions μ per bunch-crossing, i.e. in the time interval $\Delta t = 50\,\text{ns}$, is $\mu = \mathcal{L} \times \sigma \times \Delta t$; if one uses the value of $\mathcal{L} = 0.77 \times 10^{34}\,\text{cm}^{-2}\text{s}^{-1}$, $\sigma = 70\,\text{mb}$ and $\Delta t = 50\,\text{ns}$, one finds $\mu = 27$, i.e. 27 pp collisions per bunch-crossing (see last line in Table 4.1). This effect is called *event pile-up* (or just *pile-up*). Figure 4.2(b) shows the distribution of the average number of proton–proton collisions per bunch-crossing, μ, recorded by ATLAS during the 2011 and 2012 run.

Given the very small cross sections of the physics processes of interest, in a bunch-crossing there is at most one proton–proton collision characterised by a hard scattering, with all other collisions in the same bunch-crossing being soft proton–proton interactions. The particles produced in these soft processes appear together in the detector with those from the process of interest, causing some loss in the quality of the reconstruction and measurements of the physics objects in the events, in particular jets, tau leptons, b-jets, and missing transverse energy. It should be stressed, however, that powerful analysis methods were used to strongly reduce the undesired effects of multiple collisions, and hence allowed accurate studies to be done also in the presence of event pile-up.

2. Proton–proton collision physics at the LHC

2.1. *Choice of coordinates*

The first step in measuring particle momenta is defining an appropriate system of coordinates. While the laboratory frame is also the rest frame of the colliding protons, the centre-of-mass frame of the interacting partons within the proton cannot be known *a priori*, because of their internal motion. In particular, the centre-of-mass frame of the interacting partons is in general Lorentz-boosted with respect to the laboratory frame along the direction of the colliding protons; hence, the most convenient system of coordinates for expressing the momenta of particles emerging from proton–proton collisions is the one that lends itself to boosts along the beam axis.

One such system of coordinates, used widely by particle physicists, particularly in the context of hadron colliders, involves the polar angle θ of the particle with momentum p with respect to the beam line, the transverse momentum vector $p_T = p \cdot \sin\theta$, the longitudinal momentum $p_L = p \cdot \cos\theta$, the azimuthal angle in the transverse plane (ϕ) and the rapidity y, which is defined as:

$$y = \ln\left(\frac{E + p_L}{E - p_L}\right), \tag{4.1}$$

where E is the particle energy. A particle emitted along the direction of the beam line has infinite absolute value of the rapidity, while a particle emitted in a direction orthogonal to the beam line ("central production") has vanishing rapidity. The key feature of the quantity y is that differences in rapidity are invariant under Lorentz boosts along the beam axis. In the limit of zero particle mass, y can be approximated by the pseudorapidity η which depends only on the θ angle:

$$\eta = -\ln\left[\tan\left(\frac{\theta}{2}\right)\right]. \tag{4.2}$$

Consequently, the differential solid angle defined by $\Delta\eta \times \Delta\phi$ also remains approximately invariant under such boosts, thus making it convenient to express kinematic distributions in the (η, ϕ) coordinate plane.

2.2. *Particle production at the LHC*

Figure 4.3(a) shows the basic Feynman diagram describing a generic proton–proton (or proton–antiproton) collision. Two partons (quarks or gluons), which carry a fraction $x_{1,2}$ of the incoming hadron's momenta, interact with a collision energy $\hat{s} = x_1 x_2 s$, where s is the squared centre-of-mass energy of the colliding protons.

The probability of finding parton 1,2 with momentum $p_{1,2} = x_{1,2} \cdot p_{h_{1,2}}$ is described by the probability density functions (PDFs) which have been measured in (or known from) previous deep-inelastic scattering experiments. In such a collision, particles with invariant mass M up to $\sqrt{\hat{s}}$ can be produced.

The energy and longitudinal momentum of the intermediate state (which can be a single particle with mass $\sqrt{\hat{s}}$) are given by $E = (x_1 + x_2)\sqrt{s}/2$ and $p_L = (x_1 - x_2)\sqrt{s}/2$, if the parton masses are neglected. The rapidity y can be expressed by $y = \ln\sqrt{x_1/x_2}$ and $x_{1,2} = (M/\sqrt{s})e^{\pm y}$. If M is not too small with respect to the centre-of-mass energy, the two

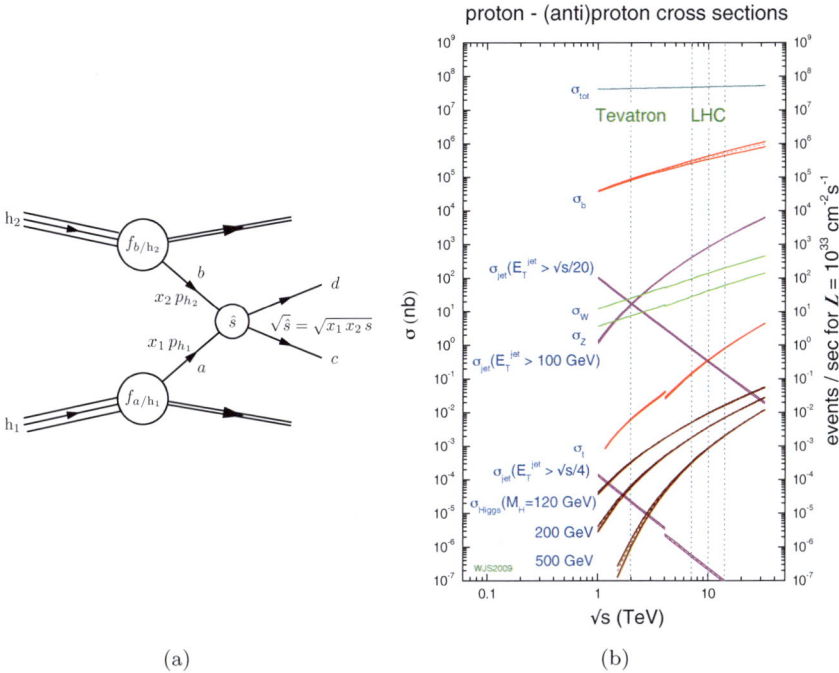

proton - (anti)proton cross sections

(a) (b)

Fig. 4.3 (a) Basic Feynamn diagram for the description of inelastic hadron–hadron collisions. (b) Inclusive production cross sections for various processes in proton–(anti)proton
collisions as a function of the centre-of-mass energy, $\sqrt{s}$. The vertical axis on the right
hand side shows the corresponding event rate assuming an instantaneous luminosity
$\mathcal{L} = 1 \times 10^{33}$ cm^{-2}s^{-1}.[10]

momentum fractions are comparable, yielding a ratio close to 1; hence, the
(pseudo)rapidity is small. Consequently, the decay products of heavy states
(or particle reactions at very large momentum transfer) tend to appear at
smaller rapidities (central production), while the bulk of softer interactions
appear at larger rapidities. A Higgs boson with mass of the order of 100 GeV
is typically produced centrally, i.e. at small rapidities.

In contrast to what happens for e^+e^- collisions, hadron–hadron interactions are characterised by large interaction cross sections. Figure 4.3(b)
shows the cross section of proton–proton (or proton–antiproton) interactions as a function of the collision energy $\sqrt{s}$ (see curve labeled "σ_{tot}"). At
$\sqrt{s} = 14$ TeV, heavy-flavour (anti)quarks are produced at the LHC with a
cross section that is of the order of 1 mb. The W and Z boson production
cross sections are about 100 nb, while the top (antitop) quark is produced
with a cross section of about 1 nb (see the corresponding typical event rate

on the right-hand scale of the figure). The Standard Model Higgs boson production cross section is predicted to be 0.2 nb assuming $m_H = 120\,\mathrm{GeV}$ for $\sqrt{s} = 8\,\mathrm{TeV}$ and 0.6 nb for $\sqrt{s} = 14\,\mathrm{TeV}$ (see Chapter 1).

The production of vector bosons W and Z in association with jets, of $t\bar{t}$ or single top events, of particle jets from parton scattering, are important backgrounds to the search for Higgs bosons. These backgrounds are due to particle mis-identification (in particular to jets faking electrons, photons, taus, or b-jets), but can be reduced to an acceptable level with a properly designed and well calibrated detector. Other backgrounds, e.g. WW and ZZ production, have final states identical to those from Higgs boson decays and are therefore classified as "irreducible" backgrounds, which can only be separated from the signal on the basis of differences in kinematic properties.

A nice overview of LHC early physics is available in Ref. [11].

Higgs boson searches at LHC take advantage of the relatively large production cross sections at this accelerator, about a factor 10 (or more) higher than at the Tevatron (see Chapter 3). Center-of-mass energy and large luminosity are crucial to the enhancement of the Higgs boson production rate; hence, decay channels that were inaccessible at the Tevatron turn out to be important at the LHC.

3. Overview of the ATLAS and CMS experiments

3.1. *ATLAS*

A schematic view of the ATLAS detector[6] is shown in Fig. 4.4.

At the core of the ATLAS detector is the inner tracking detector (ID), which is immersed in a 2 T axial field produced by a superconducting solenoidal magnet. Silicon pixel and micro-strip detectors provide measurements of charged particle trajectories in the pseudorapidity range $|\eta| < 2.5$, complemented by a straw tube tracker for $|\eta| < 2.0$, which enhances electron identification through detection of transition radiation.

The central solenoidal magnet is surrounded by a system consisting of an electromagnetic calorimeter and a hadron calorimeter. The electromagnetic calorimeter is based on a liquid argon (LAr) technology, is highly segmented, and is divided in a barrel part ($|\eta| < 1.475$) and two endcaps ($1.375 < |\eta| < 3.2$). The hadron calorimeter, based on an iron-scintillator technology, is placed around the LAr calorimeter and provides coverage in the central rapidity region ($|\eta| < 1.7$). This is complemented by a LAr hadron calorimeter in the endcap region behind the forward electromagnetic

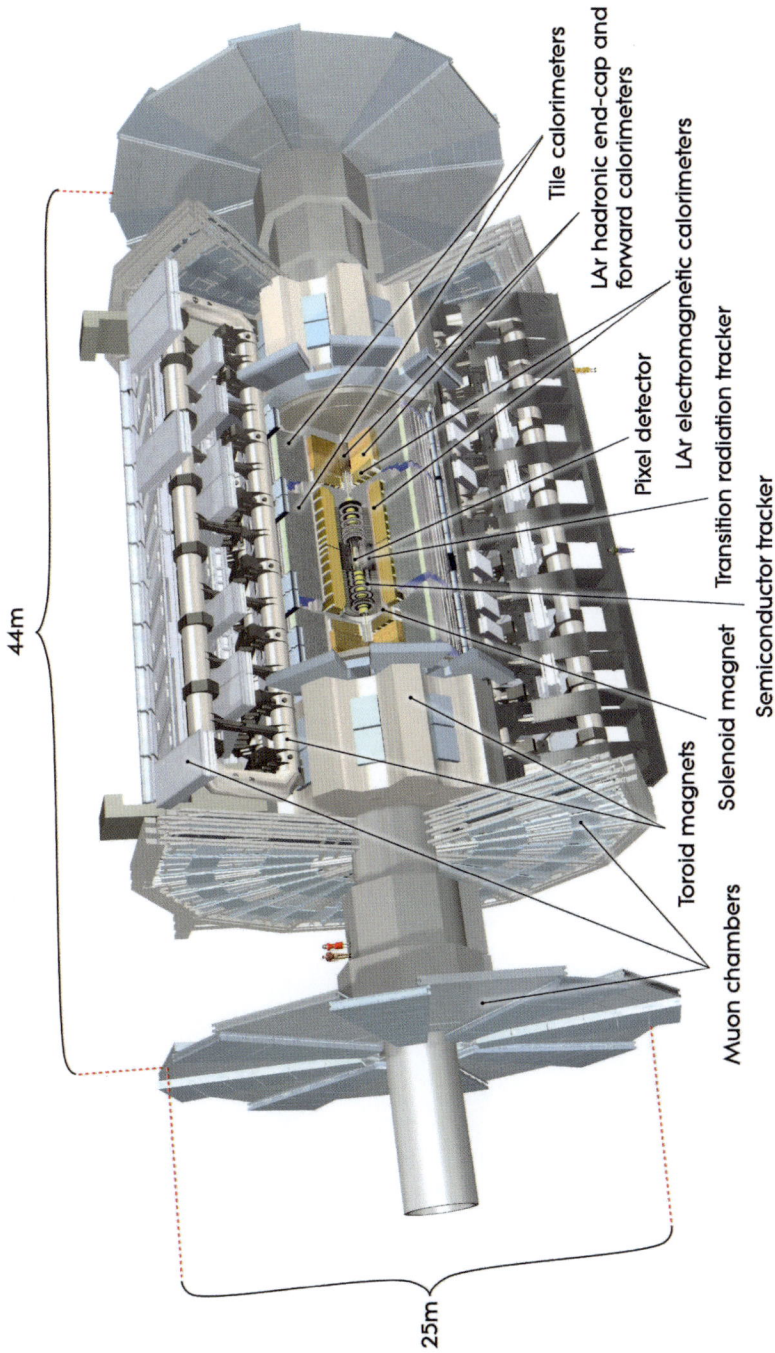

Fig. 4.4 Cut-away view of the ATLAS detector. The dimensions of the detector are 25 m in height and 44 m in length. The overall weight of the detector is approximately 7000 tons.

calorimeter ($1.5 < |\eta| < 3.2$) and by forward calorimeters covering the region $3.1 < |\eta| < 4.9$.

The muon spectrometer is composed of separate trigger and precision tracking detectors immersed in a toroidal field provided by three air-core superconducting magnets. Resistive Plate Chambers (RPCs, $|\eta| < 1.05$) and Thin Gap Chambers (CSCs, $1.05 < |\eta| < 2.4$) provide trigger information in the barrel and in the endcap parts, respectively. Monitored Drift Tubes (MDTs) provide precision measurements of muon tracks in the pseudorapidity region $|\eta| < 2.0$, together with Cathode Strip Chambers that cover the region $2.0 < |\eta| < 2.7$.

3.2. CMS

The central feature of the CMS detector[7] is a superconducting solenoid, 13 m in length and 6 m in diameter that provides an axial magnetic field of 3.8 T. The inner volume of the solenoid is instrumented with particle detection systems. The steel return yoke outside the solenoid is instrumented with a succession of particle detectors, including tracking and calorimetry, while the steel return yoke is instrumented with gas detectors used to identify muons.

Charged particle trajectories are measured by the silicon pixel and strip tracker, with full azimuthal coverage within $|\eta| < 2.5$. Unlike ATLAS, the inner tracker of CMS does not use gaseous detectors. A lead-tungstate crystal electromagnetic calorimeter (ECAL) and a brass/scintillator hadron calorimeter (HCAL) surround the tracking volume and cover the region $|\eta| < 3$. Both ECAL and HCAL are contained in the superconducting solenoid. A quartz-fibre Cherenkov calorimeter (HF) extends the coverage to $|\eta| = 5.0$.

The outermost component of the CMS detector is the muon system. It consists of gas detectors interleaved with the iron plates that act as return yoke for the central magnet, with a coverage $|\eta| < 2.4$. These detectors (RPCs, CSCs and Drift Tubes (DTs)) allow the identification and the measurement of tracks produced by muons emerging from the central volume. In the case of CMS, muon trigger information is provided by each of the three technology types. A cartoon of the CMS detector is shown in Fig. 4.5.

3.3. The ATLAS and CMS trigger systems

As shown in Fig. 4.3(a), the inelastic cross section of protons colliding at $\sqrt{s} = 14$ TeV is about 80 mb (70 mb at $\sqrt{s} = 8$ TeV). Assuming an

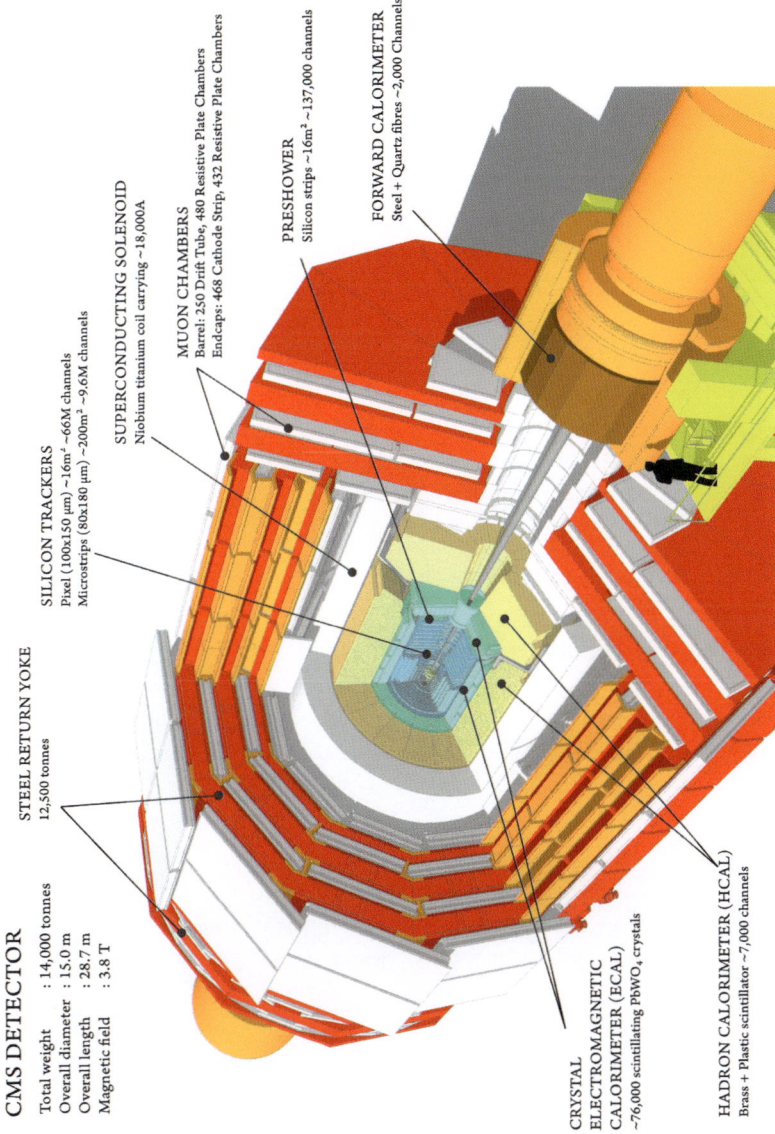

CMS DETECTOR

Total weight	: 14,000 tonnes
Overall diameter	: 15.0 m
Overall length	: 28.7 m
Magnetic field	: 3.8 T

STEEL RETURN YOKE
12,500 tonnes

SILICON TRACKERS
Pixel (100x150 μm) ~16m² ~66M channels
Microstrips (80x180 μm) ~200m² ~9.6M channels

SUPERCONDUCTING SOLENOID
Niobium titanium coil carrying ~18,000A

MUON CHAMBERS
Barrel: 250 Drift Tube, 480 Resistive Plate Chambers
Endcaps: 468 Cathode Strip, 432 Resistive Plate Chambers

PRESHOWER
Silicon strips ~16m² ~137,000 channels

FORWARD CALORIMETER
Steel + Quartz fibres ~2,000 Channels

CRYSTAL ELECTROMAGNETIC CALORIMETER (ECAL)
~76,000 scintillating PbWO₄ crystals

HADRON CALORIMETER (HCAL)
Brass + Plastic scintillator ~7,000 channels

Fig. 4.5 Cut-away view of the CMS detector. The dimensions of the detector are 15 m in height and less than 29 m in length. The overall weight of the detector is approximately 14000 tonnes.

instantaneous luminosity $\mathcal{L} = 1 \times 10^{34}\,\mathrm{cm}^{-2}\mathrm{s}^{-1}$, the rate of collisions produced by the machine is about 10^9 events per second, which corresponds to an amount of data that cannot be recorded because of real-world limitations in data collection, storage capacity and offline computing. In fact, given that the typical event size is of the order of 1 Mbyte and assuming an effective running period of 10^7 s per year, this would require 10^{10} Tbytes storage for 10^{16} events. However, only a tiny fraction of these events are of interest for new physics. For example, a SM Higgs boson with mass $m_H = 125\,\mathrm{GeV}$ would be produced with a rate of only 11 events per minute at $\sqrt{s} = 8\,\mathrm{TeV}$, and 28 events per minute at $\sqrt{s} = 14\,\mathrm{TeV}$. Hence, a fast online selection is mandatory. This is provided by the trigger system that filters the events on the basis of simple criteria and retains only those which might be interesting for physics studies (for example counting the number of leptons, photons and jets with transverse energies larger than predefined thresholds). The selected events are recorded by the data acquisition systems and reconstructed offline. The ATLAS trigger system has three levels of selections, while CMS has two. They both accept events at the first level with a rate of about 100 kHz, and record data with a rate of few hundred per second.[12–14] More detail on triggering and event selection is given below in Sec. 5.

4. Physics object reconstruction at hadron colliders

Any measurement in a particle physics experiment relies on two crucial aspects of detector performance — the capability of identifying particles produced in particle collisions and the accurate measurement of the magnitude and direction of the momentum and energy of these particles. Insofar as the search for the Higgs boson is concerned, decays into the relevant final states include photons, electrons, muons, taus, neutrinos and bottom quarks. Hence, efficient reconstruction and identification of these particles and an accurate measurement of their momentum and energy are of prime importance.

General principles of particle identification and measurement at collider experiments have already been presented in Chapters 2 and 3. In this section we focus on the aspects that specifically characterise the particle reconstruction by the ATLAS and CMS experiments.

4.1. *Muon reconstruction*

Muons traversing matter typically lose a modest amount of energy by ionisation processes that do not cause significant alteration of the momentum vector. For this reason, muons are the only particles (apart from neutrinos which go completely undetected) which pass through the heavy calorimeter systems and still produce a signal in the external detectors. Muons are therefore identified by the tracks produced in the outermost component of the detector, the muon system. Tracks in the muon system are matched to those reconstructed in the inner tracker, and a combined fit is performed to obtain the full trajectory of the muons in the whole detector.[15,16]

Low-p_T muons, however, don't have sufficient energy to traverse the full muon system and produce only track segments ("stubs"). Nevertheless, these track stubs can still be matched to tracks in the inner tracking system to reconstruct muon candidates. In addition, the fact that muons are minimum ionising particles can be exploited to identify muon candidates by requiring the calorimetric energy deposits along the muon trajectory to be consistent with that of a minimum ionising particle. This is helpful in identifying muons in regions of the detector where the muon system lacks geometric coverage, such as the centre of the barrel region ($|\eta| < 0.1$) in the case of the ATLAS experiment (these are regions where typically detector supports and subsystem services, such as electrical cables and gas/liquid pipes, are placed).

The overall geometric acceptance for muons is determined by the coverage provided by the tracking and the muon system. In the case of the CMS detector, the muon system covers up to $|\eta| = 2.4$. In the case of the ATLAS experiment, while the tracker extends up to $|\eta| = 2.5$, the coverage for muon reconstruction is extended up to $|\eta| = 2.7$ by including tracks reconstructed in the forward region of the muon system. The precision of the muon momentum measurement is driven by the reconstruction in the central tracker except for p_T larger than about 100 GeV, where the track measured in the muon system improves the overall reconstruction owing to the increased lever arm.

Muons with momentum above about 300 GeV lose energy mainly through bremsstrahlung, pair production and nuclear interactions. These processes may cause the loss of an important fraction of the muon energy.

It should be noted, however, that muons produced in decays of the discovered Higgs boson, which has a mass near 125 GeV, have energies considerably lower than the threshold mentioned above, and hence these effects are neglected in the studies presented in this book.

4.2. *Electron reconstruction*

The ATLAS and CMS electromagnetic calorimeters (ECAL) are designed such that electrons and photons are fully absorbed in them. Electrons, being electrically charged, also leave a track in the tracking system. Thus, electron candidates are reconstructed by associating tracks in the tracker with energy "clusters" in the ECAL (in contrast to photons which produce no tracks).

Electron reconstruction typically starts by searching for energy clusters in the ECAL. The algorithms searching for these clusters take into account the characteristic profile of an electron's energy deposit in the ECAL, which is narrow in the η coordinate but wide in the ϕ coordinate, due to the bending in the magnetic field as they radiate bremsstrahlung photons which interact with the tracker material. An energy cluster in the ECAL may also be produced by hadron jets; however, while electrons are absorbed in the ECAL, hadron jets are not. Hence, for a cluster to be assigned to an electron, the HCAL activity behind the cluster is required to be absent or low.

Having identified these energy clusters, the next step is to search for tracks pointing to them. Electrons tend to radiate a considerable fraction of their energy as they traverse through the tracker, unlike heavier charged particles. This fact needs to be taken into account when reconstructing electron tracks. In ATLAS, once a track is matched to a cluster, it is refitted with the Gaussian Sum Filter (GSF),[17] which takes into account the energy loss suffered by the electron. In CMS, one matches the ECAL energy deposits to hits in the innermost tracker layers and uses these hits as seeds to initiate the reconstruction of tracks using the Gaussian Sum Filter. This procedure is complemented by an "inside-out" approach wherein track seeds are used to map electron energy deposits in the ECAL. The latter approach leads to some improvement in reconstruction efficiency for low p_T electrons.

Once an electron is identified, its energy has to be determined. In ATLAS, the ECAL cluster energy is corrected taking into account the electron energy deposited before the ECAL and the estimated energy that leaks out both laterally and longitudinally.

In CMS, the ECAL cluster energy is also corrected to account for leakage losses. In addition, for all physics analyses in CMS and for Higgs boson analyses in ATLAS, the momentum measured from the electron track is also used to compute the electron energy as it improves the energy measurement at low p_T. Finally, the momentum vector of the electron and its distance of close approach to the interaction point (impact parameter) are determined for the reconstructed electron track.[18-21]

4.3. *Photon reconstruction*

The response of the ECAL to photons is similar to electrons; hence, the procedure to reconstruct electron energy from ECAL clusters applies also to photons. However, unlike electrons, photons do not leave a track in the tracking system. In some cases, though, photons may interact with the tracker material and convert to electron–positron pairs that leaves tracks in the tracking system. Such tracks from conversions can be recognised and are helpful in associating photons to a primary vertex, which in turn e.g. helps in the accurate determination of the photon direction.

In ATLAS, the procedure for associating photons to primary vertices makes use also of the "photon-pointing" feature of the ECAL where the longitudinal segmentation of the electromagnetic calorimeter helps in tracing the trajectory of the photon; details are given in Chapter 6 and in Ref. [22]. Details of photon reconstruction and identification in CMS are described in Ref. [23]. In CMS, the vertex associated to the photon cluster is identified by studying the kinematic properties of the tracks belong to it; details are reported in Chapter 6.

4.4. *Jet reconstruction*

Quarks and gluons fragment and hadronise almost immediately after being produced in hard interactions, leading to a collimated spray of high-energy hadrons, called a "jet". Such jets are relevant for both testing our understanding of their production and properties, and also for identifying the partonic component of Higgs boson final states. Jets are important for tagging Higgs boson events produced in VBF processes, as well as Higgs bosons produced in association of other particles (e.g. $t\bar{t}$ processes).

In order to reconstruct the observed hadrons into a jet (or several distinct jets) one uses a *jet definition* coded in a computer algorithm. Good jet definitions should be simple, weakly sensitive to non-perturbative effects arising in particular from soft particles produced in hadronisation processes, or present in the underlying event[a] or in the pile-up. An extensive discussion of jet definition in hadron colliders is given in Refs. [24,25].

Jets are reconstructed starting from the basic measurements that are provided by the hadron and electromagnetic calorimeters. Charged particles reconstructed in the inner tracker can also be used. In ATLAS and CMS, the elementary input for jet reconstruction consists of close-by *clusters* of calorimeter cells, each providing the energy deposit measured. Associated to the cluster is the momentum, equivalent to the sum of the cell energies (after detector calibration). In addition, CMS employs a novel Particle-Flow algorithm[26] that combines the information from all sub-detectors to identify and reconstruct all visible particles produced in a pp collision, namely charged hadrons, neutral hadrons, photons, muons, and electrons. The resulting list of particles can then be used to construct a variety of higher-level objects and observables such as jets, missing transverse energy, and taus, and is also used to quantify lepton and photon isolation, b-jet tagging, etc.

The most basic jet definition is provided by the *cone* algorithm, starting with a cone associated to a given "seed" particle (or cluster) and with fixed size R in the (η, ϕ) space. Particle vector momenta falling inside this cone are summed up. The direction of the sum serves as an updated seed, and this procedure is repeated until the cone is stable. The vector sum of the transverse momenta is the jet transverse momentum; if it is larger than a given threshold $p_{T,min}$ the jet is retained for physics analysis. The size R of the cone should be optimised for the type of physics processes under study. In general, cone-jets are not infrared-safe, meaning that if a soft gluon is emitted between two partons, the reconstruction of the jet would change. There are many variants of cone-jet algorithms; a modern version is offered by SISCone.[27]

ATLAS and CMS have chosen to use the infrared- and collinear-safe anti-k_T algorithm.[28] A distance parameter $d_{ij} = \min(k_{t,i}^{2p}, k_{t,j}^{2p})\Delta_{i,j}^2/R^2$ is evaluated using all pairs of particles (or clusters) i, j, where $\Delta_{i,j}$ is their distance in the pseudorapidity–azimuth plane ($\Delta_{i,j} = \sqrt{\Delta_{\eta;i,j}^2 + \Delta_{\phi;i,j}^2}$), $k_{t,i}$ is the transverse momentum of the particle i w.r.t. the incoming beam direction, and R is a free parameter. The smallest parameter between d_{ij} and $d_{i,B} = k_{t,i}^{2p}$, is selected. If it is d_{ij}, i and j are merged in proto-jets with

---

[a]The underlying event consists of the remnants of the colliding protons after the hard scatter takes place as well as additional, soft multi-parton interactions.

momentum corresponding to the vector sum of the two particle momenta, and used as a new particle. The procedure is then repeated. If the smallest parameter is $d_{i,B}$ the proto-jet i is added in the final collection of jets, and removed from the list of proto-jets. This procedure is iterated until the list of proto-jets is empty. The parameter p determines the type of algorithm; for the anti-k_T $p = -1$. Also in this case the value of the parameter R is optimised for the type of physics process under study. Typical values adopted by ATLAS and CMS are $R = 0.4$ and 0.6. Details of the jet performance are available in Refs. [29–31].

4.5. *Tau reconstruction*

Tau leptons are the heaviest of all leptons, and with mass of 1.776 GeV, they are the only leptons that decay into hadrons (plus a tau neutrino). As they are short-lived, with a mean lifetime of 2.9×10^{-13} seconds, most tau leptons decay before leaving the beam-pipe and have to be reconstructed from their decay products. About 35% of taus decay leptonically, producing either an electron or a muon, and two neutrinos. In the remaining cases, the taus decay hadronically, producing predominantly one or three charged hadrons, a neutrino and up to two neutral kaons or pions, of which the latter promptly decay into photons.

Hadronic decays of taus can therefore be identified with either one or three hadronic tracks along with photonic clusters in the ECAL. Such decays can be reconstructed as jets, and in fact jets serve as the starting point for identifying tau leptons in an event.

The principle algorithm for identifying taus in CMS is called the *Hadron Plus Strip* (HPS) algorithm. This algorithm considers several possibilities depending on the presence of one or three *particle flow*[32] charged hadrons in the jet. Neutral pion decays are identified in the form of fired inner tracker strips associated to photons reconstructed by the particle flow algorithm ("photon strips"). Several decay topologies are considered. These include a single charged hadron with zero, one or two strips associated, as well as three charged hadrons, all emerging from a common secondary vertex. In order to reconstruct the four-momentum of the tau lepton, the charged hadrons are assumed to be pions and, depending on the decay mode, they are required to be consistent with the masses of the intermediate π-meson resonances such as ρ and a_1.

In the ATLAS experiment, tau reconstruction starts by associating tracks to calorimetric jets in order to identify one prong and three prong tau decays. Then, in order to identify a jet as a tau candidate, several identification variables are considered. These include variables that reflects the topological distribution of tracks and energy clusters within the jet, invariant mass reconstructed from constituent energy subclusters or tracks, tracking-related variables such as the decay-length significance of a secondary vertex, and others. A multivariate discriminant constructed using these variables is further used to discriminate genuine tau decays from QCD jets.[33,34]

4.6. Missing transverse energy (E_T^{miss})

Neutrinos, and other weakly interacting particles that may possibly exist in nature, cannot be directly reconstructed in the detectors since they rarely interact with any material. Direct neutrino detection with an acceptable rate is possible only in dedicated apparatuses characterised by very large, massive systems. In a pp collision, the presence of weakly interacting particles that fly away undetected can be inferred by the momentum imbalance they create. In hadron colliders, the total energy and the total longitudinal momentum of the interacting partons is not known *a priori*. Given that the momentum of the interacting partons is along the beam direction, the total momentum in the plane transverse to the beam axis must be zero before and after the collision. Hence, undetected neutrinos imply that the transverse momentum vectors of all detected particles will not add up to zero. In fact, the negative of this vector sum gives the total transverse momentum of the escaping neutrinos and its size is known as "missing transverse energy", E_T^{miss}.

In ATLAS, E_T^{miss} is computed by adding the contributions from energy deposits in the calorimeters. Since muons do not leave much energy in the calorimeters, their contribution is separately added. Low-p_T tracks are further added to this computation to take into account particles which do not carry sufficient momentum to reach the calorimeters.[35]

In the CMS experiment, the E_T^{miss} is formed by adding up all particle flow objects in the event.[36]

4.7. Jets from b-quarks

Since Higgs bosons, if they are not too heavy, decay preferentially to heavy-flavour quarks (see Chapter 1), the detection of b-quarks is an essential ingredient of Higgs boson searches. Jets originating from b-quarks can be identified by exploiting the lifetime of the weakly decaying *b*-hadrons produced in their fragmentation. The average lifetime of a *b*-hadron[b] is $(1.57 \pm 0.01) \times 10^{-12}$ s.[37] So, in its own rest frame, the *b*-hadron, once produced at the event production vertex (the primary vertex), travels on average for about 0.5 mm and then decays. Considering the relativistic factor $\beta\gamma = p_T(b)/m_b$ and the fact that due to the hard *b*-quark fragmentation function most of the original *b*-jet energy is kept by the *b*-hadron, transverse flight lengths are observed in the detector which range on average from 3–9 mm for *b*-jets with transverse momenta between 50 GeV and 150 GeV. Charged particles arising from the decay of the *b*-hadron, which are reconstructed as tracks in the inner tracking detector, will in general be displaced with respect to the primary vertex, while most of the charged particles produced within light-quark or gluon jets will not have any significant lifetimes and therefore be compatible with the primary vertex. This allows them to be separated from *u*, *d* or *s*-quarks (usually defined as *light-quark* jets) and gluons and, to a lesser extent, from *c*-jets.

There are two different types of algorithms that exploit the lifetime signature, with different levels of efficiencies and purities. The first relies on the impact parameter of charged particle tracks reconstructed in the inner detector, i.e. an evaluation whether the tracks selected within a jet are compatible with originating from the primary vertex or not. This type of algorithm has high *b*-jet identification efficiency: often a single track with a significant impact parameter is in principle sufficient to tag the jet as a *b*-jet. The second type of algorithm uses more than one track to explicitly reconstruct a secondary vertex and, beyond the lifetime information, further properties of the decay vertex, such as the vertex mass, the charged decay multiplicity and the fraction of charged jet energy associated to the vertex, are exploited as well. The simple picture of a single secondary vertex may be blurred by the fact that, in most of cases, at least one of the decay products of the *b*-hadron is a weakly decaying *c*-hadron, which again travels significantly before decaying. However, since the *c*-hadron has a shorter lifetime and is less boosted, its vertex is in general not separated from the *b*-quark vertex; hence, in most cases a single vertex is observed. Nevertheless, some algorithms try to reconstruct the full decay chain made of the primary, the secondary (*b*-hadron), and tertiary (*c*-hadron) vertex as well. While the vertex-based algorithms have a lower efficiency than those using

---

[b]This can be a B^+, a B^0, a B_S^0 (or one of their charge conjugates) or a *b*-baryon.

impact parameters, they allow b-jets to be identified with a significantly higher purity.

In their main Higgs boson searches both ATLAS and CMS use a combination of the impact parameter and secondary vertex-based algorithms, which are called *MV1* (Multi Variate 1) and *CSV* (Combined Secondary Vertex), respectively. The former relies on a neural network algorithm to combine the inputs, while the latter uses a likelihood-based approach. In terms of inputs, the biggest difference is that ATLAS also performs a full b- to c-hadron decay chain fit while CMS adds a *pseudo-vertex* approach, where vertex-based quantities are computed from high impact parameter tracks even if the explicit reconstruction of a vertex fails. The performance of the two approaches is not too different: for a typical working point of 70% b-jet efficiency, c-jets and light jets are selected with an efficiency of $\approx 20\%$ and $0.7-1\%$, respectively. Rejecting c-jets is difficult because of the non-negligible lifetime of c-hadrons but several properties can be exploited, such as the vertex mass or the decay chain topology, to suppress c-jets further, at least in the regime of lower b-tagging efficiencies. The *MV1c* algorithm in ATLAS explicitly rejects also c-jets; in this way, for a b-jet efficiency of 50%, the c-jet mis-identification probability was reduced from 8% to 4% while keeping the light jet efficiency well below 0.1%. More details are available in Refs. [38,39].

4.8. *Lepton isolation*

Several physics processes of interest result in the production of leptons and photons that are isolated in space with little activity in their vicinity. "Isolation" is the measure of energy in a small neighbourhood of a particle. A lepton produced in the decay of a W or a Z boson is typically isolated while a particle inside a hadronic jet is typically surrounded by a considerable amount of energy and therefore is not isolated. Hence, isolation serves as an extremely important discriminant against background from QCD. The presence of pile-up (see discussion in Sec. 1) however, degrades this discrimination by spewing energy uniformly across the detector. As a result, particles which would be isolated in the absence of pile-up pick up this additional energy in their neighbourhood; hence, corrections have to be made to account for pile-up in the measure of isolation of a given particle. In the case of charged particles that enter the isolation sum, the tracks are required to be associated to the primary vertex. In the case of calorimetric energy, the overall ambient energy is computed on an event-by-event basis from the transverse energy density of low p_T jets and an appropriate amount of energy is subtracted.

Pile-up also affects the momentum determination of physics objects that rely on calorimetric information. The p_T of jets, for instance, needs to be corrected to account for additional energy from pile-up. Similarly the calibration of the energy of electrons and photons, as well as the E_T^{miss} measurements, are sensitive to the effect of pile-up, especially when running the collider at high luminosity.

5. The trigger selection for Higgs boson events

The online selection of Higgs boson production and decays at LHC poses severe challenges to the trigger system. All physics objects described above are essential to the search of this particle; moreover, selection thresholds on transverse momentum or energy of these objects have to be quite low, particularly if low-mass Higgs bosons are concerned. Selection of many different inclusive signatures with low-p_T thresholds imply a potential large rate of accepted events, to be balanced with the bandwidth of the data acquisition system, and with the capability of the computing system.

Table 4.2 shows the trigger menu implemented in the ATLAS experiment during the 2012 run at $\sqrt{s} = 8$ TeV, at the average instantaneous

Table 4.2   The trigger table of the ATLAS experiment during the 2012 run at $\sqrt{s} = 8$ TeV and at the average instantaneous luminosity $L = 5 \times 10^{33}$ cm^{-2}s^{-1}. The selection thresholds for the various physics objects and the corresponding acceptance rates are reported, including the overall data acquisition rate. The latter is compared to expected production rate of the SM Higgs boson with 125 GeV mass.

Trigger	Threshold (GeV)	Rate (Hz)
Single electron	24	70
Double electron	12	8
Double photon	35, 25	7
Single muon	24	45
Double muon	18, 8	8
Single jet	360	5
Multi-jets	5 jets, each $p_T > 55$	8
b-jets	1 b-jet + 3 jets, all with $p_T > 55$	4
Transverse momentum imbalance (E_T^{miss})	80	17
Total trigger rate		~ 400
125 GeV SM Higgs boson production rate		0.1

luminosity $L = 5 \times 10^{33}\,\mathrm{cm}^{-2}\mathrm{s}^{-1}$. Similar trigger rates characterised the CMS online event selection.

References

1. A. Asner, Y. Baconnier, O. Barbalat, M. Bassetti, C. Benvenuti, *et al.*, *Proc. ECFA-CERN Workshop on Large Hadron Collider in the LEP Tunnel*, Lausanne and CERN, Geneva, Switzerland, 21–27 Mar 1984, (1984), p. 1.
2. L. Evans, The Large Hadron Collider from conception to commissioning: A personal recollection, *Rev.Accel.Sci.Tech.* **3**, 261–280 (2010). doi: 10.1142/S1793626810000373.
3. O. Brüning, P. Collier, P. Lebrun, S. Myers, R. Ostojic, *et al.* (eds.), LHC Design Report. 1. The LHC Main Ring, CERN-2004-003-V-1, (2004).
4. O. Brüning, P. Collier, P. Lebrun, S. Myers, R. Ostojic, *et al.* (eds.), LHC Design Report. 2. The LHC Infrastructure and General Services, CERN-2004-003-V-2, (2004).
5. M. Benedikt, P. Collier, V. Mertens, J. Poole, K. Schindl (eds.), LHC Design Report. 3. The LHC Injector Chain, CERN-2004-003-V-3, (2004).
6. ATLAS Collaboration, The ATLAS experiment at the CERN Large Hadron Collider, *JINST.* **3**, S08003 (2008). doi: 10.1088/1748-0221/3/08/S08003.
7. CMS Collaboration, The CMS experiment at the CERN LHC, *JINST.* **3**, S08004 (2008). doi: 10.1088/1748-0221/3/08/S08004.
8. LHCb Collaboration, The LHCb detector at the LHC, *JINST.* **3**, S08005 (2008). doi: 10.1088/1748-0221/3/08/S08005.
9. ALICE Collaboration, The ALICE experiment at the CERN LHC, *JINST.* **3**, S08002 (2008). doi: 10.1088/1748-0221/3/08/S08002.
10. J. Stirling, private communication, 2009.
11. G. Dissertori, LHC detectors and early physics. pp. 197–230 (2010). arXiv: 1003.2222.
12. A. Nisati, Trigger systems for experiments at the Large Hadron Collider in: *Proceedings of the XVI IFAE Italian Meeting on High Energy Physics*, Torino, April 14–16, 2004. INFN (2004). URL http://www.ph.unito.it/ifae/Proceedings/Proceedings.pdf.
13. T. Schoerner-Sadenius, The trigger of the ATLAS experiment, *Mod.Phys.Lett.* **A18**, 2149–2168 (2003). doi: 10.1142/S0217732303011800.
14. CMS Trigger and Data Acquisition Group, The CMS high level trigger, *Eur.Phys.J.* **C46**, 605–667 (2006). doi: 10.1140/epjc/s2006-02495-8.
15. ATLAS Collaboration, Measurement of the muon reconstruction performance of the ATLAS detector using 2011 and 2012 LHC proton–proton collision data, *Eur.Phys.J.* **C74**(11), 3130 (2014). doi: 10.1140/epjc/s10052-014-3130-x.

16. CMS Collaboration, Performance of CMS muon reconstruction in *pp* collision events at $\sqrt{s}$ = 7 TeV, *JINST.* **7**, P10002 (2012). doi: 10.1088/1748-0221/7/10/P10002.

17. R. Frühwirth, A Gaussian-mixture approximation of the Bethe–Heitler model of electron energy loss by bremsstrahlung, *Computer Physics Communications* **154**(2), 131–142 (2003). doi: 10.1016/S0010-4655(03)00292-3.

18. ATLAS Collaboration, Electron reconstruction and identification efficiency measurements with the ATLAS detector using the 2011 LHC proton–proton collision data, *Eur.Phys.J.* **C74**(7), 2941 (2014). doi: 10.1140/epjc/s10052-014-2941-0.

19. ATLAS Collaboration, Electron and photon energy calibration with the ATLAS detector using LHC Run 1 data, *Eur.Phys.J.* **C74**(10), 3071 (2014). doi: 10.1140/epjc/s10052-014-3071-4.

20. CMS Collaboration, Performance of electron reconstruction and selection with the CMS detector in proton–proton collisions at $\sqrt{s}$ = 8 TeV (2015). arXiv: 1502.02701.

21. CMS Collaboration, Energy calibration and resolution of the CMS electromagnetic calorimeter in *pp* collisions at $\sqrt{s}$ = 7 TeV, *JINST.* **8**, P09009 (2013). doi: 10.1088/1748-0221/8/09/P09009.

22. ATLAS Collaboration, Measurements of the photon identification efficiencies with the ATLAS detector using LHC Run 1 data. Technical Report CERN-EP-2016-110, CERN, Geneva (2016). arXiv:1606.01813.

23. CMS Collaboration, Performance of photon reconstruction and identification with the CMS detector in proton–proton collisions at $\sqrt{s}$ = 8 TeV, *JINST.* **10**(08), P08010 (2015). doi: 10.1088/1748-0221/10/08/P08010.

24. G. P. Salam, Towards jetography, *Eur.Phys.J.* **C67**, 637–686 (2010). doi: 10.1140/epjc/s10052-010-1314-6.

25. S. Ellis, J. Huston, K. Hatakeyama, P. Loch, and M. Tonnesmann, Jets in hadron–hadron collisions, *Prog.Part.Nucl.Phys.* **60**, 484–551 (2008). doi: 10.1016/j.ppnp.2007.12.002.

26. CMS Collaboration, Particle-flow event reconstruction in CMS and performance for jets, taus, and MET. CMS-PAS-PFT-09-001 (Apr, 2009). URL http://cds.cern.ch/record/1194487.

27. G. P. Salam and G. Soyez, A practical seedless infrared-safe cone jet algorithm, *JHEP.* **0705**, 086 (2007). doi: 10.1088/1126-6708/2007/05/086.

28. M. Cacciari, G. P. Salam, and G. Soyez, The anti-k(t) jet clustering algorithm, *JHEP.* **0804**, 063 (2008). doi: 10.1088/1126-6708/2008/04/063.

29. ATLAS Collaboration, Jet energy measurement and its systematic uncertainty in proton–proton collisions at $\sqrt{s}$ = 7 TeV with the ATLAS detector, *Eur.Phys.J.* **C75**(1), 17 (2015). doi: 10.1140/epjc/s10052-014-3190-y.

30. ATLAS Collaboration, Single hadron response measurement and calorimeter jet energy scale uncertainty with the ATLAS detector at the LHC, *Eur.Phys.J.* **C73**(3), 2305 (2013). doi: 10.1140/epjc/s10052-013-2305-1.

31. CMS Collaboration, Determination of jet energy calibration and transverse momentum resolution in CMS, *JINST*. **6**, P11002 (2011). doi: 10.1088/1748-0221/6/11/P11002.

32. CMS Collaboration, Commissioning of the particle-flow event reconstruction in minimum-bias and jet events from *pp* collisions at 7 TeV. CMS Physics Analysis Summary. CMS-PAS-PFT-10-002 (2010). URL http://cdsweb.cern.ch/record/1279341.

33. ATLAS Collaboration, Identification and energy calibration of hadronically decaying tau leptons with the ATLAS experiment in *pp* collisions at $\sqrt{s} = 8$ TeV, *Eur. Phys. J.* **C75**(7), 303 (2015). doi: 10.1140/epjc/s10052-015-3500-z.

34. CMS Collaboration, Performance of tau-lepton reconstruction and identification in CMS, *JINST*. **7**, P01001 (2012). doi: 10.1088/1748-0221/7/01/P01001.

35. ATLAS Collaboration, Performance of missing transverse momentum reconstruction in proton–proton collisions at 7 TeV with ATLAS, *Eur. Phys. J.* **C72**, 1844 (2012). doi: 10.1140/epjc/s10052-011-1844-6.

36. CMS Collaboration, Performance of the CMS missing transverse momentum reconstruction in *pp* data at $\sqrt{s} = 8$ TeV, *JINST*. **10**(02), P02006 (2015). doi: 10.1088/1748-0221/10/02/P02006.

37. Particle Data Group, Review of particle physics, *Chin. Phys.* **C38**, 090001 (2014). doi: 10.1088/1674-1137/38/9/090001.

38. ATLAS Collaboration, Performance of b-jet identification in the ATLAS experiment, *JINST*. **11**, P04008 (2016). doi: 10.188/1748-0221/11/04/P04008.

39. CMS Collaboration, Identification of *b*-quark jets with the CMS experiment, *JINST*. **8**, P04013 (2013). doi: 10.1088/1748-0221/8/04/P04013.

Chapter 5

Higgs boson observation and measurements of its properties in the $H \to ZZ \to 4\ell$ decay mode

Andrey Korytov[*] and Konstantinos Nikolopoulos[†]

University of Florida, Gainesville, FL 32611 USA
†*University of Birmingham, Birmingham, B15 2TT, UK*

In their searches for the Standard Model (SM) Higgs boson in the mass range 115–1000 GeV, both ATLAS and CMS Collaborations observed a narrow four-lepton resonance with a mass near 125 GeV with local significances in excess of 5σ. In the combination of the ATLAS and CMS $H \to ZZ \to 4\ell$ measurements, the mass of the observed boson was found to be 125.15 ± 0.37 (stat) ± 0.15 (syst) GeV. The event rates attributed to the signal and the studied differential cross sections were compatible with the SM Higgs boson hypothesis. Kinematic properties of leptons in signal candidate events agreed with those expected for a state with spin-parity quantum numbers of the SM Higgs boson ($J^P = 0^+$) and strongly disfavoured states with alternative quantum numbers or 0^+ states with non-SM-like tensor structures of their couplings to Z bosons. The yield and kinematic properties of events in the high four-lepton mass region allowed one to probe off-shell production of the discovered boson and set model-dependent upper limits on its natural width.

1. Introduction

The $H \to ZZ \to 4\ell$ channel,[a] where ℓ denotes electrons and muons, provides excellent search sensitivity to the SM Higgs boson in a broad mass range and is often referred to as the "golden channel". The $H \to ZZ \to 4\ell$ channel was one of the key benchmarks that defined the designs of the ATLAS and CMS experiments. The main virtues of this channel are the fully reconstructed final state of four leptons; the excellent,

---

[a]In this chapter, intermediate on-shell and off-shell Z bosons as well as γ^*, when allowed, are commonly referred to as Z, unless explicitly stated otherwise.

experimentally attainable four-lepton mass resolution; and the outstanding signal-to-background ratio. Despite the small expected branching fraction, this channel was the most sensitive for discovering a SM Higgs boson in the mass ranges 120−150 and above 180 GeV.[1,2,b]

The number of $H \to ZZ \to 4\ell$ events expected to be produced in the LHC Run 1 ranged from a few to about two hundred, depending on the assumed Higgs boson mass, m_H. The branching fraction, $\mathcal{B}(H \to 4\ell)$, for a SM Higgs boson with $m_H = 125$ GeV is 1.25×10^{-4} [3], implying fewer than 70 four-lepton events per experiment. This is the lowest signal event yield among all decay channels presented in this book. After accounting for detector acceptance, lepton reconstruction and event selection efficiencies, the expected number of detectable signal events was reduced to about 20.

Thanks to the narrow intrinsic width of the SM Higgs boson in the low mass range and the excellent electron and muon momentum reconstruction of the ATLAS and CMS experiments, such a signal would manifest itself as a narrow peak in the four-lepton mass distribution. For a SM Higgs boson with a mass below 300 GeV, the instrumental four-lepton mass resolution of about 1–2% was expected to dominate over the Higgs boson natural width.

The final state with four prompt leptons also ensured small background since such a signature was not characteristic of QCD processes. The main background in this search was electroweak non-resonant diboson production ($q\bar{q} \to ZZ \to 4\ell$) with a relatively small and well-understood cross section. The narrow signal peak over the continuous distribution of the low-rate background gave rise to a good signal-to-background ratio, 2:1 or higher, for the entire mass range considered in the search. The signal-to-background ratio for the $H \to ZZ \to 4\ell$ channel is the best among all SM Higgs boson decay modes. Moreover, angular and dilepton mass distributions of four well-reconstructed leptons provided rich information on physics underlying four-lepton production processes, which allowed ATLAS and CMS to enhance the signal-vs-background separation even further.

Finally, this channel provided excellent means for studying the properties of the discovered boson. The narrow mass peak and high signal-to-background ratio facilitated precise mass measurements and studies of production-related properties, such as the Higgs boson's transverse

---

[b]In the mass ranges below 120 and between 150−200 GeV, the branching fraction for the SM Higgs boson is impractically small, as discussed in Chapter 1.

momentum, multiplicities of jets produced in association with the Higgs boson, etc. The kinematics of the four final state leptons, being sensitive to helicity amplitudes of Z bosons produced in $X \to ZZ$ decays, allowed for detailed studies of the spin-parity properties of the discovered boson. The relative production rates of off-shell $H^* \to ZZ \to 4\ell$ and on-peak $H \to ZZ \to 4\ell$ events were used for probing, albeit with some model-dependent assumptions, the natural width of the observed boson.

Unless stated otherwise, the experimental results presented in this chapter are based on Refs. [4–13].

2. Physics objects used in the analysis

The physics objects used in the $H \to ZZ \to 4\ell$ analyses are leptons, photons, and jets. The reconstruction specifics relevant to this channel are discussed below, while the general description of physics object reconstruction can be found in Chapter 4.

2.1. *Electrons and muons*

With at least four leptons in the final state, the search for a Higgs boson in the $H \to 4\ell$ decay mode and measurements of the discovered boson's properties demanded a high pseudorapidity acceptance and a high reconstruction efficiency for leptons.

Figure 5.1(a) shows the pseudorapidity, η, distribution for the highest-pseudorapidity lepton originating from decays of Higgs boson with $m_H = 125$ GeV, produced via gluon fusion at $\sqrt{s} = 8$ TeV. The probability that all four leptons would have pseudorapidity $|\eta| < 2.4$ (muon detector acceptance in CMS) is only 66%. Acceptance for muons in ATLAS was $|\eta| < 2.7$. Electrons in ATLAS and CMS were reconstructed within $|\eta| < 2.5$.

Efficient reconstruction of low-p_T leptons ($p_T < 10$ GeV) was of a particular importance. In the low Higgs boson mass range ($m_H < 2m_Z$), at least one Z boson originating from the $H \to ZZ$ decay is off-shell and gives rise to relatively low-p_T leptons. For $m_H = 125$ GeV, the typical invariant mass of such off-shell Z bosons is only about 35 GeV. Figure 5.1(b) shows the p_T distribution of leptons for $m_H = 125$ GeV, ordered by their p_T, for events with all four leptons within the detector acceptance. Less than 80% (40%) of these events would have all leptons with $p_T > 5$ (10) GeV.

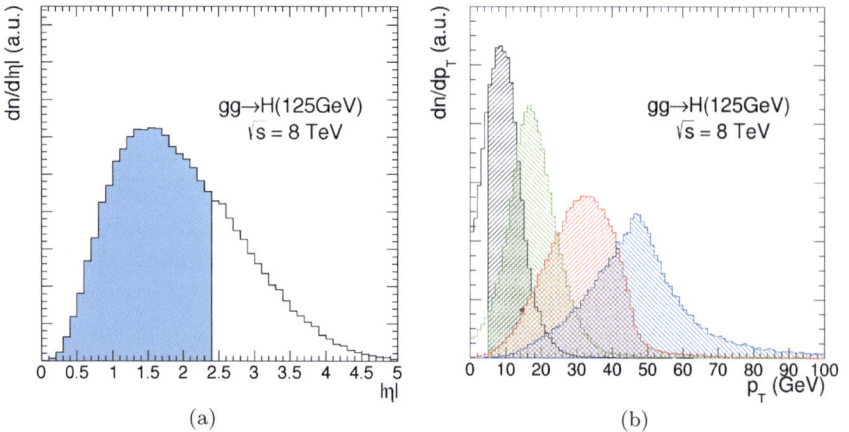

Fig. 5.1 Kinematic distributions of leptons in the $H \rightarrow ZZ \rightarrow 4\ell$ decays of a Higgs boson, $m_H = 125$ GeV: (a) pseudorapidity of the highest-$|\eta|$ lepton (the highlighted area corresponds to $|\eta| < 2.4$); (b) transverse momenta of leptons, ranked by their p_T, in events with all four leptons having $|\eta| < 2.4$ (the hatched areas correspond to $p_T > 5$ GeV). These distributions do not include experimental reconstruction efficiencies.

Thus, both ATLAS and CMS made sure that their detector designs and lepton reconstruction algorithms allowed for reconstructing muons and electrons with transverse momenta as low as 5–7 GeV. No other Higgs boson analysis described in this book had to face the experimental challenges associated with using such low-p_T leptons. The "turn-on" of the muon reconstruction efficiency, shown on example of ATLAS in Fig. 5.2(a), was near $p_T \sim 3$ GeV. The efficiency "turn-on" for electrons was at $p_T \sim 7$ GeV.

It was also important to achieve a high lepton-p_T resolution, which defined the observable four-lepton mass peak width and, hence, had a direct impact on the Higgs boson discovery sensitivity and on measurements of the discovered boson's properties. Typical lepton-p_T resolutions achieved by ATLAS and CMS were around 1–2% for $p_T \lesssim 100$ GeV. Such resolutions implied that the width of the observable four-lepton mass peak for a SM Higgs boson with $m_H < 300$ GeV would still be defined by the instrumental resolution, while for larger masses by the Higgs boson's natural width. As an example, Fig. 5.2(b) shows the electron E_T resolution achieved by CMS.

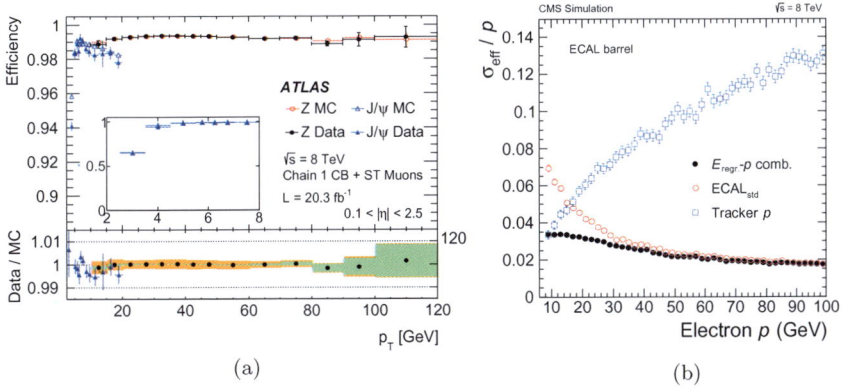

Fig. 5.2 (a) Muon reconstruction efficiency as a function of p_T obtained with the tag-and-probe method applied to the J/ψ and Z resonances (ATLAS). (b) Electron relative E_T resolution as a function of electron's energy as predicted by simulation (CMS). Contributions of the track's momentum and electromagnetic cluster's energy measurements are also indicated.

2.2. *Photons*

A few percent of the four-lepton events were expected to have a reconstructible final-state radiation (FSR) photon emitted by one of the leptons. When identified, such photons were included in the calculation of the Higgs boson candidate mass.

Generally, an FSR photon is expected to be soft and nearly collinear to the emitting lepton. Reconstruction of photons with E_T in the $1-4\,\mathrm{GeV}$ range and with small angular separation to leptons was yet another feature specific to this analysis. Such low E_T thresholds were driven by the experimental four-lepton invariant mass, $m_{4\ell}$, resolution: not accounting for an FSR photon with E_T of few GeV would result in mismeasuring the mass of a Higgs boson candidate by more than $1-2\%$ and, consequently, removing it from the signal peak.

The main challenge in FSR photon recovery was to keep a high efficiency of recovering genuine FSR photons and a high rejection factor for abundant unwanted photons mostly coming from π^0 decays. Since only a small fraction of events had reconstructible FSR, the net gain in sensitivity from the FSR photon recovery was $\mathcal{O}(1\%)$. Nevertheless, given the small expected signal yield, both ATLAS and CMS opted for recovering FSR photons in order to make the best use of every single event.

2.3. *Jets*

Both ATLAS and CMS used jets to gain sensitivities to vector boson fusion (VBF) and associated (VH) production. Jets were reconstructed using the anti-k_T algorithm, discussed in Chapter 4, with the clustering distance parameter $D = 0.4$ for ATLAS and 0.5 for CMS. In the case of ATLAS, jets were considered in the analysis only if they had pseudorapidity $|\eta| < 2.5$ and $p_T > 25\,\text{GeV}$ or pseudorapidity $2.5 < |\eta| < 4.5$ and $p_T > 30\,\text{GeV}$. CMS considered jets with $p_T > 30\,\text{GeV}$ in the pseudorapidity range of $|\eta| < 4.7$.

3. Background processes

Broadly speaking, all four-lepton backgrounds can be classified as either "irreducible" with four prompt leptons not directly associated with jets (hence, very similar to leptons from Higgs boson decays) or "reducible" with one or more leptons, real or fake, closely associated with jets. This association of leptons with jets allows one to suppress, or reduce, the latter backgrounds by very large factors, hence the name "reducible".

Feynman diagrams for the main contributions to "irreducible" background are shown in Fig. 5.3. Figure 5.4 presents the four-lepton mass distributions for each of these contributions.

The mass distribution for events produced via t-channel $q\bar{q} \to ZZ \to 4\ell$, shown in light blue in Fig. 5.4, has three characteristic regions. For $m_{4\ell} \ll 100\,\text{GeV}$, both propagators, shown in Fig. 5.3 (a), are predominantly γ^* ($q\bar{q} \to \gamma^*\gamma^* \to 4\ell$) and the cross section nearly diverges as $m_{4\ell}$ goes to

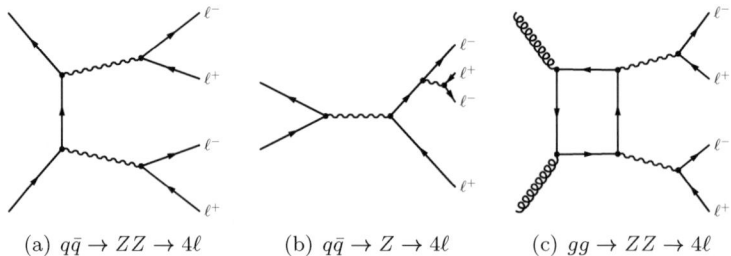

(a) $q\bar{q} \to ZZ \to 4\ell$ (b) $q\bar{q} \to Z \to 4\ell$ (c) $gg \to ZZ \to 4\ell$

Fig. 5.3 The main production modes of "irreducible" background $pp \to 4\ell$ search: (a) and (b) show LO diagrams for $q\bar{q} \to 4\ell$, while (c) shows NNLO diagram of $gg \to 4\ell$. NLO processes are not shown.

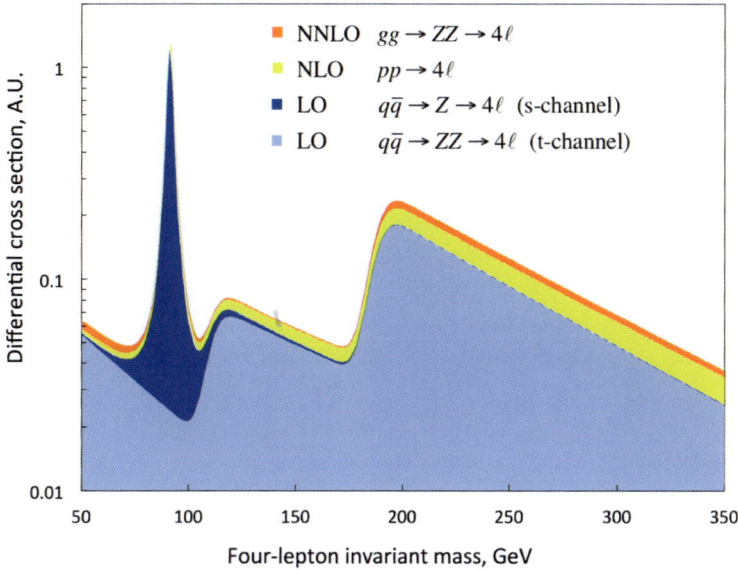

Fig. 5.4 Four-lepton mass distribution for events associated with "irreducible" background $pp \to 4\ell$, obtained at generator-level. Leptons are required to have $p_{\mathrm{T}} > 5\,\mathrm{GeV}$ and $|\eta| < 2.5$, while the invariant mass of opposite-sign dileptons has to be greater than $12\,\mathrm{GeV}$. The following contributions are identified by color and stacked on the plot: LO processes $q\bar{q} \to ZZ \to 4\ell$ (t-channel) and $q\bar{q} \to Z \to 4\ell$ (s-channel); NLO correction to these two LO processes; NNLO process $gg \to ZZ \to 4\ell$.

zero. For $m_{4\ell} \approx 100\,\mathrm{GeV}$, the cross section exhibits a step increase as one of the two propagators becomes predominantly an on-shell Z boson, while the other one still has to be γ^*: $q\bar{q} \to Z\gamma^* \to 4\ell$. Near $200\,\mathrm{GeV}$, there is another step increase in cross section as both propagators become predominantly on-shell Z bosons: $q\bar{q} \to ZZ \to 4\ell$. The s-channel $q\bar{q} \to Z \to 4\ell$ decays, shown in Fig. 5.3(b), form an $m_{4\ell}$ peak at the Z boson mass (dark blue in Fig. 5.4) with negligible spillover into the search region with $m_{4\ell} > 110\,\mathrm{GeV}$.

The NLO corrections to the t- and s-channels include $q\bar{q} \to 4\ell + g$, $qg \to 4\ell + q$, and interference of NNLO $q\bar{q} \to ZZ$ with LO $q\bar{q} \to ZZ$. They add about 30% to the LO cross section and are shown in yellow in Fig. 5.4 (mind the logarithmic scale). The NNLO $gg \to ZZ \to 4\ell$ process, shown in Fig. 5.3(c), contributes about 10% with respect to the NLO $pp \to 4\ell$ production for $m_{4\ell} > 200\,\mathrm{GeV}$. However, for four-lepton mass near $125\,\mathrm{GeV}$, this process has a negligible contribution, since diagrams with various quarks in the loop of the $gg \to \mathrm{box} \to Z\gamma^*$ process have many partial cancellations arising from sign flips of Z boson and γ couplings to left/right and up/down-type quarks. For $gg \to$ box $\to ZZ$ and $gg \to$ box $\to \gamma^*\gamma^*$, there are no such cancellations as the Z boson and γ^* couplings to the quarks in the loop are effectively squared.

The main sources of the so-called "reducible" background are Z + jets (including heavy flavour quark jets) and $t\bar{t} \to WW + 2$ b-jets, both giving rise to two prompt and two non-prompt leptons. There is also a contribution from backgrounds with three prompt and one non-prompt leptons: WZ + jets and $Z\gamma$+jets (where the photon converts into an electron–positron pair in the detector volume and is misidentified as a prompt electron). After applying all selection requirements, the sum of all "reducible" backgrounds was assessed by ATLAS and CMS from data and found to contribute about $30-40\%$ with respect to "irreducible" background in the four-lepton mass range $100-180\,\mathrm{GeV}$ and much less at higher masses.

4. Event selection and categorisation

ATLAS datasets certified as taken during periods of nominal detector operation and usable for the $H \to ZZ \to 4\ell$ analyses corresponded to an integrated luminosity of 20.3 fb^{-1} at $\sqrt{s} = 8\,\mathrm{TeV}$ and 4.5 fb^{-1} at $\sqrt{s} = 7\,\mathrm{TeV}$. The corresponding numbers for CMS were 19.6 and 5.1 fb^{-1}.

Both single-lepton and dilepton triggers were used for online event selection by ATLAS and CMS. CMS also used a tri-electron trigger for the $4e$ case. The trigger efficiency for a signal with respect to the offline selection, described below, was $97-100\%$, depending on the final state and the assumed Higgs boson mass in the range $100-1000\,\mathrm{GeV}$.

4.1. *Event selection*

For an event to be selected in the offline analysis, it had to contain at least two pairs of same-flavour and opposite-charge leptons: $e^+e^-e^+e^-$, $\mu^+\mu^-\mu^+\mu^-$ or $e^+e^-\mu^+\mu^-$. The leptons had to satisfy the criteria listed in Table 5.1, which were defined through simulation-based optimisations maximising the *expected* search sensitivity (exclusion limits). ATLAS and CMS, in general, arrived to very similar requirements; however, there were some variations as well. In some cases, these variations were a result of differences in the detector designs and performance, while in other cases

Table 5.1   Four-lepton event selection criteria in the ATLAS and CMS analyses. SF and DF stand for same-flavour (*ee* or $\mu\mu$) and different-flavour (*e*μ) lepton pairings, respectively. The $m_{Z_2}^{\min}$ cut in ATLAS is 12 GeV for $m_H < 140$ GeV, rose linearly to 50 GeV at $m_H > 190$ GeV, and then stayed constant for searches of a Higgs boson with higher masses.

Observables	ATLAS	CMS
Leptons:		
electron $|\eta|^{\max}$	2.47	2.5
muon $|\eta|^{\max}$	2.7	2.4
electron $p_T^{\min}$ (GeV)	7	7
muon $p_T^{\min}$ (GeV)	6	5
maximum relative isolation energy (see text)	0.2–0.3 ($\Delta R = 0.2$)	0.4 ($\Delta R = 0.4$)
maximum impact parameter significance $|d|/\sigma$	3.5(μ), 6.5(*e*)	4.0
1st/2nd/3rd leading lepton $p_T^{\min}$ (GeV)	20/15/10	20/10/-
Dileptons:		
$\Delta R^{\min}$ (SF/DF)	0.1/0.2	0.02/0.05
invariant mass of Z_1 pair, m_{Z_1} (GeV)	50–106	40–120
invariant mass of Z_2 pair, m_{Z_2} (GeV)	$m_{Z_2}^{\min}$–115	12–120
$m_{\ell^+\ell^-}^{\min}$ (GeV)	5 (SF)	4

the optimal selection criteria had fairly broad ranges so that seemingly different values of requirements resulted in nearly identical expected search sensitivities. Also, as often happens, the same ultimate objectives could be achieved by somewhat different means.

The pseudorapidity requirements on leptons, $|\eta|^{\max}$, were set by the detectors' geometrical acceptance. The minimal p_T requirements on leptons, $p_T^{\min}$, were a matter of careful optimisation. As discussed in Sec. 2, reconstruction of low-p_T leptons was crucial for gaining signal efficiency for a low-mass Higgs boson. However, one also had to worry about the quickly rising rate of "reducible" backgrounds with low-p_T "fake" leptons and, also, about reliable measurements of lepton reconstruction efficiencies quickly falling below $p_T \sim 5$ GeV for muons and 10 GeV for electrons.

The isolation requirements on the amount of hadronic energy flowing around leptons were the primary tool for suppressing the reducible backgrounds with non-prompt leptons originating from jets (see Sec. 3). ATLAS used tracker-based and calorimeter-based isolations, which were partially correlated as charged hadrons contributed to both. The relative tracker-based isolation, defined as the sum of the transverse momenta of all tracks, excluding those associated with leptons of interest, inside a cone of $\Delta R = \sqrt{\Delta\phi^2 + \Delta\eta^2} < 0.2$ around a lepton, divided by the lepton's E_T, had to be smaller than 0.15. The relative calorimeter-based isolation, defined as a sum of energy deposits in the electromagnetic and hadron calorimeters, excluding the lepton's calorimetric footprint, in a cone of $\Delta R < 0.2$ around the lepton, divided by lepton's E_T, was required to be smaller than 0.3 for muons and 0.2 (0.3) for electrons in 8 TeV (7 TeV) datasets. CMS used a particle flow algorithm,[14] in which all tracks and all electromagnetic/hadron calorimeter deposits were grouped to form mutually exclusive "particles" of five kinds: electrons, muons, photons, charged and neutral hadrons. To quantify the amount of energy flowing around a given lepton, the transverse energy of all such particles in a cone $\Delta R < 0.4$ around a lepton was calculated and the ratio of this energy over the lepton's E_T was required to be less than 0.4. To maintain the same isolation requirement efficiency for different pile-up conditions (and hence minimise potential systematic uncertainties), both ATLAS and CMS used isolation energy corrections calculated on the per-event basis as a function of the number of vertices found and the average amount of transverse energy flowing in the event.

Rejecting leptons not pointing to the primary vertex was another handle to suppress non-prompt leptons originating from long-lived hadrons, such as B mesons, or electrons originating from photon conversions in the beam pipe and detector material. ATLAS discarded muons (electrons), if their impact parameter in the transverse plane divided by its estimated measurement uncertainty, $|d|/\sigma$, was greater than 3.5 (6.5). CMS used a 3D-impact parameter, the closest distance between the track helix and the event primary vertex, and discarded leptons with $|d|/\sigma > 4$.

Requirements on the transverse momentum, p_T of the two leading leptons (20 and 15 GeV for ATLAS, 20 and 10 GeV for CMS) ensured that selected events had a high trigger efficiency and, hence, known with small uncertainties.

Leptons were required to be separated from each other by a minimum ΔR^{min} distance. The primary purpose of this requirement was to reject rare occasions of duplicates when one lepton was reconstructed as two. Whenever this happened, the original and "fake" leptons would tend to be nearly collinear.

In the decays of a Higgs boson with a mass in the range 110–180 GeV, one Z boson is expected to be mostly on-shell, while the other off-shell. Therefore, between four leptons in an event, the pair of opposite-sign same-flavour leptons with its invariant mass closest to the Z boson mass, to be denoted henceforth as Z_1, was required to have its mass in the range 50–106 GeV for ATLAS and 40–120 GeV for CMS. The m_{Z_1} cuts had a very high efficiency for a signal and helped reduce backgrounds without Z

bosons, such as $t\bar{t}$ and $WW + \text{jets}$. The low-end cut on m_{Z_1} was as low as 40 or 50 GeV in order to keep a high acceptance for a low-mass Higgs boson occasionally decaying into two off-shell Z bosons: $H \to Z^*Z^* \to 4\ell$.

The low-end cut on the invariant mass of the remaining pair,[c] denoted as Z_2, was a subject of detailed optimisations. Lowering this requirement would naturally increase the signal efficiency for a low-mass Higgs boson. However, the rate of backgrounds, including the "irreducible" background $q\bar{q} \to Z(Z^*/\gamma^*) \to 4\ell$ would grow as well. Both ATLAS and CMS arrived to a conclusion that the cut near 12 GeV would be close to optimal for the low-mass Higgs boson searches ($m_H < 140$ GeV). In a search for a Higgs boson of higher masses, the cut could be raised, as was done by ATLAS (see Table 5.1). CMS refrained from varying the cut value, as the expected sensitivity gains were considered to be small. The upper cut on m_{Z_2}, 115 or 120 GeV, was not consequential and was added as a safe-guard against non-ZZ background.

Finally, the cut $m^{\min}_{\ell^+\ell^-} > 5/4$ GeV (ATLAS/CMS) on all same-flavour lepton pairs helped suppress background $q\bar{q} \to \gamma^*\gamma^* \to (\ell^+\ell^-)(\ell^+\ell^-)$. The cross section for this background rises fast as m_{γ^*} decreases. The potential danger was that such events with two γ^*'s of low mass and high p_T might nevertheless pass the selection criteria should the Z_1 and Z_2 pairs be formed from leptons associated with different γ^*'s. CMS decided to extend the $m^{\min}_{\ell^+\ell^-}$ cut to include lepton pairs of different flavour, motivated by considering backgrounds with 2 b-quark jets, e.g.: $pp \to b\bar{b} \to (e^+\mu^- + X)(e^-\mu^+ + X)$. Again, same-flavour lepton pairs might pass the m_{Z_1} and m_{Z_2} cuts, but would fail the $m^{\min}_{e^\pm\mu^\mp}$ cut as $e^\pm\mu^\mp$ pairs in such events would come from B mesons with a mass of about 5 GeV.

The signal efficiency, which includes the detector acceptance, lepton reconstruction efficiency, and efficiency of event selection requirements, is shown in Fig. 5.5. The difference between efficiencies for the three final states was due to the lower reconstruction efficiency for electrons in comparison to muons. The overall decline of efficiency at lower Higgs boson masses was due to smaller efficiency for low p_T leptons (acceptance, reconstruction, and isolation) and falling efficiency of the low-end m_{Z_2} cut.

Table 5.2 shows the expected and observed event yields after all selection requirements in the wide range of four-lepton masses as well as for a

---

[c]The subtleties of selecting a Z_2 pair in rare events with more than four leptons can be found in the original papers.

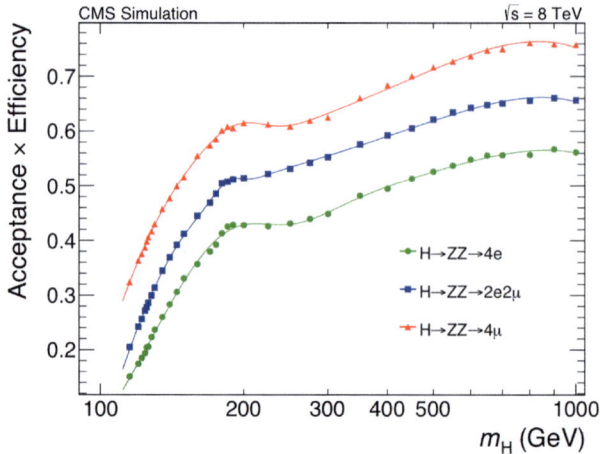

Fig. 5.5   Four-lepton event acceptance times reconstruction/selection efficiency as a function of the Higgs boson mass for CMS in the 8 TeV dataset. Efficiencies for different final states are shown separately.

Table 5.2   Expected and observed numbers of events in two four-lepton mass ranges: $m_{4\ell} > 100$ GeV and in 10/9 GeV windows near 125 GeV. The assumed SM Higgs boson mass is 125 GeV. The numbers in parentheses are not given in the original publications; they are estimates made by the authors of this chapter.

	ATLAS		CMS	
	>110 GeV	120–130 GeV	>100 GeV	121.5–130.5 GeV
"Irreducible" background	381 ± 20	7.41 ± 0.40	387 ± 31	6.8 ± 0.3
"Reducible" background	18.9 ± 2.4	2.95 ± 0.33	22.6 ± 3.6	2.6 ± 0.4
$H \to ZZ \to 4\ell$	18.2 ± 1.8	16.2 ± 1.6	(19)	17.3 ± 1.3
Total expected	418 ± 20	26.5 ± 1.7	(429 ± 31)	26.7 ± 1.4
Observed	466	37	470	25

narrow range around 125 GeV. For simplicity, all four-lepton final states are summed.

4.2. *Recovery of final state radiation*

Both ATLAS and CMS searched for FSR photons (see Sec. 2.2) and, when identified, included them in the Higgs boson candidate mass calculation. FSR photons have a spectrum $dN/dE_\gamma \sim 1/E_\gamma$ and tend to be emitted along leptons. Considering this, ATLAS defined "collinear" FSR candidates

($\Delta R_{\ell\gamma} < 0.15$ and $E_T^\gamma > 1.5\,\mathrm{GeV}$) and "non-collinear" FSR candidates ($\Delta R_{\ell\gamma} > 0.15$ and $E_T^\gamma > 10\,\mathrm{GeV}$). CMS definitions for FSR candidates were as follows: "collinear" if $\Delta R_{\ell\gamma} < 0.07$ and $E_T^\gamma > 2\,\mathrm{GeV}$, and "non-collinear" if $0.07 < \Delta R_{\ell\gamma} < 0.5$ and $E_T^\gamma > 4\,\mathrm{GeV}$. To be included in calculation of the mass, an FSR photon candidate had to be isolated and was required to bring the invariant mass of $\ell^+\ell^-\gamma$ closer to the Z boson mass in comparison to the $\ell^+\ell^-$ mass. In ATLAS, the FSR recovery procedure was applied to leptons in the leading Z_1 pair only, while in CMS it was applied to all leptons.

On average, about 5% of four-lepton events were FSR-corrected. The estimated efficiency of the procedure was about 50–70%, with respect to the number of true FSR photons with the same ΔR and E_T^γ requirements. The purity was estimated to be between 80–95%, i.e. 5–20% of photons identified as FSR were actually not FSR photons. The FSR recovery procedure was validated by applying it to $Z \to \ell^+\ell^-$ decay. FSR photons were found in a few percent of Z boson decays, in agreement with expectations, and the reconstructed invariant mass of the $\ell^+\ell^-\gamma$ system for such events formed a much improved peak around the Z boson mass.

4.3. Event categorisation

Selected events were classified into categories, aiming to improve search and measurement sensitivities, to facilitate studies of various signal properties, and help with analysis flow bookkeeping.

4.3.1. Categorisation by centre-of-mass energy

Both ATLAS and CMS treated events from the $\sqrt{s} = 7$ and $8\,\mathrm{TeV}$ datasets separately. This simplified the basic analysis bookkeeping (run conditions, calibrations, simulation samples, etc.).

4.3.2. Categorisation by flavour of the four-lepton system

CMS used three lepton-flavour-dependent event categories: $4e$, 4μ, $2e2\mu$. Keeping events of different expected signal-to-background ratio in separate groups helped to maximise the overall search and measurement sensitivities. In comparison to the $4e$ final state, 4μ events were expected to have a higher signal-to-background ratio due to better four-lepton mass resolution (see Table 5.3 and Fig. 5.6) and smaller "reducible" backgrounds (see Sec. 6).

Table 5.3   Four-lepton mass resolution in GeV (σ of the Gaussian core) for a 125 GeV Higgs boson.

	4e	2e2μ/2μ2e	4μ
ATLAS	2.2	1.8	1.6
CMS	2.0	1.6	1.2

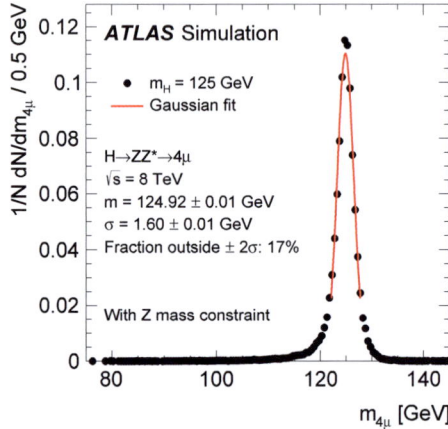

Fig. 5.6   Simulated four-muon invariant mass distributions for $H \to ZZ \to 4\mu$ (ATLAS).

In terms of the signal-to-background ratio, the $2e2\mu$ event category was in between.

ATLAS divided $2e2\mu$ events further into two separate categories, $2e2\mu$ and $2\mu2e$, depending on whether 2e or 2μ pair would be counted as Z_1. For events with $m_{4\ell} < 2m_Z$, the Z_1 pair would contribute the most to the four-lepton mass and, hence, $2\mu2e$ events were expected to have a better average mass resolution than $2e2\mu$ events. On the other hand, $2\mu2e$ events with low p_T electrons associated with the Z_2 pair were expected to have a higher relative contribution of "reducible" background in comparison to $2e2\mu$ events, since soft electrons are easier to fake. The two trends, however, had opposite effects on the signal-to-background ratio and largely compensated each other so that the net gain in sensitivities from splitting events with dilepton pairs of different flavours into $2e2\mu$ and $2\mu2e$ categories was found to be relatively small.

4.3.3. *Categorisation by production mechanism signatures*

To probe different Higgs boson production mechanisms, ATLAS and CMS introduced production-mechanism-sensitive categories. ATLAS used four categories (dijet VBF, dijet VH, leptonic VH, and untagged), while CMS used only two categories (dijet and untagged).

The categorisation employed by ATLAS was as follows. If an event had at least two jets and the two highest p_T jets (two "leading" jets) had an invariant mass $m_{jj} > 130\,\mathrm{GeV}$, the event was assigned to the VBF-tagged category. If $40\,\mathrm{GeV} < m_{jj} < 130\,\mathrm{GeV}$ and a dedicated multivariate-observable trained using a Boosted Decision Tree algorithm (BDT, see Appendix B) $\mathrm{BDT_{VH}} > -0.4$, such an event would be assigned to the dijet VH-tagged category. The $\mathrm{BDT_{VH}}$ was trained to separate VH from gluon-gluon fusion production using the following five inputs: m_{jj}, p_T of each of the two leading jets, pseudorapidity of the highest p_T jet, and absolute value of the difference between two leading-jet pseudorapidities $|\Delta\eta_{jj}|$. From the remaining events, those with at least one additional (i.e. fifth) lepton with $p_T > 8\,\mathrm{GeV}$ would form the leptonic VH category. The untagged category comprised all remaining events.

In the CMS categorisation scheme, events with two or more jets were assigned to the dijet category and all other events to the untagged category.

Table 5.4 lists the expected and observed event counts in each of the production mechanism categories. Note that VBF and VH categories are expected to contain a substantial fraction of signal events produced via gluon-gluon fusion. In comparison to the untagged category, VBF and VH categories were expected to have better signal-to-background ratio albeit with considerably smaller event yields.

Table 5.4   Expected and observed event counts in each of the production mechanism tag categories. ATLAS (CMS) results are for four-lepton mass range 120–130 (121.5–130.5) GeV and for an assumed Higgs boson mass of 125 (126) GeV.

		gF+ttH+bbH	VBF	VH	Total H	Total bkg	Observed
ATLAS	Untagged	12.8	0.57	0.35	13.7	9.8	34
	Dijet VBF	1.18	0.75	0.10	2.03	0.42	3
	Dijet VH	0.40	0.03	0.21	0.64	0.18	0
	Leptonic VH	0.013	<0.001	0.069	0.082	0.031	0
CMS	Untagged	15.4	0.70	0.49	16.6	8.5	20
	Dijet	1.7	0.87	0.37	3.0	0.9	5

5. Continuous observables

The four-lepton events were further characterised by introducing observables that helped enhance search and measurement sensitivities, but were not used for explicit cuts. Instead, the analyses took into account entire shapes of their distributions, which allowed for using maximally the discriminating information carried by these observables. In this section, we describe such observables that were selectively used in different sub-analyses to be presented in Secs. 7–11. Observables specific to measurements of the discovered boson's mass, total width via its far off-shell production, and studies of spin-parity properties are described in the corresponding sections.

5.1. *Four-lepton invariant mass*

The four-lepton invariant mass was the prime shape observable. The SM Higgs boson events were expected to form a distinct resonance peak not characteristic of any background. As mentioned earlier, for $m_H < 300\,\text{GeV}$, the intrinsic width of the SM Higgs boson is smaller than the instrumental four-lepton mass resolution. Whenever an FSR photon was identified, it was added to the calculation of the mass. ATLAS improved the average $m_{4\ell}$ resolution by about 15% by refitting the invariant mass of each four-lepton event with Z-mass constraints, on the Z_1-pair mass for events with $m_{4\ell} < 190\,\text{GeV}$ and both Z_1 and Z_2 pair masses for events with $m_{4\ell} > 190\,\text{GeV}$, in which both Z bosons would tend to be produced on-shell. The fit was taking into account the Z boson line-shape, the reconstructed dilepton mass and its instrumental uncertainty.

Figure 5.7 shows the four-lepton invariant mass distributions (separately for the full and the low-mass ranges) for all events passing the selection criteria. In this and following figures, events of all four-lepton final states and data from the 7 TeV and 8 TeV runs are shown together, although, in the actual analyses, events of different event categories were always analysed separately. In the low-mass range, signal-like peaks near 125 GeV are visually evident in both ATLAS and CMS data without any formal statistical analysis. Outside the 125 GeV peak region, the distributions were consistent with the expected background. The peak at $m_{4\ell} \sim m_Z$ corresponds to $Z \to 4\ell$ decays (Fig. 5.3(b)); it is more pronounced in CMS data owing to the lower p_T cuts.

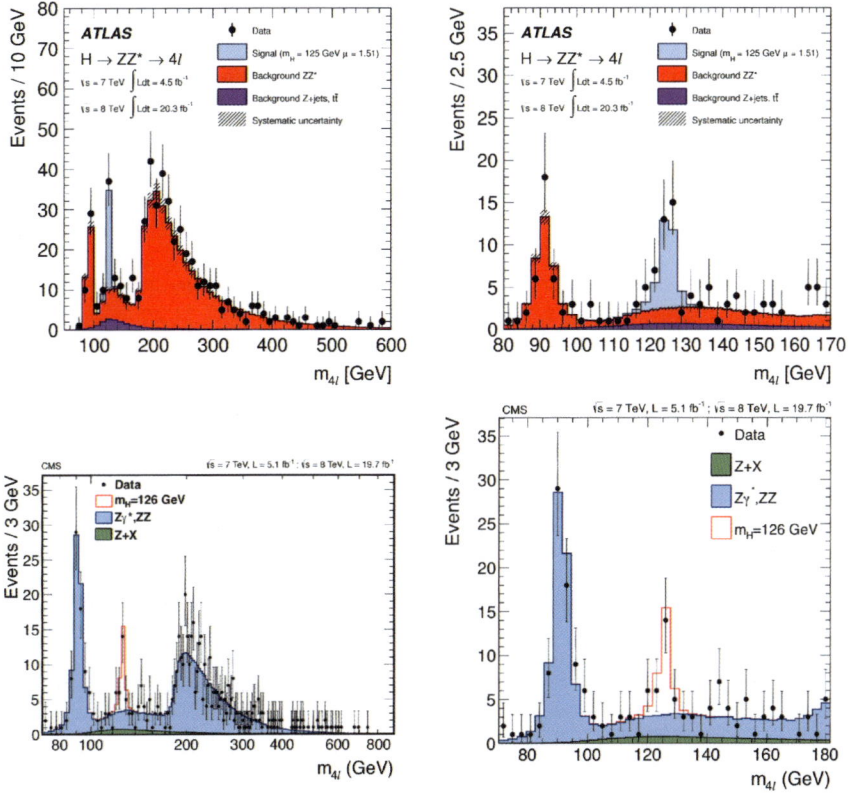

Fig. 5.7 Four-lepton invariant mass distributions for ATLAS (top) and CMS (bottom) data in the full mass range (left) and the low-mass range (right). Data are shown as points. The background contributions along with the expectation for the SM Higgs boson production are shown as histograms. For ATLAS, the expectations for SM Higgs boson with $m_H = 125\,\mathrm{GeV}$ are scaled up by factor 1.51, the best-fit signal strength obtained in data.

5.2. *Four-lepton kinematic discriminants*

To help separate signal from background, ATLAS and CMS made use of kinematic discriminants calculated for each four-lepton event as the ratio of leading-order matrix elements, $\mathcal{A}_{\mathrm{H}}(\mathcal{P})$ for the $gg \to H \to 4\ell$ signal, and $\mathcal{A}_{\mathrm{ZZ}}(\mathcal{P})$ for the prevailing $q\bar{q} \to ZZ \to 4\ell$ background:

$$d = \frac{|\mathcal{A}_{\mathrm{H}}(\mathcal{P})|^2}{|\mathcal{A}_{\mathrm{ZZ}}(\mathcal{P})|^2}, \qquad (5.1)$$

where $\mathcal{P}$ stands for momenta of the four final-state leptons. The signal matrix element $\mathcal{A}_H$ was calculated assuming $m_H = m_{4\ell}$; hence, the observable d discriminated on the basis of a kinematic configuration of four leptons in an event, rather than on the four-lepton invariant mass. Matrix elements were calculated and cross-validated by using a number of tools,[15–22] including various event generators, exact for any four-lepton final state, and explicit analytic formulas, which were available for the $2e2\mu$ final state only.[d]

In the mass range near 125 GeV, the matrix-element-based discriminant boosted the Higgs boson search sensitivity by about 20%. This discrimination power arises mostly from the m_{Z_2} mass distribution differences [18]: for a signal, the Z_2 pair is always produced via off-shell Z boson and tends to "prefer" the highest possible mass; while for background, dominated in the low four-lepton mass range by $q\bar{q} \to Z\gamma^* \to (\ell^+\ell^-)(\ell^+\ell^-)$, this lepton pair is expected to "prefer" lower invariant masses. Also, the low mass tail of the m_{Z_1} distribution is expected to be less suppressed for Higgs boson events. When Z_1 goes off-shell ($m_{Z_1} < m_Z$), the suppression is partially compensated by allowing for m_{Z_2} to get closer to m_Z. Figure 5.8 shows the m_{Z_1} and m_{Z_2} distributions for events in the 125 GeV peak. Indeed, the excess events revealed the expected pattern.

For the purposes of technical convenience, the discriminant d was monotonically transformed.[e] ATLAS simply used $D_{ZZ^*} = \log(d)$. CMS transformed the discriminant d so that the new one, denoted as $D_{\text{bkg}}^{\text{kin}}$, would range between 0 and 1: $D_{\text{bkg}}^{\text{kin}} = [1 + c_{m_{4\ell}} \cdot d]^{-1}$, where an ad-hoc constant $c_{m_{4\ell}}$ was adjusted for each bin of $m_{4\ell}$ in order to prevent distributions from being too compressed against 0 or 1.

Matrix elements calculated at LO are oblivious of the four-lepton event transverse momentum $p_T^{4\ell}$ (NLO effect) or its rapidity $y^{4\ell}$ (PDF and NLO effects). However, these observables carry some signal-vs-background separation. Since the initial state radiation from gluons is more prolific than from quarks, the four-lepton system tends to have higher transverse momentum for signal ($gg \to H \to 4\ell$) than for ZZ background ($q\bar{q} \to ZZ \to 4\ell$). Also, the four-lepton system produced in $q\bar{q} \to ZZ \to 4\ell$

---

[d]The case of the $4e$ and 4μ states is more complicated due to interference associated with permutations of identical leptons, which is particularly pronounced for events with off-shell Z bosons.
[e]Monotonic transformations do not change the discrimination power of observables as they do not change relative ranking of events.

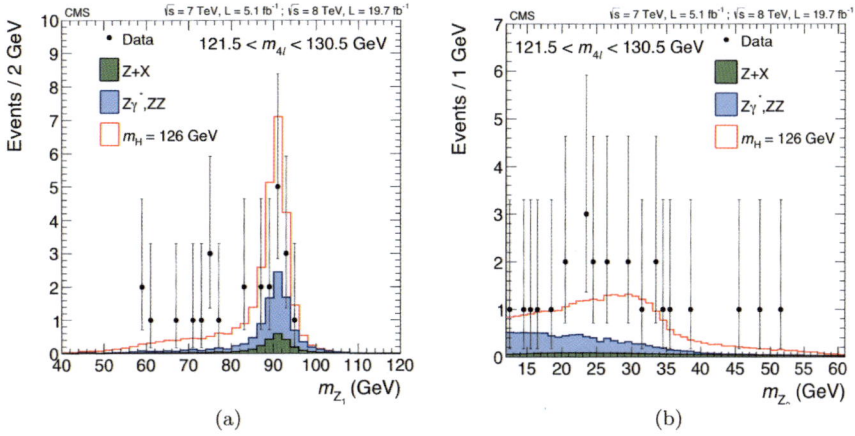

Fig. 5.8 Dilepton mass distributions for events with 121.5 GeV $< m_{4\ell} <$ 130.5 GeV, as obtained by CMS: (a) m_{Z_1} for the pair of leptons of same flavour and opposite charge with their invariant mass closest to m_Z, (b) m_{Z_2} for the remaining pair.

process tends to be more boosted along the beam line due to very different quark and antiquark parton density functions. ATLAS exploited these features by introducing a multivariate observable, BDT_{ZZ^*}, trained using three inputs: matrix-element discriminant D_{ZZ^*}, four-lepton transverse momentum $p_T^{4\ell}$, and four-lepton pseudorapidity $\eta^{4\ell}$. CMS also used $p_T^{4\ell}$, but kept it as an independent observable, aiming both to help separate signal from ZZ background as described above and to attain some sensitivity to the VBF + VH production component of the untagged event category. A Higgs boson produced in VBF or VH processes recoils already at LO against either two quarks or a W/Z boson, which leads to higher Higgs boson transverse momenta in comparison to the case of gluon–gluon fusion process. However, quantitative studies showed that gains in Higgs boson discovery sensitivity from using the four-lepton transverse momentum and rapidity observables were rather small, at the level of $\mathcal{O}(1\%)$.

Figures 5.9(a) and 5.9(b) show the BDT_{ZZ^*} (ATLAS) and $D_{\mathrm{bkg}}^{\mathrm{kin}}$ (CMS) distributions for events with $m_{4\ell}$ near 125 GeV. The expected distributions, shown by histograms, demonstrate a clear separation between the SM Higgs boson and the dominant $q\bar{q} \to ZZ$ background. Also, one can see that the observed events are skewed to the right, as one would expect in the presence of a Higgs boson signal.

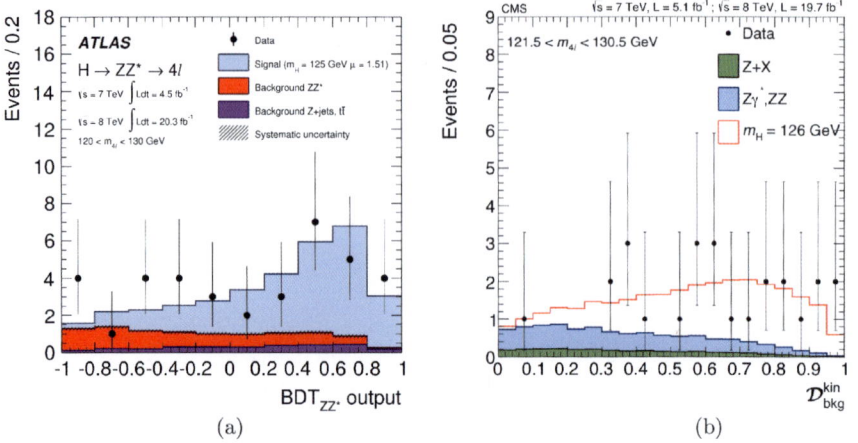

Fig. 5.9   (a) Discriminant BDT_{ZZ^*} distribution for events with four-lepton invariant mass in the range 120–130 GeV (ATLAS). (b) Discriminant D_{bkg}^{kin} distribution for events with four-lepton invariant mass in the range 121.5–130.5 GeV (CMS).

5.3. *Dijet discriminant*

The dijet tagged category was enriched in signal events; nevertheless, it still contained a large fraction of $gg \rightarrow H$ signal events and background as well. The large rapidity distance between two leading jets $|\Delta\eta_{jj}|$ and their invariant mass m_{jj} were both used to help distinguish signal events produced via VBF mechanism from gluon–gluon fusion. However, these two observables had a strong correlation. CMS consolidated them into one VBF discriminant $D_{jet} = \alpha\,|\Delta\eta_{jj}| + \beta\,m_{jj}$, where α and β coefficients were optimised to provide the maximum discrimination power between VBF and gluon fusion production mechanisms. Such a linear construct is known as a Fisher discriminant. ATLAS trained an MVA-observable BDT_{VBF} using the following five inputs: m_{jj}, $|\Delta\eta_{jj}|$, p_T of each of the two leading jets, and the pseudorapidity of the leading jet. The VBF discriminants were also efficient in separating the VBF produced Higgs boson signal events in the VBF dijet category from the background, thus improving the overall signal sensitivity.

ATLAS and CMS observed three and five events, respectively, in their VBF dijet categories with $m_{4\ell} \sim 125$ GeV. One ATLAS event had Higgs signal to ZZ background ratio of about 6, and 60% probability to arise from VBF production, still very modest for any affirmative claims. The VBF purity of all five CMS events was smaller.

6. Signal and background modelling

The signal and background modelling, including the assessment of associated systematic uncertainties, was evolving over time. The description presented in this chapter is based on the final analyses published with the full Run 1 dataset. The tools and methods used by ATLAS and CMS in setting up their models were very similar.

6.1. *Processes with four prompt leptons*

The processes with four prompt leptons were modelled using simulation. Particle-level events were simulated using appropriate event generators as shown in Table 5.5. The best-known cross sections were used to normalise simulated events to represent the expected event yields for the given integrated luminosity. Signal cross sections were taken from Refs. [3,23] (see Chapter 1 for details). The particle-level events were then processed through the ATLAS and CMS detector simulation, which propagated the generated particles through the detector volume and emulated detector's response. Pile-up pp interactions were added with their multiplicity distributions matched to those observed in data. Then, the simulated detector response was used to reconstruct events with the same algorithms that were used with the data. Finally, the simulation-based signal and background models

Table 5.5 Simulated processes with four prompt leptons. Parton showering and hadronisation, including the underlying event, are simulated with PYTHIA for all listed processes. Processes $gg/q\bar{q} \to X$ stand for production of a particle with spin-parity properties different from the SM Higgs boson. Processes marked with "Yes" had additional theory-based corrections applied, as described in the text.

Process	ME generator	Corrections	Cross section
$gg \to H$	POWHEG (NLO)	Yes	$(\text{NNLO} + \text{NNLL})_{\text{QCD}} + \text{NLO}_{\text{EW}}$
VBF	POWHEG (NLO)	Yes	$\text{NNLO}_{\text{QCD}} + \text{NLO}_{\text{EW}}$
WH and ZH	PYTHIA (LO)		$\text{NNLO}_{\text{QCD}} + \text{NLO}_{\text{EW}}$
$t\bar{t}H$	PYTHIA (LO)		NLO_{QCD}
$b\bar{b}H$ (ATLAS)	see text		$\text{NLO}_{\text{QCD(4FS)}} + \text{NNLO}_{\text{QCD(5FS)}}$
$gg/q\bar{q} \to X$	JHUGEN (LO)	Yes	Normalised to data
$q\bar{q} \to ZZ$	POWHEG (NLO)		NLO_{QCD}
$gg \to ZZ$	GG2ZZ (LO)		LO and (LO) $\times$ K-factor
DPI $(q\bar{q} \to Z) \times 2$	PYTHIA		Phenomenological

were corrected for differences between simulated and actual detector performance (lepton reconstruction efficiency, lepton p_T scale, lepton p_T resolution, jet energy scale, etc.), where the actual detector performance was obtained from dedicated analyses of data, relying on data-driven techniques such as the tag-and-probe method.

Some of the simulated samples were corrected to improve the theoretical accuracy of models implemented in corresponding generators. For example, the POWHEG-generated gluon fusion events were given additional weights so that the p_T distribution of reweighted events would match the one calculated at the NLO + NNLL level.

For a SM Higgs boson with $m_H \gtrsim 400\,\text{GeV}$, the narrow-width approximation implemented in POWHEG was not adequate. Also, for gluon–gluon fusion the effect of interference between $gg \to H \to ZZ$ and non-resonant $gg \to ZZ$ needed to be accounted for. CMS reweighted the POWHEG samples ($gg \to H$ and VBF) in this mass range to match the Higgs boson mass lineshape calculated in a complex-pole scheme.[24]

Neither POWHEG nor PYTHIA simulated the effects of interference associated with permutations of identical leptons in $H \to ZZ \to 4e$ and 4μ decays, which was important in spin-parity studies.[18] CMS opted to keep Higgs bosons stable in POWHEG $gg \to H$ and VBF samples and perform Higgs bosons' decays with JHUGEN,[15–17] which treated the interference properly. In their spin-parity studies, for simulation of $gg \to H$, ATLAS used LO JHUGEN and reweighted the simulated events to match their p_T distribution to the NLO + NNLL calculation.

The small contributions from WH, ZH, and $t\bar{t}H$ processes were simulated with PYTHIA. The SM $b\bar{b}H$ production is expected to contribute less than 1% of the total cross section. ATLAS included and treated it as a small correction to gluon fusion, while CMS chose to neglect it.

Production of bosons with exotic spin-parity properties was simulated with JHUGEN. In the absence of established contender models for such exotic bosons that would predict definitive production cross sections, event yields in those models were set to the number of observed signal events.

Background $q\bar{q} \to ZZ \to 4\ell$ was generated with NLO POWHEG, which was also used for assessing the process cross section. Background $gg \to ZZ \to 4\ell$ was simulated with GG2ZZ at LO. In analyses probing the Higgs boson off-shell production, where $gg \to ZZ$ background was of high importance, the $gg \to H \to ZZ$ K-factor was used as a proxy[25] for the $gg \to ZZ$ background.

Double-parton interactions (DPI) with two "independent" $q\bar{q} \to Z$ interactions taking place within the same proton–proton collision was simulated with PYTHIA. The overall cross section was assessed from a general phenomenological formula $\sigma_{\mathrm{DPI}} = \sigma_1 \sigma_2 / \sigma_{\mathrm{eff}}$, where σ_1 and σ_2 are cross sections for two single parton–parton processes of interest ($q\bar{q} \to Z_1$ and $q\bar{q} \to Z_2$), and σ_{eff} is a universal phenomenological parameter measured to be about 15 mb for pp collisions at $\sqrt{s} = 7$ TeV.[26] The DPI background was found to be negligible.

6.2. *Reducible background*

In the mass range 100–180 GeV, "reducible" backgrounds consisting of events with one or two leptons of non-prompt origin were estimated to contribute as much as 30–40% in addition to the "irreducible" background (see Table 5.2). Although the primary processes contributing to "reducible" backgrounds were well known (see Sec. 3), accurate simulation of rare instances of reconstructing objects of non-prompt origin as "tight" leptons was challenging. Here, "tight" refers to the final lepton selection criteria. Objects of non-prompt origin could be a lepton from B hadron decays, an electron originating from a photon conversion, a muon from a pion in-flight decay $\pi \to \mu\nu$, a charged pion faking an electron via the charge-exchange interaction (e.g., $\pi^- p \to \pi^0 n$) in the electromagnetic calorimeter, etc. Probabilities of reconstructing such objects as "tight" leptons, being very small, depended strongly in relative terms on various subtleties of jet fragmentation and particle interactions with the detector, whose accurate modelling was not guaranteed.

Therefore, to predict the net "reducible" background in the signal region, ATLAS and CMS opted for data-driven methods. The method used by ATLAS was based on unravelling the background composition and dealing with each contribution separately, while CMS treated all background sources more inclusively. In the following, we briefly describe the CMS method and then highlight the distinct features of the ATLAS approach.

The CMS method assumed that there were two distinct groups of "reducible" background processes with either two or three prompt leptons, with the unknown number of events X_2 and X_3 respectively, and that the probability of reconstructing an object of non-prompt origin as a lepton with "tight" and "loose" selection criteria were ϵ_{T} and ϵ_{L}, respectively. "Loose" selection did not include isolation requirements and certain reconstruction quality criteria. The number of observed events N_{CR2}^{obs} and N_{CR3}^{obs}

in two control regions — $CR2$ with two "tight" leptons and two "loose" leptons, failing the "tight" selection criteria, and $CR3$ with three "tight" leptons and one "loose-but-not-tight" lepton — were related to X_2, X_3, ϵ_T, and ϵ_L as follows:

$$N_{CR2}^{obs} = X_2 \cdot (\epsilon_L - \epsilon_T)^2$$
$$N_{CR3}^{obs} = X_3 \cdot (\epsilon_L - \epsilon_T) + X_2 \cdot 2\epsilon_T(\epsilon_L - \epsilon_T).$$

From these two equations, the predicted expected number of "reducible" background events in the signal region ($N_{SR} = X_2 \cdot \epsilon_T^2 + X_3 \cdot \epsilon_T$) was obtained:

$$N_{SR} = N_{CR3}^{obs} \cdot \frac{\epsilon}{1 - \epsilon} - N_{CR2}^{obs} \cdot \left(\frac{\epsilon}{1 - \epsilon}\right)^2, \qquad (5.2)$$

where ϵ is the "tight-to-loose" ratio of two probabilities, $\epsilon = \epsilon_T/\epsilon_L$, also referred to as a transfer factor. This ratio was measured in control regions of $Z(\ell^+\ell^-) + e/\mu$ events with low missing transverse momentum required in order to suppress WZ events. The non-Z lepton in such events provided a clean source of reconstructed leptons of non-prompt origin.

Potential systematic biases arising from differences in ϵ for different non-prompt objects were minimised by using isolation as the prime difference between "loose" and "tight" leptons and by measuring "tight-to-loose" ratios for electron and muons in bins of p_T and η. The validity of this method was confirmed by simulation. With ϵ being a function of p_T and η, the two terms in Eq. (5.2) then represent sums of $CR2$ and $CR3$ events reweighted by measured probabilities on an event-by-event basis. The predictions of Eq. (5.2) were obtained in bins of various four-lepton observables used in the $H \to ZZ \to 4\ell$ analyses. The method was validated on data using events with four leptons of wrong charge/flavour combinations ($\mu^+\mu^-\mu^+e^-$, $\mu^+\mu^-e^+e^+$, etc), as shown in Fig. 5.10(a).

In the ATLAS approach, separate analysis schemes were developed for $\ell\ell + \mu\mu$ and $\ell\ell + ee$ events, motivated by the fact that leptons of non-prompt origin were most likely to end up in the Z_2 dilepton pair and that processes leading to "fake" muons and electrons were notably different. Using physics considerations, "reducible" background processes were classified in a few groups. Then, a simultaneous fit to observations in several control regions with one or two "loose" leptons, enriched differently by each process, was used to extract the number of events of each type and extrapolated to the signal region, by applying transfer factors tuned for each process.

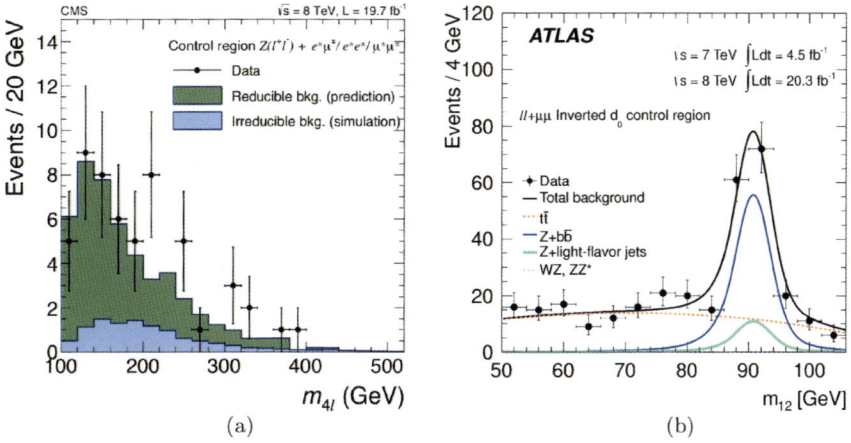

Fig. 5.10   (a) CMS: Predicted and observed "reducible" background in the control region with four leptons of wrong charge/flavour combinations and passing all "tight" selection requirements (validation of the method). (b) ATLAS: Observed m_{Z_2} mass distribution in one of the four $\ell\ell + \mu\mu$ control regions (see text). The fitted contributions from $t\bar{t}$, $Zb\bar{b}$, and $Z + $ jets are also shown.

For $\ell\ell + \mu\mu$, the processes considered were $t\bar{t}$, $Zb\bar{b}$, and $Z + $ jets (light quarks and gluons). Four control regions were defined and the fit was performed for the m_{Z_1} distribution in each of them. Figure 5.10(b) shows such a fit in a control region, where both muons from the Z_2 pair were not required to pass isolation cuts and at least one of them had to fail the impact parameter requirement. This control region was expected to be enriched with $t\bar{t}$ and $Zb\bar{b}$, as the fit results indeed confirmed.

For $\ell\ell + ee$, there was one control region with considerably relaxed requirements on the sub-leading p_T electron from the Z_2 pair. The processes considered were classified according to the origin of an object potentially faking the "loose" electron: $(\ell\ell + e)$ plus a light-flavour hadron, $(\ell\ell + e) + \gamma$, and $(\ell\ell + e) + b$. The three contributions were disentangled in a fit of events in the 2D-distribution of two observables: the number of track hits in the innermost layer of the pixel detector (electrons induced by conversions often originated past the innermost layers), and the ratio of the number of high-threshold and low-threshold hits in the Transition Radiation Tracker (for charged hadrons, this ratio tended to be low).

The simulation-based transfer factors for each process were validated/corrected in the analysis of the $Z(\ell^+\ell^-) + \ell$ data, again with selection adjustments enhancing different sources of non-prompt leptons. Results obtained in data agreed well with simulation for all processes with one exception: rates with which charged light hadrons were faking electrons had to be corrected by a p_T-dependent factor of ~ 2.

6.3. *Uncertainties*

Thanks to the narrow Higgs boson width, most uncertainties in the signal and background models could be effectively treated as uncertainties on the yield of four-lepton events with $m_{4\ell} \approx m_H$. The only important exception is the shape of the four-lepton mass distribution for the signal. Overall, the impact of systematic uncertainties on the final results obtained in the $H \to ZZ \to 4\ell$ channel in Run 1 was marginal.

6.3.1. *Normalisation uncertainties*

The normalisation uncertainties were assessed as a function of four-lepton mass and separately for each exclusive event category. Uncertainties in different exclusive categories arising from the same source were treated as correlated.

Theoretical uncertainties on the overall signal and irreducible background event yields were of the order of 10% and 5%, respectively. These uncertainties — assessed separately for each contributing process — were either taken explicitly from phenomenological papers or were obtained by following the commonly accepted phenomenological prescriptions: the PDF4LHC recommendation[27] for assessing PDF+α_s related uncertainties, varying QCD renormalisation and factorisation scales by a factor of two, and the Stewart–Tackmann prescription[28] for jet-based categorisations.

The main sources of instrumental uncertainties on the expected event yields for simulated samples (signal and irreducible background) were associated with lepton reconstruction/selection efficiencies. These efficiencies were evaluated using the "tag-and-probe" technique exercised on $Z \to \ell\ell$ and $J/\psi \to \ell\ell$ events in bins of the probe lepton's p_T and η. The finite statistics of events in each (p_T, η)-bin and the differences with respect to simulation were used to define the corresponding uncertainties. It is worthwhile noting that with four leptons in the final state, single-lepton efficiency uncertainties could have as large as a four-fold effect on the final event yields. The range of uncertainties on event rates was from 4% (4μ) to 10% ($4e$). The integrated luminosity uncertainty was less than 3%.

Systematic uncertainties on the "reducible" background yields, coming from the limited number of events in the control regions and from the imperfect knowledge of transfer factors, were in the 20−40% range.

6.3.2. *Shape uncertainties*

For a low-mass Higgs boson ($m_H < 300\,\text{GeV}$), the shape of the signal four-lepton mass distribution could be affected by uncertainties on the absolute electron/muon momentum scales and on the electron/muon momentum resolution. These uncertainties were particularly important in the measurements of the mass and width of the discovered boson and are described in detail in Sec. 8, where these measurements are presented.

For a high-mass Higgs boson ($m_H > 300\,\text{GeV}$), the observable four-lepton mass distribution was defined by the intrinsic Higgs boson mass shape, which was subject to substantial theoretical uncertainties. In the earlier ATLAS and CMS searches, the narrow-width approximation encoded in POWHEG was used. To account for the poor description of the four-lepton mass lineshape for a heavy Higgs boson by the narrow-width approximation, a large ad hoc systematic uncertainty[23] on signal event yield, parametrised as $150\% \times (m_H/\text{TeV})^3$, was used. In the more recent analyses, POWHEG samples were reweighted by using the predictions for the mass lineshape calculated in the complex-pole scheme[24] which also provided a more coherent treatment of the lineshape uncertainties.

Various sources of uncertainties on distribution shapes of other variables used in different analyses (e.g., four-lepton kinematic discriminants, VBF discriminants, etc.) were studied and found to have a negligible impact on the relevant results.

7. Observation of a new boson in the SM Higgs boson search

The SM Higgs boson was searched for in the mass range 110–1000 GeV. To quantify the search results, both ATLAS and CMS used the profile-likelihood-ratio test statistics, given by Eqs. (A.14) and (A.15) with unbinned likelihoods. A simplified view of the full likelihood is:

$$\mathcal{L}(\text{data}|b + \mu s) = \prod_k \mathcal{L}_k(\text{data}|b + \mu s), \qquad (5.3)$$

$$\text{with } \mathcal{L}_k(\text{data}|b + \mu s) = e^{-B_k - \mu S_k} \cdot \frac{(B_k + \mu S_k)^{N_k}}{N_k!} \cdot \prod_i \mathcal{F}_k(\mathcal{O}_i|b + \mu s),$$

where the index k enumerates different event categories, each with N_k observed events and the total expected background events B_k and signal events μS_k. The factor μ is a signal strength modifier common for all event categories. Index i runs over all events in each category, and $\mathcal{F}_k(\mathcal{O}|b + \mu s)$ is the probability density function (pdf) of a set of observables $\mathcal{O}$ for an event in category k under the "$b + \mu s$" hypothesis, i.e. the nominal expected background plus the SM Higgs boson signal, whose event yields were scaled by the factor μ.

In the search for the Higgs boson, ATLAS used 8 categories (two centre-of-mass energies times four final-state flavours). The categorisation by production mechanism was not used. As far as the observables $\mathcal{O}$ are concerned, the statistical analysis was performed in each of the 8 categories using 2D-pdf's: $\mathcal{F}(m_{4\ell}, \mathrm{BDT}_{ZZ^*})$. CMS used 12 event categories (two centre-of-mass energies, three final state flavours, two production tag categories) and 3D-pdf's in each category: $\mathcal{F}(m_{4\ell}, D_{\mathrm{bkg}}^{\mathrm{kin}}, D_{jet})$ for dijet-tagged events; $\mathcal{F}(m_{4\ell}, D_{\mathrm{bkg}}^{\mathrm{kin}}, p_{\mathrm{T}}^{4\ell})$ for untagged events.

In comparison to using four-lepton mass alone, adding the four-lepton kinematic discriminants against the ZZ background, BDT_{ZZ^*} (ATLAS) and $D_{\mathrm{bkg}}^{\mathrm{kin}}$ (CMS), helped increase the Higgs boson search sensitivity by as much as 20–30% in the explored range of possible Higgs boson masses.

Adding the dijet categorisation in combination with using observables sensitive to vector-boson-fusion (VBF) Higgs boson production improved the search sensitivity by as much as 40% for a 1-TeV Higgs boson. However, in the low mass range, the VBF event yield relative to gluon fusion was expected to be very small and the gain in the search sensitivity from including these observables was less than 2% near $m_H \sim$ 125 GeV. The dijet categorisation, of course, was important for probing processes responsible for production of the discovered boson, as described in Sec. 9.

The number of events in simulation and control regions for "reducible" backgrounds were sufficient to populate 2D distributions and, hence, build 2D-pdf's, but too low to build statistically accurate 3D-pdf's. Hence, the 3D probability density functions used by CMS were constructed in a factorised form as a product of a 2D-pdf $\mathcal{F}(m_{4\ell}, D_{\mathrm{bkg}}^{\mathrm{kin}})$ and conditional 1D-pdf $p(z \mid m_{4\ell})$, where z would stand for either D_{jet} or $p_{\mathrm{T}}^{4\ell}$. Using simulation, it was checked that neither D_{jet} nor $p_{\mathrm{T}}^{4\ell}$ correlated with the kinematic discriminant $D_{\mathrm{bkg}}^{\mathrm{kin}}$. This was expected since the kinematic discriminant depended on properties of four leptons in their centre-of-mass frame where there would be no information on jets or four-lepton system $p_{\mathrm{T}}^{4\ell}$ observed

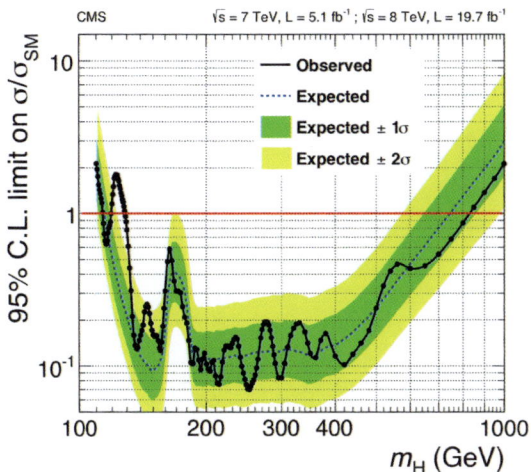

Fig. 5.11 Signal strength $\mu = \sigma_H/\sigma_H^{SM}$ excluded by CMS experiments at 95% CL as a function of the hypothesised Higgs boson mass m_H.

in the lab frame. The correlation, which might arise only from second-order acceptance effects, was found to be negligibly small.

Using the example of CMS results, Fig. 5.11 presents the 95% CL upper limits on the signal-strength modifier, $\mu = \sigma/\sigma_{SM}$, as a function of m_H. In the mass range where $\mu < 1$ (114.5–119 and 129.5–832 GeV), the SM Higgs boson was excluded at 95% CL.

The oscillations of the observed 95% CL upper limits around the median expected limits, assuming the background-only hypothesis, is a direct consequence of the high mass resolution, which results in observations for hypothesised Higgs bosons with nearby masses being statistically independent. The sensitivity to a heavier SM Higgs boson is reduced owing to the lower expected production rate in combination with the increasing natural width of the resonance. The steep loss of sensitivity at the lower end of the mass range is due to the diminishing branching fraction $\mathcal{B}(H \to ZZ^*)$ channel and reduced reconstruction and selection efficiency for low-p_T leptons. For $2\,m_W < m_H < 2\,m_Z$, $\mathcal{B}(H \to ZZ^*)$ is very small, which propagates into a considerable loss of sensitivity in this mass range as well.

In most of the explored Higgs boson mass range, the observed limits were generally within the 68% or 95% bands around the median expected values and, hence, statistically compatible with the background-only hypothesis. However, in the range $119\,\text{GeV} < m_H < 129.5\,\text{GeV}$, the limits were considerably weaker than expected in the absence of the SM Higgs boson.

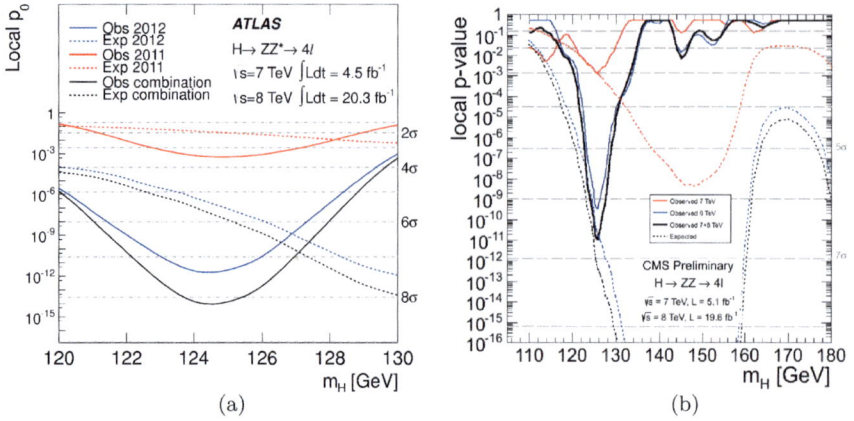

Fig. 5.12　Local p-value p_0 for (a) ATLAS and (b) CMS searches for the SM Higg boson as a function of the hypothesised Higgs boson mass.

Despite having reached the SM Higgs boson sensitivity in this low-mass range, neither experiment could exclude the SM Higgs boson there, due to the excesses of events observed near 125 GeV.

To quantify the inconsistency of the observed excesses near 125 GeV with the background-only hypothesis, Fig. 5.12 shows a scan of the local p-value as a function of m_H, as obtained by ATLAS and CMS experiments. The probabilities for background fluctuating near that mass at least as high as observed were $\sim 10^{-14}$ (ATLAS) and 5×10^{-12} (CMS), which corresponded to statistical significances of $8.2\,\sigma$ and $6.8\,\sigma$, respectively. The very low probability for the excesses to arise from statistical fluctuations of background implied that both ATLAS and CMS observed a new boson[f] with a mass near 125 GeV and decaying to four leptons.

The dashed lines in Fig. 5.12 indicate the expected significances at different values of m_H, should the SM Higgs boson at those masses have existed. ATLAS observed an excess larger than expected for a SM Higgs boson with a mass near 125 GeV, while CMS observed an excess with about the expected significance. The statistical compatibility of the observed excesses with the expectations for the SM Higgs boson are discussed in Sec. 9.

Figure 5.13 shows an event display of a representative four-electron Higgs boson candidate. No other high momentum tracks or high energy

---

[f]A particle decaying to even number of fermions must be a boson.

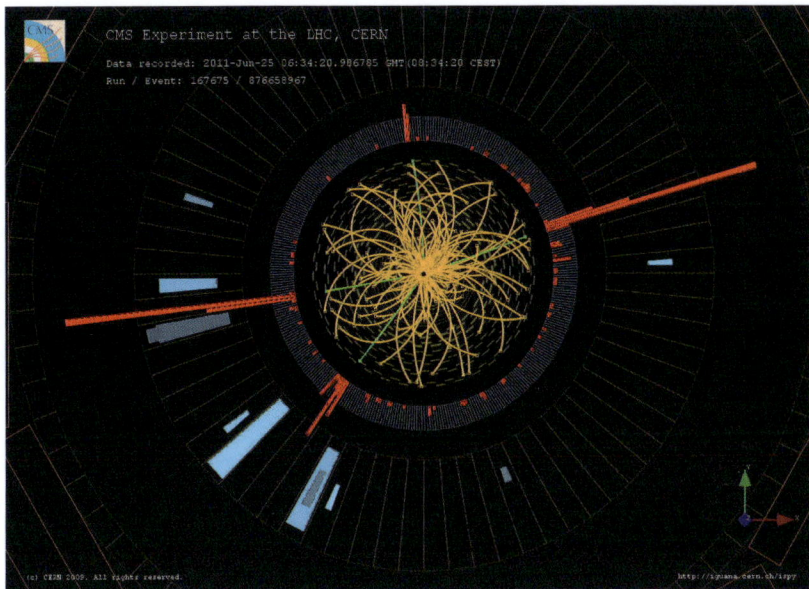

Fig. 5.13 Event display of a representative Higgs boson event candidate in the $H \to ZZ \to 4e$ decay mode observed by CMS. Tracks matched to the four pronounced electromagnetic clusters are highlighted in green. Invariant masses $m_{4\ell}$, m_{Z_1} and m_{Z_2} in this event are 125.7 GeV, 92.3 GeV and 27.2 GeV, respectively. All other tracks, highlighted in yellow, have a large curvature and correspond to low p_T charged particles. Amount of energy deposited in the hadron calorimeter (blue) is also small. No muon candidates are found in the event (muon detectors are outside of the figure box).

deposits in calorimeters are seen next to the electrons, which implies that all four electrons in the event are isolated.

8. Mass and total width of the new boson

8.1. *Mass*

The mass of the observed boson was measured from a m_H-scan of the unbinned negative log-likelihood ratio, $-2\Delta \ln \mathcal{L}(\text{data}|m_H)$. In this fit, the overall signal strength was allowed to float, while the relative event yields in the $4e$, 4μ, and $2e2\mu$ final states were assumed to be the same as for the SM Higgs boson.

With all nuisance parameters refit to maximise the likelihood at each m_H, the scan provided the total uncertainty of the measurement, as described in Appendix A. Statistical uncertainties on the measured masses were evaluated by performing the scan with all nuisance parameters, except for the overall signal strength, fixed at their best-fit values. The systematic uncertainty was evaluated as the difference in quadrature between the total and statistical uncertainties: $\sigma_{syst}^2 = \sigma_{tot}^2 - \sigma_{stat}^2$. The likelihood scans corresponding to the ATLAS, CMS, and combined ATLAS + CMS[12] data are shown in Fig. 5.14(a), while the numerical results obtained from these scans are presented in Table 5.6.

ATLAS performed the measurement using two observables: the four-lepton invariant mass $m_{4\ell}$ and the kinematic discriminant BDT_{ZZ^*} (see Sec. 5), whose role in the mass fit was to give more weight to events that were more $H \to ZZ$ signal-like than $q\bar{q} \to ZZ$ background-like. The

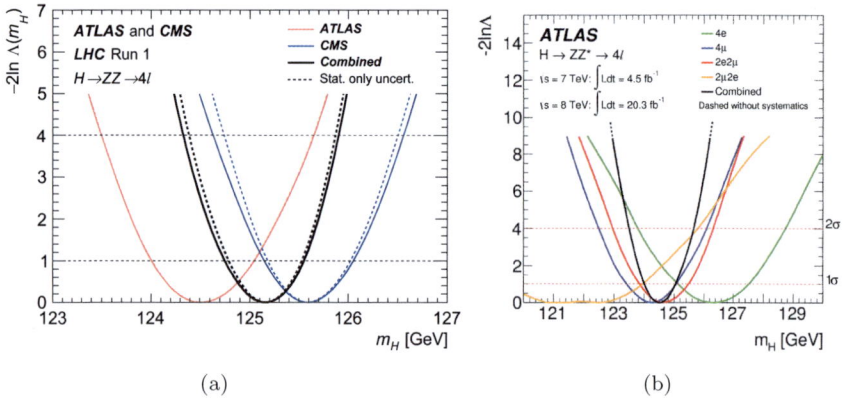

(a)   (b)

Fig. 5.14   (a) Scan of the negative log-likelihood ratio as a function of the hypothesised mass of the new boson obtained for the ATLAS (red), CMS (blue), and combined ATLAS + CMS (black) data. The solid (dashed) lines correspond to scans with (without) systematic uncertainties. (b) Scan of the negative log-likelihood ratio for the different four-lepton final states as a function of the hypothesised mass of the new boson, as obtained by ATLAS.

Table 5.6   Mass measurements for the observed boson by ATLAS, CMS, and ATLAS + CMS combination. The 95% CL limits on the intrinsic total width Γ_{tot} from the four-lepton resonance peak fit, as obtained by ATLAS and CMS, are also reported.

Dataset used	Measured mass	95% CL limit on Γ_{tot}
ATLAS	124.51 ± 0.52 (stat) ± 0.04 (syst) GeV	<2.6 GeV
CMS	125.59 ± 0.42 (stat) ± 0.17 (syst) GeV	<3.4 GeV
ATLAS + CMS	125.15 ± 0.37 (stat) ± 0.15 (syst) GeV	

use of the kinematic discriminant in the fit helped improve the statistical uncertainty of the mass measurement by about 4%. Event categorisation based on the production-specific tags was not used in the mass measurements.

Even though events with four leptons of different flavours were treated as separate categories, individual events in the same final state could have different mass resolutions varying by as much as a factor of three, depending on leptons' p_T, η, and the overall reconstruction quality of leptons in a given event. Therefore, in addition to the four-lepton mass and kinematic discriminant, CMS added to the mass fit a third observable: an estimated per-event four-lepton mass uncertainty, $\sigma_{m_{4\ell}}$, whose purpose was to give more weight to events that had smaller uncertainties on the measured four-lepton mass. Including the per-event four-lepton mass uncertainties improved the expected mass resolution by about 8%. It also allowed one to assign a more accurate uncertainty on the measured mass, given the four-lepton mass uncertainties for the observed events (instead of the expected average uncertainties). The latter was particularly important since the measurement was based on a small number of events. ATLAS used per-event four-lepton uncertainties in a cross-check analysis of the main mass measurement result and in the direct measurements of the width of the discovered boson.

The main systematic uncertainties affecting the mass measurements were associated with the absolute electron/muon momentum scale calibration affecting the position of the peak, and on the electron/muon momentum resolution affecting the peak width. The scale of such potential biases could be evaluated by analysing the reconstructed peak position and width of Z, J/ψ, and Υ resonances for events sorted in bins of lepton's (p_T, η). In CMS, the differences in resonance mass peak positions between data and simulation in different data subsets were well covered by conservative envelopes of $\pm 0.1\%$ and $\pm 0.3\%$ uncertainties on the muon and electron momentum scales, respectively. ATLAS produced a more detailed map of uncertainties, as a function of lepton's p_T and η (e.g., see Fig. 5.15(a)). Using the detailed maps of uncertainties, the net systematic uncertainty on the mass measurement of the Higgs boson in $4e$ and 4μ final states was assessed to be of the order of $\pm 0.04\%$. The data–simulation differences in the widths of the Z, J/ψ, and Υ peaks were found to be well covered by assigning 20% uncertainties on the simulated muon/electron momentum measurement resolutions. The impact of this uncertainty on the mass measurement was negligible.

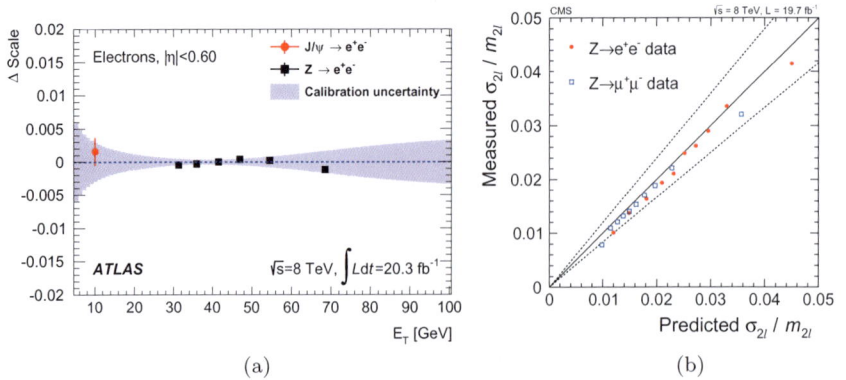

Fig. 5.15   (a) Relative difference between the measured and the nominal electron energy scales, as a function of E_T for J/ψ and Z events in $|\eta| < 0.6$. The uncertainty on the nominal energy scale is shown by the shaded band. (b) Measured-versus-predicted relative dilepton mass resolution for $Z \to e^+e^-/\mu^+\mu^-$ events in data. The dashed lines represent the $\pm 20\%$ systematic uncertainty assigned to predictions of per-event mass resolutions.

The ability to predict per-event mass resolutions was validated using $Z \to \ell\ell$ events, as shown in Fig. 5.15(b). It was found that a 20% uncertainty covers all observed differences between data and simulation. To use per-event four-lepton mass uncertainties, one had to extend the signal and background models to include appropriate $\mathcal{F}(\sigma_{m_{4\ell}}|m_{4\ell})$ pdf's. The $\mathcal{F}(\sigma_{m_{4\ell}}|m_{4\ell})$ distributions for Higgs boson and irreducible ZZ background were obtained from simulation and validated with data using $Z \to 4\ell$ events and $ZZ \to 4\ell$ events with $m_{4\ell} > 200\,\text{GeV}$, where the "reducible" background contribution was negligible. For the "reducible" background, $\mathcal{F}(\sigma_{m_{4\ell}}|m_{4\ell})$ was obtained directly from data using events from the control region with two "tight" leptons and two "loose-but-not-tight" leptons (see Sec. 6).

Table 5.6 shows that the ATLAS and CMS mass measurements were compatible, which was a necessary prerequisite for the combination of the two results. The reason why CMS result was statistically more accurate, despite of a fewer number of observed events in the peak, was due to a somewhat better average four-lepton mass resolution (see Table 5.3) and due to the use of per-event four-lepton mass uncertainties. In dedicated studies preceding the combination, the dominant systematic uncertainties on lepton momentum scales and resolutions were found to be uncorrelated between the two experiments.

Table 5.7 Measurements of the overall (μ) as well as fermionic and bosonic (μ_F and μ_V) signal strengths. The ATLAS results are evaluated at the ATLAS combined Higgs boson mass.[6] The CMS results are at the best-fit mass measured in the four-lepton channel alone.

	μ	μ_F	μ_V
ATLAS at $m_H = 125.4$ GeV	$1.44^{+0.34}_{-0.31}$(stat)$^{+0.21}_{-0.11}$(syst)	$1.7^{+0.5}_{-0.4}$	$0.3^{+1.6}_{-0.9}$
CMS at $m_H = 125.6$ GeV	$0.93^{+0.26}_{-0.23}$(stat)$^{+0.13}_{-0.09}$(syst)	$0.80^{+0.46}_{-0.36}$	$1.7^{+2.2}_{-2.1}$

8.2. *Total width*

The spread of events in the four-lepton mass peak serves as a direct, *model-independent*, probe of the total intrinsic width Γ_{tot} of the observed resonance. Although the expected sensitivity, $\delta\Gamma_{tot} \sim \mathcal{O}(1)$ GeV, was by far larger than the expected SM Higgs boson width of 4.2 MeV, the measurement was important in the context of models beyond the Standard Model. For example, an observation of a non-zero width might imply the existence of more than one Higgs boson with a mass split comparable to the instrumental resolution. Both ATLAS and CMS observed that the excess near 125 GeV was consistent with the hypothesis of a single boson of a small total width. The obtained 95% CL upper bounds on the total width are shown in Table 5.6. Limits on the total width of the observed boson obtained in studies of the high-mass four-lepton events, where the observed boson was expected to contribute via its far off-shell production, are discussed in Sec. 12.

8.3. *Cross-checks*

By exploiting the independence of the electron and muon momentum scales one could perform a cross-check of the mass measurement results by comparing measurements in the different final states. Figure 5.14(b) shows the ATLAS mass likelihood scans for the four channels used in the analysis. The results obtained by ATLAS and CMS in different channels did not display any statistically significant biases.

In addition, the four-lepton mass peak near 91 GeV (see Fig. 5.7) arising from rare $Z \rightarrow 4\ell$ decays[29,30] (see Fig. 5.3(b)) could be used as a "standard candle" for a direct validation of the mass and width reconstruction of the newly discovered boson. The Z boson mass and width measured by CMS using $Z \rightarrow 4\ell$ events were 91.16 ± 0.23 GeV and $2.98^{+0.54}_{-0.50}$ GeV, respectively, in a good agreement with the world average values.[31] Note that the Z boson mass measurement uncertainty was lower than the uncertainty on the Higgs boson mass.

9. Signal strength

The signal strength modifier, μ, acts as a common scale factor on the number of events predicted by the SM for each Higgs boson signal process, or, equivalently, it scales the SM predicted cross section times branching fraction, $\sigma_{SM} \cdot \mathcal{B}_{SM}$. The signal strength values measured by ATLAS and CMS are shown in Table 5.7. The ATLAS and CMS results were statistically compatible with each other and with the expectations for the SM Higgs boson, i.e. with $\mu = 1$.

When comparing the obtained μ values, one needs to keep in mind the assumed Higgs boson masses at which they were evaluated. The *expected* signal event yield for the SM $H \rightarrow ZZ \rightarrow 4\ell$ channel increases at a rate of 7.2%/GeV as a function of the assumed Higgs boson mass near 125 GeV.

These measurements were extended to probe signal strength factors for specific production modes. In this analysis, the production mechanisms were grouped into "fermionic" and "bosonic". The "fermionic" group consisted of the gluon–gluon fusion, $t\bar{t}H$, and $b\bar{b}H$ modes. The "bosonic" group comprised VBF, WH, and ZH production mechanisms. Respective signal strength modifiers, μ_F and μ_V, were introduced to scale the expected SM Higgs boson event yields in each group and then fitted to the data in the production-tagged categories. The fit results are presented in Fig. 5.16. The best-fit values for (μ_F, μ_V), shown in Table 5.7, were found to be consistent with the SM Higgs boson expectation, i.e. $(\mu_F, \mu_V) = (1, 1)$.

Since the line $\mu_F = 0$ lies outside the 95% CL contour, the measurement established a non-zero fermionic coupling for the observed boson in the context of the tested two-parameter model. With the current amount of data, the $H \rightarrow ZZ \rightarrow 4\ell$ analysis did not yet reach sufficient sensitivity to establish explicitly the presence of VBF $+ VH$ production. As seen from Fig. 5.16, for $\mu_V = 0$ there was a range of μ_F values statistically compatible with the data.[g] However, the observed decays to the ZZ final state imply a non-zero coupling of the discovered boson to Z bosons.

---

[g]Technically, pdfs used in the construction of unbinned likelihoods to describe the overall event probability density are checked to be positive definite, where data events are observed. This effectively limits the magnitude of possible negative values of μ_V for a

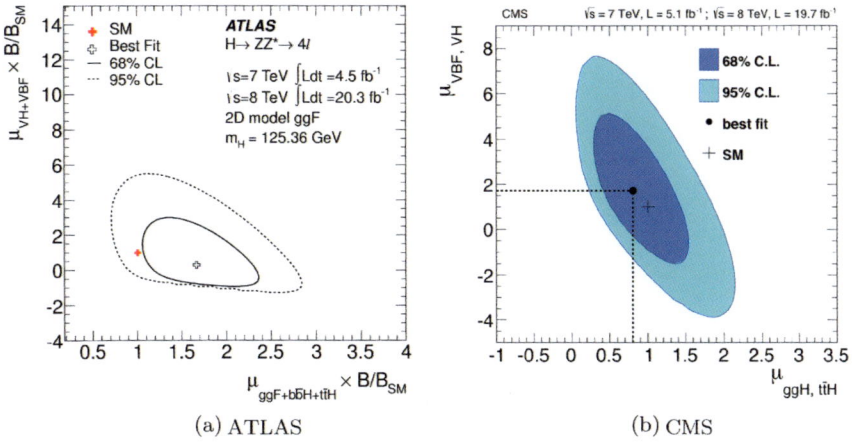

(a) ATLAS   (b) CMS

Fig. 5.16   Simultaneous fit for signal strengths μ_F (x-axis) and μ_V (y-axis) by (a) ATLAS and (b) CMS. The best-fit values and the 68% (95%) CL contours are shown. The point (1,1), indicated by +, corresponds to the SM Higgs boson expectations.

10. Fiducial total and differential cross sections

The extrapolation of event rates, measured within the detector acceptance, towards a total cross section can vary by a large factor, depending on the assumed production mechanism (e.g. by a factor of two for $gg \to H$ vs $q\bar{q} \to H$). Also, some signal selection efficiencies have a strong model-dependence (e.g. the lepton isolation efficiency is about 40% smaller for $gg \to t\bar{t}H$ in comparison to $gg \to H$). To minimise model-dependence of the obtained results, measurements of inclusive and differential cross sections[32,33] were performed within a carefully selected fiducial volume, defined at the generator level, for which theoretical predictions could be made. The fiducial volume was closely matched to the four-lepton event selection criteria summarised in Table 5.1. ATLAS did not include the lepton isolation in the fiducial volume definition, while CMS did. Hence, ATLAS results are valid for models with jet activity similar to that expected for the SM Higgs boson production. CMS results do not have such a limitation.

The inclusive cross section for a resonance to produce four leptons at 8 TeV within the fiducial volume defined by ATLAS was measured to be

---

given μ_F and vice versa. The clipping of the ATLAS contours at the bottom was due to an event with a relatively high VBF-like purity.

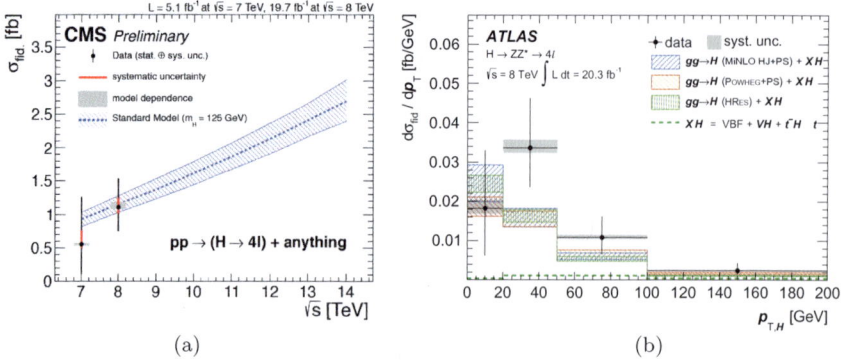

Fig. 5.17 (a) Fiducial cross sections measured at 7 and 8 TeV and theoretical predictions in the 7–14 TeV range. (b) Differential fiducial cross section for the transverse momentum $p_{T,H}$ of the observed boson. Several theoretical calculations are presented, along with their uncertainties, as hatched blocks.

$2.11^{+0.53}_{-0.47}$(stat) ± 0.08(syst) fb, while the expected value for the SM Higgs boson was 1.30 ± 0.13 fb. The CMS results, obtained in the tighter fiducial volume, are shown in Fig. 5.17(a). The measured fiducial cross section at 8 TeV was $1.11^{+0.41}_{-0.35}$(stat)$^{+0.14}_{-0.10}$(syst) fb, with the expectation for the SM Higgs boson being 1.15 ± 0.13 fb.

The differential cross section measurements were performed in several production-related observables, including: Higgs boson's transverse momentum and rapidity, associated jet multiplicity, transverse momentum of the leading jet, etc. Differential cross sections related to decay-related observables were also studied. As an example, Fig. 5.17(b) shows the differential cross section measurements for the Higgs boson's transverse momentum, $p_{T,H}$. The results were found to be consistent with predictions, while the statistical precision of the measurements was not yet accurate enough to distinguish between different theoretical calculations for the SM Higgs boson.

11. Determination of the spin-parity quantum numbers

After the discovery of the 125 GeV boson and establishing its mass and signal strength with respect to the expectations for the SM Higgs boson of that mass, it was important to test the compatibility of its spin-parity quantum numbers with those expected for the SM Higgs boson ($J^P = 0^+$). The charge conjugation parity of the observed boson was set to C = +1 by the observation of $\gamma\gamma$ decays, which also excluded J = 1 by the Landau–Yang theorem, as described in Chapter 1, Sec. 9. A boson with generic spin-parity quantum numbers is henceforth denoted as X.

Before presenting the spin-parity analyses carried out with the four-lepton final state, a brief phenomenological preamble is given in Sec. 11.1. It is followed by Sec. 11.2 describing results of pair-wise tests of data compatibility with the SM Higgs boson versus various alternative (exotic) boson models. Analyses aiming to detect admixtures of anomalous non-SM-like amplitudes in the four-lepton decays of the observed boson are described in Sec. 11.3. The presented results are based on Refs. [4,5,9,13].

11.1. *Phenomenological considerations*

The $X \to ZZ \to 4\ell$ and $X \to WW \to 2\ell 2\nu$ four-body decay modes are well suited for probing all possible J^P spin-parity states, albeit with different levels of sensitivities. The diphoton decays cannot be used for probing alternative J = 0 amplitudes. Two photons emerging from a spin-zero state decays are back-to-back in the boson's rest frame, with the diphoton axis direction uniformly distributed in stereo angle, and, unless one can measure the two photons' polarisations, provide no handles for studying the different possible decay amplitudes that can be associated with a J = 0 state. For a boson with J = 2, distribution of the diphoton axis direction with respect to the beam line, however, is not uniform and depends on the underlying structures in both production and decay amplitudes.

11.1.1. *Spin zero*

The Effective Field Theory (EFT) Lagrangian up to dimension-five operators for a spin-zero state decaying to two vector bosons, $X \to VV$, is given by Eq. (1.22) in Chapter 1, Sec. 9. The first dimension-three operator corresponds to the SM Higgs boson. The other three α-, β-, γ-terms are dimension-five operators and can be thought of as effective operators for loop-induced decays. In this convention the "effective couplings" α, β, γ absorb actual couplings, mass scales of particles in the loops (relative to the vacuum expectation value of the Higgs field), loop factors, etc. The α- and β-terms are even-parity scalars, kinematically distinguishable from the SM Higgs boson, while the γ-term corresponds to a pseudoscalar. In fact, SM particles contribute to all three terms, but at a very small level:[15,34]

$\alpha \sim \beta \sim \mathcal{O}(\alpha_{\mathrm{EW}}) \sim 10^{-2}$ and $\gamma \sim \mathcal{O}(10^{-11})$. In SM, it takes at least three loops to generate a pseudoscalar-like term.

11.1.2. *Spin one*

The spin-one hypothesis was studied using $X \to ZZ$ and/or $X \to WW$ decays. One can consider those studies either as a test carried out independently of the observation of diphoton decays, prohibiting spin-one states, or as a test for a possible conspiracy of multiple nearly mass-degenerate states with different quantum numbers. In the case of the spin-one resonance, there are two distinct decay amplitudes corresponding to a vector (1^-) and a pseudovector (1^+),[35] both of which correspond to dimension-four operators. Note that production of spin-one state via gluon fusion must be strongly suppressed; for on-shell gluons, production amplitude $\mathcal{A}(gg \to X)$ must be zero by the very same Landau–Yang theorem that forbids decays to diphoton final states.

11.1.3. *Spin two*

As mentioned in Chapter 1, there is not a self-consistent quantum field theory of elementary spin-two massive states. Moreover, within the EFT framework, one can write out a large number of XVV and Xff Lagrangian terms, associated with kinematically distinguishable spin-two states. All this ensures complications for experimental analyses aiming to assess the relative odds of the observed boson being an exotic — perhaps, composite — spin-two state or a SM-like Higgs boson.

For a colour-, weak- and electromagnetic-singlet spin-two resonance (henceforth denoted as 2_m^+, following notations introduced in Ref. [15]), the Lagrangian is unique:[36]

$$\mathcal{L} \sim -\frac{1}{\Lambda} \sum_i \kappa_i T^i_{\mu\nu} X^{\mu\nu}, \qquad (5.4)$$

where i runs over all SM particles and $T^i_{\mu\nu}$ is the energy-momentum tensor of particle i. This Lagrangian can be associated with the RS graviton,[37] in which case couplings κ_i to all SM particles are universal. However, the premise of universal couplings is obviously in a strong contradiction with the observed relative decay rates of the discovered boson (see Chapter 12).

Therefore, one is compelled to consider a 2_m^+ state with non-universal couplings. This, however, creates new problems. First, distributions of the production-related observables would strongly depend on the assumed relative couplings of X to quarks and gluons (assuming that the dominant production mechanism is $q\bar{q}/gg \to X$). Second, should the quark and gluon couplings be different, the X boson would acquire unitarity-violating behaviour manifesting itself in substantial X-boson's p_T-boosts at NLO.[36]

On the other hand, one can assume that the unitarity problems are resolved by yet-to-be-discovered new physics at higher energy scales and use the abnormal $p_T(X)$ distributions as yet another observable discriminating between 2_m^+ and the SM Higgs boson for the case when $\kappa_q \neq \kappa_g$. The observed $p_T(X)$ distributions in the $\gamma\gamma$ and ZZ decay modes (Sec. 6.5 in Chapter 6, and Sec. 10 in this chapter, respectively), constrain the allowed range of non-universality to $0 \leq \kappa_q \lesssim 2\kappa_g$.[13] Finally, one should beware that, if a spin-two boson is produced in association with jets, its production-related observables are considerably modified with respect to those at LO (the effect does not exist for $J = 0$ and is expected to be negligible for $J = 1$ bosons).[36]

As stated at the beginning of this sub-section, from the EFT stand point, one can write many more Lagrangian terms associated with XVV vertex. Ref. [15] defines 10 such terms, going up to dimension seven. In this context, the 2_m^+ state is a combination of two out of these terms, the lowest dimension of which is denoted as 2_b^+. Notations for the alternative eight possibilities, consisting of dimension-five and dimension-seven operators, are: 2_h^+, 2_h^-, 2_{h2}^+, 2_{h3}^+, 2_{h6}^+, 2_{h7}^-, 2_{h9}^-, 2_{h10}^-.[9]

11.2. *Pair-wise tests: SM Higgs vs. alternative* J^P *states*

To test alternative J^P signal (X) hypotheses against the SM Higgs boson (H), both ATLAS and CMS opted for matrix-element discriminants. Cross-checks using multivariate-observables trained using Boosted Decision Trees were also performed. The event categorisations by production mechanism were not used.

Probability density functions for an event to come from a given process could be written as follows:

$$\mathcal{F}(\mathcal{P}|\text{process}) \sim |\mathcal{A}_{\text{process}}(\mathcal{P})|^2 \cdot \epsilon(\mathcal{P}) \cdot \mathcal{F}(m_{4\ell}|\text{process}). \qquad (5.5)$$

In this equation, $\mathcal{P}$ stands for momenta of four leptons in the final state, $\mathcal{A}_{\text{process}}(\mathcal{P})$ is the LO matrix elements calculated in the same manner as described in Sec. 5.2, $\epsilon(\mathcal{P})$ is the four-lepton reconstruction efficiency, $\mathcal{F}(m_{4\ell}|\text{process})$ is the four-lepton mass probability distribution function.[h]

With three main processes relevant to this analysis (SM Higgs boson H, exotic boson X, and background dominated by $q\bar{q} \to ZZ$), there were only two independent ratios that could be formed; the two actually used were:

$$d_X = \frac{\mathcal{F}(\mathcal{P}|X)}{\mathcal{F}(\mathcal{P}|H)} = \frac{|\mathcal{A}_X|^2}{|\mathcal{A}_H|^2},$$ (5.6)

$$d_{\text{bkg}} = \frac{\mathcal{F}(\mathcal{P}|\text{bkg})}{\mathcal{F}(\mathcal{P}|H)} = \frac{|\mathcal{A}_{ZZ}|^2}{|\mathcal{A}_H|^2} \cdot \frac{\mathcal{F}(m_{4\ell}|\text{bkg})}{\mathcal{F}(m_{4\ell}|H)}.$$ (5.7)

The first one separated the alternative signal hypothesis from the SM Higgs boson, while the second discriminated against backgrounds. In these ratios the lepton reconstruction efficiencies, to a good approximation, cancel out. For convenience, the discriminants were monotonically transformed to be constrained between 0 and 1, and the analyses were performed by building likelihoods of the observed events in the space of the two transformed observables, D_X and D_{bkg}.

By using the 2D probability density functions $pdf(D_{\text{bkg}}, D_X)$ and taking into account the expected event yields for backgrounds, the observed test statistic values were calculated as follows:

$$q = -2\ln \frac{\mathcal{L}_{max}(\text{data}|\hat{\mu}_X \cdot X + \text{bkg}, \hat{\theta}_X)}{\mathcal{L}_{max}(\text{data}|\hat{\mu}_H \cdot H + \text{bkg}, \hat{\theta}_H)}.$$ (5.8)

The H and X signal event yields were not constrained by any external assumptions and treated on par with all other nuisance parameters (θ) in the fits maximising the likelihoods in the numerator and denominator.

The observed values of test statistic, q^{obs}, were compared to the expectations obtained by simulating pseudo-observations generated using the same pdfs. Numbers of signal events in pseudo-experiments were drawn from the Poisson distributions with the best-fit rates as obtained in the data. Figure 5.18 shows ATLAS and CMS results of testing the pseudoscalar against the SM Higgs boson hypotheses. The expected test statistic distributions for the two hypotheses were well separated, indicating that both experiments had reached fair sensitivity to distinguish between pure 0^- and SM-like Higgs boson states. The observed test statistic values were in the core of the distributions expected for the SM Higgs boson (and, hence, consistent with

---

[h]ATLAS used events only in a narrow mass-window around the peak and did not include $\mathcal{F}(m_{4\ell}|\text{process})$ in their spin-parity studies.

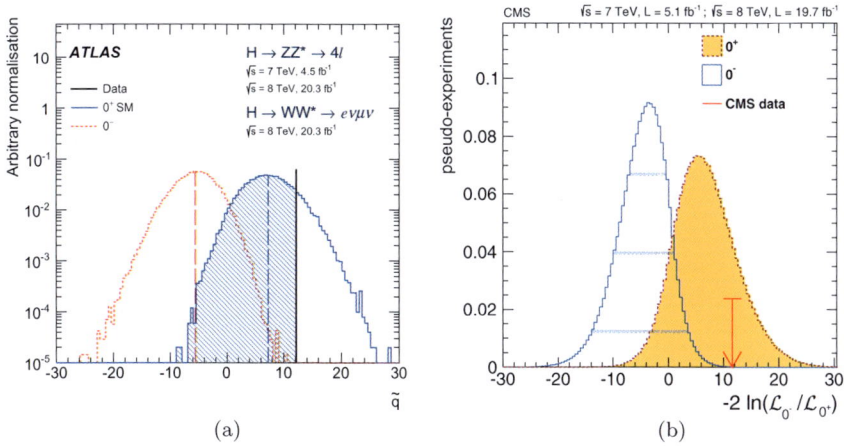

Fig. 5.18 Test statistic distributions for the SM Higgs boson and the pseudoscalar hypotheses for (a) ATLAS and (b) CMS. The observed values of the test statistic are also shown.

the SM Higgs boson hypothesis) and were far in the tails of the distributions expected for a pure pseudoscalar, thus strongly disfavouring this hypothesis.

Quantitatively, deviations of q^{obs} from the expected medians were characterized by p-values, converted then into a number of standard deviations. On the example of ATLAS, the deviation of observed q^{obs} from the median expected for the SM Higgs boson had a p-value of 88% (area of the blue histogram on the left side from the observation), which would correspond to -1.2 standard deviations ($\Delta_q = -1.2\sigma$). Following the statistical convention, deviations Δ_q from medians away from an alternative hypothesis distribution were counted as negative. The p-value for the pseudoscalar hypothesis (area of the red histogram on the right side from the observation) comprised 1.2×10^{-5}, corresponding to $\Delta_q = +4.2\sigma$. The CL_s values for ATLAS and CMS were 0.01% and 0.05%, respectively, which implied that the hypothesis of a psedoscalar was excluded by each experiment with more than 99.9% CL. ATLAS and CMS limits were somewhat tighter than expected due to the "lucky" statistical fluctuations of data in both experiments: the observed value of the test statistic was on the right from the median value expected for the SM Higgs boson, which made the observations particularly unlikely for a pseudoscalar.

Table 5.8 shows the ATLAS and CMS results of all alternative J^P states tested by both collaborations. CMS also tested nine other, more

Table 5.8   Results of testing alternative J^P state hypotheses against the SM Higgs boson (H). See text for explanations.

J^P	$xx \to X$	ATLAS $\Delta_q(H)$	$\Delta_q(J^P)$	CL_s	CMS (any production) $\Delta_q(H)$	$\Delta_q(J^P)$	CL_s
0^-		-1.2σ	$+4.2\sigma$	0.01%	-1.0σ	$+3.8\sigma$	0.05%
0_h^+		-0.8σ	$+3.4\sigma$	0.18%	-0.3σ	$+2.1\sigma$	4.5%
1^-	$q\bar{q} \to 1^-$	$+1.0\sigma$	$+1.6\sigma$	6.0%	-2.0σ	$> 5.0\sigma$	<0.001%
1^+	$q\bar{q} \to 1^+$	-0.1σ	$+3.1\sigma$	0.2%	-2.3σ	$> 5.0\sigma$	<0.001%
2_m^+	$(q\bar{q}, gg) \to 2_m^+$	-0.4σ	$+2.7\sigma$	0.97%	-1.6σ	$+3.4\sigma$	0.71%

exotic, amplitude tensor structures for spin-two boson decays $X \to VV$, corresponding to higher-dimensional operators (2_b^+, 2_h^+, 2_h^-, 2_{h2}^+, 2_{h3}^+, 2_{h6}^+, 2_{h7}^+, 2_{h9}^-, 2_{h10}^-). All tested exotic boson models were excluded at 99% CL or higher, while the data agreed well with the SM Higgs boson in each of the nine tests.

Note that spin-zero hypothesis tests are not sensitive to the mechanism responsible for the production of the observed boson, as a spin-zero state has no information on its production history in its centre-of-mass frame. For other spins, kinematical properties of decay products are not completely decoupled from the particles participating in the production.

ATLAS analyses assumed specific production mechanisms in their spin-one and spin-two tests. A spin-one particle cannot be produced via fusion of on-shell gluons (see Sec. 11.1); hence, the spin-one tests were performed assuming quark–antiquark annihilation. For the spin-two case, ATLAS probed various mixtures of quark–antiquark annihilation and gluon–gluon fusion; the most conservative limit among all of the tested mixtures is shown in the table.

CMS made their analyses nearly independent of an assumed production mechanism by integrating matrix elements for $xx \to X \to 4\ell$ over degrees of freedom connecting final state leptons and particles participating in production of X. This is equivalent to using $1 \to 4$ matrix elements $\mathcal{A}(X \to 4\ell)$ instead of $2 \to 1 \to 4$ matrix elements $\mathcal{A}(xx \to X \to 4\ell)$. The loss of sensitivity from such a generalisation was typically less than 10%.

11.3. *Search for presence of anomalous decay amplitudes*

In addition to excluding hypotheses of pure non-SM-like states, ATLAS and CMS probed the phenomenological possibility of a SM-like Higgs boson

Table 5.9   Obtained 95% CL intervals on the *allowed* couplings of alternative, not SM-like, spin-zero states with respect to those of the SM scalar state.

	α/κ	β/κ	γ/κ
ATLAS	not tested	$[-2.45, 0.75]$	$[-0.95, 2.85]$
CMS	$[-1.2, 1.5]$	$[-\infty, 0.69]$ $[1.9, 2.3]$	$[-2.2, 2.1]$

being mixed with α-, β-, and γ-terms in the spin-zero Lagrangian, given by Eq. (1.22). Testing for an admixture of a pseudoscalar (γ-term) was and will remain of a particular interest: should it be observed, it would open one more portal for CP-violating processes.

The analysis was performed in a variety of technical ways (all giving compatible results), one of which was identical to what was described above, with the only difference that in the discriminant given by Eq. (5.6), X would stand for a mixed state. Since absolute cross sections were not used in the discrimination (as in the case of pure states), the only relevant variables in these analyses were ratios of couplings. Table 5.9 summarises the allowed 95% CL intervals for alternative spin-zero state admixtures.

12. Total width determination via off-shell production

The total width for a SM Higgs boson with 125 GeV mass, $\Gamma_{\text{SM}} = 4.2$ MeV, is approximately three orders of magnitude below the sensitivity of the direct measurements described in Sec. 8. As discussed in Chapter 1, the yield and properties of high-mass ZZ events are sensitive to the width of the 125-GeV boson, with an important caveat that the actual numerical relationship is model-dependent. Following these considerations, both ATLAS and CMS pursued studies of high-mass ZZ events.[10,11] Events with $m_{4\ell} > 220$ GeV were used in both analyses. In this section, X refers to a resonance of 125 GeV with spin-parity quantum numbers identical to the SM Higgs boson and total width $\Gamma \neq \Gamma_{\text{SM}}$.

To improve the measurement sensitivity, ATLAS and CMS introduced ME-based discriminants. With multiple distinct underlying physics processes involved ($q\bar{q} \to ZZ$ (Fig. 5.3(a)), $gg \to$ "box" $\to ZZ$ (Fig. 5.3(c)), $gg \to H \to ZZ$, $gg \to X \to ZZ$, and the signal–background interference between $gg \to$ "box" $\to ZZ$ and $gg \to H/X \to ZZ$, it was not surprising that discriminant implementations chosen by the two experiments were somewhat different.

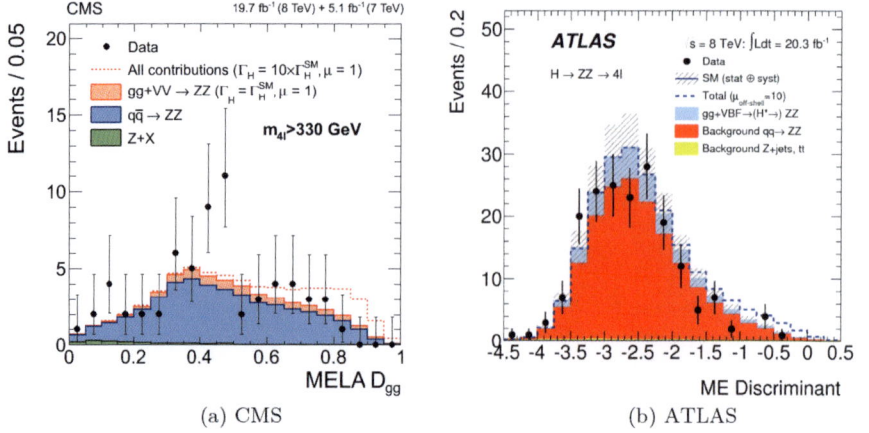

Fig. 5.19   Distributions of events for matrix-element-based observables used by (a) CMS and (b) ATLAS in the analyses probing the amount of off-shell Higgs boson production. The observables shown are monotonic transformations of the primary discriminants given by Eqs. (5.9) and (5.10). The CMS and ATLAS distributions are shown for $m_{4\ell} > 330$ and >220 GeV, respectively.

CMS used the following discriminant:

$$d_{gg} = \frac{|\mathcal{A}(gg \to (X_{10}/\text{``box''}) \to ZZ)|^2}{|\mathcal{A}(q\bar{q} \to ZZ)|^2}. \tag{5.9}$$

Here, X_{10} denotes a boson with width $\Gamma_{10} = 10\,\Gamma_{SM}$. The factor of 10 was picked based on the expected experimental sensitivity to the width with the Run 1 data set. By construction, this discriminant helped separate the gg-induced processes from the $q\bar{q} \to ZZ$ background. Moreover, its shape was also sensitive to the boson's width Γ, as can be seen in Fig. 5.19(a). For models with an X boson of a different width, this discriminant was somewhat sub-optimal. To get the maximum sensitivity, CMS performed its analysis in a 2D plane of two observables: four-lepton mass and the ME-based discriminant.

ATLAS defined its discriminant as follows:

$$d = \frac{|\mathcal{A}(gg \to H \to ZZ)|^2}{|\mathcal{A}(gg \to (H/\text{``box''}) \to ZZ)|^2 + c \cdot |\mathcal{A}(q\bar{q} \to ZZ)|^2}, \tag{5.10}$$

where c is an empirical constant, chosen to be 0.1, to approximately balance the two contributions in the denominator. The analysis was performed using the ME discriminant alone with a simple cut on four-lepton mass $m_{4\ell} > 220$ GeV. As seen from Fig. 5.19(b), in addition to the overall event yield,

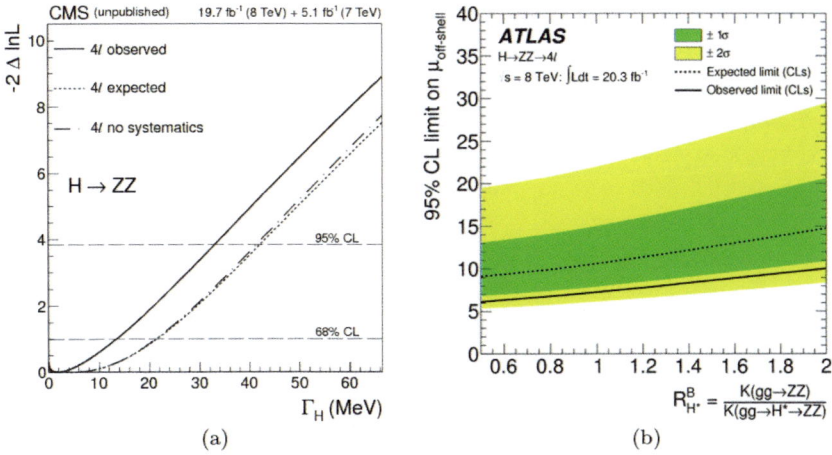

Fig. 5.20   (a) Likelihood scan as a function of the total width of the observed boson from CMS. The solid line corresponds to the observation, the dashed line to the expectation, while the dash-dotted line is the expectation neglecting systematic uncertainties. (b) The observed and expected 95% CL$_s$ upper limits on the off-shell signal strength, $\mu_{\text{off-shell}}$, as a function of R^B_{H*} (see text) from ATLAS.

the distribution of the ATLAS ME discriminant had a clear sensitivity to the width of the underlying boson.

The CMS scan of likelihood $\mathcal{L}(data|\Gamma)$, shown in Fig. 5.20(a), reveals that the best-fit width was close to zero and the upper bound on the total width could be set at $8.0 \times \Gamma_{\text{SM}}$, or 33 MeV, at 95% CL. When the $H \rightarrow ZZ \rightarrow 4\ell$ analysis was combined with an analysis assessing the off-shell signal event rate in the $ZZ \rightarrow 2\ell2\nu$ channel,[11] the limit at 95% CL on the width of the discovered boson became $5.4 \times \Gamma_{\text{SM}}$, or 22 MeV.

Figure 5.20(b) shows the expected and observed results obtained by ATLAS in the 4ℓ analysis for the off-shell signal strength, $\mu_{\text{off-shell}}$, a parameter defined as the ratio of the number of off-shell signal events to the number expected for the SM Higgs boson. Interference with $gg \rightarrow$ "box" $\rightarrow ZZ$ plays no role in the definition of $\mu_{\text{off-shell}}$. The figure explicitly shows that the inferred limit on $\mu_{\text{off-shell}}$ has a non-negligible dependence on the assumed K-factor for $gg \rightarrow$ "box" $\rightarrow ZZ$ background, expressed in units of the K-factor for $gg \rightarrow H^*$: $R^B_{H*} = K(gg \rightarrow$ "box" $\rightarrow ZZ)/K(gg \rightarrow H^*)$.[i] To reinterpret the obtained results in the context of the observed boson's

---

[i]The CMS results were obtained in assumption of $R^B_{H*} = 1$ with a 10% systematic uncertainty.

width, limits on $\mu_{\text{off-shell}}$ need to be combined with measurements of the on-shell signal strength $\mu_{\text{on-shell}}$. ATLAS did this exercise in the combination of the $H \to ZZ \to 4\ell$, $ZZ \to 2\ell 2\nu$, and $WW \to e\nu\mu\nu$ analyses.[11] Assuming $R_{H^*}^B = 1$, the obtained 95% CL limit was $\Gamma < 5.5 \times \Gamma_{\text{SM}}$, or 23 MeV.

13. Summary and outlook

The ATLAS and CMS experiments searched for a Standard Model Higgs boson with a mass in the range 110–1000 GeV and decaying to the $ZZ \to 4\ell$ final state. Both experiments observed a narrow four-lepton resonance with a mass near 125 GeV with local statistical significances of 8.2σ and 6.8σ, respectively. This unambiguously established the existence of a new boson decaying to four leptons. The mass of the discovered boson was measured to be 124.5 ± 0.5 GeV (ATLAS), 125.6 ± 0.5 GeV (CMS), giving a combined mass measurement of 125.15 ± 0.37 (stat) ± 0.15 (syst) GeV. The best-fit signal strengths with respect to the expected event yield for the SM Higgs boson were $1.44^{+0.40}_{-0.33}$ (ATLAS) and $0.93^{+0.29}_{-0.25}$ (CMS). The Run 1 dataset was not yet sufficient to establish the presence of sub-leading production mechanisms such as VBF, VH, and $t\bar{t}H$. Inclusive differential distributions for a number of observables, such as transverse momentum and rapidity of the discovered boson, number of jets produced in association with the boson and leading jet E_T, were found to be consistent with those expected for the SM Higgs boson, albeit with limited statistical accuracy. The angular distributions of leptons were statistically consistent with the new boson having spin-parity quantum numbers $J^P = 0^+$, as expected for the SM Higgs boson, and were inconsistent at 95% CL or higher with alternative J^P hypotheses, which included pseudoscalar, vector, pseudovector, and ten distinct spin-two tensor hypotheses. By comparing the on-shell event yield in the four-lepton final state to the ZZ and WW event rates at high masses, where the new boson would contribute via off-shell production, ATLAS and CMS set model-dependent upper limits on the total Breit–Wigner width of the resonance at 23 and 22 MeV at 95% CL. Within the experimental uncertainties, the ATLAS and CMS measurements were compatible with each other and, also, consistent with the expectations for the SM Higgs boson.

Over the next two decades, LHC is expected to deliver about 3000 fb^{-1} at a centre-of-mass energy near 14 TeV. The production cross section for the 125 GeV Standard Model Higgs boson at $\sqrt{s} = 14$ TeV increases by a factor of 2.6 with respect to 8 TeV. Therefore, the ultimate Higgs boson dataset will be about 300 times larger than the Run 1 dataset, which will allow for much more precise measurements than those described in this chapter and, also, provide opportunities for conducting completely new searches and measurements. The statistical uncertainty on the mass measurement will decrease to $(0.4\,\text{GeV})/\sqrt{300} \sim 25$ MeV, or 0.02%. Reducing the corresponding systematic uncertainties to levels much lower than 0.02% will be challenging. Fiducial cross sections $\sigma(pp \to H \to ZZ \to 4\ell)$ can potentially be measured with about $(30\%)/\sqrt{300} \sim 2\%$ statistical precision, challenging the accuracy of theoretical predictions, while differential measurements will provide a wealth of information on the dynamics associated with the Higgs boson production. All main Higgs boson production modes should become detectable and the prevailing ones will be studied in detail. With such a large dataset, the expected 2σ-sensitivity of a search for an admixture of a CP-odd component in the HZZ coupling is estimated to be about $\gamma/\kappa \sim 0.5$[17] and ~ 0.25.[38] The decay mode $H \to \gamma^*\gamma^* \to 4\ell$ is expected to be observed and will allow one to probe the tensor structure of the $H\gamma\gamma$ coupling as well.[39] Studies assessing the ultimate precision of off-shell production measurements at LHC and the associated inference of the total Higgs boson width are under way. And, of course, searches for exotic 4ℓ decays of the 125 GeV Higgs boson (e.g. $H \to ZZ_{dark} \to 4\ell$[40]) and for additional heavier or lighter bosons decaying to the 4ℓ final state[41,42] will continue and, one may hope, bear fruit.

References

1. ATLAS Collaboration, Observation of a new particle in the search for the Standard Model Higgs boson with the ATLAS detector at the LHC, *Phys.Lett.* **B716**, 1–29, (2012). doi: 10.1016/j.physletb.2012.08.020.
2. CMS Collaboration, Observation of a new boson at a mass of 125 GeV with the CMS experiment at the LHC, *Phys.Lett.* **B716**, 30–61, (2012). doi: 10.1016/j.physletb.2012.08.021.
3. LHC Higgs Cross Section Working Group, Handbook of LHC Higgs Cross Sections: 1. Inclusive Observables (2011). doi: 10.5170/CERN-2011-002. arXiv: 1101.0593.
4. CMS Collaboration, Measurement of the properties of a Higgs boson in the four-lepton final state, *Phys.Rev.* **D89**, 092007, (2014). doi: 10.1103/PhysRevD.89.092007.

5. ATLAS Collaboration, Evidence for the spin-0 nature of the Higgs boson using ATLAS data, *Phys.Lett.* **B726**, 120–144, (2013). doi: 10.1016/j.physletb.2013.08.026.

6. ATLAS Collaboration, Measurement of the Higgs boson mass from the $H \to \gamma\gamma$ and $H \to ZZ^* \to 4\ell$ channels with the ATLAS detector using 25 fb^{-1} of pp collision data, *Phys.Rev.* **D90**, 052004, (2014). doi: 10.1103/PhysRevD.90.052004.

7. ATLAS Collaboration, Measurements of Higgs boson production and couplings in the four-lepton channel in pp collisions at center-of-mass energies of 7 and 8 TeV with the ATLAS detector, *Phys.Rev.* **D91**(1), 012006, (2015). doi: 10.1103/PhysRevD.91.012006.

8. ATLAS Collaboration, Measurement of the muon reconstruction performance of the ATLAS detector using 2011 and 2012 LHC proton–proton collision data, *Eur.Phys.J.* **C74**(11), 3130, (2014). doi: 10.1140/epjc/s10052-014-3130-x.

9. CMS Collaboration, Constraints on the spin-parity and anomalous HVV couplings of the Higgs boson in proton collisions at 7 and 8 TeV, *Phys. Rev.* **D92**(1), 012004, (2015). doi: 10.1103/PhysRevD.92.012004.

10. CMS Collaboration, Constraints on the Higgs boson width from off-shell production and decay to Z-boson pairs, *Phys.Lett.* **B736**, 64, (2014). doi: 10.1016/j.physletb.2014.06.077.

11. ATLAS, Constraints on the off-shell Higgs boson signal strength in the high-mass ZZ and WW final states with the ATLAS detector, *Eur.Phys.J.* **C75**(7), 335, (2015). doi: 10.1140/epjc/s10052-015-3542-2.

12. ATLAS and CMS Collaborations, Combined measurement of the Higgs boson mass in pp collisions at $\sqrt{s} = 7$ and 8 TeV with the ATLAS and CMS experiments, *Phys.Rev.Lett.* **114**, 191803, (2015). doi: 10.1103/PhysRevLett.114.191803.

13. ATLAS Collaboration, Study of the spin and parity of the Higgs boson in diboson decays with the ATLAS detector, *Eur.Phys.J.* **C75**(10), 476, (2015). doi: 10.1140/epjc/s10052-015-3685-1.

14. CMS Collaboration, Particle-flow event reconstruction in CMS and performance for jets, taus, and MET. (CMS-PAS-PFT-09-001) (Apr, 2009). URL http://cds.cern.ch/record/1194487.

15. Y. Gao, A. V. Gritsan, Z. Guo, K. Melnikov, M. Schulze, *et al.*, Spin determination of single-produced resonances at hadron colliders, *Phys.Rev.* **D81**, 075022, (2010). doi: 10.1103/PhysRevD.81.075022.

16. S. Bolognesi, Y. Gao, A. V. Gritsan, K. Melnikov, M. Schulze, *et al.*, On the spin and parity of a single-produced resonance at the LHC, *Phys.Rev.* **D86**, 095031, (2012). doi: 10.1103/PhysRevD.86.095031.

17. I. Anderson, S. Bolognesi, F. Caola, Y. Gao, A. V. Gritsan, *et al.*, Constraining anomalous HVV interactions at proton and lepton colliders, *Phys.Rev.* **D89**(3), 035007, (2014). doi: 10.1103/PhysRevD.89.035007.

18. P. Avery, D. Bourilkov, M. Chen, T. Cheng, A. Drozdetskiy, *et al.*, Precision studies of the Higgs boson decay channel $H \to ZZ \to 4\ell$ with MEKD, *Phys.Rev.* **D87**(5), 055006, (2013). doi: 10.1103/PhysRevD.87.055006.

19. J. Alwall, M. Herquet, F. Maltoni, O. Mattelaer, and T. Stelzer, MadGraph 5: Going Beyond, *JHEP*. **1106**, 128, (2011). doi: 10.1007/JHEP06(2011)128.

20. A. Belyaev, N. D. Christensen, and A. Pukhov, CalcHEP 3.4 for collider physics within and beyond the Standard Model, *Comput.Phys.Commun.* **184**, 1729–1769, (2013). doi: 10.1016/j.cpc.2013.01.014.

21. J. S. Gainer, K. Kumar, I. Low, and R. Vega-Morales, Improving the sensitivity of Higgs boson searches in the golden channel, *JHEP*. **11**, 027, (2011). doi: 10.1007/JHEP11(2011)027.

22. Y. Chen, N. Tran, and R. Vega-Morales, Scrutinizing the Higgs signal and background in the $2e2\mu$ golden channel, *JHEP*. **01**, 182, (2013). doi: 10.1007/JHEP01(2013)182.

23. S. Dittmaier, S. Dittmaier, C. Mariotti, G. Passarino, R. Tanaka, *et al.*, Handbook of LHC Higgs Cross Sections: 2. Differential Distributions (2012). doi: 10.5170/CERN-2012-002. arXiv: 1201.3084.

24. LHC Higgs Cross Section Working Group, Handbook of LHC Higgs Cross Sections: 3. Higgs Properties (2013). doi: 10.5170/CERN-2013-004. arXiv: 1307.1347.

25. M. Bonvini, F. Caola, S. Forte, K. Melnikov, and G. Ridolfi, Signal–background interference effects for $gg \to H \to W^+W^-$ beyond leading order, *Phys.Rev.* **D88**(3), 034032, (2013). doi: 10.1103/PhysRevD.88.034032.

26. ATLAS Collaboration, Measurement of hard double-parton interactions in $W(\to l\nu)+$ 2 jet events at $\sqrt{s} = 7\,\mathrm{TeV}$ with the ATLAS detector, *New J.Phys.* **15**, 033038, (2013). doi: 10.1088/1367-2630/15/3/033038.

27. M. Botje, *et al.*, The PDF4LHC Working Group Interim Recommendations (2011).

28. I. W. Stewart and F. J. Tackmann, Theory uncertainties for Higgs and other searches using jet bins, *Phys.Rev.* **D85**, 034011, (2012). doi: 10.1103/PhysRevD.85.034011.

29. CMS Collaboration, Observation of Z decays to four leptons with the CMS detector at the LHC, *JHEP*. **1212**, 034, (2012). doi: 10.1007/JHEP12(2012)034.

30. ATLAS Collaboration, Measurements of four-lepton production at the Z resonance in pp collisions at $\sqrt{s} = 7$ and $8\,\mathrm{TeV}$ with ATLAS, *Phys.Rev.Lett.* **112**(23), 231806, (2014). doi: 10.1103/PhysRevLett.112.231806.

31. Particle Data Group, Review of Particle Physics, *Chin.Phys.* **C38**, 090001, (2014). doi: 10.1088/1674-1137/38/9/090001.

32. ATLAS Collaboration, Fiducial and differential cross sections of Higgs boson production measured in the four-lepton decay channel in pp collisions at $\sqrt{s} = 8\,\mathrm{TeV}$ with the ATLAS detector, *Phys.Lett.* **B738**, 234–253, (2014). doi: 10.1016/j.physletb.2014.09.054.

33. CMS Collaboration, Measurement of inclusive and differential fiducial cross sections for Higgs boson production in the $H \to 4\ell$ decay channel in pp collisions at $\sqrt{s} = 7$ and 8 TeV, *JHEP*. **04**, 005, doi: 10.1007/JHEP04(2016)005. arXiv: 1512.08377.

34. A. Soni and R. M. Xu, Probing CP violation via Higgs decays to four leptons, *Phys.Rev.* **D48**, 5259–5263, (1993). doi: 10.1103/PhysRevD.48.5259.

35. W.-Y. Keung, I. Low, and J. Shu, Landau–Yang theorem and decays of a Z' boson into two Z bosons, *Phys.Rev.Lett.* **101**, 091802, (2008). doi: 10.1103/ PhysRevLett.101.091802.

36. P. Artoisenet, P. de Aquino, F. Demartin, R. Frederix, S. Frixione, *et al.*, A framework for Higgs characterisation, *JHEP.* **1311**, 043, (2013). doi: 10.1007/ JHEP11(2013)043.

37. L. Randall and R. Sundrum, A Large mass hierarchy from a small extra dimension, *Phys.Rev.Lett.* **83**, 3370–3373, (1999). doi: 10.1103/Phys-RevLett.83.3370.

38. M. Chen, T. Cheng, J. S. Gainer, A. Korytov, K. T. Matchev, *et al.*, The role of interference in unraveling the ZZ-couplings of the newly discovered boson at the LHC, *Phys.Rev.* **D89**(3), 034002, (2014). doi: 10.1103/Phys-RevD.89.034002.

39. Y. Chen, R. Harnik, and R. Vega-Morales, Probing the Higgs couplings to photons in $h \to 4\ell$ at the LHC, *Phys.Rev.Lett.* **113**(19), 191801, (2014). doi: 10.1103/PhysRevLett.113.191801.

40. ATLAS Collaboration, Search for new light gauge bosons in Higgs boson decays to four-lepton final states in pp collisions at $\sqrt{s}$ = 8 TeV with the ATLAS detector at the LHC, *Phys.Rev.* **D92**(9), 092001, (2015). doi: 10.1103/PhysRevD.92.092001. arXiv: 1505.07645.

41. CMS Collaboration, Search for a Higgs boson in the mass range from 145 to 1000 GeV decaying to a pair of W or Z bosons, *JHEP.* **10**, 144, (2015). doi: 10.1007/JHEP10(2015)144. arXiv: 1504.00936.

42. ATLAS Collaboration, Search for an additional, heavy Higgs boson in the $H \to ZZ$ decay channel at $\sqrt{s}$ = 8 TeV in pp collision data with the ATLAS detector, *Eur.Phys.J.* **C76**(1), 45, (2016). doi: 10.1140/epjc/s10052-015-3820-7. arXiv: 1507.05930.

Chapter 6

Observation of the diphoton decay of the Higgs boson and measurements of its properties

Josh Bendavid[*] and Kerstin Tackmann[†]

*CERN
CH-1211 Geneva 23, Switzerland
†Deutsches Elektronen-Synchrotron (DESY)
D-22607 Hamburg, Germany

The diphoton decay channel contributed strongly to the discovery of the new particle in the summer of 2012, and has since then played an important role in the measurement of its properties. This chapter describes the reconstruction of this final state in the CMS and ATLAS detectors, the strategies employed for the Higgs search and property measurements, and the results obtained with the data collected at $\sqrt{s} = 7$ and 8 TeV.

1. Introduction

The Higgs boson decay into two photons was expected to be one of the promising decay channels at the LHC for a low-mass SM Higgs boson, i.e. with a mass below about 130 GeV. Indirect constraints from electroweak precision measurements pointed to a relatively light SM Higgs boson (see Fig. 2.7 and Refs. [1,2] for results incorporating the recently obtained precise measurements of the top-quark and W masses), which made the diphoton decay channel an important channel to be studied at the LHC.

In the SM, Higgs boson production at the LHC proceeds dominantly through the gluon fusion process (ggH, see Fig. 1.6), which is mediated by a top-quark loop and, at leading order, produces only the Higgs boson in the final state. The cross sections for production through vector boson fusion (VBF) and associated production with a vector boson (VH) or with a $t\bar{t}$ pair ($t\bar{t}H$) are significant, but much smaller than for ggH. Their characteristic signatures with two forward jets (VBF), a vector boson (VH), and a $t\bar{t}$ pair ($t\bar{t}H$) in the final state allow for dedicated studies of these production modes

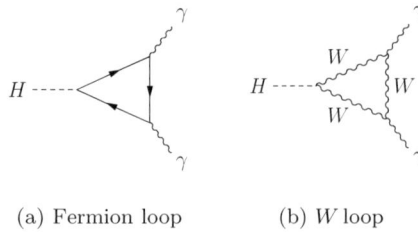

(a) Fermion loop   (b) W loop

Fig. 6.1   Higgs boson decay into two photons. In the SM, the decay into two photons proceeds through (a) fermions (predominantly top quarks) and (b) W bosons. The main contribution comes from the W loop, with a destructive interference between the two contributions.

by defining specific categories of events with a given signature. Through this categorisation, the couplings of the Higgs boson to various SM particles can be measured. VBF and VH production are sensitive to the coupling of the Higgs boson to vector bosons, while ggH and $t\bar{t}H$ production are sensitive to the couplings of the Higgs boson to the fermions.

The decay of a Higgs boson into two photons proceeds via a loop process (see Fig. 6.1) and as such has a very small branching fraction (about 0.23% for $m_H = 125$ GeV).

The $H \to \gamma\gamma$ decay has a simple final state, consisting of two energetic photons. This allows it to be reconstructed with high efficiency, resulting in a substantial signal yield, despite the small branching fraction. Around 500 events were expected to be reconstructed per experiment for a SM Higgs boson with a mass around 125 GeV in the data sample collected by ATLAS and CMS at centre-of-mass energies of 7 TeV (about $5\,\mathrm{fb}^{-1}$) in 2011 and 8 TeV (about $20\,\mathrm{fb}^{-1}$) in 2012 (for a description of the LHC running conditions, see Chapter 4).

The background consists predominantly of events with two photons (see Fig. 6.2 for examples). The production cross sections for these processes are significantly larger than for $H \to \gamma\gamma$ events. The separation of the small $H \to \gamma\gamma$ signal from the large backgrounds relies on the invariant mass $m_{\gamma\gamma}$ of the diphoton system. The signal is visible as an enhancement in the diphoton invariant mass distribution, smeared by detector resolution, above a smoothly falling background. Achieving a clearly visible signal relies on an excellent experimental diphoton mass resolution.

The diphoton decay channel strongly contributed to the discovery in the summer of 2012.[3,4] The focus of the experiments then shifted to the measurement of the properties of the new particle and their comparison

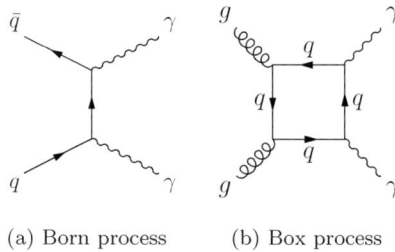

(a) Born process   (b) Box process

Fig. 6.2 Nonresonant production of diphotons in the quark-initiated "Born" process and the gluon-initiated "box" process, which proceeds through quark loops.

with the expectations for the SM Higgs boson. The reconstruction of the $H \to \gamma\gamma$ decay, the discovery of the Higgs boson decaying into two photons and the measurements of its properties are described in this chapter.

2. Detecting and measuring the diphoton signal

The $H \to \gamma\gamma$ decay can be fully reconstructed: a precise energy measurement can be made in the electromagnetic calorimeter, and the photons' directions can be determined using the electromagnetic calorimeter and the tracker. From these measurements, the diphoton invariant mass can be determined with an excellent resolution.

The primary challenge for the discovery and subsequent measurements of the $H \to \gamma\gamma$ decay at the LHC is the overwhelming amount of background. The backgrounds fall into two categories: reducible and irreducible. Reducible backgrounds consist of events where one or two hadronic jets are misidentified as photons. As the production cross sections for events with jets at LHC are several orders of magnitude larger than for events with two photons, a very strong suppression of these backgrounds by the photon identification procedure is needed. Irreducible backgrounds are due to nonresonant diphoton production.

2.1. *Photon reconstruction and identification*

Photons can have several signatures in the ATLAS and CMS detectors: they can convert into electron–positron pairs in the material in front of the electromagnetic calorimeter, or reach the calorimeter without converting. Because of the substantial amount of material in the tracking detectors (more than one radiation length in large regions of the detector, as depicted

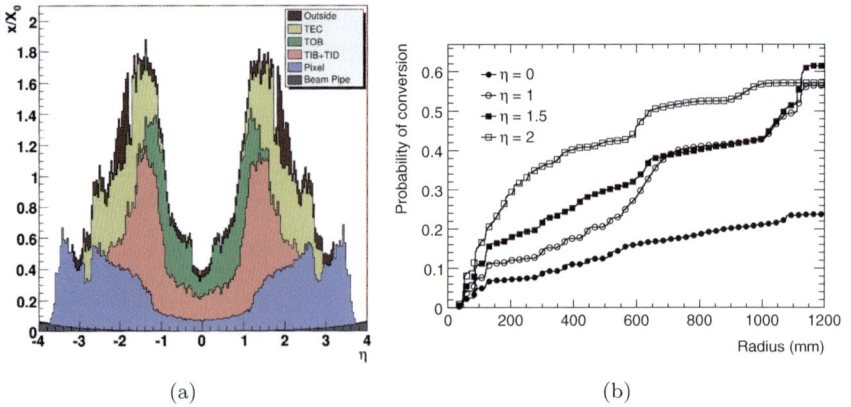

(a)   (b)

Fig. 6.3   (a) Radiation length of the material in front of the CMS electromagnetic calorimeter. The different colours represent the contributions of the various elements from the beam pipe to the material just in front of the calorimeter.[5] (b) Probability for a photon to convert within a given radius on its way to the ATLAS calorimeter for photons at different pseudorapidity η.[6]

in Fig. 6.3(a) for the CMS detector), a significant fraction of the photons convert in the tracker into electron–positron pairs (see Fig. 6.3(b)). Among $H \to \gamma\gamma$ events, close to 45% are expected to have two unconverted photons, and about the same amount to have one unconverted and one converted photon. Since electrons and positrons emit bremsstrahlung in the tracker material, the energy measurement is better for unconverted photons. The best invariant mass resolution is achieved for events with two unconverted photons in the barrel regions of the detectors. Less than 20% of events fall into this class.

Photon reconstruction

The photon reconstruction starts from electromagnetic clusters in the calorimeter. For photon conversions that happen in the tracker, the corresponding tracks and vertices can be reconstructed and matched to the electromagnetic clusters. In ATLAS, the presence and properties of the reconstructed tracks are used to classify the photons as converted, and a dedicated identification and energy calibration is applied to unconverted and converted photons. The reconstruction, identification, and classification of photons in CMS use the transverse profile of the electromagnetic shower to discriminate between converted and unconverted photons. While unconverted photons tend to produce narrower showers, for converted photons the electrons and positrons are deflected in the magnetic field, such that the positions of electrons, positrons and bremsstrahlung photons from the resulting shower are spread over an extended region in the electromagnetic calorimeter. This broadening occurs primarily in the direction perpendicular to the magnetic field, corresponding to the ϕ direction in a solenoidal field, and is more pronounced in CMS due to the stronger magnetic field of 3.8 T in the inner tracker, compared to the 2 T field in ATLAS.

Photon identification

A hadronic jet can be mistakenly reconstructed as a photon, especially when most of the jet's energy is carried by a π^0 (or η) meson with little other hadronic activity around it. The identification of photons is based on the shape of the electromagnetic showers, making use of the segmentation of the calorimeters. Even though the two photons from a π^0 decay merge into a single electromagnetic shower, the small opening angle tends to produce a wider shower than for a single photon. As converted photons have a shower which is broader in the ϕ direction, the discrimination between single photons and $\pi^0 \to \gamma\gamma$ relies more strongly on the shower width in the η direction. Both CMS and ATLAS use the measured width of the shower to discriminate between single photons and hadronic background. In addition, the fine transverse segmentation of the first layer of the ATLAS calorimeter helps to separate showers from π^0 decaying into two close-by photons from single photon showers by resolving the substructure in the $\pi^0 \to \gamma\gamma$ showers. Figure 6.4 shows an event selected by the $H \to \gamma\gamma$ analysis and the pattern of energy deposits of the photon candidates in the electromagnetic calorimeter. In order to avoid any significant systematic uncertainty on the final results, it was required to determine the efficiency of the photon identification with percent-level accuracy. It was measured using (a) electrons from $Z \to e^+e^-$ decays, making use of the similarity between electromagnetic showers from electrons and photons, (b) photons from radiative Z decays, $Z \to \ell\ell\gamma$, and (c) isolated photon candidates.[7,8]

Photon isolation

As hadronic jets deposit a larger fraction of their energy in the hadron calorimeter than photons, the photon identification also rejects photon candidates with too much energy deposited in the hadron calorimeter. Similarly, since hadronic jets consist of many close-by particles, even a photon

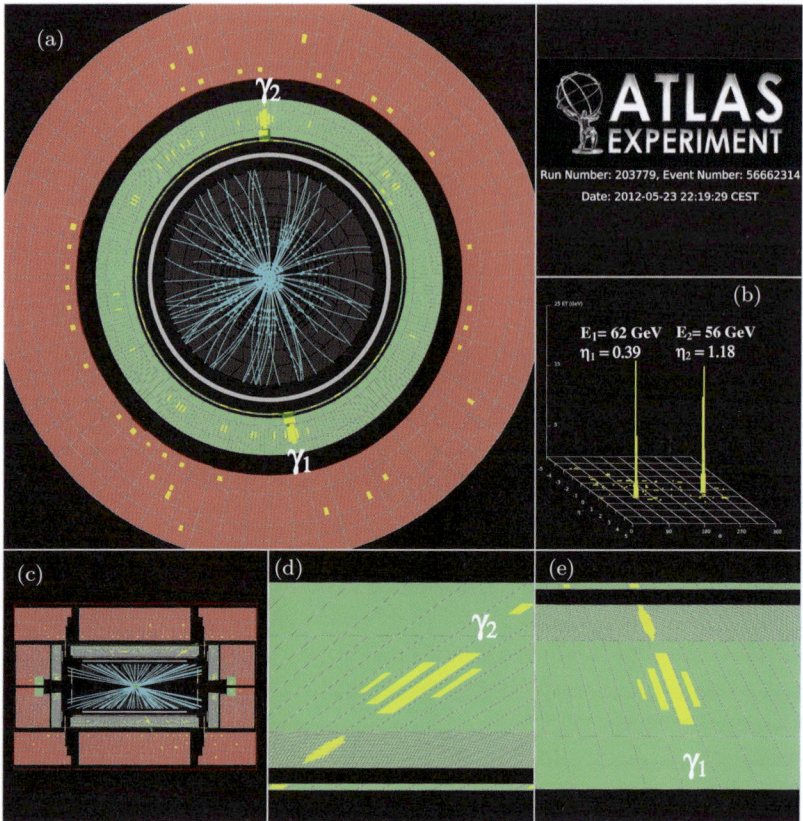

Fig. 6.4   Event display of a $H \to \gamma\gamma$ candidate in the ATLAS detector.[9] The invariant mass of the diphoton system is 126.9 GeV. (a) The view along the beam axis. The two photons are visible as energy depositions (yellow) in the electromagnetic calorimeter (green). (b) Deposited energy in the calorimeter in the (η, ϕ) plane, where the photons are visible as the largest energy depositions. (c) The view parallel to the beam line. The collimated electromagnetic showers of the (d) subleading (γ_2) and (e) leading (γ_1) photons. The very fine segmentation of the first longitudinal layer does not give any indication of a substructure in the shower, which would be the signature of a π^0 rather than a single photon.

candidate reconstructed from an energetic π^0 is typically surrounded by additional activity. For this reason, the photon candidates were required to be well isolated, that is, only small amounts of energy were allowed to be deposited in the calorimeter or reconstructed as tracks around the photon. Typically, the threshold of the transverse energy or transverse momentum sum around the photon was chosen to be a few GeV.

Photon identification in CMS was implemented using a Boosted Decision Tree (BDT) classifier (see Appendix B for an introduction) trained to distinguish prompt photons from mis-reconstructed jets, and combining shower profile and isolation information. After applying a loose preselection based on shower profile and isolation variables, the classifier value for each of the selected photons was used as one of the inputs for the subsequent categorisation of $\gamma\gamma$ events, allowing a finer-grained usage of the identification information.

Adaptation to increasing event pileup

The photon reconstruction and identification strategies had to be adapted to the increase of event pileup during 2011 and 2012. For the tracker-based reconstruction of converted photons in ATLAS, the increased track density in the detector would result in the reconstruction of fake converted photons from spurious associations between clusters of unconverted photons and pileup tracks if no measures were taken to suppress spurious tracks and track–cluster associations. Tighter quality requirements were imposed on the tracks and the track–cluster associations to keep the number of fake converted photons small. In the electromagnetic calorimeter, the showers became wider in the presence of pileup from the showers of pileup particles overlapping the photon showers. The requirements on the shape of the showers were adjusted to keep a high identification efficiency.

2.2. *Selection of diphoton events*

All $H \to \gamma\gamma$ measurements started with events containing at least two reconstructed photon candidates, with transverse momenta $p_T^{\gamma 1} > 0.35m_{\gamma\gamma}$ and $p_T^{\gamma 2} > 0.25m_{\gamma\gamma}$ (ATLAS), and $p_T^{\gamma 1} > \frac{1}{3}m_{\gamma\gamma}$ and $p_T^{\gamma 2} > \frac{1}{4}m_{\gamma\gamma}$ (CMS). The transverse momentum requirements were scaled with the diphoton invariant mass in order to keep the background distribution smooth, slowly varying, and free from threshold effects. The candidates were required to be identified as photons and well isolated. Dedicated triggers with a high efficiency for the $H \to \gamma\gamma$ event selection were deployed. They required two energy deposits in the calorimeter above certain energy thresholds, and with shapes consistent with those of electromagnetic showers. Photons with wider transverse showers are more difficult to distinguish from the background, and therefore CMS triggers also imposed loose isolation requirements on photon candidates with wider showers to suppress jet background.

2.3. *Measurement of the diphoton invariant mass*

A precise measurement of the invariant mass of the diphoton system, $m_{\gamma\gamma} = \sqrt{2E_1 E_2(1 - \cos\alpha)}$, where $E_{1,2}$ are the energies of the two photons and α is the opening angle between them, requires precise measurements of the photon energies and the angle α.

Photon energy measurement

The $H \to \gamma\gamma$ decay channel served as one of the benchmark processes for the design of the electromagnetic calorimeters of the ATLAS and the CMS detectors. The two experiments arrived at two different designs. The ATLAS electromagnetic calorimeter is a lead–liquid argon sampling calorimeter with accordion-shaped absorbers and electrodes. The calorimeter is finely segmented in the transverse and the longitudinal direction, allowing for a high jet background rejection. The expected resolution is $\sigma_E/E = 10\%/\sqrt{E} \oplus 0.7\%$[6] (with E measured in GeV). The longitudinal segmentation of the calorimeter also allows the directions of the photons to be measured by the calorimeter alone. The CMS calorimeter consists of lead tungstate crystals with an excellent intrinsic energy resolution of $\sigma_E/E = 2.8\%/\sqrt{E} \oplus 0.12/E \oplus 0.3\%$ as determined in test beam measurements,[5] but without longitudinal segmentation.

The calibration of the calorimeters is of crucial importance. The inter-calibration (relative calibration) of the individual channels of the CMS calorimeter was based on a combination of the invariant mass measured for π^0 and η decays to two photons, ϕ-symmetry of energy flow in minimum bias events, and energy–momentum matching for electrons from W and Z decays.[10] During the LHC running periods, the transparency of the crystals in the CMS calorimeter degrades due to irradiation, and recovers during periods with no proton beams. This change in transparency has a complex time-dynamic, and varies substantially as a function of pseudo-rapidity, given the higher radiation dose in the forward region, and from crystal to crystal due to manufacturing differences and impurities. A laser and LED system was therefore used to monitor and correct for the change of transparency on a per-crystal basis on a timescale of approximately 30 minutes.[11]

For the calibration of the ATLAS calorimeter, dedicated corrections derived from $W \to e\nu$, $Z \to e^+e^-$, inclusive γ events and minimum-ionising energy deposits from $Z \to \mu^+\mu^-$ were applied to correct non-uniformities of the calorimeter and to intercalibrate the longitudinal layers. The stability of the electron energy response with time and with pileup was measured using $Z \to e^+e^-$ decays and found to be approximately within 0.05%.[12]

To calibrate the photon energy and to obtain an optimal energy resolution, both ATLAS and CMS used multivariate regressions (see Appendix B) to correct the photon energy on an event-by-event basis. These algorithms used the shower profile and position to correct for energy losses in the material in front of the calorimeters and in the non-active regions of the calorimeters. For CMS, the multivariate regression also corrected for pileup effects on the energy measurement and provided a per-photon estimate of the energy resolution, used as part of the subsequent categorisation of $\gamma\gamma$ events.

There is no heavy narrow resonance decaying into photons that could be used to calibrate the absolute energy scale of the electromagnetic calorimeters. For this reason, both experiments used the mass distribution from $Z \to e^+e^-$ decays to set the absolute energy scale and to derive corrections to the energy resolution modelling in simulations (see Fig. 6.5). This procedure relies on the fact that the response of the calorimeter to electrons and photons is very similar, and that small, simulation-based corrections can be applied for the remaining differences.

CMS was able to reduce the systematic uncertainties associated with the extrapolation from electrons to photons by subdividing the electrons from $Z \to e^+e^-$ according to the same transverse shower width variables used to distinguish converted and unconverted photons. The electrons with narrow shower profile are less sensitive to the detector material and have a closer correspondence to unconverted photons, whereas those with wider shower profiles have a closer correspondence to converted photons.

To verify the applicability of the energy scale derived from $Z \to e^+e^-$ to photons, the ATLAS experiment used photons selected from radiative Z decays, $Z \to \ell^+\ell^-\gamma$. After applying the calibration derived from $Z \to e^+e^-$ to the clean photon sample from $Z \to \ell^+\ell^-\gamma$, the correction to the photon energy scale determined from $m_{\ell^+\ell^-\gamma}$ was consistent with unity.

$H \to \gamma\gamma$ vertex selection and photon direction measurement

The measurement of the opening angle between the photons can either be performed from a direct measurement of the photon directions, or can use the knowledge of the position of the pp interaction vertex from which the two photons originated. Given the resolution in the energy measurement, for the second case the vertex position needs to be measured with a

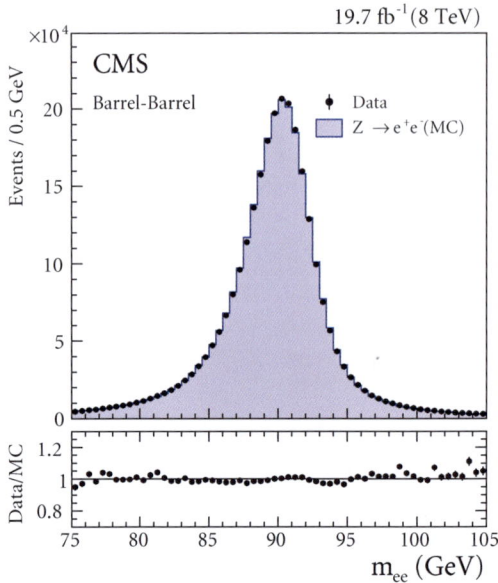

Fig. 6.5 The $Z \to e^+e^-$ peak for CMS data after application of energy scale corrections and in simulation after application of the resolution corrections to ensure a good modelling of the resolution observed in data. Both electrons were requited to be in the barrel of the electromagnetic calorimeter.[13]

precision of about 1 cm in order to remain a negligible contribution to the $m_{\gamma\gamma}$ resolution. A difficulty arose in distinguishing the $pp \to H \to \gamma\gamma$ interaction vertex from the on-average 20 additional pileup interaction vertices (see Fig. 4.2(b)), which were distributed over the luminous region of about 6 cm. The standard technique at LHC to identify the pp interaction vertex of the interesting, high-energy collision, such as the collision producing the $H \to \gamma\gamma$, is based on the transverse momenta p_T of the tracks associated with the vertex. In $gg \to H \to \gamma\gamma$ events where no hard jets are emitted with the Higgs boson, the pp interaction vertex has little track p_T associated with it. A sketch of the pp collision region containing a $H \to \gamma\gamma$ event is shown in Fig. 6.6(a). In these events, the standard, purely track-based vertex identification algorithms selected the correct vertex in only 60% of events in typical data taking conditions.

In CMS, the identification of the hard-scattering pp interaction vertex relied on a BDT which used as input the recoil momentum of tracks from the vertex, the direction and p_T of the diphoton system, and, for converted photons, the direction measurement from the associated tracks and the

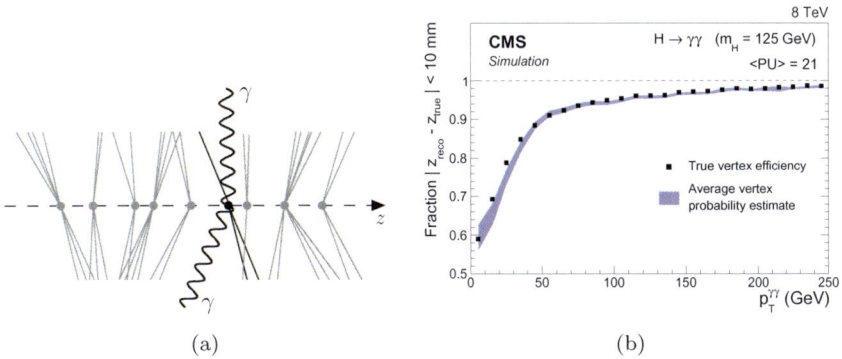

Fig. 6.6 (a) Sketch of the pp collision region in a $H \to \gamma\gamma$ event. The grey dots signify pileup vertices, from which soft jets from low-energy QCD interactions emerge, and the charged particles (shown as lines) are recorded by the inner tracker. The black dot shows the $H \to \gamma\gamma$ vertex, from which two photons and the jets produced in conjunction with the Higgs boson emerge (the charged particles in the jets are shown as black lines). (b) The fraction of events in which the correct pp interaction vertex was expected to be selected in the CMS detector for Higgs events in the 8 TeV dataset as a function of the Higgs boson transverse momentum, as well as the fraction predicted by the BDT and used as input to the event categorisation.[13]

conversion vertex. The combined vertex identification method had a probability of about 80% to select a vertex with a position within 1 cm of the true vertex in $H \to \gamma\gamma$ events. Since events where the vertex was not selected within this precision had poor α and thus $m_{\gamma\gamma}$ resolution, an additional BDT was trained to estimate the probability of a good vertex position measurement. This probability was used as one of the inputs for the subsequent categorisation of $\gamma\gamma$ events. The fraction of events in which a pp collision vertex was expected to be selected within 1 cm of the correct vertex in the CMS detector is shown in Fig. 6.6(b) as a function of the Higgs boson p_T, illustrating the dependence on the hadronic recoil momentum from the collision in which the Higgs boson was produced.

The longitudinal segmentation of the ATLAS calorimeter allows a direct measurement of the photon direction from the electromagnetic shower. In contrast to the photon direction measurement based on the identification of the correct pp interaction vertex, the direction measurement from the ATLAS calorimeter is independent of the amount of event pileup. For converted photons, the location of the conversion vertex was also used to improve the precision of the photon direction measurement. When both photons converted, the resolution on the longitudinal (z) position of the pp interaction vertex was about 0.6 cm, and in other cases, it was about

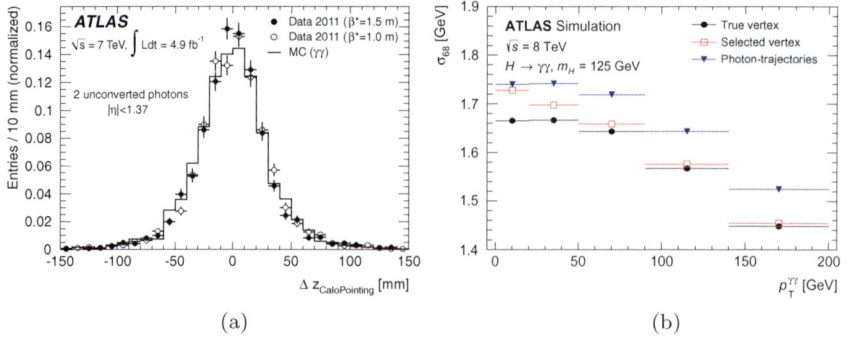

Fig. 6.7 (a) Difference in estimated position of the high-energy primary vertex along the beam direction based on the two photons' directions measured with the ATLAS calorimeter.[3] (b) Invariant mass resolution σ_{68}, defined as half of the smallest $m_{\gamma\gamma}$ interval containing 68% of the signal events in simulation, using the true primary vertex position, the measured photon trajectories, and the primary vertex selected in the analysis to measure the photon direction, as a function of the p_T of the diphoton system.[15]

1.5 cm[14] (see Fig. 6.7(a)). While this precision would have been sufficient to ensure a good $m_{\gamma\gamma}$ resolution, the selection of jets produced together with the Higgs boson required the identification of the correct pp interaction vertex to a greater precision in order to use the tracks originating from the selected $pp \rightarrow H \rightarrow \gamma\gamma$ vertex to distinguish jets produced in association with the Higgs boson from those produced in pileup interactions. To this end, ATLAS deployed a more sophisticated method for the determination of the pp interaction vertex. The track-based measurements, similar to the information used by CMS, were combined with the measurements of the photon momenta, the direction measurement from the calorimeter and, if available, the conversion vertex location, using a neural network (see Appendix B for an introduction). In the typical pileup conditions during the 8 TeV data taking, the ATLAS algorithm had an efficiency of about 85% to select a reconstructed primary vertex within Δz of 0.3 mm of the $pp \rightarrow H \rightarrow \gamma\gamma$ vertex. Using the vertex selected using the neural network in the measurement of the photon directions, the invariant mass resolution in ATLAS also improved for events with high diphoton p_T, where the hadronic recoil allowed for an improved measurement of the vertex position with respect to the purely calorimeter-based determination (see Fig. 6.7(b)).

3. Signal modelling

The $H \to \gamma\gamma$ signal properties were modelled using simulation. The main production modes, ggH, VBF, VH and $t\bar{t}H$, were simulated, including a full simulation of the detector response based on GEANT.[16] Several corrections were applied to the simulation to model the expected signal in data as closely as possible. This included corrections to the photon energy resolution derived from $Z \to e^+e^-$ events as described in Sec. 2.3. Another important correction was applied to improve the simulation of the photon identification efficiency derived from the identification efficiency measurements described in Sec. 2.1. The shape of the invariant mass distribution of the signal was then modelled by a parametric function in each analysis category, consisting of sums of Gaussian and/or Crystal Ball functions.[17]

4. Background composition and subtraction

Irreducible backgrounds are due to nonresonant diphoton production. This includes tree-level processes with $q\bar{q}$ initial states, as well as the loop-induced processes with gluon–gluon initial states, Feynman diagrams for which are shown in Fig. 6.2.

Reducible backgrounds include both processes with a photon and one or more jets in the final state, with one jet reconstructed as a photon, as well as QCD multi-jet processes in which two jets are reconstructed as photons. Although multi-jet production at the LHC has a much larger cross section, the relatively small rate for jets to be misidentified as photons meant that the γ+jet background was more important than the multi-jet background.

There is an additional smaller reducible background resulting from processes in which electrons are misreconstructed as photons, mainly in cases where the electron track is not reconstructed, or the electron is misidentified as a converted photon. This background includes dielectron events where both electrons are misreconstructed as photons, as well as $W\gamma$ events with one misreconstructed electron.

The ATLAS experiment determined the composition of the event sample by studying the shower shapes and isolation of the two photon candidates in the selected events, and demonstrated[15] that the sample was indeed dominated by events with two real photons as shown in Fig. 6.8. For example, in the range 105 GeV $< m_{\gamma\gamma} <$ 160 GeV, the relative contributions from $\gamma\gamma$ and γ+jet events were 77% $\pm$ 3% and 20% $\pm$ 2%, respectively, in the 8 TeV data.

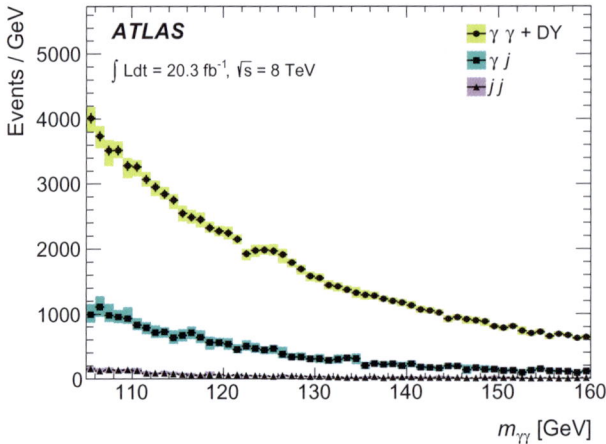

Fig. 6.8 Measured diphoton invariant mass spectrum in the 8 TeV data, and contributions of the different types of events (diphoton and dielectron (DY) events, one photon and one misreconstructed jet, and two misreconstructed jets).[15] The depletion near $m_{\gamma\gamma} = 120$ GeV is a fluctuation in the background.

Predictions from simulation of the different kinds of background were useful for optimising several aspects of the analyses, such as the photon identification and for testing the background parametrisations, as well as training BDT analyses and neural networks. However, simulation predictions were not directly used to derive the final results.

In order to minimise the dependence on any a priori background prediction and simulation, and exploiting the fact that all background contributions had a smoothly falling diphoton mass distribution, the background contribution was determined from the data itself. This was achieved by parametrising the background shape by an analytic function in an invariant mass window 100 GeV $< m_{\gamma\gamma} <$ 180 GeV (CMS) and 105 GeV $< m_{\gamma\gamma} <$ 160 GeV (ATLAS). The background parametrisations and procedures were tested on simulated samples to ensure that any potential biases from the background subtraction were smaller than 20% of the statistical uncertainties on the estimated number of background events (ATLAS) or 14% of the expected statistical uncertainties on the signal yield in each category (CMS). ATLAS assigned the remaining potential bias as systematic uncertainty, while the criteria employed by CMS ensured that neglecting this uncertainty did not underestimate the total signal yield uncertainty by more than 1%.

5. Event categorisation

Selected diphoton events were further divided into categories in order to separate the different Higgs production modes, as well as to group together events with similar signal-to-background ratios. This was accomplished by imposing additional requirements on the properties of the photons and by tagging additional objects in the event. In order to ensure non-overlapping sets of events, the selection proceeded sequentially, with events selected first for exclusive categories targeting the more topologically distinctive production modes, and assigning the remaining events to the subsequent categories.

First, events with topologies typical of the different production modes were selected. Two categories were enriched in $t\bar{t}H$ events, based on the presence of jets, b-tagged jets, leptons and missing transverse momentum E_T^{miss}. Four more categories were enriched in VH events, making use of the vector-boson decays into charged leptons and neutrinos, as well as jets. Two (ATLAS) and three (CMS) further categories were enriched in VBF events, taking advantage of its typical signature of two forward jets with large $\Delta\eta$. All events not falling into any of these categories were collected in the "untagged" categories, dominated by ggH events.

Naturally, such topology-based categories yield event samples that are mixtures of the production modes. Simulation-based studies were used to estimate the contributions of the production modes to each of the categories. Global fits to all categories were used in the measurements described in Sec. 6.

In addition to their use in separating the various Higgs production modes, the categories differed in their expected $m_{\gamma\gamma}$ resolution and signal-to-background ratios, which enhanced the sensitivity to production mode independent measurements, such as the Higgs mass and the inclusive signal yield. Since the additional objects selected for exclusive categories are quite rarely produced in conjunction with diphotons, these categories had a relatively high signal-to-background ratio (larger than 10%). The untagged events were categorised according to their signal-to-background ratio and their $m_{\gamma\gamma}$ resolution (see Sec. 5.4 for more details). The categorisation prevented the good resolution and signal-to-background ratios in the "best" categories from being diluted by the "worse" categories. The "best" categories dominated the results, but the remaining categories also contributed to the measurements.

5.1. *t̄tH categories*

As production through $t\bar{t}H$ is expected to be rare, the categories targeting this production mode were defined to have a high efficiency for $t\bar{t}H$ events. Two event categories targeting semileptonic and fully hadronic decays of the $t\bar{t}$ pair were defined. All selections required one or two jets to be tagged as b-jets. The leptonic $t\bar{t}$ category required the presence of at least one electron or muon, while the hadronic $t\bar{t}$ category required the presence of at least four additional jets. The approximate number of background events in the smallest mass window expected to contain 90% of the signal, the number of expected signal events, as well as $t\bar{t}H$ events in the different categories enriched in $t\bar{t}H$ production are given in Table 6.1[13,15] for the $\sqrt{s} = 8$ TeV datasets.

5.2. *VH categories*

Categories enriched in VH events were defined by a fairly loose selection of the decay products of the vector bosons. The exact definitions of the categories differed between the two experiments, but followed the same concepts. The ATLAS dilepton and the CMS tight lepton categories, which were sensitive to $Z(\to \ell\ell)H$ production, contained events with two electrons or two muons in the final state. Events with one electron or muon and fulfilling a loose requirement on $E_{\mathrm{T}}^{\mathrm{miss}}$ entered the ATLAS one lepton and the CMS tight lepton categories, sensitive to $W(\to \ell\nu)H$ production. The CMS loose lepton category, sensitive to both $Z(\to \ell\ell)H$ and $W(\to \ell\nu)H$ production, was defined by the presence of an electron or muon in the event. Both experiments defined an event category with significant $E_{\mathrm{T}}^{\mathrm{miss}}$ (sensitive to $Z \to \nu\nu$ and $W \to \ell\nu$ if the lepton escaped detection), as well

Table 6.1 Approximate number of background events (B_{90}) in the smallest mass window expected to contain 90% of the signal events, expected number of total signal events N_S, and of $t\bar{t}H$ events $N_S^{t\bar{t}H}$ in the different categories enriched in $t\bar{t}H$ production in the ATLAS and CMS analyses of the $\sqrt{s} = 8$ TeV dataset.

Category	B_{90}	N_S	$N_S^{t\bar{t}H}$
ATLAS leptonic	0.53	0.6	0.5
ATLAS hadronic	1.8	0.5	0.4
CMS leptonic	0.89	0.5	0.5
CMS hadronic	2.6	0.6	0.6

Table 6.2   Approximate number of background events (B_{90}) in the smallest mass window expected to contain 90% the signal events, expected number of total signal events N_S and of VH events N_S^{VH} in the different VH categories in the ATLAS and CMS analyses of the $\sqrt{s} = 8$ TeV dataset.

Category	B_{90}	N_S	N_S^{WH}	N_S^{ZH}
ATLAS dilepton	0.27	0.3	0.0	0.3
ATLAS one lepton	4.4	1.7	1.5	0.1
ATLAS missing E_T significance	3.2	1.1	0.4	0.6
ATLAS dijet	18	3.2	1.0	0.6
CMS tight lepton	2.1	1.4	1.1	0.3
CMS loose lepton	6.3	0.9	0.7	0.2
CMS missing E_T	7.2	1.8	0.6	0.6
CMS dijet	4.3	1.6	0.6	0.4

as a category with two jets with an invariant mass consistent with that of a W or Z boson. Only a small number of signal events was expected to be selected for each of these event categories. The approximate number of background events in the smallest mass window expected to contain 90% of the signal, the number of expected signal events, as well as VH events in the categories enriched in VH production, are given in Table 6.2[13,15] for the $\sqrt{s} = 8$ TeV datasets. While the number of events is small, the purity of VH events is very high. With the data collected at $\sqrt{s} = 7$ and 8 TeV, the sensitivity to this production mode is low (see Sec. 6), but these categories improve the sensitivity of the measurement of the coupling strength to vector bosons when combined with the (more sensitive) categories enriched in VBF events.

5.3.   *VBF categories*

The topology of events produced through VBF is characterised by two jets with relatively large p_T emitted with a large separation in rapidity, and little hadronic activity in between. The definition of categories enriched in VBF events was hence based on the pseudorapidity separation of the two highest p_T jets, the invariant mass of the dijet system and the separation and angular balance between the dijet and the diphoton system. Figure 6.9 shows the first two quantities for simulated background and signal events, and for events selected on data, which were dominated by background events. Figure 6.10 shows an event with VBF topology selected by the CMS analysis.

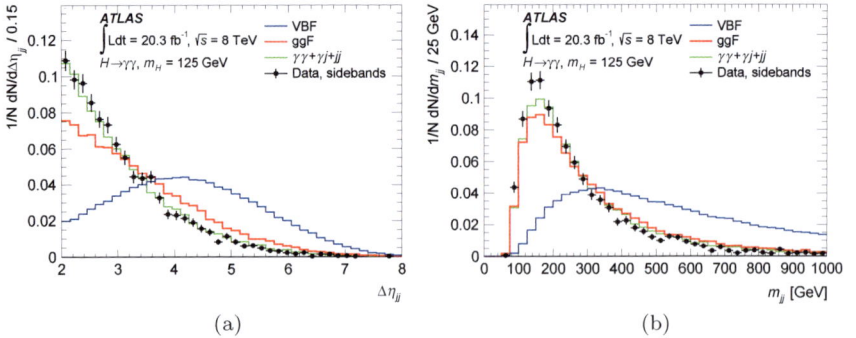

Fig. 6.9 (a) Separation in η and (b) invariant mass of the two highest p_T jets in events selected in data outside of the $H \rightarrow \gamma\gamma$ signal region, in simulated $\gamma\gamma$, γ+jet and jet+jet background events, and in signal events produced through ggH and VBF.[15]

A high purity in signal events produced in VBF is needed for a good sensitivity to this production mode. Both the ATLAS and CMS experiments used BDTs to combine the information from the different discriminating variables. The categories were then defined by ranges in the output values of the BDT. CMS defined three ("tight", "medium", "loose") categories, while ATLAS defined two ("tight" and "loose"). The analogous information to Table 6.1 is given in Table 6.3[13,15] for the categories enriched in VBF events.

5.4. "Untagged" categories

The remaining events, which were not selected for any of the production-mode-based categories described above, were further divided into categories according to different expected $m_{\gamma\gamma}$ resolution and signal-to-background ratio. This increased the overall sensitivity for the measurements of the mass, the inclusive signal yield and the signal yield in the ggH production mode.

The ATLAS experiment followed a cut-based approach, where the events were categorised according to the pseudorapidity of the two photons in the detector ("central" if both photons had $|\eta| < 0.95$ and "forward" otherwise) and the transverse momentum ("high" and "low") of the diphoton system. In general, events with both photons in the central region of the detector have a better mass resolution and a better signal-to-background

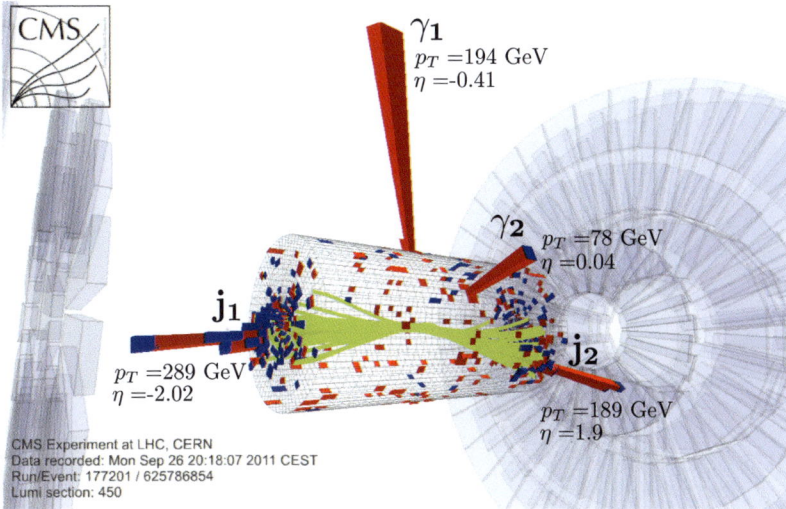

Fig. 6.10   Display of a diphoton event in the CMS detector, which is a candidate for a Higgs produced in the VBF topology and decaying to two photons.[18] Red bars represent energy in the electromagnetic calorimeter, blue bars represent energy in the hadron calorimeter, and the yellow lines represent tracks. The isolated red bars in the central part of the event represent the two photon candidates γ_1 and γ_2, and the large groups of red and blue bars and associated tracks represent the two forward jets j_1 and j_2.

Table 6.3   Approximate number of background events (B_{90}) in the smallest mass window expected to contain 90% of the signal events, expected number of total signal events N_S and of signal events produced in VBF N_S^{VBF} in the different categories enriched in VBF in the ATLAS and CMS analyses of the $\sqrt{s} = 8$ TeV dataset.

Category	B_{90}	N_S	N_S^{VBF}
ATLAS tight	6.7	5.7	4.6
ATLAS loose	44	9.3	5.3
CMS tight	3.4	4.5	3.7
CMS medium	13	5.6	3.9
CMS loose	115	13.7	7.3

ratio than events where one or both photons are not central. The separation in transverse momentum served two purposes: events with high transverse momentum have a better signal-to-background ratio and, due to Higgs bosons produced by VBF being more boosted than Higgs bosons produced in ggH, these categories also had a larger fraction of events produced in

VBF but not assigned to the dedicated VBF categories. Hence, this categorisation also improved the separation of the different production modes. The categorisation used the p_{Tt}[19,20] of the diphoton system. The p_{Tt} of the diphoton system is defined as the orthogonal component of the diphoton p_T when projected on the axis given by the difference of the transverse momenta of the two photons. While p_{Tt} is correlated with the p_T of the Higgs boson, it has a better experimental resolution (analogously to the measurement of the Z boson transverse momentum distribution described in Ref. [20]) and is less correlated with the diphoton invariant mass.

CMS opted for an inclusive categorisation based on a BDT. The inputs were the kinematic properties of the diphoton system, an estimate of the per-event mass resolution constructed from the per-photon energy resolution and vertex selection probability, and the output scores of the multivariate photon identification. The BDT score was then used to define six event categories for the 8 TeV dataset, and five event categories for the 7 TeV dataset. The category with the lowest score in each dataset was discarded from the analysis. This approach made optimal use of all available information on the diphoton mass resolution and signal-to-background ratio, while still allowing the analysis to be performed in a reasonable number of categories. The category with the highest score was found to be populated almost exclusively by events with a high diphoton p_T, while the category with the second highest score was found to be dominated by events with two central unconverted photons. The BDT output distribution for events passing a loose pre-selection is shown in Fig. 6.11 together with the distributions for the four production modes as well as the Monte Carlo background prediction.

Figure 6.12 illustrates the contribution of the different production modes to the different event categories in the ATLAS analysis.

6. Results

6.1. *Discovery of a new particle*

With all the ingredients in hand, the experiments measured the invariant mass distribution of the selected diphoton events.[13,15] The measured spectra from the combined $\sqrt{s} = 7$ and 8 TeV dataset for all categories combined are shown in Fig. 6.13. An excess of events can clearly be seen around 125 GeV — the sign of a new resonant state in this decay channel. The overlaid fit fixed the width of the resonance to the expected width, which is

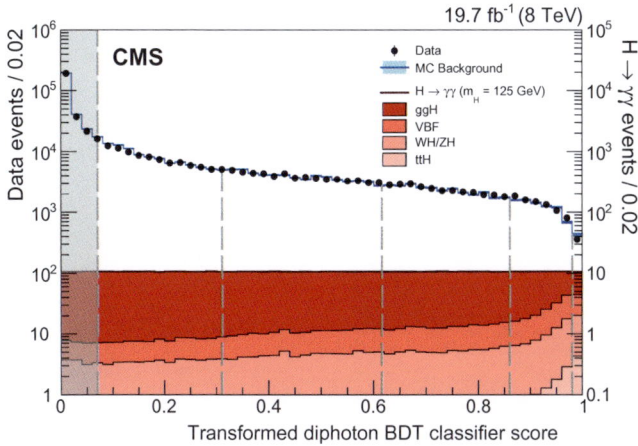

Fig. 6.11 BDT classifier score used to categorise untagged events in the 8 TeV CMS dataset for events passing a loose pre-selection.[13] The distribution for data (points with error bars, left axis) is compared to the distributions for simulated background (outlined histogram) and the four simulated Higgs production modes (solid histograms, right axis). The vertical dashed lines show the boundaries of the untagged event classes, with the left-most dashed line representing the score below which events were discarded and not used in the analysis. The classifier score was transformed such that the expected distribution for the SM Higgs boson is a uniform distribution.

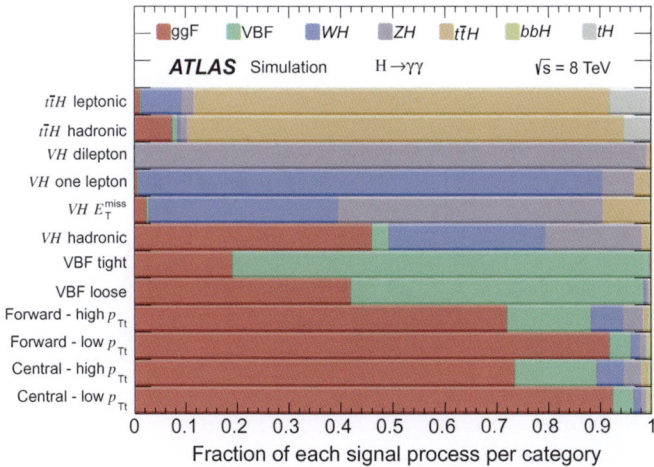

Fig. 6.12 Expected contribution of the different Higgs boson production modes to the categories defined in the ATLAS analysis in the $\sqrt{s} = 8$ TeV data. The definition of the categories is described in the text, and in further detail in Ref. [15]. The high purity of the production mode categories that are enriched by dedicated selections can be seen for the $t\bar{t}H$, VH and VBF categories.

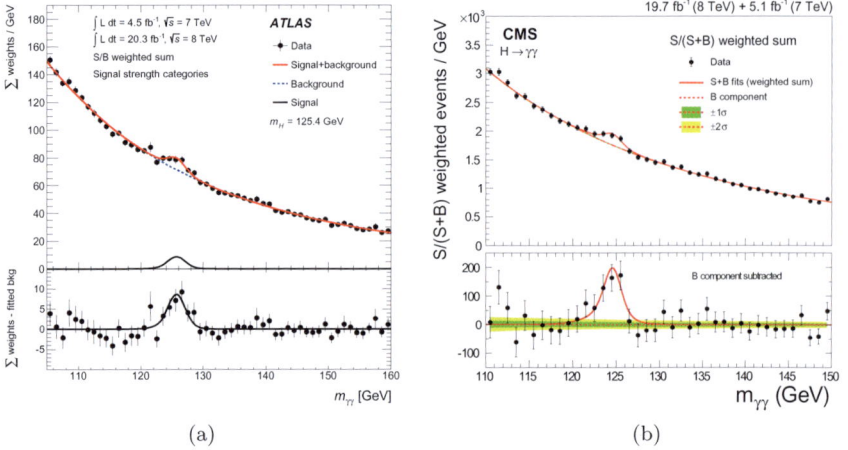

Fig. 6.13   The measured $m_{\gamma\gamma}$ spectra by the (a) ATLAS[15] and (b) CMS[13] experiments with the full 7 and 8 TeV datasets. Every event was weighted according to the value of the expected signal-to-background ratio (ATLAS) and the signal fraction (CMS) under the mass peak in its category, to illustrate approximately the effective $m_{\gamma\gamma}$ distribution resulting from the categorisation. The lower panels of the figures show the data after the subtraction of the fitted background.

completely dominated by the detector resolution. Good agreement between the expected width from detector resolution and the data is observed. Thus, the width of the newly observed state is consistent with the predicted width of a SM Higgs boson of about 4 MeV.

The p-value gives the probability of the observed excess to be produced by a fluctuation of the background (see Appendix A) as a function of the Higgs boson mass hypothesis and is shown in Fig. 6.14. In the full dataset, the ATLAS experiment observed a p-value that corresponded to a $5.2\,\sigma$ excess at about 125.4 GeV, when about $4.6\,\sigma$ were expected for a SM Higgs boson of the same mass. The CMS experiment observed an excess of $5.7\,\sigma$ at 124.7 GeV, when about $5.2\,\sigma$ were expected for a SM Higgs boson of that mass. In high-energy physics, the convention is to require at least a $5\,\sigma$ excess to claim a discovery of a new particle or phenomenon. Thus, both CMS and ATLAS can claim a discovery of the new boson in its diphoton decay channel.

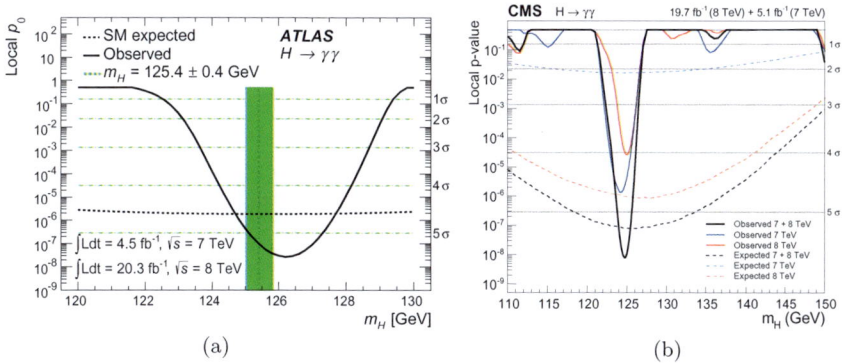

Fig. 6.14 Expected and observed p-values (p_0), which quantify the probability of an observed excess to be produced by a fluctuation of the background, as observed by the (a) ATLAS[15] and (b) CMS[13] experiments. The expected p-values for a SM Higgs boson are shown as a function of the hypothetical Higgs boson mass. The p-values observed in data show a clear excess around 125 GeV. The value of $m_H = 125.4$ GeV indicated in the ATLAS plot is the mass measured from $H \rightarrow \gamma\gamma$ and $H \rightarrow 4\ell$ (see Chapter 12).

6.2. Mass measurement

As one of the two decay channels with an excellent mass resolution, the diphoton decay was important for the measurement of the mass of the new state.[13,21] The mass measurement was dominated by the event categories with the best $m_{\gamma\gamma}$ resolution, while the event categories with the worst resolution had practically no impact.

While CMS used the same event categorisation for the measurement of the production modes and the mass, ATLAS opted for a dedicated event categorisation for the mass measurement. It was based on the pseudo-rapidity of the two photons in the detector, whether or not they had converted, and the p_{Tt} of the diphoton system. The precision of the measurements was dominated by the statistical uncertainties. The masses measured by the two experiments were

$$m_H = 125.98 \pm 0.42(\text{stat}) \pm 0.28(\text{syst}) \text{ GeV (ATLAS) and}$$
$$m_H = 124.70 \pm 0.31(\text{stat}) \pm 0.15(\text{syst}) \text{ GeV (CMS)}.$$

The systematic uncertainties were dominated by the uncertainties on the electromagnetic energy scales of the calorimeters. Since both experiments determined the absolute energy scale using $Z \to e^+e^-$ decays, the most important systematic uncertainties on the photon energy scale were related to differences between the electromagnetic showers of electrons and photons in the calorimeters, and differences in kinematics between $Z \to e^+e^-$ events and $H \to \gamma\gamma$ events, mainly the higher p_T spectrum of the photons from the Higgs decay compared to electrons from the Z, due to the larger Higgs mass. The extrapolation from the electron to the photon energy scale relied mainly on the detector simulation. For CMS, the main associated systematic uncertainties were related to possible imperfections in the simulation, the most important was the description of the detector material in front of the calorimeters, and the modelling of the electromagnetic showers. For ATLAS, the largest contributions to the systematic uncertainties were due to an observed non-linearity of the electromagnetic calorimeter cell response, the modelling of the material between the tracking detectors and the calorimeter, and the intercalibration of the calorimeter layers.

The measurements from the two experiments were found to be compatible with each other within $2.1\,\sigma$.[22] Combining the ATLAS and CMS measurements yielded

$$m_H = 125.07 \pm 0.25(\text{stat}) \pm 0.14(\text{syst}) \text{ GeV}.$$

As the experimental systematic uncertainties were related to the specifics of the detectors and calibration procedures, they were treated as uncorrelated between the two experiments. The excellent precision of the measurement of 0.2% was possible thanks to the careful calibration of the electromagnetic calorimeters.

6.3. *Measurement of the signal strength and separation of production modes*

A quantitative measure of how well the number of measured signal events agrees with the expectation is the signal strength μ, defined as the ratio of the measured to the expected number of events for a SM Higgs boson, $\mu = \frac{(\sigma \cdot B)_{\text{meas}}}{(\sigma \cdot B)_{\text{SM}}}$ (with σ the production cross section and B the branching ratio into two photons). The signal strengths determined by the two experiments were[13,15]

$$\mu = 1.17 \pm 0.23(\text{stat})^{+0.10}_{-0.08}(\text{syst})^{+0.12}_{-0.08}(\text{theory}) \text{ (ATLAS) and}$$

$$\mu = 1.14 \pm 0.21(\text{stat})^{+0.09}_{-0.05}(\text{syst})^{+0.13}_{-0.09}(\text{theory}) \text{ (CMS)},$$

consistent with the SM expectation. The largest contributions to the experimental systematic uncertainties were different in CMS and ATLAS. For CMS the modelling of the electromagnetic shower shapes in the simulation gave the largest contribution since the distinction between unconverted and converted photons was largely based on the shower shapes and directly entered the resolution estimate used in the event categorisation. For ATLAS, the uncertainty on the photon energy resolution used in the modelling of the signal shape gave the largest contribution.

A central prediction of the SM is the contribution of the different production modes to the Higgs boson production. Therefore, a separation of the different production modes and the measurement of the individual signal strengths is a crucial test of the nature of the new particle. The categories enriched in events with the signature of an event produced in VBF, VH or $t\bar{t}H$ production, as described in Sec. 5, served exactly this purpose. Figure 6.15 shows the signal strengths measured in various production modes. The measurements were still dominated by statistical uncertainties, especially for the rare production modes. CMS performed a combined measurement for WH and ZH (VH), while ATLAS measured the signal strength for WH and placed an upper limit on the ZH signal strength. These measurements were an important input to the coupling measurements described in Chapter 12. Within the uncertainties, all measured signal strengths were consistent with the SM expectations. These

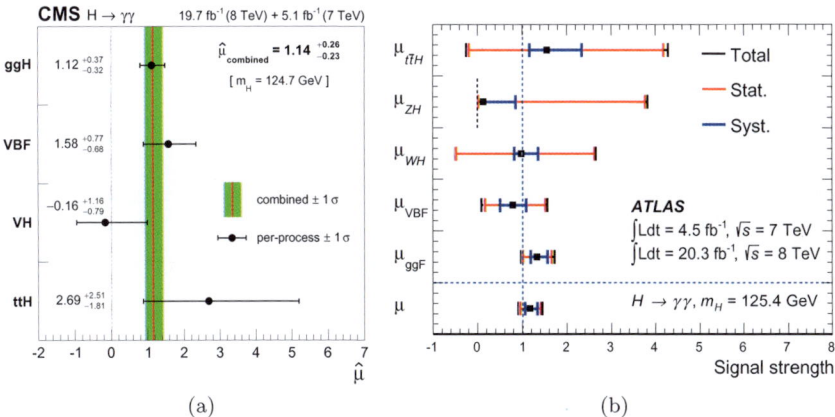

Fig. 6.15   Measured signal strengths ($\hat{\mu}$) for the different production modes from the (a) CMS[13] and (b) ATLAS[15] analyses.

measurements also provided evidence for VBF production at the level of 2σ for each experiment.

6.4. *Spin study*

According to the Landau–Yang theorem,[23,24] the observation of the decay into two photons already rules out that the new particle has spin one (as explained in Chapter 1). As a two-body final state, kinematic distributions of the diphoton system are not sensitive to the CP value of the intermediate resonance, but the angular distribution of the two photons carries information about its spin. Both experiments made use of the distribution of the polar angle θ^* of the two photons with respect to the z-axis in the Collins–Soper frame[25] to study the spin properties of the new particle. A spin-0 particle decays isotropically in its rest frame, which implies a flat distribution of $\cos\theta^*$ before it is distorted by the experimental requirements on the p_T of each of the two photons. Spin-2 models are challenging from a theoretical point of view (see Chapter 1). The models chosen for this study are described in detail in Chapter 5. CMS chose a leading-order model, where the spin-2 particle is produced as the only particle in the final state.[26] The distributions of $\cos\theta^*$ in this model for pure gluon- and pure quark-induced production are shown for signal events before and after selection cuts in Fig. 6.16, illustrating the effect of the acceptance cuts in reducing the discriminating power between different models.

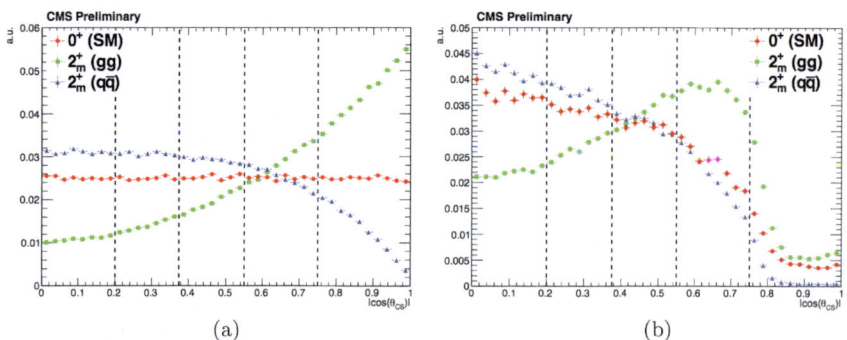

Fig. 6.16 Distribution of $\cos\theta^*$ for spin-0, and for a spin-2 model with pure gluon-induced or pure quark-induced production,[26] with distributions shown before (a) and after (b) selection cuts, illustrating the effect of the acceptance cuts on the distributions.[27]

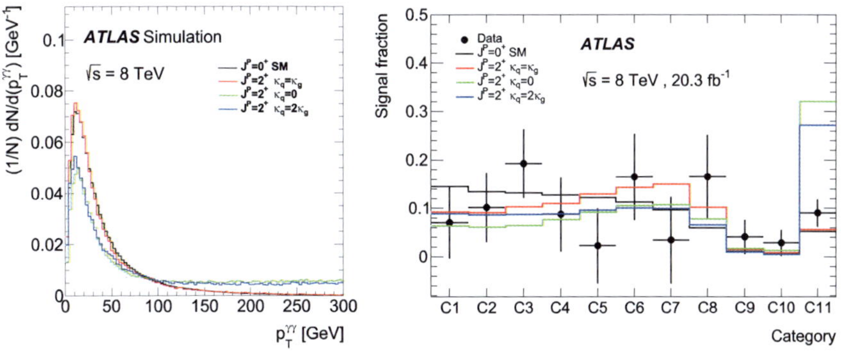

Fig. 6.17 (a) The p_T spectrum predicted for a SM Higgs boson, and several alternative spin-2 models with different couplings to gluons (κ_g) and quarks (κ_q). (b) The measured $\cos\theta^*$ distribution in bins of 0.1 (labelled "C1" to "C10") for events with $p_T < 125$ GeV and the signal yield for 125 GeV $< p_T < 300$ GeV (labelled "C11").[30]

ATLAS chose an Effective Field Theory (EFT) approach,[28,29] which allows more complex processes with the emission of one or more additional partons to be modelled. The predicted $\cos\theta^*$ distributions offer less separation from the SM, but if the strength of the coupling of the Higgs boson to gluons and to quarks is not identical, the Higgs p_T spectrum is visibly enhanced at high p_T, see Fig. 6.17(a). This was exploited by defining a separate event category for events with 125 GeV $< p_T < 300$ GeV, where the upper cutoff was motivated by the fact that the EFT is only valid up to a high energy scale where new physics effects would appear. Events with $p_T > 300$ GeV, which constitute a negligible fraction of the total number of events, were therefore not used in the analysis when testing models where the couplings to vector bosons and to fermions are not identical.

The compatibility of the data with the alternative spin models (see the distributions in Fig. 6.17(b) as an example from the ATLAS study) was determined through hypothesis tests. For this, a likelihood ratio q was constructed as the ratio of the likelihood of the SM hypothesis and the likelihood of the alternative hypothesis, as explained in detail in Appendix A, taking into account the background and expected signal distributions in $m_{\gamma\gamma}$ and $\cos\theta^*$ (and p_T) for the SM and the spin-2 model. The distribution of q for the SM and for the spin-2 models was obtained from pseudo-experiments.

By the nature of the test performed, spin-2 cannot be ruled out in general, but rather only for the particular models which were tested. The models tested by CMS were disfavoured with confidence levels varying from 71% to 94% depending on the relative fraction of gluon and quark-induced production, with the pure gluon fusion case being excluded at 94% confidence level. The models tested by ATLAS were ruled out at 99% confidence level for both the pure gluon fusion production as well as a scenario with enhanced coupling to quarks.

6.5. *Fiducial and differential cross section measurements*

The measurement of fiducial and differential cross sections provides results which are nearly model-independent. They can be compared to predictions from simulation and analytical calculations. They serve as a test of the predictions and deviations from the SM. When more precise theoretical predictions become available in the future, they can easily be compared to the measured cross sections. The differential cross sections are sensitive to different aspects of Higgs boson production and decay. For example, the p_T and rapidity of the Higgs boson are sensitive to the production kinematics, while a measurement of the number of jets produced in association with the Higgs boson and their p_T are a measure of the jet activity in Higgs boson production, and allow tests of higher-order theoretical predictions. Several other distributions offer particularly good separation between ggH and VBF production, such as the dijet invariant mass spectrum, or to the spin of the new particle, such as the azimuthal angle between the two highest p_T jets.

The large signal yield in the diphoton decay channel allowed ATLAS and CMS to perform differential and fiducial cross section measurements.[31,32] In ATLAS, the common fiducial region for all measurements was chosen to limit the dependence on the simulation of $H \to \gamma\gamma$ events, and was given by $|\eta_\gamma| < 2.37$ and $p_T/m_{\gamma\gamma} > 0.35$ and > 0.25 for the two photons. For event signatures with jets, the definition of the fiducial region was given by $p_T^j > 30$ GeV and $|y^j| < 4.4$. Fiducial cross sections were measured for several event signatures and are shown in Fig. 6.18.

The differential measurements were performed by subtracting the background in each bin of the observable, and correcting the signal event yield for efficiency and detector resolution effects. As examples, Fig. 6.19(a) presents the p_T spectrum of the diphoton system and Fig. 6.19(b) shows the p_T of the highest p_T jet in the event.[31] Similar results were obtained by CMS.[32] Within the uncertainties all measurements were consistent with the theoretical predictions for a 125 GeV SM Higgs boson.

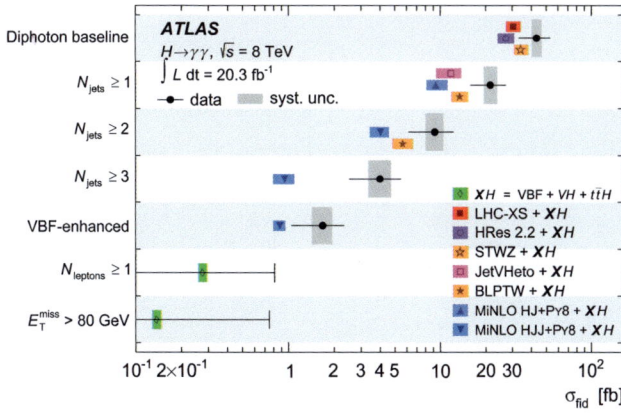

Fig. 6.18   Fiducial cross sections for several event signatures: fully inclusive, with a minimum number of jets, in a phase space region with enhanced contribution from VBF, with an identified electron or muon, and with large E_T^{miss}. The measurements are compared to several state-of-the-art predictions.[31]

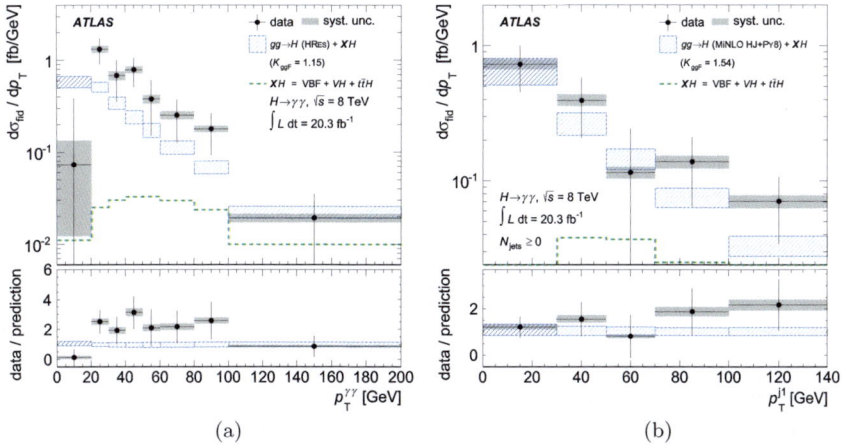

Fig. 6.19   Differential cross sections measured in $H \to \gamma\gamma$ events: p_T of the (a) Higgs boson, $p_T^{\gamma\gamma}$, and (b) the leading jet, p_T^{j1}.[31] For the predictions, the bins are treated as fully correlated. None of the deviations between the data and the predictions are statistically significant.

7. Summary and outlook

The $H \to \gamma\gamma$ decay channel saw an exciting time during the last years. After the important contribution to the discovery of the new particle, it quickly became one of the primary channels to study its properties. Not only did it allow for a precision mass measurement, it also allowed a study of the production modes and the production and decay kinematics. Within the uncertainties, all measured properties were consistent with the expectations for a 125 GeV SM Higgs boson. The results from the $H \to \gamma\gamma$ channel also contributed strongly to the combined, and therefore more precise measurements of the Higgs boson mass, the signal strength and couplings as discussed in Chapter 12.

Most of the measurements performed with the 7 and 8 TeV data were dominated by statistical uncertainties, and none were limited by systematic uncertainties. The larger datasets to be accumulated over the next years at the LHC will allow for more detailed measurements of the Higgs boson properties. The expected increase in luminosity and production cross section at the higher centre-of-mass energy should increase the number of $H \to \gamma\gamma$ events by close to a factor of ten over the next few years. This dataset will allow more precise measurements of the Higgs boson couplings and differential cross sections, and more detailed comparisons with precise theoretical predictions. As a process which contains a loop at leading order, the $H \to \gamma\gamma$ decay mode is particularly sensitive to potential contributions from physics beyond the SM. A particular focus will also be on establishing the existence of the rarer production modes. On the other hand, the higher intensity and therefore higher pileup will also be a significant challenge for the photon reconstruction and identification, the trigger strategy, as well as the vertex selection for $H \to \gamma\gamma$.

References

1. The ALEPH, DELPHI, L3, OPAL, SLD Collaborations, the LEP Electroweak Working Group, the SLD Electroweak and Heavy Flavour Groups, Precision electroweak measurements on the Z resonance, *Phys. Rept.* **427**, 257–454 (2006). doi: 10.1016/j.physrep.2005.12.006.
2. The ALEPH, DELPHI, L3, OPAL Collaborations, the LEP Electroweak Working Group, Electroweak measurements in electron–positron collisions at W-boson-pair energies at LEP, *Phys. Rept.* **532**, 119–244 (2013). doi: 10.1016/j.physrep.2013.07.004.
3. ATLAS Collaboration, Observation of a new particle in the search for the Standard Model Higgs boson with the ATLAS detector at the LHC, *Phys. Lett.* **B716**, 1–29 (2012). doi: 10.1016/j.physletb.2012.08.020.

4. CMS Collaboration, Observation of a new boson at a mass of 125 GeV with the CMS experiment at the LHC, *Phys. Lett.* **B716**, 30–61 (2012). doi: 10.1016/j.physletb.2012.08.021.

5. CMS Collaboration, The CMS experiment at the CERN LHC, *JINST.* **3**, S08004 (2008). doi: 10.1088/1748-0221/3/08/S08004.

6. ATLAS Collaboration, The ATLAS experiment at the CERN Large Hadron Collider, *JINST.* **3**, S08003 (2008). doi: 10.1088/1748-0221/3/08/S08003.

7. ATLAS Collaboration, Measurement of the photon identification efficiencies with the ATLAS detector using LHC Run 1 data. Technical Report CERN-EP-2016-110, CERN, Geneva (2016). arXiv:1606.01813.

8. CMS Collaboration, Performance of photon reconstruction and identification with the CMS detector in proton–proton collisions at $\sqrt{s} = 8$ TeV, *JINST.* **10**(08), P08010 (2015). doi: 10.1088/1748-0221/10/08/P08010.

9. ATLAS Collaboration, Observation of an excess of events in the search for the Standard Model Higgs boson in the $\gamma\gamma$ channel with the ATLAS detector. Technical Report ATLAS-CONF-2012-091, CERN, Geneva (2012). URL http://cds.cern.ch/record/1460410. https://atlas.web.cern.ch/Atlas/GROUPS/PHYSICS/CONFNOTES/ATLAS-CONF- 2012-091/.

10. CMS Collaboration, Energy calibration and resolution of the CMS electromagnetic calorimeter in pp collisions at $\sqrt{s} = 7$ TeV, *JINST.* **8**, P09009 (2013). doi: 10.1088/1748-0221/8/09/P09009.

11. CMS Collaboration, Electromagnetic calorimeter calibration with 7 TeV data. Technical Report CMS-EGM-10-003 (2010). URL http://cds.cern.ch/record/1279350.

12. ATLAS Collaboration, Electron and photon energy calibration with the ATLAS detector using LHC Run 1 data, *Eur. Phys. J.* **C74**(10), 3071 (2014). doi: 10.1140/epjc/s10052-014-3071-4.

13. CMS Collaboration, Observation of the diphoton decay of the Higgs boson and measurement of its properties, *Eur. Phys. J.* **C74**(10), 3076 (2014). doi: 10.1140/epjc/s10052-014-3076-z.

14. ATLAS Collaboration, Search for the Standard Model Higgs boson in the diphoton decay channel with 4.9 fb^{-1} of pp collisions at $\sqrt{s} = 7$ TeV with ATLAS, *Phys. Rev. Lett.* **108**, 111803 (2012). doi: 10.1103/PhysRevLett.108.111803.

15. ATLAS Collaboration, Measurement of Higgs boson production in the diphoton decay channel in pp collisions at center-of-mass energies of 7 and 8 TeV with the ATLAS detector, *Phys. Rev.* **D90**(11), 112015 (2014). doi: 10.1103/PhysRevD.90.112015.

16. GEANT4 Collaboration, Geant4: A simulation toolkit, *Nucl. Instrum. Meth.* **A506**, 250–303 (2003). doi: 10.1016/S0168-9002(03)01368-8.

17. M. Oreglia. A Study of the reactions $\psi' \to \gamma\gamma\psi$. Technical Report SLAC-0236, UMI-81-08973, SLAC-R-0236, SLAC-R-236 (1980). URL http://www.slac.stanford.edu/cgi-wrap/getdoc/slac-r-236.pdf.

18. CMS Collaboration, Search for the Standard Model Higgs boson decaying into two photons in pp collisions at $\sqrt{s} = 7$ TeV. URL https://twiki.cern.ch/twiki/bin/view/CMSPublic/Hig11033TWiki.

19. OPAL Collaboration, Search for anomalous production of dilepton events with missing transverse momentum in e^+e^- collisions at $\sqrt{s} = 161$-GeV and 172-GeV, *Eur. Phys. J.* **C4**, 47–74 (1998). doi: 10.1007/PL00021655.

20. M. Vesterinen and T. Wyatt, A novel technique for studying the Z boson transverse momentum distribution at hadron colliders, *Nucl. Instrum. Meth.* **A602**, 432–437 (2009). doi: 10.1016/j.nima.2009.01.203.

21. ATLAS Collaboration, Measurement of the Higgs boson mass from the $H \to \gamma\gamma$ and $H \to ZZ^* \to 4\ell$ channels with the ATLAS detector using 25 fb^{-1} of pp collision data, *Phys. Rev.* **D90**, 052004 (2014). doi: 10.1103/PhysRevD.90.052004.

22. ATLAS and CMS Collaborations, Combined measurement of the Higgs boson mass in pp collisions at $\sqrt{s} = 7$ and 8 TeV with the ATLAS and CMS experiments, *Phys. Rev. Lett.* **114**, 191803 (2015). doi: 10.1103/PhysRevLett.114.191803.

23. L. Landau, On the angular momentum of a two-photon system, *Dokl. Akad. Nauk Ser. Fiz.* **60**, 207–209 (1948).

24. C.-N. Yang, Selection rules for the dematerialization of a particle into two photons, *Phys. Rev.* **77**, 242–245 (1950). doi: 10.1103/PhysRev.77.242.

25. J. C. Collins and D. E. Soper, Angular distribution of dileptons in high-energy hadron collisions, *Phys. Rev.* **D16**, 2219 (1977). doi: 10.1103/PhysRevD.16.2219.

26. Y. Gao, A. V. Gritsan, Z. Guo, K. Melnikov, M. Schulze, *et al.*, Spin determination of single-produced resonances at hadron colliders, *Phys. Rev.* **D81**, 075022 (2010). doi: 10.1103/PhysRevD.81.075022.

27. CMS Collaboration, Properties of the observed Higgs-like resonance using the diphoton channel. Technical Report CMS-PAS-HIG-13-016, CERN, Geneva (2013). URL http://cds.cern.ch/record/1558930.

28. LHC Higgs Cross Section Working Group, Handbook of LHC Higgs Cross Sections: 3. Higgs Properties (2013). doi: 10.5170/CERN-2013-004. arXiv:1307.1347.

29. P. Artoisenet, P. de Aquino, F. Demartin, R. Frederix, S. Frixione, *et al.*, A framework for Higgs characterisation, *JHEP.* **1311**, 043 (2013). doi: 10.1007/JHEP11(2013)043.

30. ATLAS Collaboration, Study of the spin and parity of the Higgs boson in diboson decays with the ATLAS detector, *Eur. Phys. J.* **C75**(10), 476 (2015). doi: 10.1140/epjc/s10052-015-3685-1.

31. ATLAS Collaboration, Measurements of fiducial and differential cross sections for Higgs boson production in the diphoton decay channel at $\sqrt{s} = 8$ TeV with ATLAS, *JHEP.* **1409**, 112 (2014). doi: 10.1007/JHEP09(2014)112.

32. CMS Collaboration, Measurement of differential cross sections for Higgs boson production in the diphoton decay channel in pp collisions at $\sqrt{s} = 8$ TeV, *Eur. Phys. J.* **C76**(1), 13 (2016). doi: 10.1140/epjc/s10052-015-3853-3. arXiv:1508.07819.

Chapter 7

Observation of the Higgs boson in the $H \to W^+W^{-(*)} \to \ell^+\nu\ell^-\nu$ final state

Emanuele Di Marco[*] and Jonas Strandberg[†]

[*]*Istituto Nazionale di Fisica Nucleare, Roma*
P. le Aldo Moro, 2 - 00185 Roma, Italy
[†]*Physics Department, Royal Institute of Technology, Stockholm, Sweden*

After analysing the full LHC Run 1 data set, the ATLAS and CMS experiments both observed an excess of events above the expected background in the $H \to W^+W^{-(*)} \to \ell^+\nu\ell^-\nu$ channel ($\ell = e, \mu$) consistent with the expectations from the SM Higgs boson with a mass of 125 GeV. The probability to observe an excess equal to or larger than the one seen, under the background-only hypothesis, corresponded to a significance of 6.1 standard deviations for ATLAS (5.8 standard deviations expected) and 4.3 standard deviations for CMS (5.8 standard deviations expected). The observed $\sigma/\sigma_{\mathrm{SM}}$ value for $m_H = 125$ GeV was $1.09^{+0.23}_{-0.21}$ for ATLAS and 0.72 ± 0.20 for CMS. The 0^+ spin-parity hypothesis of the SM was favoured by the data over a narrow resonance with $J^P = 2^+$ or $J^P = 0^-$ decaying into W^+W^-. The results were consistent between the two experiments and provided strong evidence for a SM-like Higgs boson decaying into a W boson pair.

1. Introduction

The coupling between the Higgs boson and the W boson is a cornerstone of the electroweak theory in the Standard Model, providing mass to the W boson and consequently breaking the symmetry between the weak and the electromagnetic forces at low energies. Searches for the Higgs boson through the decay $H \to W^+W^{-(*)}$ was instrumental in finding the Higgs boson, and this channel remains one of the most important for exploring properties such as the spin and parity of this new particle.

The SM Higgs boson decay to two W bosons is characterised by a large branching fraction, for a Higgs boson mass of 125 GeV it is 21.5%. The experimental signature is determined by the subsequent decays of the two W bosons. This chapter concerns the fully leptonic decay mode where both W bosons decay to a charged lepton and a neutrino, $H \rightarrow W^+W^{-(*)} \rightarrow \ell^+\nu\ell^-\nu$ (the search for the Higgs boson at high mass in the semi-leptonic final state, $H \rightarrow W^+W^{-(*)} \rightarrow \ell\nu qq$, is described in Chapter 10). The presence of charged leptons and missing transverse energy from the neutrinos in the detector separates the signal from the large QCD background at the LHC. Other sources of background are more problematic. In particular, the production of top quark pairs that decay via $t\bar{t} \rightarrow W^+b W^-\bar{b}$ yield the same set of leptons and missing transverse energy as the Higgs boson events and the reduction of this background is what largely determines the design of the analysis. After all selections, the SM continuum production of W boson pairs constitutes the largest background but many other smaller sources of background events, mainly from production of single W or Z bosons, also have to be considered. A precise determination of all the backgrounds, in regions of phase space where the simulation suffers from large systematic uncertainties, is one of the biggest challenges in this analysis. To reduce the uncertainties on the background predictions, dedicated control regions in data were used to estimate the majority of the backgrounds.

Another challenge in the $H \rightarrow W^+W^{-(*)} \rightarrow \ell^+\nu\ell^-\nu$ analysis stems from the presence of neutrinos in the final state, which makes it impossible to reconstruct the Higgs boson mass from the visible decay products. Quantities such as the invariant mass distribution of the two charged leptons, or the invariant mass distribution in the transverse plane of the leptons and the missing energy in the event, do depend on the Higgs boson mass however and they can be used to isolate a signal region and reduce the backgrounds. The sensitivity to the Higgs boson mass, m_H, from the observable quantities in the $H \rightarrow W^+W^{-(*)} \rightarrow \ell^+\nu\ell^-\nu$ decay is rather poor though, resulting in a resolution of about 20% in this channel alone.

The $H \rightarrow W^+W^{-(*)} \rightarrow \ell^+\nu\ell^-\nu$ channel was used in combination with the $H \rightarrow \gamma\gamma$ (Chapter 6) and $H \rightarrow ZZ^{(*)} \rightarrow llll$ (Chapter 5) decay modes to establish the existence of the Higgs boson in 2012. Searches for the SM Higgs boson in the $H \rightarrow W^+W^{-(*)} \rightarrow \ell^+\nu\ell^-\nu$ final state at the LHC were carried out in 2011 using data at $\sqrt{s} = 7$ TeV by the CMS[1-3] and ATLAS[4] experiments. The presence of a SM Higgs boson was then excluded at 95% CL in the mass ranges 129–270 GeV and 133–261 GeV, respectively. The results in this channel with the complete Run 1 dataset (7 TeV and 8 TeV data) can be found in the CMS and ATLAS publications from 2014 and 2015.[5,6]

This chapter is organised as follows. Section 2 describes the signal and its characteristic features. Section 3 introduces the sources of background events and the important properties of these processes. Section 4 then describes the event selections and Sec. 5 discusses how the backgrounds are estimated and introduces the various control regions. Section 6 describes the results of the analysis, including the signal strength, the Higgs boson mass sensitivity and the test of spin and CP properties of the discovered particle. Finally, Sec. 7 provides a quick summary of the results and outlines future prospects in the $H \to W^+W^{-(*)} \to \ell^+\nu\ell^-\nu$ analysis.

2. Signal characteristics

To understand the motivation for the event selections in the $H \to W^+W^{-(*)} \to \ell^+\nu\ell^-\nu$ analysis it is instructive to look at the characteristics of the Higgs boson signal events and the various backgrounds. Here we will discuss the important features of the signal process, while the backgrounds are discussed in Sec. 3.

No jets are expected at leading-order when the Higgs boson is produced through gluon-fusion (shown in Fig. 7.1(a)), which is the principal production mode at the LHC. Higher order contributions lead to a non-negligible fraction of the signal events being reconstructed with one or more jets. Higgs bosons created through the vector-boson fusion production mode, depicted in Fig. 7.1(b), naturally contain two jets from the outgoing quarks in the final state. To obtain maximum sensitivity to the signal, events with 0, 1 or ≥ 2 jets were treated as separate channels due to the differences in the expected background compositions.

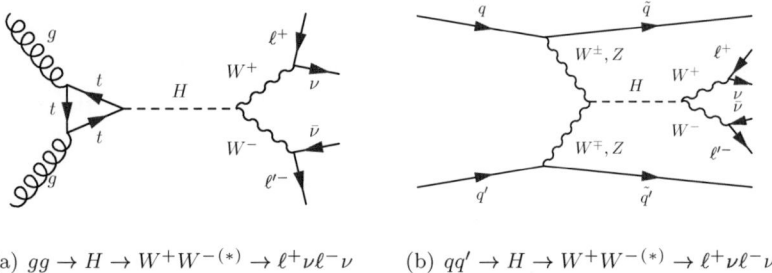

(a) $gg \to H \to W^+W^{-(*)} \to \ell^+\nu\ell^-\nu$ (b) $qq' \to H \to W^+W^{-(*)} \to \ell^+\nu\ell^-\nu$

Fig. 7.1   Feynman diagrams for the gluon-fusion Higgs boson production (a) and vector-boson-fusion Higgs production (b) followed by the decay of the Higgs boson into a W boson pair that subsequently decay into a fully leptonic final state.

The Higgs boson decay into two W bosons is the favoured decay mode for Higgs boson masses above 140 GeV, while at $m_H = 125$ GeV it is the second most likely decay mode with a branching fraction of 21.5% (as discussed in Chapter 1). Only the direct decay of a W boson into a muon or an electron plus the corresponding neutrino gives a significant contribution to the selected dilepton sample. While the decay chain $W \to \tau \bar{\nu}_\tau$ followed by $\tau \to \ell \nu_\tau \bar{\nu}_\ell$ gives the same experimental signature, the typical momentum of the final electron or muon is lower. This makes the selection efficiency almost negligible compared to the direct decays. Combining the decay probabilities for the two W bosons, about 7% of all W-pair events will result in two oppositely charged leptons and at least two neutrinos. Since the neutrinos do not interact with any components of the detector they will escape undetected, resulting in a large transverse momentum imbalance.

For a Higgs boson mass of 125 GeV, there is not enough energy available in the decay to create two on-shell W bosons. The favoured mass configuration is for one of the two W bosons to be off-shell with a mass of around 45 GeV (denoted as W^*) while the other W boson is on-shell. This results in an asymmetric probability distribution for the momenta of the two charged leptons in the final state. The less energetic of the two leptons, typically coming from the off-shell W boson decay, has a transverse momentum of less than 20 GeV in approximately 75% of the $H \to W^+W^{-(*)} \to \ell^+\nu\ell^-\nu$ decays and less than 15 GeV in approximately 30% of the decays. These kinematic characteristics imply that the experimental analysis must be able to efficiently trigger, reconstruct and identify pairs of relatively low-p_T leptons.

Due to the spin-0 nature of the Higgs boson in the SM, the two W bosons originating from the decay must have opposite spin orientations. Because of the V-A nature of the W boson decay, the left-handed ℓ^- is preferentially emitted opposite to the W^- spin direction. For the same reason the right-handed ℓ^+ is preferentially emitted along the W^+ spin direction. A schematic of the spin correlations in the $H \to W^+W^{-(*)} \to \ell^+\nu\ell^-\nu$ decay is shown in Fig. 7.2. For the Higgs boson signal, the distributions of the lepton opening angle in the transverse plane, $\Delta\phi(\ell, \ell)$, and the invariant mass of the two leptons, $m_{\ell\ell}$, are expected to be enhanced at low values compared to the continuum WW background, where the spin orientations of the two W bosons are not strongly correlated.[7,8]

Event displays of two candidate events selected in the $H \to W^+W^{-(*)} \to \ell^+\nu\ell^-\nu$ analysis are shown in Figs. 7.3 and 7.4. Both events contain one electron and one muon, with a small opening angle between

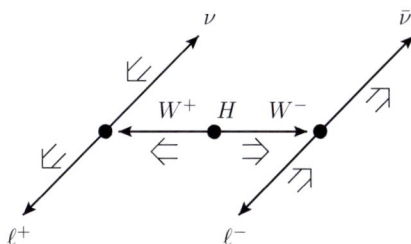

Fig. 7.2 Schematic representation of the spin correlations for a scalar Higgs boson decaying into two leptons and two neutrinos. The small arrows indicate the particles' directions of motion and the large double arrows indicate their spin projections. The spin-0 Higgs boson decays to W bosons with opposite spins, and the spin-1 W bosons decay into leptons with aligned spins. The H and W decays are shown in the decaying particle's rest frame. Because of the V-A decay of the W bosons, the charged leptons have a small opening angle in the laboratory frame.

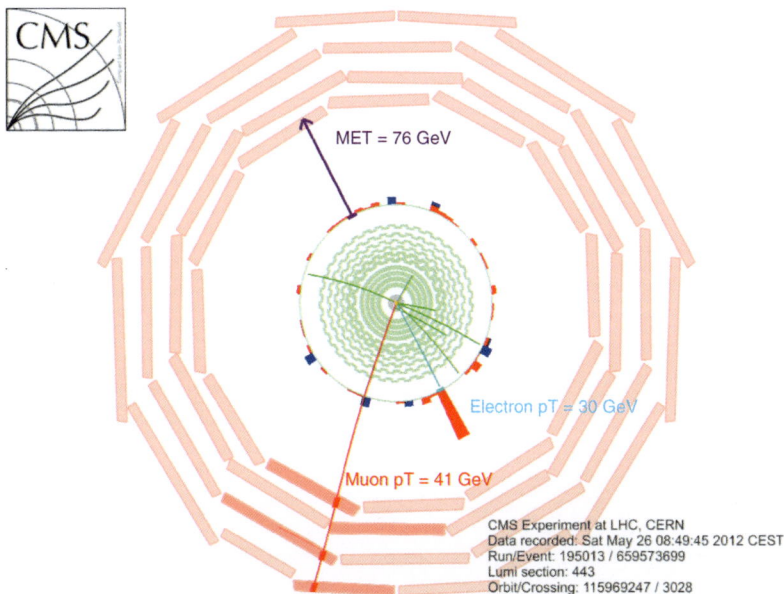

Fig. 7.3 Event display $((x,y)$ view) of one selected event in the 8 TeV data with a Higgs boson candidate decaying into $e\nu_e\mu\nu_\mu$ from the CMS experiment. The electron with $p_T \approx 30$ GeV can be seen as the one depositing the energy in the electromagnetic calorimeter (red towers) while the muon with $p_T \approx 41$ GeV interacts only in the external muon chambers. The opening angle between them in the (x,y) plane is $\Delta\phi(e,\mu) = 0.6$. Both are isolated from any major activity in the calorimeter or the tracker. Opposite to the direction of the two leptons, the purple arrow indicates the direction of the large missing transverse energy in the event, $E_T^{\text{miss}} = 76$ GeV. No jets above a p_T threshold of 30 GeV are observed in this event, typical of the gluon-fusion production mechanism.

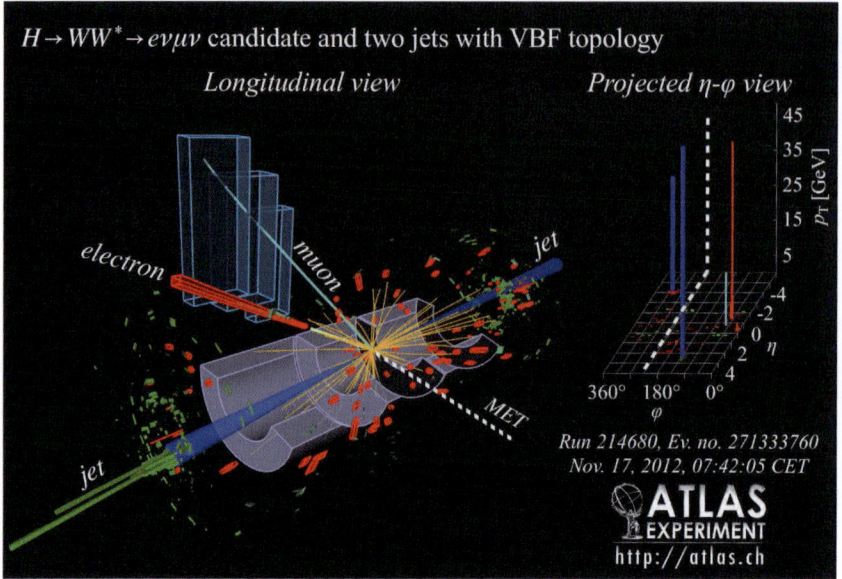

Fig. 7.4   Event display of one selected event in the 8 TeV data with a Higgs boson candidate decaying into $e\nu_e\mu\nu_\mu$ from the ATLAS experiment. The electron and the muon can be seen to the left in the detector, with a small opening angle between them $\Delta\phi(e,\mu) = 0.1$. The electron has transverse momentum 51 GeV and the muon 15 GeV. Both are isolated from any major activity in the calorimeter or the tracker. The dashed white line indicates the large missing transverse energy in the event, $E_{\mathrm{T}}^{\mathrm{miss}} = 59$ GeV, pointing in the direction opposite from the two leptons. Two jets, with transverse momenta 67 GeV and 41 GeV, travelling almost parallel to the beam-pipe can be seen. The invariant mass of the pair of jets is 1.4 TeV. The location of energy deposits in the calorimeter, projected on the transverse plane, is shown in the (η, ϕ) view in the right panel. The electron (red) and muon (cyan) are seen in the central part of the detector. The two jets (dark blue) are visible at large values of η. Energetic jets in the forward directions are characteristic of the VBF production mechanism.

the two leptons. The leptons are not balanced by anything visible in the opposite hemisphere, presumably because of the two neutrinos travelling in that direction. The first event, observed in the CMS detector, is devoid of any reconstructed jets. The absence of jets is typical for Higgs bosons produced through the gluon-fusion mechanism. In the second event, from the ATLAS detector, two energetic jets travelling almost parallel to the direction of the beam-pipe can be observed. Such jets are typical for Higgs bosons produced through the vector-boson-fusion mechanism.

3. Main backgrounds

Dominant sources of background events in the Run 1 $H \to W^+W^{-(*)} \to \ell^+\nu\ell^-\nu$ analysis were processes that produce two W bosons, mainly SM continuum WW production, $t\bar{t}$ production and to a smaller extent tW production. Another process that resulted in leptons and neutrinos in the final state is the decay of a Z boson to a pair of tau leptons, $Z \to \tau^+\tau^-$, with the sub-sequent decays $\tau \to \ell\nu_\tau\bar{\nu}_\ell$. All other background processes involved some amount of mis-identification of an object. The dominant sources were from production of single W bosons in conjuncture with jets, from Drell–Yan production of e^+e^- or $\mu^+\mu^-$ with large $E_{\mathrm{T}}^{\mathrm{miss}}$ from a mis-measurement of the momentum of one of the leptons or jets in the event, and from $W\gamma^{(*)}$ processes where a photon was mis-identified as a charged lepton.

3.1. *The WW background*

The Standard Model production of continuum W^+W^- pairs proceeds mainly through $q\bar{q} \to W^+W^-$. The lowest order diagrams are the s-channel contribution, with a Z boson or a virtual photon as intermediate particle, and the t-channel contribution with an intermediate quark. The corresponding Feynman diagrams are shown in Fig. 7.5(a) and 7.5(b), respectively. Pair production of W bosons also occurs via gluon-fusion, where the two W bosons are produced through a quark loop (Fig. 7.5(c)). The latter production mode accounts for less than 5% of the total WW rate at $\sqrt{s} = 7$ and 8 TeV. The main characteristics of the background WW events which allowed them to be separated from the Higgs boson signal were the decay angles and the invariant mass of the leptons. As was discussed in Sec. 2, the spin correlations in the signal events lead to different expected distributions for these variables for the signal compared to the continuum background.

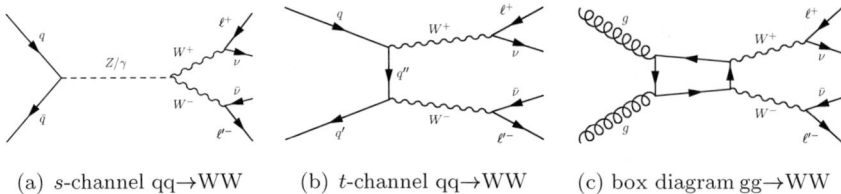

(a) s-channel qq→WW   (b) t-channel qq→WW   (c) box diagram gg→WW

Fig. 7.5   Feynman diagrams for the LO WW production at the LHC through (a) the $q\bar{q}$ initial state in the s-channel, (b) $qq\prime$ scattering in the t-channel and (c) via the gluon-fusion channel.

The production cross section of non-resonant WW pairs in the fully leptonic final state at 7 TeV has been measured by both the CMS and ATLAS experiments[9,10] and at 8 TeV by the CMS experiment.[11]

3.2. *Other diboson backgrounds*

Diboson processes, other than WW production, also yield final states with multiple leptons and missing transverse energy. The $W\gamma$ and $W\gamma^*$ processes constituted the bulk of this background. The W boson decay yields one lepton plus missing transverse energy, and the second lepton required to mimic the Higgs signal was selected from a pair of leptons arising either from photon conversion in the material of the detector or from prompt production through $\gamma^* \to \ell^+\ell^-$ (Fig. 7.6(b)). Very small background contributions to the $H \to W^+W^{-(*)} \to \ell^+\nu\ell^-\nu$ analysis came from WZ or ZZ production. These processes have low cross sections, and when both vector bosons decay leptonically the final states comprise more than two high momentum leptons. Such events only entered the analysis in the rare cases where only two of the leptons were successfully identified.

3.3. *The top quark backgrounds*

Other sources of W boson pairs, in conjunction with jets, are decays of top quarks (the top quark almost always decays into a W boson and a b quark). Production of $t\bar{t}$ pairs leads to a final state of two W bosons and two b quarks, the latter two nominally reconstructed as two jets. Single top quarks can also be produced via the weak force, where the main contribution to the background in this analysis came from tW production. Top quark events were mainly rejected by looking at the number of reconstructed jets in the event, and by rejecting events where any jet was tagged as coming from a b quark.

3.4. *The W+jets, Z/γ^*+jets and QCD backgrounds*

Another set of background sources were processes with high production cross sections, such as the associated production of single W or Z bosons with jets or QCD multijet events. The leptonic decays of W bosons yield one charged lepton plus missing transverse energy. This signature would normally not pass the dilepton selection criterion, the exception is on rare occasions when an additional jet in the event gets mis-identified as the

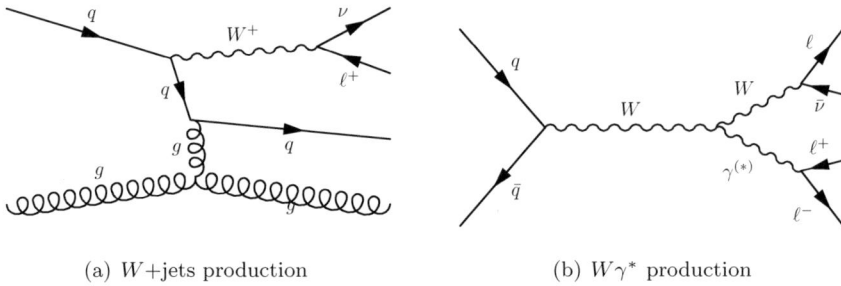

(a) W+jets production   (b) $W\gamma^*$ production

Fig. 7.6 Feynman diagrams for (a) W+jets production (one jet from a gluon and another jet from a quark in this example) and (b) for $W\gamma^*$ production with the subsequent decays of the W boson and the γ^* into leptons.

second lepton. The Feynman diagram for the production of W+jets is shown in Fig. 7.6(a). The decay $Z \rightarrow \tau^+\tau^- \rightarrow \ell^+\bar{\nu}_\tau\nu_\ell\ell'^-\nu_\tau\bar{\nu}_{\ell'}$ can yield charged leptons of the same or of different flavours in the final state. Due to the high mass of the Z boson, the tau leptons are usually emitted at a large opening angle and the neutrinos are boosted in the directions of the decaying tau leptons. This results in events with typically low momentum charged leptons, since the tau lepton momentum has to be shared between three decay products, and on average low missing transverse energy due to the cancellations of neutrinos travelling in opposite directions. The direct decays of Z/γ^* bosons into pairs of electrons or muons (Drell–Yan production) constituted an important background in the same-flavour lepton final states. There is no natural source of missing transverse energy for this process, but due to the finite E_T^{miss} resolution a small fraction of events passed the event selections. Since the initial cross section is very large compared to the signal cross section, the Z/γ^*+jets background was sizeable in the same-flavour final states. Multijet production has no natural sources of high-p_T leptons nor of missing transverse energy, and was therefore a very minor background despite its enormous cross section.

4. Event selection

The event selection consisted of three main stages. Events with two isolated leptons and high missing transverse energy were selected in the first step. This reduced the contributions from high cross section processes, such as multijet production and single W and Z boson production, and resulted in a sample dominated by the WW and $t\bar{t}$ backgrounds. The second step involved dividing the sample according to the number of reconstructed jets in the events and their properties. For events with one or more jets, additional selections such as a b-tagging veto were applied to reduce the $t\bar{t}$ background. The final step attempted to isolate the Higgs boson signal from the WW and $t\bar{t}$ backgrounds by looking at the kinematics of the two identified leptons. The distributions of variables that differentiate between the signal and the background, such as the opening angle between the two leptons or the invariant mass of the leptons, were used in multidimensional fits with signal and background templates to extract the signal, as described later in Sec. 6.1.

4.1. *The preselection*

All details regarding the CMS and ATLAS event selections are given in Refs. [5] and [6], here the main requirements imposed in the event selections are summarised:

(1) The events were required to contain two leptons with a minimal p_T of 10 GeV. At least one of the two leptons had to have a p_T of more than 22 (20) GeV in the ATLAS (CMS) analysis. Events where more than two leptons were identified were discarded, to reduce contributions from diboson processes decaying to multiple charged leptons. The asymmetric lepton p_T requirements ensured a high efficiency for low Higgs boson masses where the leptons originating from the off-shell W boson in the decay have lower transverse momentum on average. The leptons had to pass strict identification and isolation requirements, to reduce the impact of the W+jets background. Single lepton selection efficiencies were in the range 70–80% depending on the p_T and η of the leptons, typically lower for leptons with $p_T < 20$ GeV due to the stricter identification requirements applied to those leptons.

(2) The invariant mass of the two leptons, $m_{\ell\ell}$, had to be larger than 12 GeV to reject events from low-mass Drell–Yan production or resonances decaying to a pair of leptons. In the e^+e^- and $\mu^+\mu^-$ final states, events were also vetoed if the invariant mass was within 15 GeV of the nominal Z boson mass. Due to the Z boson background, both ATLAS and CMS treated the different-flavour ($e\mu$) and the same-flavour ($e^+e^-/\mu^+\mu^-$) final states as separate analysis channels.

(3) The transverse momentum of the dilepton four-vector, $p_T^{\ell\ell}$, was required to be above 30 (45) for ATLAS (CMS). The $p_T^{\ell\ell}$ variable is highly correlated with a small opening angle $\Delta\phi(\ell, \ell)$ of the two leptons, and it

mainly reduces the background from $Z \to \tau^+\tau^-$ and Drell–Yan production where the two leptons are produced back-to-back. If the decaying Z/γ^* boson is boosted in the transverse plane it can produce leptons with a high $p_T^{\ell\ell}$ and thus this selection is less performant in events with high-p_T jets.

(4) Several different definitions of the E_T^{miss} were used in the analyses to obtain optimal rejection of backgrounds without neutrinos in the final state. The main variable used was the magnitude of the E_T^{miss} relative the direction of the closest lepton or jet in the transverse plane, since for backgrounds without neutrinos the E_T^{miss} tends to align with a lepton or a jet. A quantity $E_{T,\mathrm{rel}}^{\mathrm{miss}}$ was defined as $E_{T,\mathrm{rel}}^{\mathrm{miss}} = E_T^{\mathrm{miss}} \sin\Delta\phi_{\mathrm{near}}$ if $\Delta\phi_{\mathrm{near}} < \pi/2$, where $\Delta\phi_{\mathrm{near}}$ is the azimuthal angle between the E_T^{miss} and the nearest lepton or jet, and as $E_{T,\mathrm{rel}}^{\mathrm{miss}} = E_T^{\mathrm{miss}}$ otherwise. The $E_{T,\mathrm{rel}}^{\mathrm{miss}}$ distribution after the requirement of two isolated leptons is shown in Fig. 7.7. The selection criteria on the $E_{T,\mathrm{rel}}^{\mathrm{miss}}$ varied substantially depending on the lepton flavours and the number of reconstructed jets, with additional small variations between the two experiments. Typically the $E_{T,\mathrm{rel}}^{\mathrm{miss}}$ was required to be larger than 20 GeV in all channels (the ATLAS analysis targeting VBF production in the $e\mu$ final state omitted the $E_{T,\mathrm{rel}}^{\mathrm{miss}}$ cut altogether) . In the e^+e^- and $\mu^+\mu^-$ final states, a more stringent requirement of $E_{T,\mathrm{rel}}^{\mathrm{miss}} > 40$ GeV was necessary to suppress the background from Drell–Yan production. Another

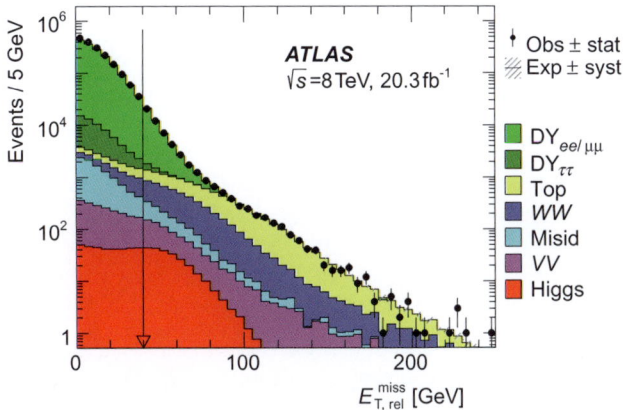

Fig. 7.7 Distribution of $E_{T,\mathrm{rel}}^{\mathrm{miss}}$ for events with exactly two selected leptons in the ATLAS detector. The arrow indicates the selection threshold for events with same-flavour leptons in ATLAS, events below the threshold were rejected. The selection reduces the dominant Drell–Yan background, but due to the finite E_T^{miss} resolution a small fraction of Drell–Yan events survive the selection.

important variable for rejecting the Drell–Yan background was the p_T^{miss}, calculated using only the tracks originating from the primary vertex of the event. This variable is insensitive to the pile-up interactions, but neglects all neutral particles and the charged particles outside the tracker acceptance. The $E_{T,rel}^{miss}$ and the p_T^{miss} estimates were largely uncorrelated for background events with no real E_T^{miss}, while they were strongly correlated for events with high-momentum neutrinos where a real energy imbalance is expected. Combining the track-based p_T^{miss} estimate with some calorimeter information (ATLAS) or using it in an MVA discriminator together with other kinematic variables (CMS) further enhanced the Drell–Yan background rejection.

(5) To further improve the rejection of the Drell–Yan background in the e^+e^- and $\mu^+\mu^-$ channels, ATLAS required that the amount of hadronic energy measured in the quadrant opposite to the dilepton direction, f_{recoil}, was small. The hadronic recoil is expected to be higher for processes without neutrinos, such as Drell–Yan, compared to processes where the E_T^{miss} comes from undetected neutrinos.

4.2. *Jet multiplicity selections*

Events were classified according to their jet multiplicity (jets with p_T above 30 GeV were considered for both experiments, except for $|\eta| < 2.4$ where ATLAS used a threshold of $p_T > 25$ GeV). The primary purpose of this selection was to define a sample of events, without reconstructed jets, where the contribution from the top quark background was small. For events comprising exactly one jet, the top quark background and the WW background were of approximately equal magnitude. Events with two or more jets were dominated by the top quark background. Additional selection criteria such as requiring a large pseudorapidity gap, $\Delta\eta(j,j)$, and large invariant mass of the two jets, m_{jj}, suppressed this background and resulted in a rather pure class of signal events dominated by VBF production (described in Sec. 5 of Chapter 1). ATLAS defined a special class of events with two or more jets that failed the VBF specific selections, which targeted Higgs bosons produced in the gluon-fusion production mode with two additional jets. Figure 7.8 summarises all the various jet multiplicity and lepton flavour channels used in the ATLAS analysis.

To reduce the background contribution from top quark processes, events with jets identified as coming from b quarks were rejected. These jets tend to contain tracks with a large impact parameter with respect to the primary

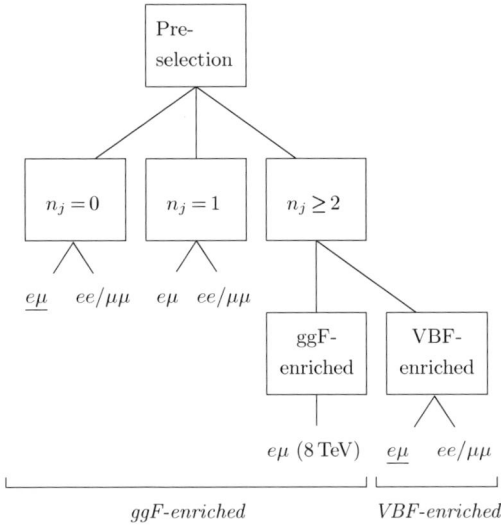

Fig. 7.8   Analysis divisions in categories based on jet multiplicity (n_j) and lepton-flavour samples ($e\mu$ and $ee/\mu\mu$) for the ATLAS analysis. The most sensitive signal region for gluon-fusion production (ggF) was $n_j = 0$ in $e\mu$, while for VBF production it was $n_j \geq 2$ in $e\mu$. These two samples are underlined. CMS used the same channel definitions, but omitted the gluon-fusion enriched $n_j \geq 2$ channel.

vertex, coming from the decay of the b-hadron. Furthermore, b-hadrons decay semi-muonically in approximatively 11% of the cases, and therefore the jets may contain non-isolated, soft ($p_T \approx 3\,\text{GeV}$) muons with large impact parameter. Multivariate techniques were used to identify b-jets using a combination of track and muon information as input.

The jet multiplicity distributions after the lepton and E_T^{miss} selections had been applied are shown in Fig. 7.9, separately for events with same-flavour leptons and different-flavour leptons. Also shown in Fig. 7.9 is the number of identified b-jets in the different-flavour final state. The noticeable changes in background composition as a function of the lepton flavours and jet multiplicity is what prompted treating each combination as a separate analysis channel in order to obtain the best sensitivity for the combined result.

4.3. *Kinematic shape analysis*

The composition of the event sample after all selections had been applied was dominated by backgrounds with two W bosons, from SM continuum

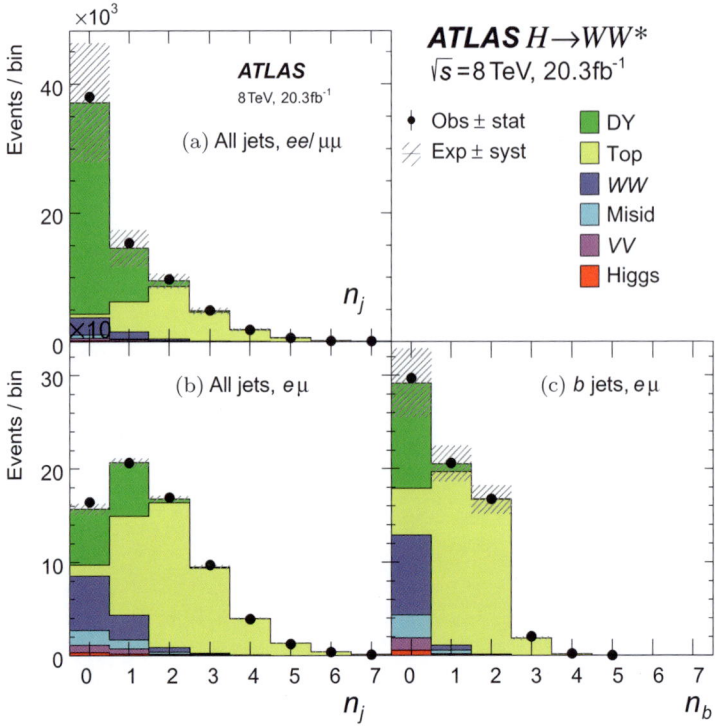

Fig. 7.9   Distributions of jet multiplicity for all jets, n_j, and for jets identified as coming from b quarks, n_b, after applying the lepton and E_T^{miss} selections in ATLAS. Additional selections which suppressed the Drell–Yan background, for example the requirement of large $p_T^{\ell\ell}$, were still to be applied. Backgrounds labelled "Misid" include contributions from W+jets and other backgrounds with mis-identified leptons.

production or from top quark decays. As discussed in Secs. 2 and 3, the spin correlation between the two W bosons originating from a scalar Higgs boson decay produces different angular distributions between the final state leptons compared to W boson pairs from the backgrounds. Two (CMS) or four (ATLAS) variables that depend on the lepton kinematics were used for extracting the signal, using shape fits to the data that are described in more detail in Sec. 6. One variable used was the so-called transverse mass:[12]

$$m_T = \sqrt{2p_T^{\ell\ell}E_T^{\text{miss}}(1 - \cos(\Delta\phi(\ell\ell, E_T^{\text{miss}})))}, \qquad (7.1)$$

where $\Delta\phi(\ell\ell, E_T^{\text{miss}})$ is the angle between the dilepton direction and the E_T^{miss} in the transverse plane. In the approximation that the two neutrinos balance the $p_T^{\ell\ell}$, which is mostly true in events with zero jets, and if the invariant mass of the two neutrinos is on average equal to the invariant mass of the two charged leptons, the distribution of m_T is proportional to the true Higgs mass. The distribution has a large r.m.s. of around 30–40 GeV, due to deviations from these approximations in the presence of jets from initial state radiation and due to the finite resolution of the missing transverse energy. The m_T distributions in each combination of jet multiplicity and lepton flavours from ATLAS are shown in Fig. 7.10. Another variable used in the shape fit was the invariant mass of the two charged leptons, $m_{\ell\ell}$, which is highly correlated with $\Delta\phi(\ell,\ell)$ and better at discriminating between signal and background. This variable is also well described by the simulation, since both the angular and momentum resolutions of electrons

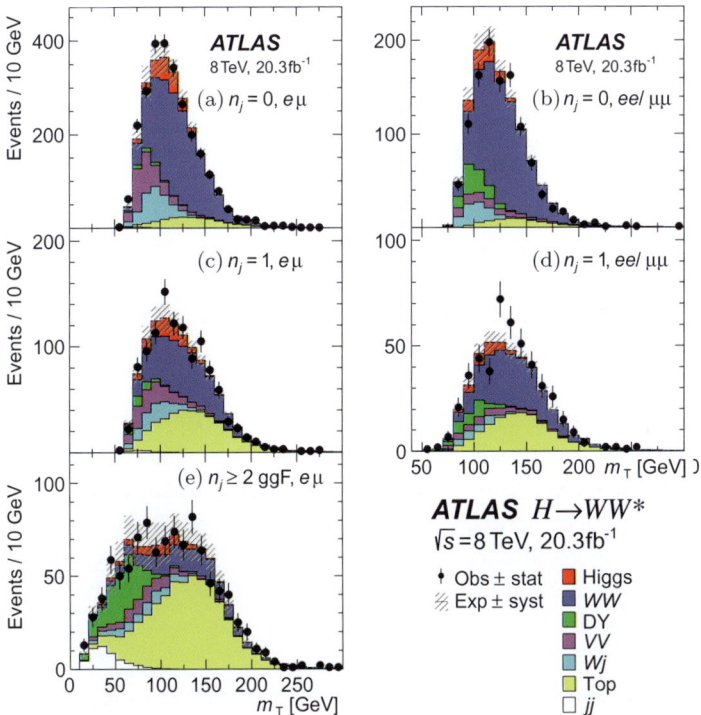

Fig. 7.10   Distributions of transverse mass, m_T, in the various jet multiplicity and lepton flavour channels in the ATLAS analysis targeting the gluon-fusion production mode. The expected contributions from all the SM backgrounds and from a $m_H = 125$ GeV SM Higgs boson are shown. All the selections up to the transverse mass have been applied. Shaded areas represent the total uncertainty on the background prediction.

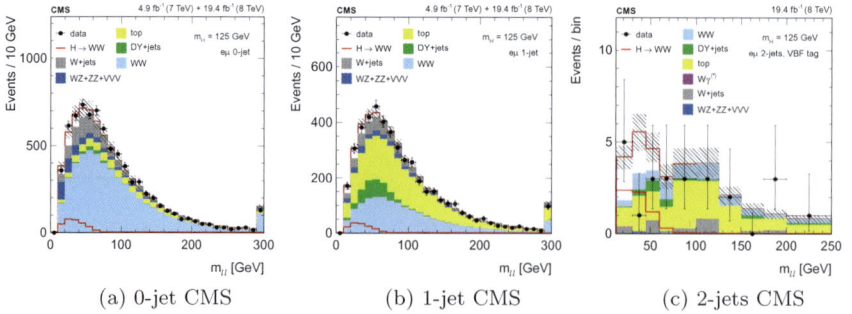

Fig. 7.11 Invariant mass distributions for the two leptons in the CMS analysis for the $e\mu$ final state in events with zero jets (a), with one jet (b), and with two jets and the VBF selection (c). The expected contributions from all the SM backgrounds and from a $m_H = 125$ GeV SM Higgs boson are shown (the signal is shown both superimposed and stacked on top the backgrounds). All selections up to the final kinematic fit have been applied. The shaded areas represent the total uncertainty on the background prediction.

and muons are well measured in CMS and ATLAS. The $m_{\ell\ell}$ distributions in events with zero, one and two jets from CMS are shown in Fig. 7.11. Given the absence of a narrow mass peak in the $H \to W^+W^{-(*)} \to \ell^+\nu\ell^-\nu$ analysis, the main handle to infer the presence of a Higgs boson was by observing an overall excess of events in the two-dimensional plane spanned by the m_T and $m_{\ell\ell}$ variables (ATLAS also used the p_T and flavour of the sub-leading lepton as a third and fourth dimension in the fit to extract the signal).

Figure 7.12 shows the reduction in sample size and the evolution of the signal and background composition after each stage of the CMS event selection. The selection up to the *anti b-tag* defined the so-called WW sample, used for the final significance estimate and other measurements. Also shown, for illustration purposes, is the effect of applying further selections on the p_T of the two leptons, on $m_{\ell\ell}$, on m_T and on $\Delta\phi(\ell,\ell)$ instead of using the full shape information of these variables as was done in the default analyses from ATLAS and CMS. An analysis which imposes selections on the final variables instead of using them in a shape fit is called a "counting analysis" and both experiments used counting analyses as cross checks for the default methods.

At the end of the selection process, the Higgs boson events surviving in the 0-jet and 1-jet samples were mainly produced through the gluon-fusion process. In events with two jets, most of the sensitivity came from the

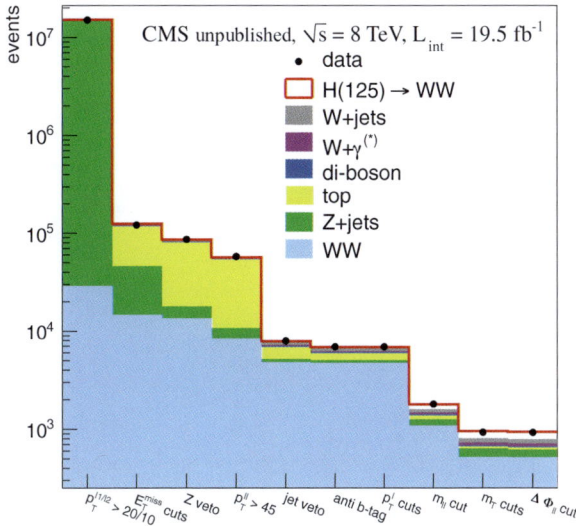

Fig. 7.12 The background composition after each stage of the event selection for the sum of same-flavour and different-flavour final states for events with fully identified leptons in the CMS detector. Shown here, for illustration purposes, is the effect of applying selections on the $m_{\ell\ell}$, m_T and $\Delta\phi(\ell,\ell)$ distributions instead of using the full shape information of these variables as was done in the default analyses from ATLAS and CMS.

selection targeting the VBF production of the signal. The expected signal yields, broken down into the different categories and production modes, is given in Table 7.1 (for CMS). The expected yields in the same-flavour channels are lower than in the $e\mu$ channel largely due to the differences in the E_T^{miss} selections. The 2-jet category, after all selections, contained only about 2% of the total yield[13] but made a significant contribution to the overall sensitivity due to the large signal-to-background ratio. In the ATLAS analysis, the contribution from the VBF selected sample to the overall sensitivity of the analysis was second only to the 0-jet $e\mu$ channel.

5. Background estimation

The $H \to W^+W^{-(*)} \to \ell^+\nu\ell^-\nu$ analysis had to account for a diverse set of backgrounds. It was preferable to use estimates obtained from data control samples instead of simulation, since the selected $\ell^+\nu\ell^-\nu$ events typically were in the tails of the kinematic distributions (for example in the high E_T^{miss} and low $m_{\ell\ell}$ tails for Drell–Yan events). The simulation does not

Table 7.1 Summary of the expected signal fractions per production mode for a SM Higgs boson with a mass of 125 GeV in the CMS analysis for each combination of lepton flavour and jet multiplicity, and the total number of $H \to W^+W^{-(*)}$ events expected at $\sqrt{s} = 7$ and 8 TeV.

| | | | | Total $H \to W^+W^{-(*)}$ yield | |
Category	ggH (%)	VBF (%)	VH (%)	$\sqrt{s} = 7$ TeV	$\sqrt{s} = 8$ TeV
0-jet eμ	95.7	1.2	3.1	52.6	245
0-jet ee, $\mu\mu$	98.1	0.9	1.0	10.4	58.5
1-jet eμ	81.6	10.3	8.1	19.8	111
1-jet ee, $\mu\mu$	83.6	11.2	5.2	3.1	19.6
2-jet eμ	22.3	77.7	0.0	1.3	6.4
2-jet ee, $\mu\mu$	14.2	85.8	0.0	0.3	2.3

always provide an accurate description of the backgrounds in these regions. This is especially true regarding the probability of a jet being mis-identified as a lepton, which determines the rate of the W+jets background. Therefore control regions were selected in data, which ideally were dominated by a single background process. The normalisation, and in some cases also the kinematic distributions, of these backgrounds were determined from the event yields in the control regions, N^{CR}_{bkg}, scaled by a factor, α, determined from simulation that relates the yield in the control region to the yield in the signal region N^{SR}_{bkg}:

$$N^{SR}_{bkg} = \alpha \cdot N^{CR}_{bkg}. \tag{7.2}$$

Since α is determined from the ratio of two event yields in the simulation, it is unaffected by the dominant systematic uncertainties having to do with the jet selections. Except for some small contributions to the total background yield, all processes in the zero-jet and one-jet channels were estimated using the control region method. In the following, the background estimation techniques and the control samples used for the different backgrounds are described (the full details of each background estimation can be found in the CMS[5] and ATLAS[6] publications).

5.1. *The WW background*

The WW control region was defined by altering the requirement on the invariant mass of the two leptons in the selection. The signal region was defined by requiring $m_{\ell\ell} < 55$ GeV, while control regions contained the events with $m_{\ell\ell}$ above 100 GeV for CMS[5] and $m_{\ell\ell}$ between 55 and 110 GeV for ATLAS.[6] Residual backgrounds from other SM processes were subtracted. The resulting number of events in the control region was multiplied by the extrapolation factor to obtain the expected contribution of non-resonant WW in the signal region. The main systematic uncertainty on this estimate was from the knowledge of the extrapolation factor, which is based on the accuracy of the description of the $m_{\ell\ell}$ distribution by the WW generators. The uncertainty on the extrapolation factor was estimated from a comparison of results obtained with alternative generators, and resulted in an uncertainty of approximatively 10% and 30% for the $q\bar{q} \to WW$ and $gg \to WW$ processes, respectively. Because the WW process was the dominant background in the category with the highest sensitivity (0-jet, different flavour), this systematic uncertainty was among the most important ones for the final results.

5.2. *The top quark backgrounds*

The top-quark background processes mostly affected the channels with jets in the final state. In both the ATLAS and CMS analyses, the normalisation of these backgrounds was estimated from a high-purety control sample of top-quark events obtained by requiring a b-tagged jet. In the zero-jet channels a b-tagged jet with p_T greater than 20 GeV, but below the jet identification threshold of 30 GeV, was required. The extrapolation to the signal region, where events with b-tagged jets are vetoed, was done by means of the b-tagging efficiency of a single jet. This efficiency could be measured directly in the data, thereby reducing the uncertainty on the extrapolation between the control and signal regions. The distribution of the transverse mass in the control region for events with zero jets is shown in Fig. 7.13(a) (for CMS). The dominant systematic uncertainty on this estimate was from the limited statistics in the control region for events with no jets, amounting to approximately 20%.

5.3. *The W + jets background*

Events with leptonic decays of W bosons may pass the analysis selections when an additional jet is mis-identified as an isolated lepton. The mis-identified lepton may be either a true electron or muon from the decay of a heavy quark, or a product of the fragmentation of a quark or a gluon incorrectly being reconstructed as an isolated lepton. The estimation of this process can only be done reliably in data. It was obtained by using a control

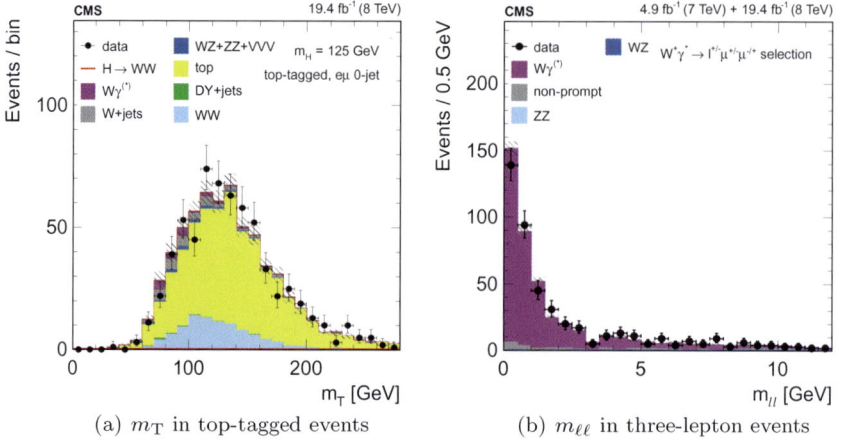

(a) m_T in top-tagged events   (b) $m_{\ell\ell}$ in three-lepton events

Fig. 7.13   The Higgs boson transverse mass m_T in the top-tagged control region in the zero-jet events used to estimate the top background (a) and the invariant mass of the two leptons $m_{\ell\ell}$ in the three-lepton control region with zero jets used to estimate the $W\gamma^{(*)}$ contribution in CMS (b). Shaded areas represent the total uncertainty on the simulation prediction.

sample of events in which one of the two leptons was fully identified, and the other lepton failed these criteria but satisfied a loosened selection. The normalisation of the background in the signal region was obtained by scaling the number of events in the control sample by the probability for the loosely identified lepton to satisfy the signal identification criteria (denoted as the *fake rate*). The fake rate was estimated in a statistically independent sample, either an inclusive dijet data sample or a Z+jets data sample, after subtracting the contributions from leptonic W and Z boson decays. The systematic uncertainties on the fake rate dominated the total uncertainty on the background estimate. The uncertainty on the fake rate incorporated the dependence on the sample composition, since the fake rate depended on whether the mis-identified lepton originated from a quark-induced jets, a gluon-induced jets, a heavy flavour decay, from charge exchange in the electromagnetic calorimeters or from a photon conversion. The uncertainty was determined by modifying the jet p_T threshold of the dijet sample, which modified the jet composition. Another large uncertainty on the W + jets estimate came from the validity of the estimation method itself. The uncertainty on the method was obtained from a closure test using simulated events, where the fake rate was derived from simulated QCD multijet events and applied to simulated samples of W + jets events to predict the number of background events in the signal region. The total uncertainty on the W + jets background estimate, including the statistical precision of the control sample, was of the order of 40%.

5.4. *The* $W\gamma/W\gamma^*$, *WZ and ZZ backgrounds*

The $W\gamma$ and the $W\gamma^*$ processes can pass the event selections when the photon (either real or virtual) produces a lepton pair (through pair production or prompt production, respectively), where one of the leptons is lost either because of low momentum or other inefficiencies in the reconstruction. The CMS analysis used a control sample with either three muons or one electron and two muons to normalise this background. The events in the control region were also required to have high E_T^{miss}, and for the events with one electron and two muons the invariant mass of the two muons was required to be low. Figure 7.13(b) shows the $m_{\ell\ell}$ distribution of the selected three-lepton events, compared with the expected distribution, dominated by $W\gamma^{(*)}$ and W+jets events. The ATLAS analysis used a control sample with same-charge leptons, since the contributions from the $W\gamma$ and $W\gamma^*$ backgrounds are charge symmetric. The total background yields expected from the $W\gamma$ and $W\gamma^*$ processes were small but not negligible, since they closely mimic the kinematic features of the Higgs boson signal.

The WZ and ZZ backgrounds were minor, and they were estimated from simulation and normalised to their theoretical cross sections.

5.5. *The* $Z/\gamma^* \to e^+e^-, \mu^+\mu^-$ *backgrounds*

Drell–Yan production was a major background for the same-flavour final states, because of its large cross section. It was reduced by several orders of magnitude from the high E_T^{miss} and low $m_{\ell\ell}$ requirements, but a small fraction of events passed the stringent selections applied in the same-flavour channels. CMS estimated the number of surviving events in the signal region by an extrapolation from the number of observed events at intermediate E_T^{miss} values within a narrow mass range around the Z boson pole mass. The central value for the extrapolation factor was estimated from simulation and validated on data. The largest uncertainty, between 20% to 50%, on the Drell–Yan background estimate came from the dependence of the extrapolation factor on the E_T^{miss}. ATLAS estimated the Drell–Yan background with a method based on the different efficiencies for Drell–Yan and non-Drell–Yan processes to pass the f_{recoil} selection. These efficiencies were measured in data, and the number of Drell–Yan events passing the f_{recoil} selection could be determined from the number of events failing the selection. The largest uncertainties on the background prediction, between 30% and 40%, came from the measurements of the efficiencies.

5.6. *The $Z \to \tau^+\tau^- \to \ell^+\nu_\ell\bar{\nu}_\tau\ell'^-\bar{\nu}_{\ell'}\nu_\tau$ background*

Cascade decays of Z bosons to two leptons and four neutrinos via a pair of tau leptons can contribute to both the same-flavour and the different-flavour final states. It was a negligible background in the zero-jet category, where the two tau leptons travel in opposite directions and the contributions to the observed $E_{\text{T}}^{\text{miss}}$ from the neutrinos in their decays largely cancel, while it accounted for roughly 10% of the total background in the one-jet category. ATLAS normalised this background in a control region at high $\Delta\phi(\ell, \ell)$ and low $m_{\ell\ell}$, while CMS estimated it with a hybrid data–simulation approach. A sample of $Z \to \mu^+\mu^-$ events in data was selected, and then the reconstructed muons were substituted with the decay products of a simulated τ lepton decaying leptonically. In this way the reconstructed quantities involving the $E_{\text{T}}^{\text{miss}}$, which are hard to model correctly by the simulation in a high-pileup environment, are correctly estimated.

5.7. *Summary of background estimates*

Table 7.2 summarises the estimation method of each major background in the 0-jet and 1-jet channels (any lepton combination) for the CMS analysis. Other minor sources of background events were estimated by simulation. In the 2-jet category, where all backgrounds except $t\bar{t}$ and tW were smaller, dedicated control regions could no longer be found for many background processes. For example, the WW process was estimated from simulation because of the lack of events in the control region.

6. Results

Table 7.3 lists the signal prediction, observed number of events in data and background estimates in the $e\mu$ final state for the $\sqrt{s} = 8$ TeV data from CMS for the default shape analysis (before fitting the shapes) and for the alternative counting analysis mentioned in Sec. 4 where explicit selections were made on the $m_{\ell\ell}$, m_{T} and $\Delta\phi(\ell, \ell)$ variables instead of fitting the shapes. The reference Higgs boson mass, when not stated explicitly, is

Table 7.2   Summary of the estimation methods of the background processes, for the CMS experiment, when data events were used to estimate either the normalisation or the shape of the discriminant variables. The given examples are valid for the 0-jet and 1-jet categories.

Process	Control/template sample
WW	events at high $m_{\ell\ell}$ and m_{T}
Top-quark	top-tagged events
$W + \mathrm{jets}$	events with loosely identified leptons
$W\gamma$	events with an identified photon
$W\gamma^*$	$W\gamma^* \rightarrow 3\mu$ sample
$Z/\gamma^* \rightarrow \mu^+\mu^-$ & $Z/\gamma^* \rightarrow e^+e^-$	events at low $E_{\mathrm{T}}^{\mathrm{miss}}$
$Z/\gamma^* \rightarrow \tau^+\tau^-$	τ embedded sample

125 GeV. The corresponding predictions and observed number of events in the ATLAS analysis were similar.

The m_{T} distribution, after normalisation of the signal to the likelihood fit outcome, is shown in Fig. 7.14(a) for events with zero or one jet in the ATLAS analysis. In the bottom panel of Fig. 7.14(a), the data is shown with the background subtracted together with the predicted shape of the signal for a Higgs boson with a mass of 125 GeV. In Fig. 7.14(b), the m_{T} distribution for the VBF selected events is shown together with the expected distributions for signal and background. The bottom panel shows how the selected events were positioned in the (m_{jj}, m_{T}) plane, and what the expected signal-to-background ratios were in the different regions of the plane. The plots visualise the discrepancies between the predictions from the SM without a Higgs boson and the data of the LHC experiments, and how well the data agreed with the additional presence of a resonance with m_{T} distributions consistent with the Higgs boson. These distributions were used in the statistical analysis to determine quantitatively the statistical significance for the signal and the measurements of the cross section, spin and parity.

6.1. *Measurement of the signal strength*

The statistical interpretation of the data was obtained with a binned likelihood $\mathcal{L}(\mu, \theta)$, where the signal strength $\mu = \sigma/\sigma_{\mathrm{SM}}$ was the parameter of interest and θ represents all parameters which were integrated over (a more in-depth discussion of the likelihood function is given in Appendix A). In the CMS analyses with zero or one jet, 4×3 bins in $m_{\mathrm{T}} \times m_{\ell\ell}$ were used for the $e\mu$ final state (referred to as the "shape analysis") while only a single

Table 7.3 Signal predictions, background estimates and the observed numbers of events in data for the $\sqrt{s} = 8$ TeV CMS analysis after applying the requirements used for the counting analysis and for the shape-based analysis (before fitting the shapes). The reported uncertainties are a combination of statistical, experimental and theoretical uncertainties. The $Z/\gamma^* \to \ell\ell$ process includes the e^+e^-, $\mu^+\mu^-$ and $\tau^+\tau^-$ final states. The shape-based selections correspond to the $m_H = 125$ GeV selection.

m_H [GeV] (Shape)	ggH	VBF+VH	Data	All bkg.	WW	$WZ+ZZ$ $+Z/\gamma^* \to \ell\ell$	$t\bar{t}+tW$	W+jets	$W\gamma^{(*)}$
8 TeV $e\mu$ final state 0-jet category									
125	88 ± 19	2.19 ± 0.22	506	429 ± 34	310 ± 29	11.4 ± 1.0	19.9 ± 4.3	48 ± 13	39 ± 13
160	370 ± 80	8.75 ± 0.71	285	239 ± 19	196 ± 18	5.94 ± 0.61	24.9 ± 5.4	5.9 ± 2.0	6.3 ± 3.5
shape-analysis	227 ± 46	10.27 ± 0.41	5747	5760 ± 210	4185 ± 63	178.3 ± 9.5	500 ± 96	620 ± 160	282 ± 76
8 TeV $ee/\mu\mu$ final state 0-jet category									
125	55 ± 12	1.10 ± 0.14	423	361 ± 37	207 ± 19	106 ± 31	9.4 ± 2.2	29.0 ± 7.8	9.3 ± 3.8
160	319 ± 69	6.78 ± 0.58	258	214 ± 19	164 ± 15	28.5 ± 9.7	14.0 ± 3.2	5.7 ± 1.9	1.72 ± 0.92
8 TeV $e\mu$ final state 1-jet category									
125	37 ± 12	6.53 ± 0.53	228	209 ± 14	80 ± 11	12.9 ± 1.2	79.2 ± 4.6	25.9 ± 6.9	11.2 ± 4.6
160	180 ± 57	30.6 ± 2.5	226	174 ± 11	73.3 ± 9.6	7.98 ± 0.83	83.2 ± 4.7	8.7 ± 2.8	1.07 ± 0.69
shape-analysis	88 ± 28	19.83 ± 0.81	3281	3242 ± 90	1268 ± 21	193 ± 11	1443 ± 46	283 ± 72	55 ± 14
8 TeV $ee/\mu\mu$ final state 1-jet category									
125	15.8 ± 5.1	3.09 ± 0.28	141	111.9 ± 8.6	39.9 ± 5.4	21.2 ± 5.4	40.8 ± 3.1	6.6 ± 2.0	3.3 ± 1.7
160	103 ± 33	16.8 ± 1.5	134	113.8 ± 8.2	46.8 ± 6.2	13.8 ± 3.9	48.0 ± 3.2	3.9 ± 1.5	1.3 ± 1.0

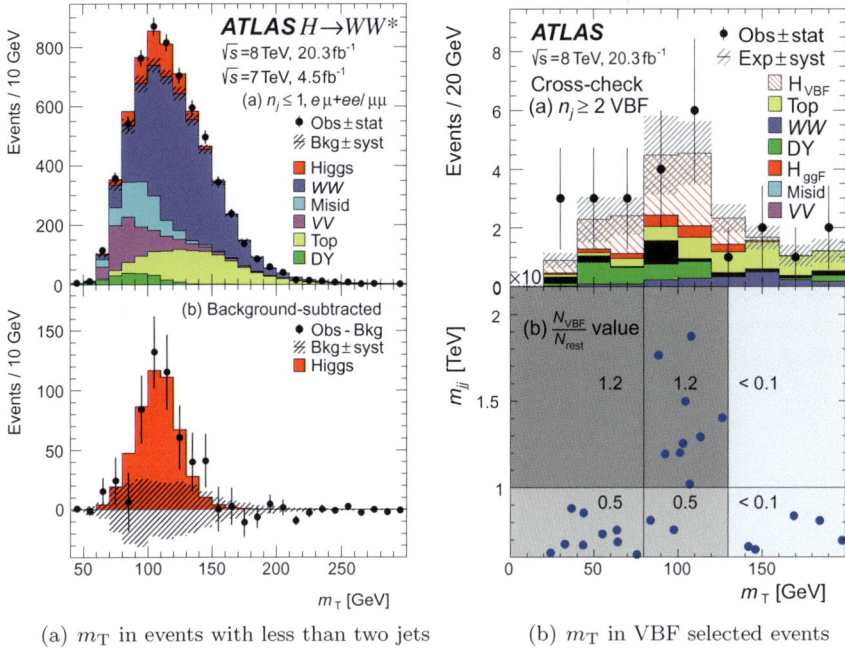

(a) m_T in events with less than two jets (b) m_T in VBF selected events

Fig. 7.14   The expected and observed m_T distribution from ATLAS in the $e\mu$ final state for the 0-jet and 1-jet categories combined (a) is shown in the top panel. The bottom panel of (a) shows the background-subtracted data with the best-fit signal component. Shaded areas represent the total uncertainty on the background prediction. The plot (b) depicts the m_T distribution in the VBF-enriched category in the 8 TeV data analysis from ATLAS (top). The bottom panel shows the m_{jj} versus m_T scatter plot for the selected events. For each region in the bottom panel, the expected ratio of signal-to-background is stated in the plot, where the background includes all processes other than the VBF signal.

bin was used for the final states with same-flavour leptons (a "counting analysis"). The choice to use a counting analysis for the same-flavour final states was made because the Drell–Yan process makes it difficult to model the $E_\mathrm{T}^{\mathrm{miss}}$, which is an input to the m_T variable.[5] For the analyses with zero or one jet in ATLAS, 10 bins in m_T for each of 12 regions was used for the $e\mu$ final state and 10 bins in m_T for a single region was used for the same-flavour final states. The 12 regions in the $e\mu$ channel were obtained from two ranges in $m_{\ell\ell}$, three ranges in p_T of the sub-leading lepton and two choices depending on the flavour of the sub-leading lepton ($2 \times 3 \times 2 = 12$).[6] Due to the small number of expected events for the VBF category, CMS used an un-binned counting analysis while ATLAS used a fit to four bins in an output distribution of a boosted decision tree trained to separate the signal from the background.

The control regions described in Sec. 5 were modelled by Poisson probability terms in the likelihood. The signal strength parameter μ multiplied the expected Standard Model Higgs boson signal in each bin. Signal and background predictions had systematic uncertainties that were parametrised by nuisance parameters θ. The expected signal and background event counts in each bin were functions of θ. The parametrisation for them was chosen such that the rates in each channel were log-normally distributed for a normally distributed θ (see Sec. A.2.1 of Appendix A).

A test statistic q_μ was constructed using the profile likelihood as defined in Eq. A.6. This test statistic was used to compute the probability (p-value) that the background could fluctuate to the level of the observed data or higher, and to calculate the exclusion limits following the modified frequentist method known as CL_s. The p-value was also used to compute the statistical significance of the excess over the background-only hypothesis.

For a Higgs boson with mass $m_H = 125$ GeV, ATLAS found an excess of events with respect to the background-only hypothesis corresponding to a significance of 6.1 standard deviations while 5.8 standard deviations was expected. For CMS, the significance of the excess at $m_H = 125$ GeV was 4.3 standard deviations, while the expected significance was 5.8 standard deviations. The observed $\sigma/\sigma_{\mathrm{SM}}$ value for $m_H = 125$ GeV was $1.09^{+0.23}_{-0.21}$ for ATLAS and 0.72 ± 0.20 for CMS. Considering the magnitude of the uncertainties, both measurements of the signal strength were consistent with the expectations for a Higgs boson with a mass of approximately 125 GeV. The most important uncertainties were the theoretical uncertainties on the signal coming from the categorisation of the events based on jet multiplicity and due to the imperfect knowledge of the factorisation and renormalisation scales.[14,15] The second most important uncertainty was the limited accuracy with which the WW cross section was known. Other important sources of systematic uncertainties were the lepton energy scale, the E_T^{miss}, and the jet energy scale experimental uncertainties, as well as the limited knowledge of the $W + $ jets and $W\gamma^{(*)}$ background processes.

Figure 7.15(a) shows the observed p-value for ATLAS and Fig. 7.15(b) shows the significance of the excess for CMS as functions of the Higgs boson mass. For comparisons, the dashed lines show the expected results

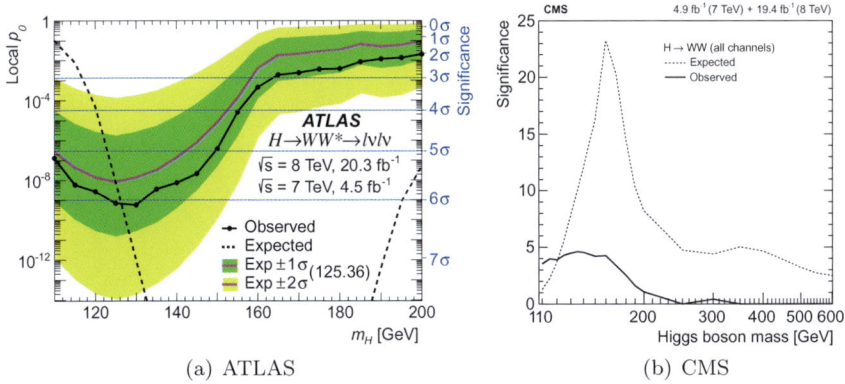

(a) ATLAS   (b) CMS

Fig. 7.15 Expected and observed local p-values from ATLAS (a) for the background-only hypothesis (the so-called p_0). The solid line without points represent the expected p-values as a function of the tested Higgs boson mass for a Higgs boson with true mass $m_H = 125$ GeV and the shaded inner (darker) and outer (lighter) regions represent the one and two standard deviations around the expected value. The dashed black line indicates the expected p-values for the presence of a signal at the value of the x-axis, tested at that same value. The solid line with points represents the observed p-values, showing good agreement with the expectation for a $m_H = 125$ GeV Higgs signal. The observed and expected significances for the combined $7 + 8$ TeV analysis of CMS data is shown in (b). The dashed line shows the expected signal significance for the value of m_H on the x-axis, while the solid line shows the observed significance as a function of the tested Higgs boson mass. Good agreement between observed and expected significance can be observed around a Higgs boson mass of $m_H = 125$ GeV.

for varying Higgs boson masses when they are only tested at their true mass. For both ATLAS and CMS the expected and observed results agree at a Higgs mass around $m_H \approx 125$ GeV. Due to the poor mass resolution of the $H \to W^+W^{-(*)} \to \ell^+\nu\ell^-\nu$ analysis, the excess of the observed narrow resonance at $m_H = 125$ GeV extended over a large range of tested Higgs boson masses (approximatively ±30 GeV around the 125 GeV mass). For ATLAS, the minimum p-value was found in a region around $m_H = 130$ GeV, and corresponded to the same significance of 6.1 standard deviations as was found at $m_H = 125$ GeV.

The value of μ could also be measured in each category: zero-jet, one-jet (sensitive mainly to the gluon-fusion production), and two-jets (sensitive to VBF production) and per lepton flavour combination. All measurements were found to be consistent with the SM expectations.[5,6] Table 7.4 shows the measured signal strengths when the analyses were parametrised according to the production modes.

Table 7.4   Summary of the measured values of the signal strength μ for the CMS and ATLAS experiments, separately in the gluon-fusion categories and for the VBF tag category, and for the combination of all the categories considered in the analysis. Also shown are the measured signal strengths in the analyses targeting the VH (where V can be either a W or a Z boson) and WH production modes from CMS. The uncertainties comprise both statistical and systematic components, including theoretical uncertainties.

Category	CMS	ATLAS
Gluon-fusion	0.76 ± 0.21	$1.02^{+0.29}_{-0.26}$
VBF tag (2-jets)	$0.62^{+0.58}_{-0.47}$	$1.27^{+0.53}_{-0.45}$
VH tag (2-jets)	$0.39^{+1.97}_{-1.87}$	—
WH tag ($3\ell3\nu$)	$0.56^{+1.27}_{-0.95}$	—
All categories	0.72 ± 0.20	$1.09^{+0.23}_{-0.21}$

6.2. *Mass sensitivity*

The effective mass resolutions of the four-lepton channel and the two-photon channel were 1–2 GeV, while the m_T distribution for the Higgs boson signal in the $H \to W^+W^{-(*)} \to \ell^+\nu\ell^-\nu$ analysis was much broader. The m_T variable peaked close to the correct m_H, but with an effective resolution of about 30 GeV. Additional sensitivity came from the compatibility of the measured excess with the expected SM Higgs $\sigma\times$BR, given the rapidly varying branching ratio to WW as a function of m_H (about 4%/GeV for $m_H \approx 125$ GeV). The contour of the two-dimensional likelihood in a simultaneous scan of μ and m_H for $H \to W^+W^{-(*)} \to \ell^+\nu\ell^-\nu$ is shown in Fig. 7.16. The ATLAS and CMS results were compatible among them, and compatible with the more precise mass measurements in the $H \to ZZ^{(*)} \to llll$ and $H \to \gamma\gamma$ channels.

6.3. *Spin measurement*

As in the $H \to ZZ^{(*)} \to llll$ final state, the kinematic correlations between the leptons can be exploited to determine the spin of the decaying resonance. The sensitivity is reduced in the $H \to W^+W^{-(*)} \to \ell^+\nu\ell^-\nu$ channel due to the fact that two of the leptons are not visible, so their directions and momenta are not measurable separately. Still, the combination of the three-momenta of the two charged leptons and the missing transverse energy carry some information that can be correlated to the Higgs field tensor structure.

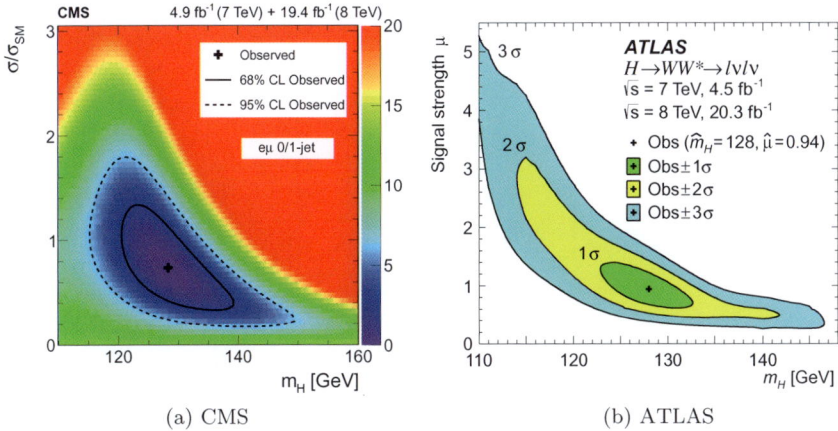

(a) CMS   (b) ATLAS

Fig. 7.16   Confidence intervals in the (μ, m_H) plane for the different-flavour final states combining the 0-jet and 1-jet categories for the shape analysis at 7 and 8 TeV in the CMS analysis (a) and the ATLAS analysis (b). The lines in (a) indicate the 68% and 95% CL contours.

6.3.1. *Spin-parity quantum numbers constraints*

The ATLAS and CMS experiments have tried to distinguish alternative J^P signal hypotheses[16] from the SM Higgs boson using the $H \to W^+W^{-(*)} \to \ell^+\nu\ell^-\nu$ channel, using a similar strategy as the one described in Sec. 11.2 of Chapter 5. Both experiments have tested the 0^+ and 2^+ hypotheses by using a two-dimensional shape fit, but with different distributions. CMS used the same variables adopted for the couplings measurements: $m_{\ell\ell}$ and m_T. ATLAS used the outputs of two different boosted decision trees (BDT), trained separately against all backgrounds to identify 0^+ and 2^+ events, respectively. The BDT for 2^+ events, where the production can go via qq annihilation through a p-wave instead of gluon-fusion, was retrained for each f_{qq} fraction (the fraction of events produced through qq annihilation).

In general the expected event yields depend on the model (SM spin 0^+ or the alternative 2^+). When the expected significance of the analysis was estimated the assumption was made that the yields were the same. The observed significance was independent of this assumption however, since the signal yields were obtained from the data simultaneously with the other parameters of interest in a binned maximum likelihood fit. The signal rates were allowed to float independently for each signal type. The test statistic was defined as $q = -2\,\ln(L_{J^P}/L_{0^+})$ where L_{0^+} and L_{J^P} were the best-fit likelihood values for the SM Higgs boson and the alternative hypothesis.

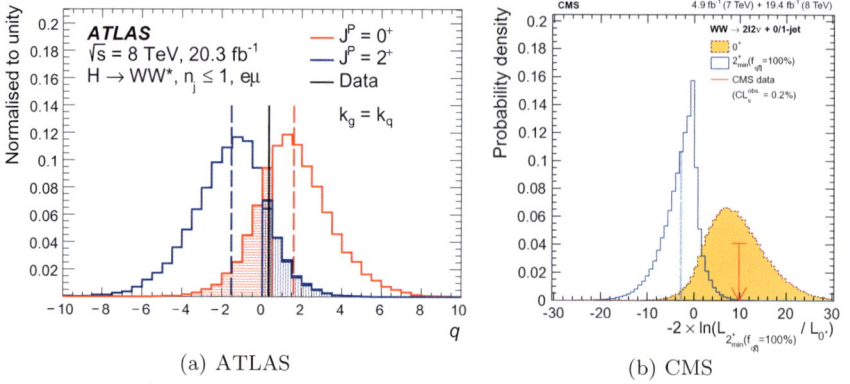

(a) ATLAS   (b) CMS

Fig. 7.17   Distributions of $q = -2 \ln(L_{JP}/L_{0^+})$ for the $e\mu$ final states combining the 0-jet and 1-jet categories, for the 0^+ and 2^+ hypotheses, produced assuming a μ equal to the one fitted on data, for the point $f_{qq}=100\%$. For the ATLAS plot (a) the median of the expected distributions for the SM (dashed red line) and the spin-2 Higgs-boson signal (dashed blue line) are also shown, together with the observed result (solid black line) from the fit to the data. In the CMS plot (b), the red arrow indicates the value observed in data.

The expected separation between the two hypotheses was computed for each considered f_{qq} fraction. Figure 7.17(a) and Fig. 7.17(b) show the result for pure $q\bar{q}$ production, for which the separation was maximal, for ATLAS[17] and CMS[5,18] respectively. The data favoured the SM hypothesis, and this resulted in an exclusion of the 2^+ hypothesis with a significance in the $[1.2$–$3.1]\sigma$ range for CMS and the 2^+ hypothesis excluded at 84.5% confidence level for ATLAS.

In a similar fashion, both collaborations tested a wide range of anomalous spin and parity states against the observed resonance. The alternative CP-odd 0^- hypothesis was excluded at 96.5% confidence level by ATLAS and at 65.3% confidence level by CMS. An overall picture of various spin-2 models which have been studied by the CMS experiment is shown in Fig. 7.18. As in the $H \to ZZ^{(*)} \to llll$ case (see Sec. 11 in Chapter 5), the alternative J^P states were disfavoured by the data.

6.3.2. *Search for anomalous decay amplitude admixtures*

Since pure states with alternative J^P quantum numbers were excluded, ATLAS and CMS also performed searches for possible admixtures of anomalous spin-zero decay amplitudes. These correspond to the α-, β-,

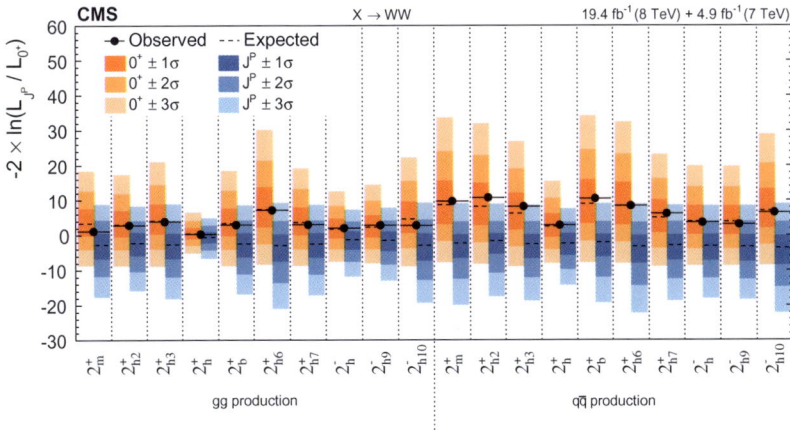

Fig. 7.18 Distribution of the test statistic $q = -2 \ln(L_{JP}/L_{0+})$ for the spin-two J^P models tested against the SM Higgs boson hypothesis in the CMS 8 TeV data. The expected median and the 68%, 95%, and 99.7% CL regions for the SM Higgs boson (orange) and for the alternative J^P hypotheses (blue) are shown. The observed q values are indicated by the black dots.

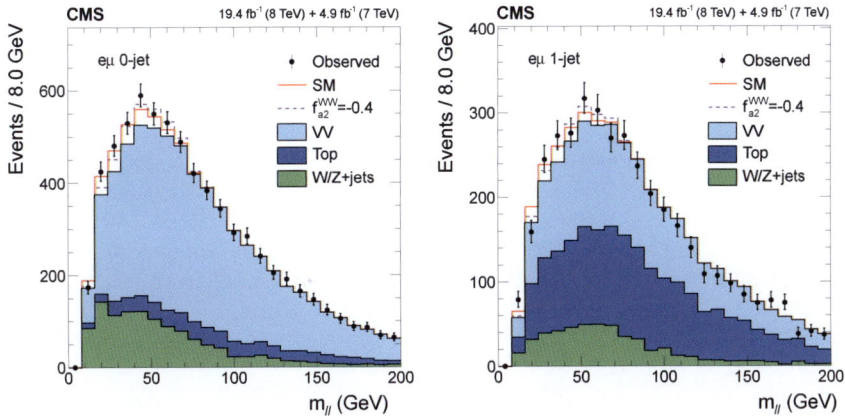

Fig. 7.19 Distributions of $m_{\ell\ell}$ for events with 0 jets (left) and 1 jet (right) for the CMS 8 TeV data. The observed data, the expectations for the SM backgrounds, the expectation for the SM Higgs boson signal, and the expectation from the alternative spin-zero resonance are shown. The mass of the resonance was taken to be 125 GeV and the SM Higgs boson cross section was used.

Table 7.5   The 95% CL intervals obtained on the allowed amounts of alternative, not SM-like, spin-zero states admixed to the primarily SM-Higgs-boson-like state.

	α/κ	β/κ	γ/κ
CMS allowed range at 95% CL	[-1.2, 1.5]	$[-\infty, 0.71]$ [1.22, $+\infty$]	not tested
ATLAS allowed range at 95% CL	not tested	[-0.4, 0.85] [1, 2.2]	[−5, +6]

and γ-terms in the spin-zero Lagrangian, given by Eq. (1.22). The observables used were the m_T and $m_{\ell\ell}$ variables for CMS and BDTs trained for the BSM CP-even and BSM CP-odd Higgs bosons for ATLAS. Figure 7.19 shows the expected shifts in the $m_{\ell\ell}$ distribution between the Standard Model Higgs boson and one of the alternative hypotheses in the 0-jet and 1-jet categories. Comparisons with the data then allowed for the characterisation of the admixture of anomalous decay amplitudes.[18] The resulting allowed intervals at 95% CL for the three higher-dimension operators in Eq. (1.22) are summarised in Table 7.5.

7. Conclusions and outlook

A search for the SM Higgs boson decaying to W boson pairs at the LHC was carried out by the CMS and ATLAS experiments. The event samples used in the analyses corresponded to the full LHC Run 1 dataset, using both the $\sqrt{s} = 7$ TeV and 8 TeV run periods. The expected signal yield in the $H \to W^+W^{-(*)} \to \ell^+\nu\ell^-\nu$ channel was comparable to the $H \to \gamma\gamma$ channel and about ten times higher than in $H \to ZZ^{(*)} \to llll$. The expected background contamination was larger than in the $H \to ZZ^{(*)} \to llll$ channel but lower than in the $H \to \gamma\gamma$ channel. The limited mass resolution in the $H \to W^+W^{-(*)} \to \ell^+\nu\ell^-\nu$ channel, compensated by the high expected signal yield and relatively small background, resulted in a comparable sensitivity in this channel to a Higgs boson with mass m_H=125 GeV as in the two high-mass-resolution channels.

Excesses of events above the expected backgrounds were observed by both the CMS and ATLAS experiments, consistent with the expectations from the SM Higgs boson with a mass of approximately 125 GeV. The probability to observe an excess equal to or larger than the one seen, under the background-only hypothesis, corresponded to a significance of 4.3 standard deviations for $m_H = 125$ GeV for CMS and 6.1 standard deviations for ATLAS. The observed signal strength for $m_H = 125$ GeV was 0.72 ± 0.20 for CMS and $1.09^{+0.23}_{-0.21}$ for ATLAS. The 0^+ spin-parity hypothesis was favoured by the data against a narrow resonance with $J^P = 2^+$ or $J^P = 0^-$ decaying into W^+W^-. The results were consistent between the two experiments and provided strong evidence for a SM-like Higgs boson decaying to W boson pairs.

When data taking resumes in Run 2 of the LHC, at a centre-of-mass energy of 13 TeV, the exploration of the Higgs boson will continue and the predictions of the SM regarding this particle will be put to stringent tests. The integrated luminosity is expected to reach over 100 fb^{-1} in Run 2, which together with the increased collision energy should lead to approximately a factor of 10 increase in Higgs boson statistics. This particularly favours channels with a small but clean signal, such as the VBF channel in the $H \to W^+W^{-(*)} \to \ell^+\nu\ell^-\nu$ analysis. The VBF channel probes the λ_{HWW} coupling between the Higgs boson and the W boson in both the production and the decay, and an increased precision in the measurement of this coupling can be expected in Run 2. The increase in statistics will lead to a better population of control regions in the $H \to W^+W^{-(*)} \to \ell^+\nu\ell^-\nu$ analysis that before were statistics limited, which will increase the precision on the measured $\sigma/\sigma_{\rm SM}$ value. The larger dataset should also facilitate observing Higgs boson decays to two W bosons in rarer production modes such as in associated production with a W or Z boson.

References

1. CMS Collaboration, Measurement of W^+W^- production and search for the Higgs boson in pp collisions at $\sqrt{s} = 7$ TeV, *Phys. Lett. B* **699**, 25–47 (2011). doi: 10.1016/j.physletb.2011.03.056.
2. CMS Collaboration, Search for the Standard Model Higgs boson decaying to a W pair in the fully leptonic final state in pp collisions at $\sqrt{s} = 7$ TeV, *Phys. Lett. B* **710**, 91–113 (2012). doi: 10.1016/j.physletb.2012.02.076.
3. CMS Collaboration, Search for the Standard Model Higgs boson produced in association with W and Z bosons in pp collisions at $\sqrt{s} = 7$ TeV, *JHEP.* **1211**, 088 (2012). doi: 10.1007/JHEP11(2012)088.
4. ATLAS Collaboration, Search for the Standard Model Higgs boson in the $H \to WW(*) \to \ell\nu\ell\nu$ decay mode with 4.7/fb of ATLAS data at $\sqrt{s} = 7$ TeV, *Phys. Lett. B* **716**, 62–81 (2012). doi: 10.1016/j.physletb.2012.08.010.
5. CMS Collaboration, Measurement of Higgs boson production and properties in the WW decay channel with leptonic final states, *JHEP.* **1401**, 096 (2014). doi: 10.1007/JHEP01(2014)096.
6. ATLAS Collaboration, Observation and measurement of Higgs boson decays to WW^* with the ATLAS detector, *Phys. Rev. D* **92**(1), 012006 (2015). doi: 10.1103/PhysRevD.92.012006.

7. M. Dittmar and H. K. Dreiner, How to find a Higgs boson with a mass between 155-GeV–180-GeV at the LHC, *Phys. Rev. D* **55**, 167–172 (1997). doi: 10.1103/PhysRevD.55.167.

8. V. D. Barger, G. Bhattacharya, T. Han, and B. A. Kniehl, Intermediate mass Higgs boson at hadron supercolliders, *Phys. Rev. D* **43**, 779–788 (1991). doi: 10.1103/PhysRevD.43.779.

9. CMS Collaboration, Measurement of the W^+W^- cross section in pp collisions at $\sqrt{s} = 7$ TeV and limits on anomalous $WW\gamma$ and WWZ couplings, *Eur. Phys. J. C* **73**(10), 2610 (2013). doi: 10.1140/epjc/s10052-013-2610-8.

10. ATLAS Collaboration, Measurement of the WW cross section in $\sqrt{s} = 7$ TeV pp collisions with ATLAS, *Phys. Rev. Lett.* **107**, 041802 (2011). doi: 10.1103/PhysRevLett.107.041802.

11. CMS Collaboration, Measurement of W^+W^- and ZZ production cross sections in pp collisions at $\sqrt{s} = 8$ TeV, *Phys. Lett. B* **721**, 190–211 (2013). doi: 10.1016/j.physletb.2013.03.027.

12. A. J. Barr, B. Gripaios, and C. G. Lester, Measuring the Higgs boson mass in dileptonic W-boson decays at hadron colliders, *JHEP.* **07**, 072 (2009). doi: 10.1088/1126-6708/2009/07/072.

13. G. Bozzi, S. Catani, D. de Florian, and M. Grazzini, Transverse-momentum resummation and the spectrum of the Higgs boson at the LHC, *Nucl. Phys. B* **737**, 73–120 (2006). doi: 10.1016/j.nuclphysb.2005.12.022.

14. LHC Higgs Cross Section Working Group, LHC HXSWG interim recommendations to explore the coupling structure of a Higgs-like particle. Technical Report. arXiv:1209.0040. LHCHXSWG-2012-001. CERN-PH-TH-2012-284 (Sep, 2012). URL `http://cds.cern.ch/record/1475887`.

15. I. W. Stewart and F. J. Tackmann, Theory uncertainties for Higgs and other searches using jet bins, *Phys. Rev. D* **85**, 034011 (2012). doi: 10.1103/PhysRevD.85.034011.

16. S. Bolognesi, Y. Gao, A. V. Gritsan, K. Melnikov, M. Schulze, *et al.*, On the spin and parity of a single-produced resonance at the LHC, *Phys. Rev. D* **86**, 095031 (2012). doi: 10.1103/PhysRevD.86.095031.

17. ATLAS Collaboration, Determination of spin and parity of the Higgs boson in the $WW^* \to e\nu\mu\nu$ decay channel with the ATLAS detector, *Eur. Phys. J. C* **75**(5), 231 (2015). doi: 10.1140/epjc/s10052-015-3436-3.

18. CMS Collaboration, Constraints on the spin-parity and anomalous HVV couplings of the Higgs boson in proton collisions at 7 and 8 TeV, *Phys. Rev. D* **92**(1), 012004 (2015). doi: 10.1103/PhysRevD.92.012004.

Chapter 8

Evidence for Higgs boson decays to τ leptons

Michail Bachtis[*] and Jürgen Kroseberg[†]

[*]CERN, CH-1211 Geneva 23, Switzerland
[†]Physics Institute, University of Bonn
Nussallee 12, D-53115 Bonn, Germany

The Higgs boson discovery in 2012 was based primarily on the analyses of Higgs boson decays to vector bosons. In the Standard Model (SM) the fermions acquire their masses via Yukawa couplings to the Higgs field, and therefore it was of fundamental importance to also confirm Higgs boson decays to fermions. In spite of significant experimental challenges, this was eventually achieved in the analyses of $\tau\tau$ final states. Using the full LHC Run 1 data set, ATLAS and CMS established $H \to \tau\tau$ signals with local significances above three standard deviations and with properties consistent with SM predictions. These results provided the first evidence for a direct coupling of the 125 GeV Higgs boson to fermions.

1. Introduction

During the LHC Run 1, the $H \to \tau\tau$ decay was one of two Standard Model Higgs boson decays into fermions that could be expected to be experimentally accessible and the only one potentially providing evidence for the Higgs boson coupling to leptons. Due to the relatively large τ lepton mass of about 1.8 GeV, the branching fraction to τ pairs is sizeable for Higgs masses significantly below $m_H = 2m_W$. For $m_H = 125$ GeV, $H \to \tau\tau$ accounts for approximately 6% of the decays.

The analysis of $\tau\tau$ final states poses several experimental challenges. The τ leptons have a mean lifetime of 291 fs corresponding to a decay length of 87 μm, therefore only the τ decay products are observed in the detector. Because of the large τ mass, in addition to the decays to electrons and muons, τ leptons also decay into light hadrons. The hadronic decays have a combined branching fraction of about 65% and typically result in collimated jets of a small number of particles. The reconstruction of the mass of the τ pairs is complicated by the presence of at least one neutrino in all τ decays. This typically results in a significant missing transverse momentum (E_T^{miss}), which can be used as an experimental signature but does not directly provide the four-momenta of the neutrinos from the individual τ decays. Therefore, neither the initial τ leptons nor their invariant mass can be fully reconstructed without additional assumptions.

The relevant experimental signatures are not restricted to the Higgs boson decay products. The simultaneous analysis of several event categories, defined mainly by the presence of additional hadron jets and their kinematic properties, improves the signal sensitivity and is also important for distinguishing between different Higgs production processes. In addition, recent analyses combine $\tau\tau$ mass information with various other final state properties into multivariate classifiers to extract the Higgs boson signal.

Thus, the $H \to \tau\tau$ search was based on detailed analyses of complex final states, in general involving leptonic and hadronic τ decays, E_T^{miss} and additional hadron jets. A candidate event observed by the ATLAS experiment is shown in Fig. 8.1. Here, two jets compatible with the signatures expected from hadronic τ decays recoiled against an additional high-p_T hadron jet. A significant E_T^{miss} was measured in the event and the reconstructed $\tau\tau$ mass was found to be close to 125 GeV.

In such analyses, $Z/\gamma^* \to \tau\tau$ events constitute an irreducible background, with a production cross section about 2000 times larger than that expected for the SM Higgs boson. Significant background contributions arise also from other sources, in particular from leptonic decays of W and Z bosons produced in association with jets, top quark production or QCD processes, when at least one of the final state objects is misidentified as a τ decay.

This chapter describes the ATLAS and CMS analyses[1,2] which provided the first evidence for $H \to \tau\tau$ decays based on the LHC Run 1 data. The integrated luminosities at pp centre-of-mass energies of 7 and 8 TeV were about 5 and 20 fb^{-1}, respectively. After discussing the reconstruction and identification of τ decays in Sec. 2 and the selection of events with $\tau\tau$ pairs in Sec. 3, Sec. 4 describes the reconstruction of the invariant mass of the $\tau\tau$ system. The event categorisation according to different final state topologies is covered in Sec. 5. The description of the methods to estimate the relevant background contributions in Sec. 6 is followed by a discussion of the signal extraction procedure in Sec. 7. Section 8 summarises the results, and an outlook is given in Sec. 9.

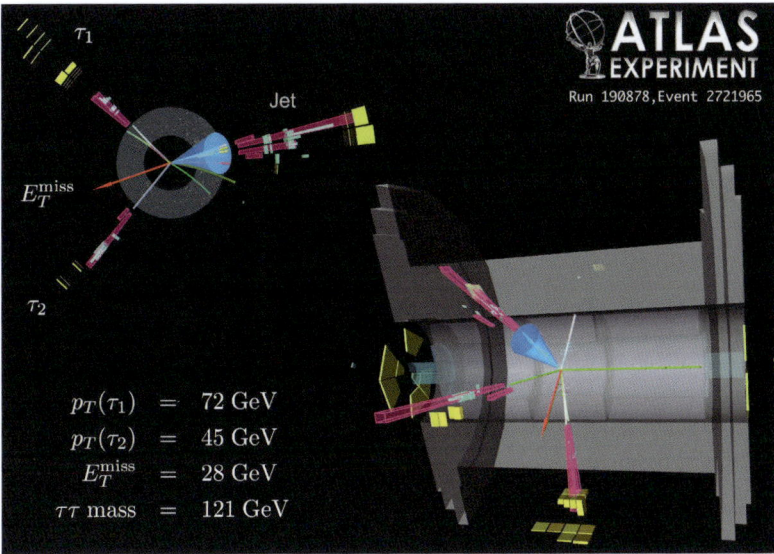

Fig. 8.1   Display of an ATLAS candidate event for Higgs boson production (in association with a hadronic jet) with subsequent decay into two hadronically decaying τ leptons. Projections into planes transverse to (left) and along the colliding proton beams (right) are shown, together with the measured transverse jet and τ momenta, $E_{\mathrm{T}}^{\mathrm{miss}}$ and invariant $\tau\tau$ mass.

2. Reconstruction and identification of hadronic τ decays

The τ lepton decays via the weak interaction to a τ neutrino and a pair of light fermions as depicted in the Feynman diagram in the left part of Fig. 8.2; the corresponding branching fractions[3] are shown on the right.

While the leptonic decays are observed in the detector as isolated muons or electrons, dedicated algorithms are required to reconstruct and identify the hadronic modes corresponding to about 65% of all τ decays. These result in signatures of typically one or three light charged hadrons and possibly additional neutral hadrons or their decay products. The dominant contributions are pions from $\tau^+ \rightarrow \pi^+ \nu_\tau$ decays or processes involving intermediate $\rho(770)$ or $a_1(1200)$ resonances with subsequent decays $\rho \rightarrow \pi^- \pi^0$ and $a_1 \rightarrow \pi^- \pi^+ \pi^-$ or $a_1 \rightarrow \pi^+ \pi^0 \pi^0$, respectively, where the neutral pions decay to photons. For τ leptons produced in high-energy collisions and/or decays of heavy particles such as the Z or the Higgs boson, the τ energy is much higher than its mass, resulting in a boosted topology with

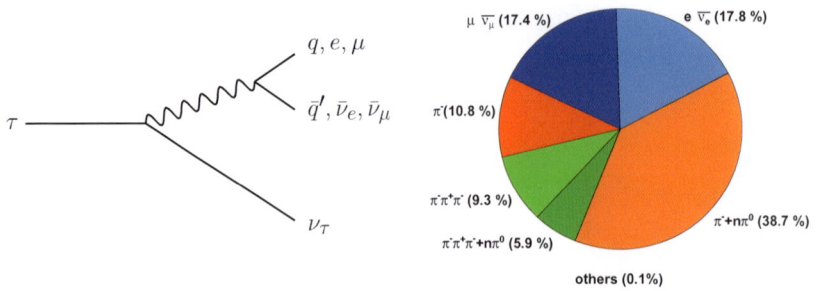

Fig. 8.2 Left: τ decay Feynman diagram. Right: τ decay branching fractions.

collimated τ decay products. In addition, the τ leptons that are produced in heavy boson decays are expected to be isolated in the detector.

As a result, hadronic τ decays are characterised by an isolated, collimated, low-multiplicity jet of particles. The main challenge in identifying these decays is to distinguish them from the ubiquitous hadron jets originating from quarks or gluons, which are henceforth referred to as *QCD jets*. Electrons can constitute a relevant background as well, especially for the τ decays to charged pions and at least one π^0.

In the following, τ_{lep} denotes a leptonically decaying τ lepton ($\tau_{\text{lep}} \to \ell \bar{\nu}_\ell \nu_\tau$). A hadronic τ decay is denoted by $\tau_{\text{had}} \to \tau_{\text{h}} \nu_\tau$, where τ_{h} corresponds to the visible decay products.

The reconstruction and identification of hadronic τ decays with the ATLAS detector[4] was seeded by jets (with $p_T > 10$ GeV) reconstructed from clusters of energy in the calorimeters using the anti-k_t algorithm[5] with radius parameter $R = 0.4$. Two regions around the direction of the τ_{h} candidate were defined as indicated in Fig. 8.3 (left): a *jet cone* with radius $\Delta R = \sqrt{\Delta\eta^2 + \Delta\phi^2} < 0.4$ and a core region (τ *cone*) with radius 0.2. The outer part of the jet cone with $0.2 < \Delta R < 0.4$, i.e. excluding the core region, was referred to as the *isolation region*. Well-reconstructed tracks with $p_T > 1$ GeV within the core region were associated to the τ_{h} candidate and were used to define its charged-particle multiplicity and net charge. In addition a π^0 reconstruction algorithm was applied. Track and calorimeter information within the jet cone but outside the core region was used to define the isolation of the candidate. In a subsequent identification step, calorimeter and track information from the core and the isolation regions were combined into boosted decision trees (BDTs) to reject backgrounds from QCD jets and electrons, respectively. The rejection of QCD jets was based on the distribution of tracks and calorimeter energy within

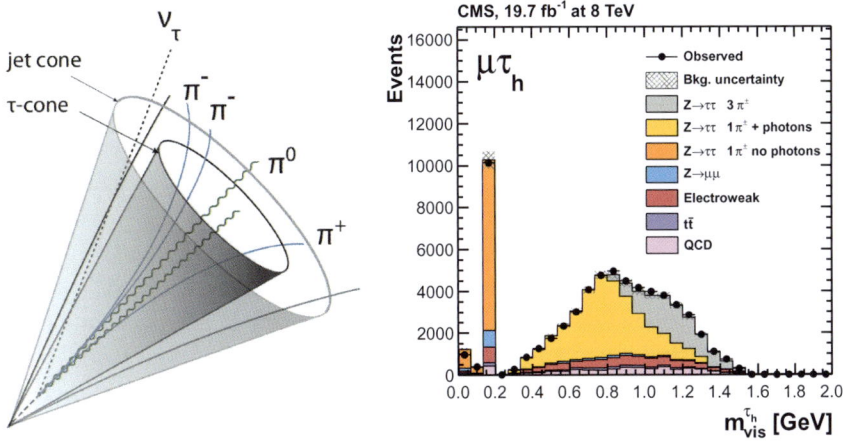

Fig. 8.3   Left: Illustration of a hadronic τ decay and the definition of the so-called τ and jet cones used in the ATLAS τ identification (see text). Right: Comparison between CMS data and simulation of the visible mass of the hadronically decaying tau lepton using $Z \to \tau\tau$ events with a muon and a hadronic τ decay.

the core and isolation regions, the fraction of the transverse momentum carried by the leading track, track impact parameter and secondary vertex information and the invariant masses of the track and/or the π^0 system. Different working points, corresponding to different τ identification efficiencies and jet misidentification rates, were defined. All identified τ_h candidates underwent a τ-specific energy calibration. The calibration constants were obtained from a combination of simulated events and data measurements.[4]

The basis for the identification of hadronic τ decays with the CMS detector[6] was the particle flow (PF) reconstruction,[7] which describes each recorded collision by a set of mutually exclusive particle candidates (electrons, muons, photons, charged and neutral hadrons). The hadronic τ decays result in charged hadrons and photons from decays of neutral pions. The charged hadron momenta are measured with the tracker. Photons from π^0 decays can be measured down to very low energies in the EM calorimeter but the material in the tracking system often causes the photons to convert into e^+e^- pairs. Since the calorimeter is situated inside the magnet, the conversion tracks bend, resulting in broad shower profiles in the bending direction. Individual clusters originating from such showers were collected in rectangular clusters (strips) that were narrow in the η direction but were allowed to grow dynamically in the ϕ direction (bending plane). The

Hadrons Plus Strips (HPS) tau identification algorithm formed combinations of hadrons and strips to build τ_h candidates for the most important decay modes: single charged hadron, a charged hadron + EM strips and three charged hadrons. For the decay modes involving more than one particle, the invariant mass of the visible decay products was required to be compatible with the ρ or a_1 resonances. If several candidates passed the quality criteria, the most isolated one was chosen. Isolation requirements for τ leptons were defined based on all particles within a cone of radius $\Delta R = 0.5$, excluding the particles associated to the reconstructed τ decay. This approach provided additional discrimination against boosted QCD jets containing collimated particles mimicking the signature of a hadronic τ decay. Figure 8.3 (right) shows the visible τ mass spectrum for different decay modes, demonstrating compatibility with the targeted resonances and an excellent agreement between data and simulation.

While the ATLAS and CMS experiments followed somewhat different approaches to reconstruct and identify hadronic τ decays, mainly attributed to different detector designs, the resulting performance within the $H \to \tau\tau$ searches was very similar.

3. Selection of $\tau\tau$ events

After reconstructing and identifying hadronic τ decay candidates, leptons and other final state objects, the next step was the selection of events with opposite-charge τ pairs. As the branching fractions and experimental signatures are quite different for hadronic and leptonic τ decay modes, the background composition and consequently the trigger and selection strategies differ as well. Therefore, the following discussion is divided into the cases where one, two, or none of the two tau leptons decay hadronically.

3.1. *Final states with a lepton and a hadronic τ decay*

The $\tau_{lep}\tau_{had}$ mode is the most prolific, accounting for about 46% of all $\tau\tau$ decays. The muons and electrons provide a clear experimental signature that is particularly valuable for the selection at the trigger level and for the reduction of the QCD jet background.

The ATLAS analysis was based on single-lepton triggers with p_T thresholds of about 25 GeV, while CMS used combined lepton–τ_h triggers with lepton and τ_h thresholds around 20 GeV. The offline thresholds were tighter for both experiments to ensure trigger efficiencies close to 100%. The τ_h candidates were required to have a minimum p_T in the range of 20–35 GeV.

Sources of major backgrounds include W+jets and $t\bar{t}$ production. Here, an electron or muon originates from the decay of a W boson or a top quark and a jet is misidentified as a hadronic τ decay; an example diagram is shown in Fig. 8.4 (left). The W+jets background was suppressed using the transverse mass computed from the lepton and the missing transverse momentum:

$$m_T(\ell, E_T^{\mathrm{miss}}) = \sqrt{2p_T^\ell E_T^{\mathrm{miss}}\left(1 - \cos\Delta\phi\right)}. \qquad (8.1)$$

In addition, there is a large QCD jet background, which was reduced by applying stringent identification and isolation requirements on the lepton and τ_h candidates.

Figure 8.4 (right) shows the transverse mass distribution for events selected by CMS in the $\mu\tau_h$ final state. The maximum of the m_T distribution is at low values for events with τ pairs and at the W boson mass for events with a W boson in the final state. Background processes with a W boson were suppressed by requiring a maximum transverse mass of 30 GeV in the CMS analysis and 70 GeV in the ATLAS analysis. The ATLAS requirement was much looser because m_T is one of the input quantities for the boosted decision trees (BDT) used in the extraction of the Higgs boson signal, see Sec. 7. In addition, ATLAS rejected events with at least one identified b jet with $p_T > 30$ GeV in order to suppress the $t\bar{t}$ background.

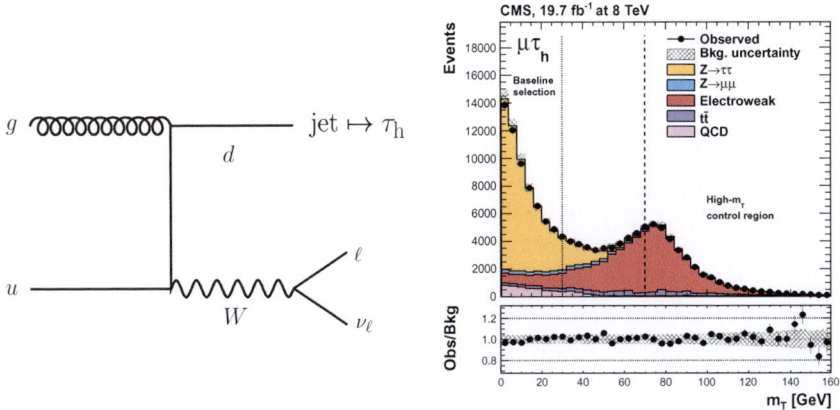

Fig. 8.4   Left: Example Feynman diagram for W production in association with a jet, which can enter the $\tau_{\mathrm{lep}}\tau_{\mathrm{had}}$ selection if the jet is misidentified as a hadronic τ decay. Right: Transverse mass distribution of the lepton and E_T^{miss} in $\mu\tau_h$ events collected with the CMS detector at $\sqrt{s} = 8$ TeV.

Final states with two leptons

The final states with two light leptons are characterised by a clean experimental signature that includes two leptons that can have the same ($\mu^+\mu^-$, e^+e^-) or different flavours ($e^+\mu^-$). They correspond, however, to about only 12% of all $\tau\tau$ decays. CMS collected events with a trigger requiring a pair of leptons, while ATLAS used both di-lepton and single-lepton triggers. Both experiments required thresholds between 10−20 GeV for the leading and sub-leading leptons, above which the trigger selection efficiency is close to 100%.

The different-flavour final state is more sensitive than the same-flavour one, benefiting from the absence of $Z/\gamma^* \to \ell\ell$ background contributions. The dominant background in this final state stems from leptonic decays of top quark pairs. The large mass of the top quark leads to typically large transverse momenta of the charged leptons and a large E_T^{miss} from the associated neutrinos. In addition, top quark decays are characterised by final-state jets originating from b quarks. ATLAS suppressed the top quark background by applying tight requirements on the E_T^{miss} as well as the invariant mass and the azimuthal angle of the two leptons. In addition, an upper limit on the scalar sum of the transverse momenta of the leptons was applied, and events with an identified b jet with $p_T > 25$ GeV were rejected. CMS suppressed the same background with a BDT that made use of the kinematic variables related to the $e\mu$ system, the E_T^{miss}, the compatibility of the leptons with the primary vertex and the probability that additional jets in the event originated from a b quark.

The selection of the same-flavour final state is challenging due to the contamination from $Z/\gamma^* \to \ell\ell$ events. Compared to this background, the muons and electrons from $H \to \tau\tau$ decays tend to have less energy, and more E_T^{miss} is expected because of the neutrinos from the τ decays. In addition, due to the large τ lifetime, final state leptons originating from τ pairs are expected to be more displaced. In the ATLAS analysis, tight requirements were applied on the invariant mass of the two leptons and the E_T^{miss}. In addition, a complementary missing transverse momentum was defined using only the high-energy objects in the event. This alternative and the nominal E_T^{miss} are strongly correlated for the $H \to \tau\tau$ signal but only loosely correlated for $Z/\gamma^* \to \ell\ell$ events. In the CMS selection, a BDT discriminated against the $Z/\gamma^* \to \ell\ell$ background based on the compatibility of the leptons with the primary vertex and the kinematics of the di-lepton system.

3.2. *Final states with two hadronic τ decays*

Final states with two hadronic τ decays ($\tau_{\mathrm{had}}\tau_{\mathrm{had}}$) are, with a branching fraction of about 42%, almost as common as the $\tau_{\mathrm{lep}}\tau_{\mathrm{had}}$ mode. With only two decay neutrinos in the final state, the resolution of the $m_{\tau\tau}$ reconstruction is less affected by undetected particles than in the other final states. However, the large background from QCD jet processes poses a significant experimental challenge.

Events were collected with the ATLAS detector based on a di-τ_{h} trigger with p_{T} thresholds of 29 and 20 GeV for the leading and sub-leading τ, respectively; the corresponding offline thresholds were 35 and 25 GeV. CMS implemented a combination of di-τ_{h} triggers (with p_{T} thresholds of 45 GeV) and triggers selecting a τ_{h} pair and an additional jet with $p_{\mathrm{T}} > 30$ GeV. Tighter τ identification requirements were implemented in the offline data analysis.

4. Mass reconstruction

The invariant mass of the $\tau\tau$ pair, $m_{\tau\tau}$, is a key observable for the extraction of the $H \to \tau\tau$ signal. A good mass resolution is particularly important for the separation between the $H \to \tau\tau$ signal and the irreducible $Z/\gamma^* \to \tau\tau$ background.

However, the $m_{\tau\tau}$ reconstruction is necessarily based on incomplete information, since each τ decay produces at least one undetected neutrino. The neutrino kinematics enter the measured $E_{\mathrm{T}}^{\mathrm{miss}}$ but the $E_{\mathrm{T}}^{\mathrm{miss}}$ resolution is modest and any information on the neutrino momenta along the beam direction is lost. Using the *visible mass* $m_{\tau\tau}^{vis}$ — defined as the invariant mass of the visible products of the τ decays — avoids any direct dependence on the missing transverse momentum reconstruction. However, the mass estimation can be improved by including also the measured $E_{\mathrm{T}}^{\mathrm{miss}}$ information.

A priori, the $E_{\mathrm{T}}^{\mathrm{miss}}$ measurement provides no information on the momenta of the individual neutrinos, and separately identifying $E_{\mathrm{T}}^{\mathrm{miss}}$ components that are associated with each of the two τ decays is not possible either without additional assumptions. Assuming the same directions for visible and invisible products for each of the τ decays, however, the measured $E_{\mathrm{T}}^{\mathrm{miss}}$ can be projected onto the visible τ axes as sketched in Fig. 8.5. This *collinear approximation* is motivated by the large mass difference between the Higgs boson and τ leptons, typically resulting in a large

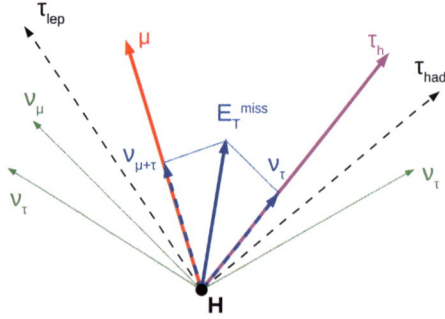

Fig. 8.5 The collinear approximation: sketch of a $H \to \tau\tau$ decay with one τ decaying to a muon (red) and the other hadronically. The directions of the two tau leptons, the neutrinos and the visible part of the hadronic decay are indicated by the dashed, green, and pink arrows, respectively. The blue arrows indicate how the the reconstructed $E_{\mathrm{T}}^{\mathrm{miss}}$(solid arrow) is projected onto the axes defined by the visible τ decay products, thus approximating the neutrino kinematics. The approximation is exact if for both tau decays the vector sum of the visible products is collinear to the original τ direction.

boost of the $H \to \tau\tau$ decay products. Neglecting the mass of the τ leptons, the corresponding *collinear mass* is defined as $m_{\tau\tau}^{coll} = m_{\tau\tau}^{vis}/\sqrt{x_1 x_2}$. Here, $x_{1,2}$ are the τ momentum fractions carried by the visible decay products, which in turn can be calculated from the measured momenta of the visible decay products and the missing transverse momentum.

For a large fraction of the experimentally relevant phase space, the collinear mass improves over the visible mass but it fails to provide a proper measurement in cases of small boosts and even breaks down if the τ leptons are back-to-back in the transverse plane. In addition, experimental resolution effects, in particular for $E_{\mathrm{T}}^{\mathrm{miss}}$, can lead to reconstructed x_i values outside the physical range of $0 < x_i \leq 1$ and correspondingly increased tails in the $m_{\tau\tau}^{coll}$ distribution.

A full $\tau\tau$ mass reconstruction without the shortcomings of a specific approximation can be achieved by a likelihood-based approach, where the under-determined event kinematics enter via free parameters which are then scanned or fitted. The measured $E_{\mathrm{T}}^{\mathrm{miss}}$ components and visible τ decay products as well as models of the decay topology and the experimental $E_{\mathrm{T}}^{\mathrm{miss}}$ resolution are used as constraints. This concept was used in the later $H \to \tau\tau$ analyses by ATLAS and CMS, albeit in somewhat different implementations denoted *missing mass calculator* (MMC)[8] and SVFit,[9] respectively. The achieved mass resolution was similar in both experiments. The improved separation of the SVFit mass between the Z and a Higgs boson at 125 GeV with respect to the visible mass is demonstrated in Fig. 8.6.

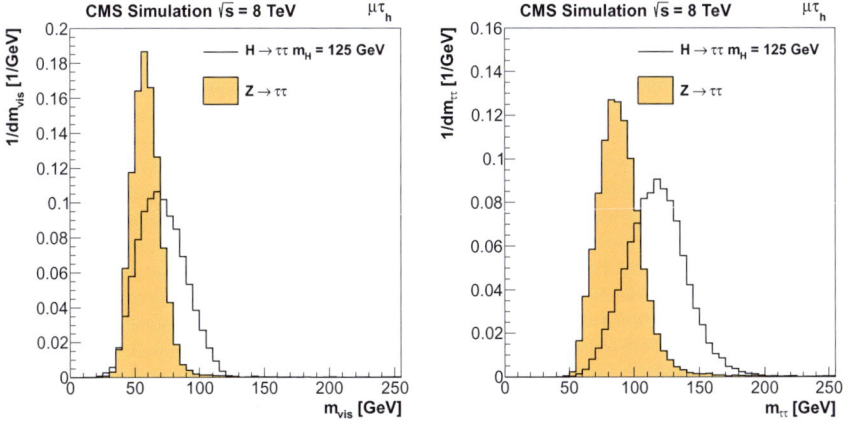

Fig. 8.6 Normalised distributions, obtained by CMS in the $\mu\tau_h$ final state, of the visible mass (left) and SVFit mass (right) from simulated samples of $Z/\gamma^* \to \tau\tau$ (filled histogram) and SM $H \to \tau\tau$ ($m_H = 125$ GeV) events (open histogram).

5. Event categorisation

The search sensitivity was substantially improved by separating the selected data into categories according to event and/or object characteristics. Here, the most important handle was the number and the kinematics of additional hadronic jets, based on two insights:

(1) The VBF Higgs production process typically results in two high-energy jets with a large invariant mass and a large separation in pseudorapidity. This provides a distinct experimental signature which can be used to significantly increase the signal-to-background ratio.
(2) Topologies with boosted Higgs bosons typically make it easier to reconstruct the missing transverse momentum and improve the experimental $\tau\tau$ mass resolution. Due to momentum conservation, these events in general include at least one additional high-p_T jet, c.f. Fig. 8.7.

ATLAS analysed events passing the selection described in Sec. 3 separately in the following categories:

- For each $\tau\tau$ decay mode, a *VBF category* was defined by requiring two high-p_T jets with a large pseudorapidity separation $\Delta\eta$. The mode-dependent selection requirements ranged from 40 to 50 GeV and 30 to 35 GeV for the minimum p_T of the leading and sub-leading jet, respectively, and from 2.0 to 3.0 for the minimum $|\Delta\eta|$. For the $\tau_{\text{lep}}\tau_{\text{lep}}$ mode,

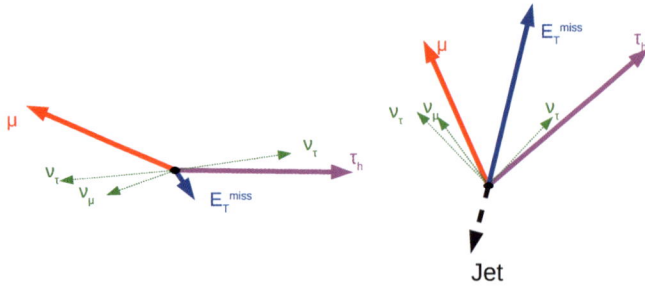

Fig. 8.7   Illustration of the effect of an additional jet on the decay topology, boosting the $\tau\tau$ candidate. The presence of the additional jet aligns the neutrinos from the two tau decays, resulting in a more precise measurement of the $E_{\mathrm{T}}^{\mathrm{miss}}$ and better estimation of the invariant mass.

the visible $\tau\tau$ mass was required to exceed 40 GeV. While signal events selected by these requirements are indeed dominantly VBF-produced, smaller contributions from ggF and *VH* production are also expected.

- For each $\tau\tau$ decay mode, events failing the VBF selection were included in *boosted categories* if the transverse momentum of the Higgs boson, as reconstructed from the vector sum of the $E_{\mathrm{T}}^{\mathrm{miss}}$ and the visible τ decay products, exceeded 100 GeV. For the $\tau_{\mathrm{lep}}\tau_{\mathrm{lep}}$ mode the presence of at least one jet with $p_T > 40$ GeV was required in addition.

Figure 8.8 shows the invariant $\tau\tau$ mass distributions obtained with the MMC in each of the two ATLAS event categories for simulated $H \to \tau\tau$ events and the $Z/\gamma^* \to \tau\tau$ background modelled with τ-embedded data as described in Sec. 6. The mass resolution, defined as the ratio between the full width at half maximum (FWHM) and the peak value of the mass distribution, was found to be of the order of 30% for both categories and all $\tau\tau$ decay modes.

CMS used a more involved categorisation, based on the presence of additional objects in the final state as well as on the transverse momenta of the di-τ system and the visible τ decay products. The selection is summarised in the following; the detailed categorisation can be found elsewhere.[2]

- Initially, three event categories were defined requiring zero, one and two additional jets in the final state, respectively.
- The *two-jets* category targeted the VBF production and was split further into so-called *tight* and *loose VBF* categories. The loose VBF category required a pseudorapidity difference of $\Delta\eta > 3.5$ between the two jets

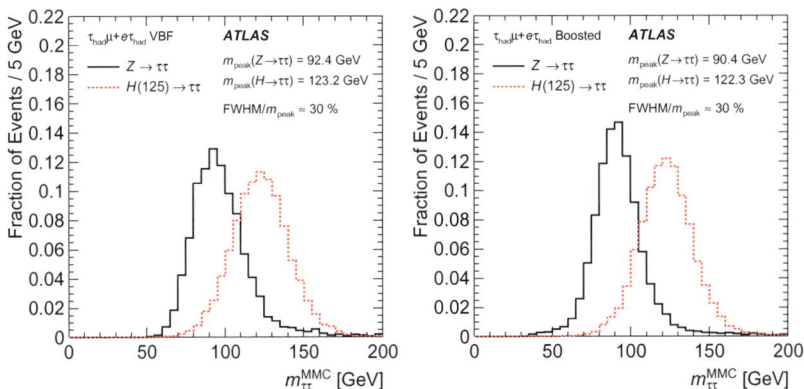

Fig. 8.8 Invariant $\tau\tau$ mass distributions obtained by ATLAS with the MMC for simulated $H \to \tau\tau$ events (red) and the $Z/\gamma^* \to \tau\tau$ background modelled with τ-embedded data (black), after the VBF category selection (left) and the boosted category selection (right) for the $\tau_{\text{lep}}\tau_{\text{had}}$ mode. A Higgs boson mass of $m_H = 125$ GeV was assumed in the signal event simulation.

and a di-jet mass of $m_{jj} > 500$ GeV, while the tight category required $\Delta\eta > 4.0$ and $m_{jj} > 700$ GeV.

- The *one-jet* category was split into a *boosted* category, where the $\tau\tau$ system was required to have $p_T > 100$ GeV, and a *non-boosted* one with the rest of the one-jet events.
- Both zero and one-jet categories were further split into two categories denoted *low-p_T* and *high-p_T*, respectively, based on the transverse momentum of the reconstructed τ or lepton candidates. This additional categorisation was motivated by the strong p_T dependence of the τ_h misidentification probability.

The different level of complexity in the event categorisation chosen by ATLAS and CMS is complementary to their signal extraction strategies that will be discussed in Sec. 7.

6. Background estimation

The selected and categorised event samples were still dominated by background contributions, which needed to be carefully modelled before any results on the Higgs boson signal content could be obtained. For example, Fig. 8.9 shows the invariant mass distribution measured by CMS in $\mu\tau_h$ events before categorisation, together with the estimated background

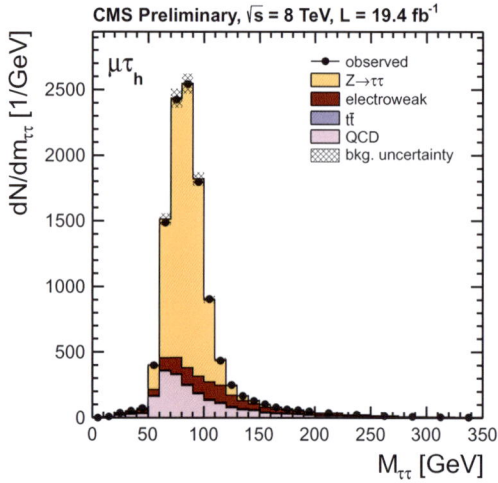

Fig. 8.9 Invariant mass distribution of $\mu\tau_h$ events from CMS in the 8 TeV data set before categorisation. The Higgs signal contribution was negligible at this stage of the selection. Most of the events originated from the irreducible $Z/\gamma^* \to \tau\tau$ background. Electroweak contributions (dominated by events with W bosons and jets) and QCD processes were other major sources of background events.

composition; the Higgs signal contribution was negligible in comparison at this stage of the analysis. The largest background was irreducible and arose from $Z/\gamma^* \to \tau\tau$ decays, while reducible contributions were dominated by electroweak and QCD jet events. Electroweak processes corresponded mainly to W boson production in association with jets, with additional contributions from Z boson decays to electrons or muons. Top quark pair production and di-boson events contributed as well but not as significantly.

The relative background contributions and the methods used to estimate the individual components varied between different decay modes and event categories; the most important concepts are summarised below.

6.1. *Irreducible backgrounds*

The dominant background source was events with $Z/\gamma^* \to \tau\tau$ decays, resulting in the same final state objects with very similar properties as the $H \to \tau\tau$ signal. The mass differences between the Z boson and the 125 GeV Higgs boson is not large compared to the experimental $m_{\tau\tau}$ resolution, so there was a significant overlap between the reconstructed Z and Higgs boson mass distributions. While the relative cross section for VBF production is much larger for the signal, which was exploited to suppress the Z background, even in the VBF-enriched analysis categories a significant fraction of the selected events originated from $Z/\gamma^* \to \tau\tau$ decays because of the much larger total Z production cross section. Therefore, a reliable model of this background, including the complex signatures that drive the $H \to \tau\tau$ search sensitivity, was of key importance. While it was desirable to rely as little as possible on simulated information, selecting a sufficiently pure $Z/\gamma^* \to \tau\tau$ control sample directly from the collision data was difficult and, more importantly, it would have been impossible to completely remove the $H \to \tau\tau$ signal events.

In order to estimate the $Z/\gamma^* \to \tau\tau$ background in a largely data-driven way, a τ-embedding technique was implemented by both ATLAS[10] and CMS. $Z/\gamma^* \to \mu\mu$ events were selected from the collision data and the muons from the Z/γ^* decay were replaced by τ leptons from simulated $Z/\gamma^* \to \tau\tau$ decays, deriving the τ kinematics from the original muons, see Fig. 8.10 (left). Thus, only well-understood features such as the decays of the τ leptons and the detector response to the τ decay products were

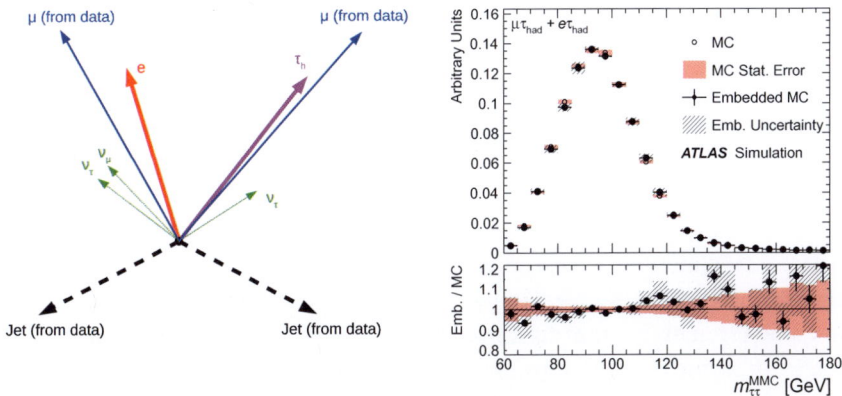

Fig. 8.10 Left: Illustration of the τ-embedding technique. A $Z/\gamma^* \to \mu\mu$ event with two jets is selected in data and the muons are replaced with simulated tau leptons that decay into an electron, a τ_h and three neutrinos, retaining the original jets in the data. Right: The reconstructed $m_{\tau\tau}$ distribution obtained by ATLAS in the $\tau_{\text{lep}}\tau_{\text{had}}$ mode for simulated $Z/\gamma^* \to \tau\tau$ events, compared to the one obtained from simulated $Z/\gamma^* \to \mu\mu$ events after τ embedding.

modelled by the simulation. All other aspects of the event, including pileup contributions and the kinematics of additional jets, were obtained from the collision data. For validation purposes, also simulated $Z/\gamma^* \to \mu\mu$ events were used as input to the procedure; the resulting embedded samples could be directly compared to simulated $Z/\gamma^* \to \tau\tau$ events, see Fig. 8.10 (right).

Sources of smaller irreducible background contributions included the production of vector boson pairs. These processes were estimated using simulation corrected for the lepton efficiencies and recent cross section measurements from the ATLAS and CMS experiments.

6.2. *Reducible backgrounds*

Major reducible backgrounds arose from QCD jet processes, W+jets, Z+jets and top quark pair production, where in general at least one final state object was misidentified as a τ decay. These background contributions were estimated from a combination of event samples obtained either from simulation or from dedicated collision data sets, where certain requirements were inverted or otherwise modified in order to enrich a particular background process. A selection of relevant concepts is sketched in the following.

CMS exploited the fact that the QCD jet background is characterised by a ratio of opposite-sign (OS) and like-sign (LS) τ pair candidates compatible with unity. This ratio was measured in data by inverting the lepton isolation requirements. In the $\mu\tau_\mathrm{h}$ final state, this ratio is expected to be slightly above unity due to OS semileptonic heavy flavour decays, while in the $e\tau_\mathrm{h}$ final state this ratio is dominated by electromagnetic fluctuations in jets resulting in a value closer to unity. To measure the contributions from W+jets and top quark pair production, the kinematics of the W decays were used. The m_T requirement was inverted, providing a W-enriched region where the simulated background was normalised to the yield in data. The same procedure applied to the Z+jets background where one lepton was not reconstructed, resulting in a W-like event topology. The background estimation started by determining the expected number of W+jets and diboson events in the LS and OS control regions. The remaining events in the LS control region were assumed to originate from QCD jet production and were scaled by the OS/LS factor to estimate the QCD jet contribution in the OS region.

In the analysis of the $\tau_\mathrm{lep}\tau_\mathrm{had}$ mode, ATLAS accounted for the misidentification of QCD jets as hadronic τ decays by measuring so-called fake factors. These were defined as the ratio of the number of jets passing the

analysis τ_h selection to the number of jets satisfying looser τ_h identification criteria but failing the analysis τ_h selection. Since τ_h candidates from
misidentified jets are not restricted to a single background physics process
and the misidentification probability depends on the kinematics and composition of the jet, the fake factors were obtained from a combination of
different data control samples, as a function of the τ_h p_T and separately
for τ_h candidates with one and three associated tracks. In addition, the
fake factors were measured separately for data control samples that were
expected to be dominated by QCD jets originating from quarks and glu-
ons, respectively, because the τ_h identification algorithms make use of the
jet shape, which is typically different for quark and gluon-induced jets. CMS
used a similar technique to estimate the reducible fake-lepton background
in the analysis of the $e\mu$ final state.

Another important background for all $\tau\tau$ decay modes are Z decays
to leptons. In final states with electrons and muons, this background is
the dominant one if the two leptons have the same flavour. For hadronic
τ decays this background is relevant as well since an electron or muon
can also mimic a τ_h signature. In both experiments the $Z/\gamma^* \to \ell\ell$ background was estimated by simulation. The simulated events were corrected
for effects related to lepton reconstruction efficiencies or τ_h misidentification

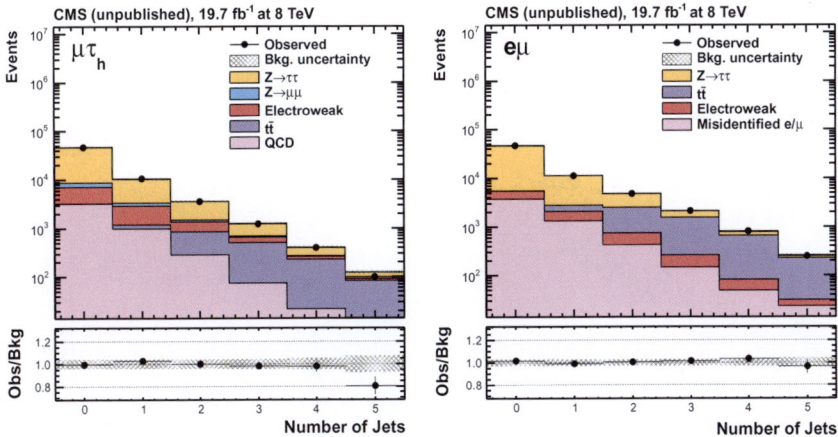

Fig. 8.11 Comparison of the distribution of the number of jets with $p_T > 30$ GeV
between CMS data and the sum of the estimated backgrounds in the $\mu\tau_\mathrm{h}$ (left) and $e\mu$
(right) final states after applying the $\tau\tau$ selection described in Sec. 3. The contribution
of the Higgs signal in these distributions is negligible.

probabilities. The corrections were based on tag-and-probe measurements in Z events selected from the collision data.

Figure 8.11 shows the comparison between data and background predictions of the distribution of the number of jets with $p_T > 30$ GeV obtained by CMS in the $\mu\tau_{\mathrm{h}}$ and $e\mu$ final states, respectively. Excellent agreement is observed for both final states.

7. Signal extraction

ATLAS and CMS followed somewhat different strategies for the extraction of the Higgs boson signal from the data events populating the different categories.

CMS used the invariant $\tau\tau$ mass in all categories except for the ee and $\mu\mu$ modes, for which the output of two multivariate (BDT) discriminants served as input to the statistical analysis.

In the ATLAS analysis, after a somewhat looser event selection and a significantly simpler categorisation, boosted decision trees were used everywhere, combining the $\tau\tau$ MMC mass with additional information depending on the $\tau\tau$ decay mode and event category. In order to take into account the different event characteristics and background compositions, the selection of input quantities and the training were performed separately for each of the $\tau\tau$ decay modes and analysis categories. The input information used by ATLAS in constructing the BDTs were:

- Invariant masses: The $\tau\tau$ invariant mass, obtained from the MMC, was used for all decay modes and categories. For the $\tau_{\mathrm{lep}}\tau_{\mathrm{lep}}$ boosted category, the invariant masses of the two leptons as well as of the two leptons and the leading jet also entered the BDT. In addition, the transverse mass of the lepton and the $E_{\mathrm{T}}^{\mathrm{miss}}$ was a BDT input in both $\tau_{\mathrm{lep}}\tau_{\mathrm{had}}$ categories.
- Transverse momenta: The p_T of the visible τ decay products and their ratios were used for some of the categories. In the modes involving hadronic τ decays, other combinations of transverse momenta entered the BDT in addition: VBF Higgs production is expected to result in a p_T balance between the Higgs boson decay products and the two leading jets. Therefore, the $\vec{p}_T$ sum of the τ_{h} candidate, the lepton, the $E_{\mathrm{T}}^{\mathrm{miss}}$ and the two leading jets is expected to be small for signal events, which is not necessarily the case for background processes. For the boosted category, the scalar p_T sum of the visible τ decay products and additional jets is typically small for the $H \to \tau\tau$ signal compared to QCD jet or top production background processes.

- Final state topology: In addition to the distances between pairs of final state objects, more complex topological information were also considered in the BDT construction. For example, for sufficiently boosted Higgs decays the $E_{\mathrm{T}}^{\mathrm{miss}}$ direction in the plane transverse to the beams is expected to be found in between the visible decay products of the two τ leptons, and in VBF events the Higgs decay products are expected to be enclosed by the two leading jets in pseudorapidity.

The most discriminating input variable throughout was the $\tau\tau$ MMC mass. Examples for other observables with significant contributions were $\Delta R(\tau_1, \tau_2)$ and — for the VBF categories — $\Delta\eta(j_1, j_2)$, see Fig. 8.12.

After defining the quantities used for signal extraction, both ATLAS and CMS performed a likelihood fit based on the methods described in Appendix A.1, combining all $\tau\tau$ decay modes and event categories. Correlations across different modes and categories were taken into account, which was particularly relevant, for example, to the lepton and τ identification efficiencies and energy scale.

All systematic uncertainties were introduced as nuisance parameters in the fit. The most important experimental systematic uncertainties were related to the τ ID efficiency ($\mathcal{O}(5\%)$), the τ energy scale ($\mathcal{O}(3\%)$), and the background yields (5–50%). Uncertainties on the gluon fusion cross

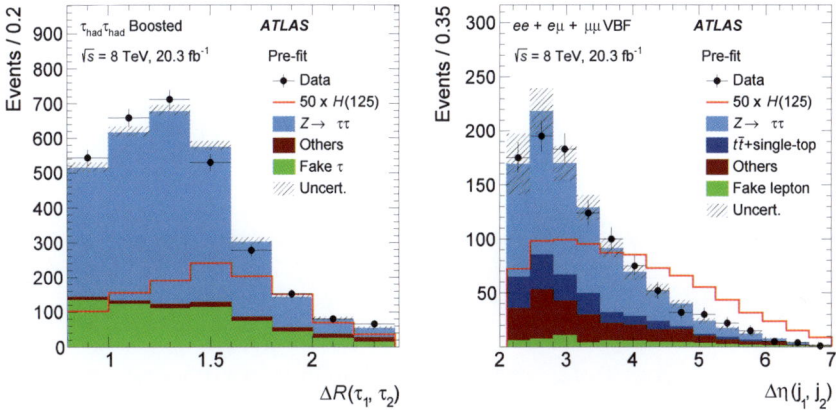

Fig. 8.12   Distributions of two of the BDT input variables used by ATLAS. Left: distance between the two τ_{h} candidates in the boosted $\tau_{\mathrm{had}}\tau_{\mathrm{had}}$ category. Right: pseudorapidity separation of the two leading jets in the $\tau_{\mathrm{lep}}\tau_{\mathrm{lep}}$ VBF category. The contributions from a Standard Model Higgs boson with $m_H = 125$ GeV are superimposed, multiplied by a factor of 50. These figures use background predictions made without the results of the global likelihood fit. The error band includes statistical and pre-fit systematic uncertainties.

section and on the shape of the Higgs p_T spectrum are the leading sources of theoretical errors. The yields for each $\tau\tau$ mode and category are correlated via the signal strength modifier μ defined as the signal yield in data measured in units of the efficiency-corrected SM $H \to \tau\tau$ yield.

8. Results

The statistical analysis of the categorised data according to the procedures described in Sec. 7 yielded firm evidence for a $H \to \tau\tau$ signal consistent with the predictions of the Standard Model.

ATLAS measured a signal strength of $\mu = 1.43^{+0.43}_{-0.37}$, corresponding to a signal significance of 4.5σ, while a sensitivity of 3.4σ was expected. CMS measured a signal strength of $\mu = 0.78 \pm 0.27$, corresponding to a significance of 3.2σ while a sensitivity of 3.7σ was expected. These results constituted the first direct evidence for a Higgs–fermion coupling and are discussed in more detail below.

Example distributions for certain $\tau\tau$ decay modes and event categories are shown in Fig. 8.13, where the expected signal and background contributions are obtained from the likelihood fit. Several mild excesses over the pure background expectation are visible in the data. Figure 8.14 illustrates the combined results from all $\tau\tau$ decay modes and categories. Figure 8.14 (left) combines the CMS $m_{\tau\tau}$ distributions for the $\mu\tau_h$, $e\tau_h$, $\tau_h\tau_h$, and $e\mu$ final states, weighted by the signal-to-background ratios for each mode, while on the right the ATLAS BDT output as a function of the signal-to-background yield is shown. Both distributions demonstrate a clear excess over the background-only expectation, compatible with the SM prediction for a 125 GeV Higgs Boson.

The signal strength results for the individual $\tau\tau$ decay modes and analyses categories as well as their combination are collected and compared in Figs. 8.15 and 8.16 for CMS and ATLAS, respectively, yielding a coherent picture for both experiments. Within uncertainties, the measured signal strengths are consistent with each other, their combination and the SM expectation of $\mu = 1$. The sensitivity is dominated by final states involving at least one hadronic τ decay and additional hadronic jets, where the $\tau_{\mathrm{lep}}\tau_{\mathrm{had}}$ modes and VBF-like jet topologies are of particular relevance.

In addition, ATLAS performed an alternative analysis, also documented in Ref. [1], using a tighter selection and a somewhat more refined categorisation, followed by a signal extraction directly from the $m_{\tau\tau}$ distribution. This analysis was restricted to the 8 TeV data, corresponding

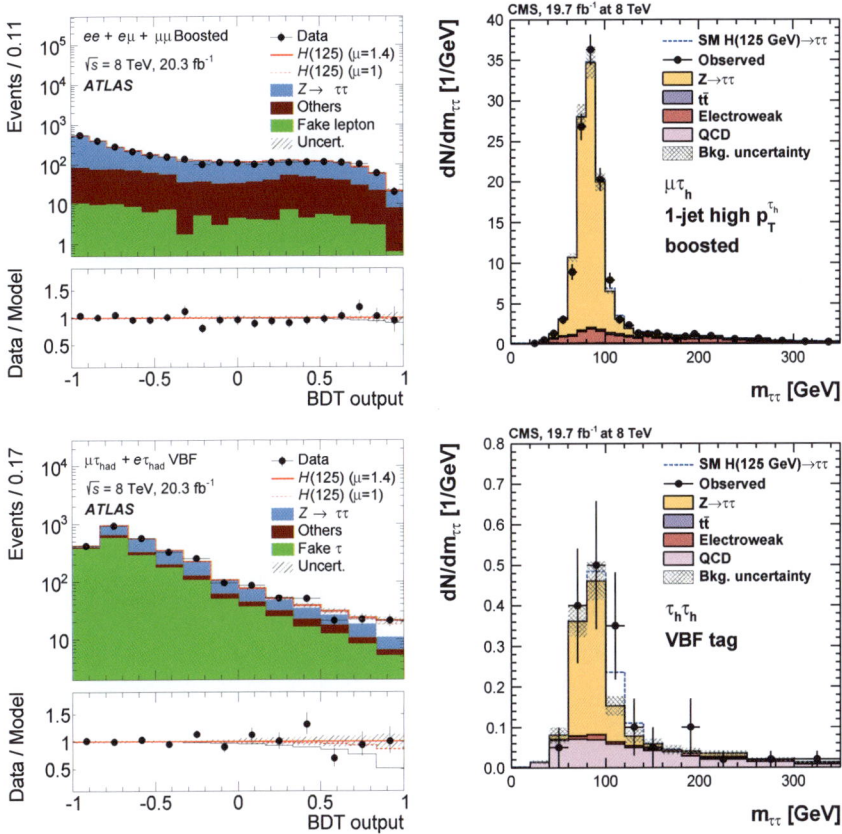

Fig. 8.13 Distributions of the BDT discriminant for ATLAS (left) and $m_{\tau\tau}$ for CMS (right) for selected analysis categories and $\tau\tau$ decay modes. The upper row contains plots with the lower-sensitivity boosted category while the lower row shows results in the VBF higher-sensitivity category.

to an integrated luminosity of 20.3 fb^{-1}, and yielded a signal strength of $\mu = 1.43^{+0.55}_{-0.49}$, in good agreement with the result $\mu = 1.53^{+0.47}_{-0.41}$ obtained from the BDT analysis of the same data. The alternative approach offered more direct access to the Higgs boson mass information, finding good consistency with $m_H = 125$ GeV, and, while not quite as sensitive as the multivariate analysis, also provided evidence for $H \to \tau\tau$ decays, with an observed (expected) significance of 3.2σ (2.5σ), to be compared with 4.5σ (3.3σ) obtained from the multivariate signal extraction. These results added further confidence to the $H \to \tau\tau$ evidence.

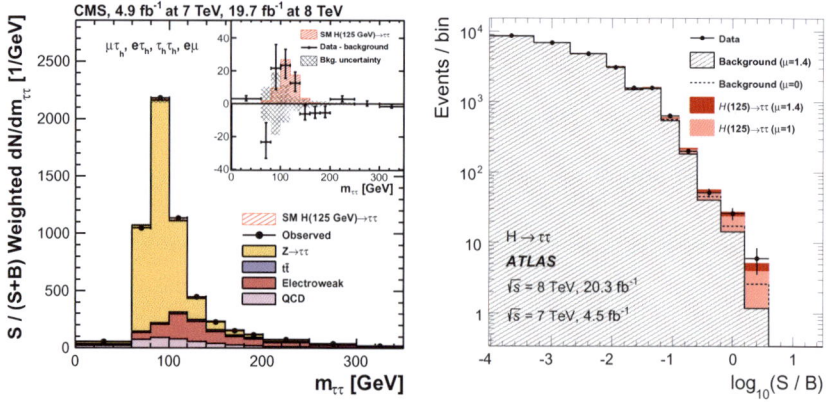

Fig. 8.14   Left: Combined observed and predicted $m_{\tau\tau}$ distributions obtained by CMS for the $\mu\tau_h$, $e\tau_h$, $\tau_h\tau_h$, and $e\mu$ final states. The distributions obtained in each category for each $\tau\tau$ decay mode are weighted by the ratio between the expected signal and signal-plus-background yields in the category. The inset shows the corresponding difference between the observed data and expected background distributions, together with the signal distribution for a SM Higgs boson at $m_H = 125$ GeV. Right: ATLAS event yields as a function of $\log_{10}(S/B)$, where S (signal yield) and B (background yield) are taken from the BDT output bin of each event, assuming the best-fit signal strength of $\mu = 1.4$. Events from all categories are included. The predicted background is obtained from the global fit (with $\mu = 1.4$) and signal yields are shown for $m_H = 125$ GeV, at $\mu = 1$ and $\mu = 1.4$.

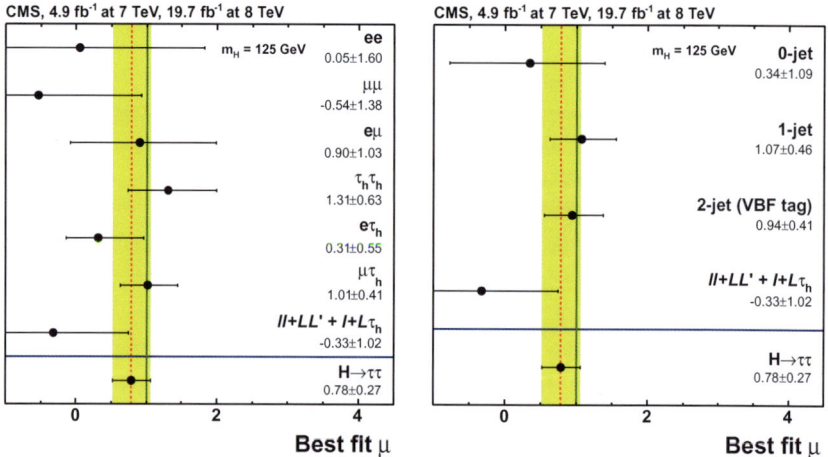

Fig. 8.15 $H \to \tau\tau$ signal strength, obtained by CMS for independent $\tau\tau$ decay modes (left) and analysis categories (right), for $m_H = 125$ GeV. In both plots, the combined $H \to \tau\tau$ value corresponds to the result of the global likelihood fit combining all categories and all decay modes; the dashed line indicates the best-fit value.

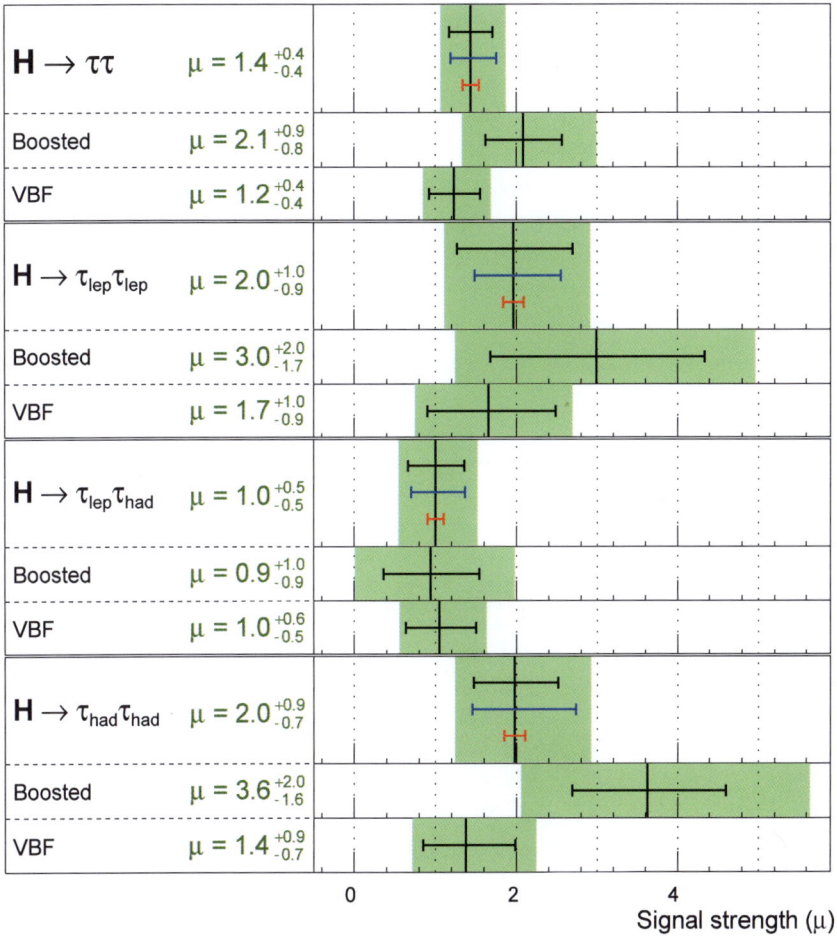

Fig. 8.16 Best-fit value for the signal strength μ in the individual $\tau\tau$ decay modes and their combination for the full ATLAS Run 1 data set, using $m_H = 125$ GeV throughout. The total $\pm 1\sigma$ uncertainty is indicated by the shaded green band, with the individual contributions from the statistical uncertainty (top, black), the total (experimental and theoretical) systematic uncertainty (middle, blue), and the theoretical uncertainty (bottom, red) on the signal cross section.

CMS investigated the mass information further by performing a likelihood scan as a function of m_H, profiling the signal strength and all nuisance parameters. This fit yielded a measured Higgs boson mass of $m_H = (122 \pm 7)$ GeV which was far from being competitive with the results obtained from the high-resolution di-photon and four-lepton Higgs decay channels but illustrated further the compatibility of the observed $\tau\tau$ signal with the 125 GeV Higgs boson.

Additional likelihood scans were performed by both experiments in order to investigate other aspects of the Higgs boson production and decay; the results are shown in Fig. 8.17. The left plot shows ATLAS results from a two-dimensional scan of separate signal strength parameters $\mu_{ggF}^{\tau\tau}$ and $\mu_{VBF+VH}^{\tau\tau}$, related to vector-boson-mediated (VBF and *VH*) and gluon-mediated (ggF) production processes, respectively. The right plot shows a CMS two-dimensional scan testing the compatibility of the observed Higgs boson couplings to fermions and bosons with the SM, where two scale factors κ_f and κ_V parametrised possible deviations of the Higgs boson couplings to fermions and vector bosons, respectively, with respect to the SM. In both cases, agreement with the SM expectation was found within the uncertainties. The underlying framework and assumptions are described in Chapter 12, where the impact of the $H \to \tau\tau$ analyses on the combined analyses of the Higgs boson couplings is also discussed.

A preliminary combination[11] of the ATLAS and CMS analyses yielded a $H \to \tau\tau$ signal with a significance of more than 5σ.

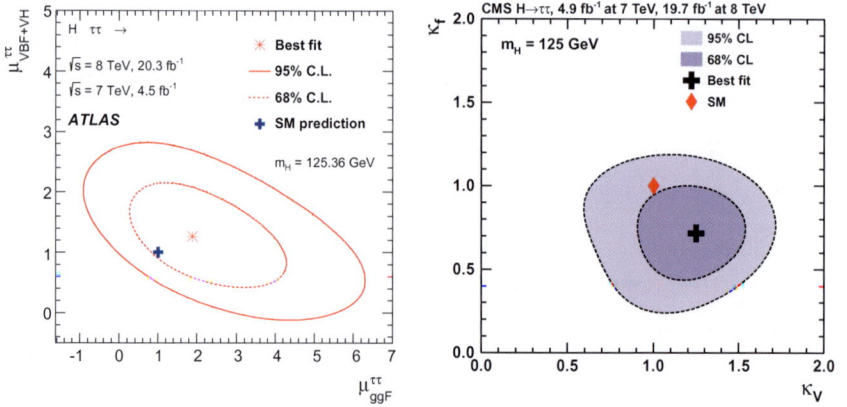

Fig. 8.17   Left: Likelihood contours in the $(\mu_{ggF}^{\tau\tau}, \mu_{VBF+VH}^{\tau\tau})$ plane, obtained by ATLAS from a combination of all $\tau\tau$ decay modes. The 68% and 95% CL contours are shown as dashed and solid lines respectively, for $m_H = 125$ GeV. The SM expectation is indicated by the filled plus symbol. The best fit to the data is shown as a star. Right: Likelihood scan as a function of κ_V and κ_f from CMS. For each point, all nuisance parameters are profiled. The observation (black cross) is compared to the expectation (red diamond) for a SM Higgs boson with mass $m_H = 125$ GeV.

9. Summary and outlook

The experimental evidence for $H \to \tau\tau$ decays was arguably one of the most important milestones of the LHC Run 1 Higgs physics program following the Higgs boson discovery in 2012. The Higgs boson coupling mechanism is conceptually different for fermions and bosons, so directly establishing a Higgs–fermion interaction was a fundamental test of the SM. With signal significances still at the 4σ level for the individual experiments, the $H \to \tau\tau$ results have already made a significant contribution to understanding the Higgs boson couplings.

Key ingredients to this success included a high-quality reconstruction of the relevant final state objects, in particular a good discrimination of hadronic τ decays against QCD jets, sophisticated $\tau\tau$ mass reconstruction algorithms, various methods for a largely data-driven estimation of the main background contributions, and event categorisation concepts exploiting the presence and kinematics of additional jets. Another important aspect, at least within the experimental approach chosen by ATLAS, was the application of multivariate techniques for the signal extraction.

For the second LHC run, the pp collision energy is increased to 13 TeV initially and possibly 14 TeV later on. In addition, the integrated luminosity is expected be at least four times higher compared to the Run 1 dataset. The necessary increase in instantaneous luminosity is accompanied by significant changes of the experimental conditions, in particular due to a reduction of the proton bunch spacing from 50 ns to 25 ns. Increased pileup affects the lepton isolation and the reconstruction and identification of hadronic τ decays. In addition, jets originating from pileup contributions could further increase the background in jet-based categories and degrade the efficiency of the VBF signal selections based on the rapidity separation of the leading jets. This is a concern especially for the trigger selection, where the higher data rates and correspondingly increased single-particle p_T thresholds could make it necessary to rely on more complex event signatures. However, with improved detectors, revised trigger strategies and improved reconstruction and identification algorithms, the experiments are expected to be well prepared to meet these challenges.

The initial goal for the $H \to \tau\tau$ physics program in Run 2 is to establish the $H \to \tau\tau$ signal with a significance exceeding 5σ in each experiment and then to increase the precision of the measurements of the couplings in combination with the other Higgs boson decay modes. The sensitivity of the Run 1 analyses were dominated by VBF production and gluon–gluon fusion with additional jets. For the larger Run 2 data sets, in addition to accurate measurements of the VBF production, the associated production with vector bosons (VH) or top quark pairs $(t\bar{t}H)$ can also be probed. The latter is of particular interest because it is dominated by purely fermionic Higgs interactions and provides direct access to the Higgs–top quark Yukawa coupling. Differential measurements of the Higgs boson decay products or additional jets will be performed as well. All the envisaged Run 2 measurements listed above will complement similar studies based on other Higgs decay channels. There are, however, parts of the Higgs physics program for which $\tau\tau$ final states add unique possibilities, in particular for future Higgs CP mixing studies, as discussed, e.g., in Ref. [12] and references therein.

References

1. ATLAS Collaboration, Evidence for the Higgs-boson Yukawa coupling to tau leptons with the ATLAS detector, *JHEP.* **1504**, 117 (2015). doi: 10.1007/JHEP04(2015)117.
2. CMS Collaboration, Evidence for the 125 GeV Higgs boson decaying to a pair of τ leptons, *JHEP.* **1405**, 104 (2014). doi: 10.1007/JHEP05(2014)104.
3. K. A. Oline *et al.*, Review of Particle Physics, *Chin. Phys.* **C38** 09001 (2014). doi: 10.10881/1674-1137/38/9/090001.
4. ATLAS Collaboration, Identification and energy calibration of hadronically decaying tau leptons with the ATLAS experiment in pp collisions at $\sqrt{s} = 8$ TeV, *Eur. Phys. J. C* **75**(7), 303 (2015). doi: 10.1140/epjc/s10052-015-3500-z.
5. M. Cacciari, G. P. Salam, and G. Soyez, The anti-k(t) jet clustering algorithm, *JHEP.* **0804**, 063 (2008). doi: 10.1088/1126-6708/2008/04/063.
6. CMS Collaboration, Performance of tau-lepton reconstruction and identification in CMS, *JINST.* **7**, P01001 (2012). doi: 10.1088/1748-0221/7/01/P01001.
7. CMS Collaboration, Particle-flow event reconstruction in CMS and performance for jets, taus, and MET. (CMS-PAS-PFT-09-001) (Apr, 2009). URL http://cds.cern.ch/record/1194487.
8. A. Elagin, P. Murat, A. Pranko, and A. Safonov, A new mass reconstruction technique for resonances decaying to di-tau, *Nucl. Instrum. Meth. A* **654**, 481 (2011). doi: 10.1016/j.nima.2011.07.009.
9. L. Bianchini, J. Conway, E. K. Friis, and C. Veelken, Reconstruction of the Higgs mass in $H \to \tau\tau$ events by dynamical likelihood techniques, *J. Phys. Conf. Ser.* **513**, 022035 (2014). doi: 10.1088/1742-6596/513/2/022035.
10. ATLAS Collaboration, Modelling $Z \to \tau\tau$ processes in ATLAS with τ-embedded $Z \to \mu\mu$ data, *JINST.* **10**, P09018 (2015). doi: 10.1088/1748-0221/10/09/P09018.

11. ATLAS and CMS Collaborations, Measurements of the Higgs boson production and decay rates and constraints on its couplings from a combined ATLAS and CMS analysis of the LHC pp collision data at $\sqrt{s} = 7$ and 8 TeV, ATLAS-HIGG-2015-07, CMS-HIG-15-002 (2016). Submitted to *JHEP*. URL http://cds.cern.ch/record/2158863. arXiv: 1606.02266.

12. S. Berge, W. Bernreuther, and S. Kirchner, Determination of the Higgs CP mixing angle in the tau decay channels at the LHC including the Drell–Yan background, *Eur. Phys. J. C* **74**(11), 3164 (2014). doi: 10.1140/epjc/s10052-014-3164-0.

Chapter 9

Search for the Higgs boson
in the $b\bar{b}$ final state at the LHC

Jacobo Konigsberg[*] and Giacinto Piacquadio[†,‡]

University of Florida
Gainesville, FL 32611, USA
†*SLAC National Accelerator Laboratory*
Menlo Park, CA 94025, USA
‡*State University of New York*
Stony Brook, NY 11794, USA

In Run 1 at the LHC, the search for the Higgs boson decay into $b\bar{b}$ by the ATLAS and CMS experiments has resulted in the observation of a small excess of events consistent with the expectation from the production of a Standard Model Higgs boson with a mass of ≈ 125 GeV. The significance of this excess is below the 3σ level, compatible with the expected sensitivity. This chapter summarizes the essential elements of these challenging analyses: the production processes studied, the techniques used in the analyses, and the prospects for the observation of $H \to b\bar{b}$ in Run 2 of the LHC.

1. Introduction

The detection of Higgs boson decays into $b\bar{b}$ is an important ingredient in the determination of the true nature of the discovered Higgs boson. The $H \to b\bar{b}$ decay directly tests its coupling to fermions, and more specifically to bottom-type quarks. In the Standard Model (SM), and for a Higgs boson mass of about $m_H = 125$ GeV, the branching ratio for $H \to b\bar{b}$ is approximately 58%, which is by far the largest branching ratio. Without the unequivocal detection of this process a consistent picture between the observed properties of the new boson and those predicted by the SM remains incomplete.

In spite of the large branching ratio, the detection of the $H \to b\bar{b}$ decay is extremely challenging at hadron colliders. Chapter 3 describes the search for this decay at the Tevatron. At the LHC the inclusive b-quark production cross section, dominated by processes mediated by the strong interaction (QCD), is about seven orders of magnitude larger than the largest cross section for Higgs boson production via gluon–gluon fusion; which is roughly 20 pb in proton–proton collisions at 8 TeV, as shown in Fig. 4.3(b) in Chapter 4. The impossibly small signal-to-background ratio in a final state with only two b-jets, together with the corresponding extremely large trigger rate, render this channel ineffectual for the $H \to b\bar{b}$ search.

Nonetheless, the $H \to b\bar{b}$ search was performed successfully during Run 1 of the LHC through the study of different production processes and the application of many sophisticated techniques that enhance the sensitivity of the analyses. Excesses of events consistent with what is expected from SM production of a Higgs boson with a mass of about 125 GeV have been found. The significance of this excess above the background expectation is smaller than the 3 standard deviations needed to claim evidence for this decay. Very importantly, however, the Run 1 analyses have demonstrated that the observation of $H \to b\bar{b}$ will likely be possible at the LHC once a large enough amount of data is collected during Run 2.

This chapter is arranged as follows. In Sec. 1.1 we present an overview on how this result was achieved, and in Secs. 2 to 4 we describe the searches in the different production processes. These descriptions aim to give the reader a general review of the analyses, describing the main features of the strategy and methodology employed in the searches, and to give a flavour of the complexity involved. These sections also include references to the publications in which more details can be found. We conclude this chapter with a summary of the LHC Run 1 results. We also present the challenges to overcome during Run 2 in order to unequivocally establish the $H \to b\bar{b}$ process and to measure the branching fraction of the Higgs boson to $b\bar{b}$ with as much precision as possible.

1.1. *Overview of $H \to b\bar{b}$ searches*

As mentioned above, the overwhelming QCD production of b-quark pairs preempts a search for $H \to b\bar{b}$ when the Higgs boson is produced through the gluon–gluon fusion process (see Fig. 1.6(a) in Sec. 5 of Chapter 1) in which two b-jets appear alone in the final state. In order to reduce the backgrounds to more manageable levels the presence of additional objects in the final state is needed. The next largest Higgs boson production cross section (approximately 1.6 pb at 8 TeV) is the vector boson fusion process (VBF) in which the Higgs boson is produced in association with two jets.

In this process two forward jets with high rapidity separation are produced, with little extra hadronic activity between them, except for the Higgs boson decaying to two b quarks. Dedicated triggers were developed to record Higgs boson events in this all-jet final state with sufficient efficiency for this process to contribute in a meaningful way to the $H \to b\bar{b}$ search. Nonetheless, such an analysis also suffers from copious QCD multijet background. Details on how this search is performed can be found in Sec. 3.

Through the "Higgsstrahlung" process the Higgs boson is produced in association with a W or a Z vector boson (V) in the final state (VH production). At 8 TeV the cross sections for the WH and the ZH processes are approximately 0.7 pb and 0.4 pb, respectively (see Table 1.2 in Sec. 5 of Chapter 1 for a complete tabulation of the Higgs boson production cross sections). In order to reduce the large multijet backgrounds that plague final states containing only jets, the analyses targeting this production mode focus on all channels in which the vector bosons decay leptonically: $W \to \ell\nu$, $Z \to \ell\ell$ or $Z \to \nu\bar{\nu}$. In these modes the main background processes are vector bosons produced in association with jets (from all quark flavors and from gluons), singly and pair-produced top quarks ($t\bar{t}$), dibosons (VV) and QCD multijet processes. Figure 9.1 shows how a

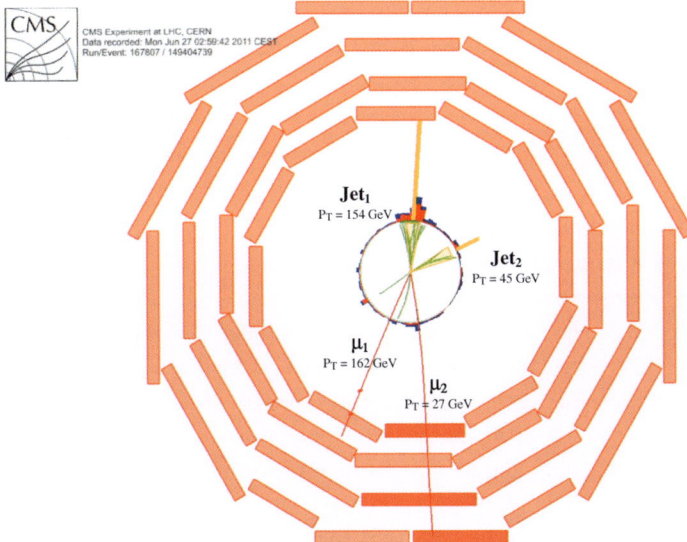

Fig. 9.1 Event display for a $ZH \to \ell\ell bb$ signal candidate event from data as seen in the CMS detector. All reconstructed objects are shown projected on the plane transverse to the colliding beams of protons.

typical $ZH \to \ell\ell b b$ signal candidate event from data looks in the CMS detector. In the innermost circle the picture shows charged particles reconstructed in the inner detector, then, just outside of it, the coloured bars indicate the amount of energy deposited in the electromagnetic and hadron calorimeters (in red and blue, respectively), while the muon system is the outermost detector. The event display nicely shows the two jets from the Higgs boson candidate corresponding to two sprays of charged particles in the inner detector, with two subsequent large energy deposits in the calorimeter (summarized by the yellow bars), while the red curved lines represent two muons from the Z boson decay, reconstructed as tracks in the inner tracking detector and in the muon system.

Due to the presence of two jets originating from b quarks in the final state, it is critical in all $H \to b\bar{b}$ searches to identify these b-jets with high efficiency and with as small a fake rate as possible. Fake b-tagged jets originate from misidentification of light-flavour and c-jets, all of which are abundant in background processes. Light-flavour jets are defined as jets originating from u, d, s quarks or gluons, while c-jets originate from c quarks. As discussed in Sec. 4.7 of Chapter 4, both ATLAS and CMS use optimized b-tagging algorithms for this purpose. The efficiency and fake rates ultimately depend on the algorithm and on the operational working point used in the different analyses in order to optimize their sensitivity. As an example, it is possible to achieve a typical b-tagging efficiency of 70% for b-jets produced within the inner tracking detector acceptance ($|\eta| < 2.5$), with corresponding c- and light-jet efficiencies of approximately 20% and 1%, respectively.

Another important feature that helps distinguish the $H \to b\bar{b}$ signal from the backgrounds is the invariant mass of the two b-quark jets, $m_{b\bar{b}}$. For the Higgs boson signal the $m_{b\bar{b}}$ distribution is approximately a Gaussian that peaks at the value of m_H, with a width of 10–15 GeV, determined by the b-jet energy resolution. Conversely, for most of the backgrounds the $m_{b\bar{b}}$ distribution lacks sharp features. Improving the resolution of the $m_{b\bar{b}}$ peak for the Higgs boson, by measuring the momentum of the b-jets as precisely as possible, helps directly improve the sensitivity of the searches. Improvements of 10–20% can be achieved as discussed in Sec. 2.

At the LHC, just like at the Tevatron, the most sensitive mode in the $H \to b\bar{b}$ search is VH production. However, at the LHC this search is even more challenging than at the Tevatron due to the different parton luminosity ratios of gluon–gluon and quark–antiquark initiated processes.[1] In going from the 2 TeV proton–antiproton collisions at the Tevatron to the

8 TeV proton–proton collisions at the LHC the rate of gluon–gluon initiated processes increases by about a factor of 20, whereas the rate for quark–antiquark initiated processes increases only by about a factor of 4. For VH production, which is a quark–antiquark initiated process, this means that the ratio of signal to background, for gluon–gluon initiated background processes, worsens significantly at the LHC. In addition, at the LHC a much higher fraction of the signal produces Higgs and vector boson decay signatures that are outside the rapidity region of the inner tracker and muon system acceptance, thus preventing the reconstruction of leptons and b-jets. In order to compensate for such signal dilution and worsened acceptance, the LHC VH searches have been optimized to explicitly rely on events in which the vector boson and the Higgs boson have a relatively large p_T. In this region of phase space the signal-to-background ratio and the detector acceptance is improved. Details on all of the features of the VH search can be found in Sec. 2.

Searches for $H \to b\bar{b}$ also look for signatures where the Higgs boson is produced in association with a top quark pair ($t\bar{t}H$ production), with a cross section of roughly 0.13 pb at 8 TeV. The only $t\bar{t}H$ final states that are studied are those that contain one or two leptons from the decays of the W bosons in the $t \to Wb$ decays.[a] These are very complicated final states with at least four jets. This leads to inefficiencies in the jet reconstruction due either to overlapping jets or due to the high likelihood of having at least one jet below the reconstructible p_T threshold. In addition, the presence of four b-jets in the final state and their limited energy resolutions make it very difficult to properly assign the b-jets to the Higgs boson on an event-by-event basis, significantly diluting the possibility of using the di-jet invariant mass (peaking at the Higgs boson mass) to discriminate against the non-resonant backgrounds. The main background to these unique signatures is $t\bar{t}$ production with additional jets from all quark flavours and gluons, but mainly from b quarks. Despite the lower sensitivity with respect to other $H \to b\bar{b}$ searches, the $t\bar{t}H$ production mode has a special interest, as it gives direct access to the Higgs boson Yukawa coupling to the top quark. Details on the $t\bar{t}H$ searches can be found in Sec. 4.

A major challenge in all these searches is the precise estimation of the background yields after the specific selection requirements applied to

---

[a] At the time of the release of this book, a search for $t\bar{t}H$ production in a a fully hadronic final state has also been published by ATLAS.[2]

enhance the $H \to b\bar{b}$ signal contribution. In most cases the simulation is not sufficiently reliable and the backgrounds must be estimated from carefully constructed control samples from data.

We have introduced here, in very broad strokes, the different $H \to b\bar{b}$ searches that have been pursued at the LHC and the challenges faced. The intricacies of these rather complex analyses can be found in the sections that follow. Particular emphasis will be placed on the $H \to b\bar{b}$ search in VH production, because of its higher signal sensitivity.

2. Search for $VH \to Vb\bar{b}$

2.1. Characterization of the final state

This search looks for the production of a Higgs boson decaying to a pair of b quarks in association with a vector boson V, later denoted as $VH \to Vb\bar{b}$, where the vector boson can be either a W or a Z boson. Three channels corresponding to the different leptonic decay modes of the W or Z bosons are explored: $ZH \to \nu\bar{\nu}b\bar{b}$, $WH \to \ell\nu b\bar{b}$ and $ZH \to \ell\ell b\bar{b}$. These are depicted in Fig. 9.2 through their leading-order Feynman diagrams. The $Z \to \nu\bar{\nu}$ decay gives rise to high transverse missing energy ($E_{\mathrm{T}}^{\mathrm{miss}}$) in the final state, while for $W \to \ell\nu$ and $Z \to \ell\ell$ the main signature is the presence of exactly one or two leptons. The presence of either high $E_{\mathrm{T}}^{\mathrm{miss}}$ or leptons allows the trigger selection to keep almost all signal events while rejecting the large multi-jet backgrounds where no leptons are expected in the final state. This allows to keep the trigger rate under control and store all interesting events for further offline data analysis. Once the events are stored, they can be analysed in much more detail and the selection can be significantly refined. In final states with neutrinos, only the overall transverse momentum of the neutrinos can be measured in terms of $E_{\mathrm{T}}^{\mathrm{miss}}$, as described in Sec. 4.6 of Chapter 4.

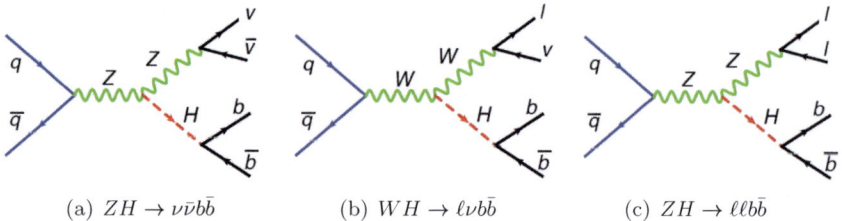

(a) $ZH \to \nu\bar{\nu}b\bar{b}$ (b) $WH \to \ell\nu b\bar{b}$ (c) $ZH \to \ell\ell b\bar{b}$

Fig. 9.2 Feynman diagrams illustrating the three channels of the $VH \to Vb\bar{b}$ analysis.

This means that the four momentum of W and Z bosons decaying to $\ell\nu$ and $\nu\bar{\nu}$, respectively, can't be fully reconstructed. Conversely, in $Z \to \ell\ell$ events, the four momentum of Z boson can be reconstructed from the sum of momenta of its oppositely-charged decay products: the invariant mass of the di-lepton system can therefore be required to be compatible with the Z boson mass ($m_Z \approx 91$ GeV). While the subdivision into three channels is driven by the leptonic signature of the vector boson, some contamination across channels is unavoidable. For example a $ZH \to \ell\ell b\bar{b}$ event where one lepton is not reconstructed or is outside the detector acceptance can easily be misinterpreted as a $WH \to \ell\nu b\bar{b}$ event: these effects are studied in simulations and are taken into account in the analysis.

2.2. *Strategies for background suppression*

There are several backgrounds that can mimic the $VH \to Vb\bar{b}$ signal, with cross sections up to several orders of magnitude larger than the signal. The analysis strategy exploits distinct features of the signal that can be used to suppress the backgrounds. A few important experimental handles are:

- Presence of two b-tagged jets
- Di-jet invariant mass $m_{b\bar{b}}$
- The leptonic signature of the associated W or Z boson
- Few or no additional jets

2.2.1. *Identification of b-jets*

The b-tagging requirement efficiently suppresses backgrounds where one or both jets forming the Higgs boson candidate stem from light-flavour or c-jets. The main background without b-jets is the subset of the associated W and Z boson production with two jets made of combinations of c and light flavours. Despite the significantly higher cross sections for such processes compared to $W/Z + b\bar{b}$ production, after b-tagging only a small residual contamination is left. Dedicated control regions with relaxed b-tagging requirements are used to estimate the contribution of these backgrounds from data.

2.2.2. *Di-jet mass requirement*

As mentioned in Sec. 1.1, the invariant mass of the two b-jets, $m_{b\bar{b}}$, is crucial to help distinguish the $H \to b\bar{b}$ signal from the backgrounds.

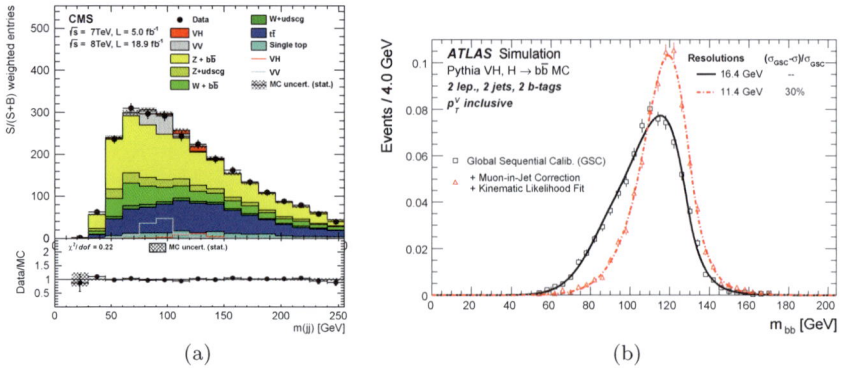

Fig. 9.3   (a) The $m_{b\bar{b}}$ distribution for the $ZH \to \nu\bar{\nu}b\bar{b}$ channel, with the histograms representing the distributions predicted by simulation for the VH signal and the main backgrounds, and the points representing the data (from Ref. [3]). (b) The $m_{b\bar{b}}$ distribution for the $ZH \to \ell\ell b\bar{b}$ signal showing the improvement in the $m_{b\bar{b}}$ resolution expected from applying a kinematic likelihood fit (from Ref. [4]).

Figure 9.3(a) shows the $m_{b\bar{b}}$ distribution in the specific case of the $VH \to Vb\bar{b}$ analysis, where the $H \to b\bar{b}$ signal is expected to be a small peak on top of a large continuum background. The width of the signal peak is of the order of 10–15 GeV and is determined by the b-jet energy resolution. A narrower peak enhances the sensitivity of the analysis, since, given a fixed $m_{b\bar{b}}$ interval around the m_H mass, it enhances the number of accepted signal events. Therefore significant effort has been devoted to improving the precision of the b-jet energy measurement. In CMS, multivariate regression techniques,[3] first developed at the CDF experiment,[5] are used to recalibrate the b-jet energy by training an algorithm to estimate the most likely value of the b-quark p_T based on the details of the measured jet properties, including information on possible soft leptons reconstructed within the jet. The latter is particularly important to include. Semi-leptonic decays of b-hadrons occur in around 40% of the b-jets, either directly through the b-hadron decay or indirectly through the decay of an intermediate c-hadron. These b-jets contain a neutrino and either a muon or an electron. While electrons contribute to the jet energy, neutrinos escape detection and muons deposit only a small fraction of their energy in the calorimeter, reducing the reconstructed b-jet p_T and thus degrading the $m_{b\bar{b}}$ resolution. ATLAS applies a simpler correction,[4] which is based directly on the four momentum of the muon (if present within the jet) and on the average b-jet energy response R in the signal ($R = p_T(\text{reco}) \, / \, p_T(\text{true})$).

Through these techniques it is possible to improve the $m_{b\bar{b}}$ resolution by 10–20%, depending on the p_T boost of the reconstructed $H \to b\bar{b}$ state. In the case of the $ZH \to \ell\ell bb$ signal a more significant improvement in the $m_{b\bar{b}}$ resolution can be obtained by exploiting the kinematic constraint that the p_T of the two b-jet system is approximately equal to the p_T of the Z boson, where the Z boson p_T can be reconstructed from the leptons with a significantly better resolution than the Higgs boson p_T. Here the use of regression techniques (CMS) or a kinematic likelihood fit (ATLAS) leads to an improvement in the $m_{b\bar{b}}$ resolution of up to 30%, as shown in Fig. 9.3(b).

The di-jet mass is ultimately the best handle to reduce the large contamination from $Z + b\bar{b}$ and $W + b\bar{b}$ production, which are among the dominant backgrounds to the VH search, as can be seen in Fig. 9.3(a) for the $ZH \to \nu\bar{\nu}b\bar{b}$ channel. The final states of these processes resemble the signal, except that the resonant $H \to b\bar{b}$ contribution is replaced with a continuum $m_{b\bar{b}}$ spectrum. The region where the value of $m_{b\bar{b}}$ is far from the Higgs boson peak, usually denoted as the "mass sideband", defines a good control region that can be used to control the normalization of the $W + b\bar{b}$ and $Z + b\bar{b}$ yield from data. The background normalization is thus derived from data, and the simulation is used only to extrapolate the yield from the sidebands to the signal region. This typically results in a more accurate estimate of the background rates in proximity of the signal peak. The specific case of the $V + b\bar{b}$ background can be generalized to a larger number of backgrounds, provided there are sufficient control regions to constrain all backgrounds simultaneously.

In addition to the Higgs boson signal, there is an additional resonant contribution: the diboson $VZ \to Vb\bar{b}$ backgrounds, which peak at $m_{b\bar{b}} \approx m_Z$. This contribution can be seen in Fig. 9.3(a) as a larger peak on the left of the $H \to b\bar{b}$ signal. The rate for diboson production is approximately five times larger than for the VH signal, but the direct contamination of this process into the signal-rich region near $m_{b\bar{b}} \sim 125$ GeV is small. Nevertheless, these resonant backgrounds need to be considered when trying to estimate the yields of the other backgrounds in the $m_{b\bar{b}}$ signal sideband region. Given the similarity to the signal but the higher expected rate, the diboson production is also ideal for testing the analysis techniques and the statistical procedure to extract a Higgs boson signal. By just treating the Z boson signal as an unknown to be measured like the Higgs boson signal, it is possible to test if the analysis is able to extract the level of diboson signal as predicted by the Standard Model.

2.2.3. *Leptonic signature*

The $ZH \to \ell\ell b\bar{b}$ channel allows a very clean VH selection, because requiring two leptons and $m_{\ell\ell}$ to be around m_Z suppresses any non-Z background. Due to the very good momentum resolution of the leptons, this requirement can significantly suppress backgrounds characterized by a continuous $m_{\ell\ell}$ spectrum, such as the di-leptonic $t\bar{t}$ background ($t\bar{t} \to Wb W\bar{b} \to \ell\nu b \ell\bar{\nu}\bar{b}$). Since jets are misidentified as isolated leptons with rates of about 0.1% or below, after requiring two leptons in the events the multijet background is also negligible. In the $\ell\nu b\bar{b}$ and $\nu\bar{\nu}b\bar{b}$ final states it is more difficult to suppress backgrounds where the $\ell\nu$ and $\nu\bar{\nu}$ signatures do not correspond to an actual W or Z boson. In the $ZH \to \nu\bar{\nu}b\bar{b}$ channel it is not possible to apply any constraint on the Z boson candidate beyond a requirement on $E_{\mathrm{T}}^{\mathrm{miss}}$ (which corresponds to $p_T(Z)$ being above a certain threshold). One of the main backgrounds in this channel is from di-leptonic $t\bar{t}$ events, with both leptons outside the p_T or η acceptance and with sizeable $E_{\mathrm{T}}^{\mathrm{miss}}$ from the two neutrinos. Mismeasured jets in multijet events can also give rise to large fake $E_{\mathrm{T}}^{\mathrm{miss}}$, and thus this background is not entirely suppressed by the $E_{\mathrm{T}}^{\mathrm{miss}}$ requirement alone. Further selection requirements are applied to reduce this background to a very small level, as described in Sec. 2.4.1. In the case of the $WH \to \ell\nu b\bar{b}$ channel, in addition to the transverse momentum of the W boson candidate the transverse mass can also be reconstructed. The transverse mass is defined as $m_T = \sqrt{p_{T,\ell}E_{\mathrm{T}}^{\mathrm{miss}}\left(1 - \cos\left(\Delta\phi\right)\right)}$, where $\Delta\phi$ is the azimuthal angle between the lepton and $E_{\mathrm{T}}^{\mathrm{miss}}$. This variable provides some discrimination against di-leptonic $t\bar{t}$ events where a second lepton is missed. In this case the reconstructed values of m_T can extend well beyond the mass of the W boson, which can be explained by a second neutrino contributing to the $E_{\mathrm{T}}^{\mathrm{miss}}$ of the event, and so rejecting events with m_T beyond the nominal W mass reduces the $t\bar{t}$ background. In the $\ell\nu b\bar{b}$ channel the multijet background, where at least one jet is misidentified as a lepton, is small but not negligible. Whenever needed, it is estimated using data-driven techniques, with both the yield and differential distributions estimated from dedicated control regions.

2.2.4. *Few or no additional jets*

In $VH \to Vb\bar{b}$ production, beyond the two b-jets, few or no additional jets are expected, while there are several important backgrounds whose leading contributions contain additional jets in the final state. The most significant of these is $t\bar{t}$ production, which arises mainly from the partonic process

$gg \to t\bar{t} \to WbW\bar{b}$. This background is particularly severe in the $WH \to \ell\nu b\bar{b}$ channel. It enters mainly in the form of $t\bar{t} \to l\nu bq\bar{q}\bar{b}$, with one top quark decaying leptonically and one hadronically, and thus with two additional jets. Another background of this type is the single-top production, either in the Wt channel $(gb \to Wt(\bar{b}) \to l\nu q\bar{q}b(\bar{b}))$,[b] with one or two extra jets, or in the t channel $(qb \to qt(\bar{b}) \to Wbq(\bar{b}))$, with typically at most one extra jet. A jet veto consists of requiring no extra jets (in addition to the two jets paired to the Higgs boson candidate) to be present in the event. When a jet veto is applied, the aforementioned backgrounds are drastically reduced. The reason for the residual inefficiency of the jet veto is two-fold: (a) jets around or below p_T threshold (20 or 25 GeV) can escape the veto, (b) a jet might not be reconstructed because it is overlapping with another jet or lepton in the event. The latter is quite frequent in events with high p_T top quark decays, which result in collimated decay products that are not always resolved by the anti-k_t clustering algorithm with radius parameter $R = 0.4$. The jet veto also helps in reducing the $t\bar{t}$ and single-top backgrounds in the $ZH \to \nu\bar{\nu}b\bar{b}$ channel and it also suppresses the contribution from hadronically decaying τ leptons. Analogously to the case of the $m_{b\bar{b}}$ sidebands, events with higher jet multiplicities which do not pass the jet veto can be used to estimate the top quark related backgrounds from data.

2.3. *The "boosted" regime*

While the requirements outlined in the previous section can enhance the expected signal-to-background ratio (S/B) significantly, with these alone it is difficult to achieve S/B greater than 1%. ATLAS and CMS have therefore worked hard to try to find additional handles to suppress the backgrounds. With respect to the CDF and D0 experiments at the Tevatron, the higher Higgs boson production rate at the LHC allows a tighter selection that focuses on regions of phase space with enhanced signal purity, while maintaining signal rates sufficient for the measurement.

As described in Sec. 5 of Chapter 1, the VH production process is characterized, up to NLO in perturbative QCD theory, by a virtual vector boson (V^*), which then radiates a Higgs boson through *Higgs-strahlung*,

---

[b]The $(\bar{b})$ in parenthesis indicates a $\bar{b}$-quark from gluon splitting in the initial state, which is often soft and forward and therefore does not always appear in the reconstructed final state.

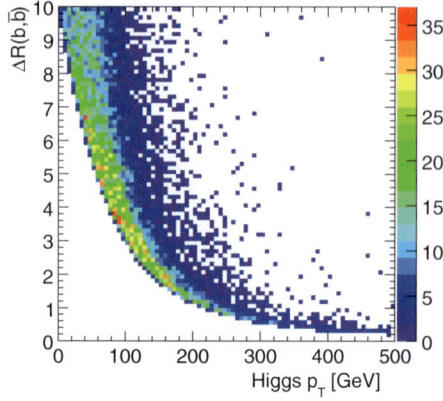

Fig. 9.4   Two-dimensional relationship of $\Delta R(b_1, b_2)$ and the Higgs boson transverse momentum, derived from $VH \to Vb\bar{b}$ simulation. The color scale illustrates the relative density of each region, using arbitrary units.

resulting in a final state vector boson. The mass of the V^* (m_{VH}) is directly related to the energy scale of the hard scattering process. At the Tevatron most of the Higgs boson events produced were just above the threshold ($m_{VH} > m_V + m_H$), resulting in Higgs and vector boson decays almost at rest. At the LHC, due to the higher collision energy, a significant number of events are produced with $p_T(V)$ and $p_T(H) > 150$ GeV, where the Higgs and the vector bosons are produced with large boost and azimuthal separation in the laboratory frame. This boosted p_T regime is of vital importance for the $VH \to Vb\bar{b}$ analysis at the LHC due to smaller backgrounds and better $m_{b\bar{b}}$ resolution.

As shown in Fig. 9.4, with higher Higgs boson p_T, the two b-jets have increasingly small opening angle, as expected from a two-body decay:

$$\Delta R(b_1, b_2) = \sqrt{\Delta \eta^2 + \Delta \phi^2} \approx \frac{2m_H}{p_T(H)}.$$

As first explored in Ref. 6, by requiring the Higgs boson to have high p_T and small $\Delta R(b_1, b_2)$, the backgrounds to $VH \to Vb\bar{b}$ can be significantly suppressed. There are several reasons why this happens, mostly related to the characteristics of the background processes. The $p_T(V)$ spectrum in the $W/Z + b\bar{b}$ backgrounds is softer with respect to the Higgs signal, and so at high $p_T(V)$ the signal purity is larger. A stronger suppression is obtained for the $t\bar{t}$ background, especially in the $WH \to \ell\nu b\bar{b}$ channel, as schematically depicted in Fig. 9.5. The two top quarks are produced mostly

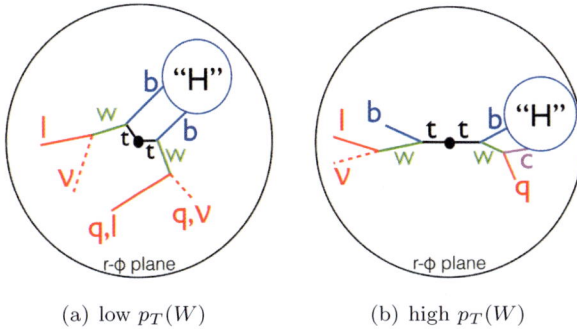

(a) low $p_T(W)$ (b) high $p_T(W)$

Fig. 9.5 Sketches representing the typical topological configurations in which the $t\bar{t}$ background mimics the $WH \to \ell\nu b\bar{b}$ signal selection for low (a) and high (b) $p_T(W)$. The circle with the "H" inside denotes the jet pairing that satisfies the $H \to b\bar{b}$ candidate selection.

back-to-back in the transverse $r\phi$ plane.[c] At low $p_T(V)$ (Fig. 9.5(a)), even if the two top quarks are most often produced back-to-back, the large top mass ($m_{\text{top}} \sim 173$ GeV) allows each b-quark to be almost equally distributed in the $r\phi$ plane, such that combinations with almost any value of angular separation of the two resulting b-jets and where $m_{b\bar{b}}$ happens to be around m_H are easily possible. Since b-tagging suppresses contributions where jets not corresponding to real b-jets are picked up as decay candidates of the Higgs boson, this turns out to be the leading background component. At high $p_T(V)$ (Fig. 9.5(b)) a new regime is entered. Once a high p_T W boson is selected, with p_T comparable or above the top quark mass, both top quarks are forced into a configuration where they are highly boosted and recoil against each other. In this kinematic configuration it is very difficult for the two real b-jets originating from the two back-to-back and boosted top decays to be produced with small angular separation, and so an upper selection cut on the angular separation $\Delta R(b\bar{b})$ can efficiently suppress the leading component of the background. As a consequence, at high $p_T(V)$ the largest residual contribution to the background from $t\bar{t}$ events is due to $b+c$-jet combinations entering the Higgs boson candidate selection, where both jets originate from the decay of the same top quark, the c-jet through the decay of the intermediate W boson ($t \to Wb$, with $W \to c\bar{s}$), and two additional jets escape the jet veto. The main advantage is that this contribution is reduced by the b-jet

---

[c]This feature is enhanced by the additional jet veto, which suppresses configurations with an additional quark or gluon either from initial or final state radiation.

tagging requirement and directly proportional to the c-jet mistagging rate. To improve the c-jet rejection further, and therefore reduce the residual top background contamination at high $p_T(V)$, without making compromises on signal efficiency, several b-tagging working points can be used simultaneously, by either separating the events in categories or using the b-tagging discriminant directly in a multivariate analysis. In fact, by tightening the b-jet tagging requirement, e.g. from 70% to 50% efficiency for real b-jets, a significant improvement in the c-jet mistagging rate, from ~20% to ~6%, can be obtained. A further improvement, from ~6% to ~3%, has been obtained in ATLAS by re-optimizing the b-tagging algorithms specifically to reject c-jets.

2.4. *Di-jet mass and MVA analyses*

Two different methods of data analysis for isolating signal events have been considered. In the *di-jet mass analysis* the signal yield is extracted from data through a maximum likelihood fit to the $m_{b\bar{b}}$ distribution. Events entering the likelihood fit must satisfy stringent kinematic and b-jet tagging requirements. To further enhance the sensitivity and simultaneously constrain the background yields from data, events are separated in categories with different expected S/B ratios and a simultaneous fit is performed, with signal and background yields correlated across categories. The categories are defined by channel, jet multiplicity and b-tagging requirements.

Alternatively a *multivariate analysis* (MVA) is performed, where a maximum likelihood fit to a multivariate discriminant is performed on the data. This discriminant, in order to obtain the best possible separation of signal and background, summarizes the information from a set of observables (describing event topology, object kinematics, b-jet likelihood, etc.) into one single variable, represented by a boosted decision tree (BDT) (as implemented in the TMVA package[7]). By using this technique the observables typically used to select events prior to the final likelihood fit are also directly used in the fit itself. Therefore significantly looser selection criteria are applied compared to the *di-jet mass analysis*. Multiple event categories are defined also in this case, for two main reasons: (a) to allow the use of category specific BDTs, which in most cases yield a better separation of signal and background; (b) to reduce the impact of systematic uncertainties on the background prediction. A more detailed description of multivariate analysis techniques can be found in Appendix B.

In the following the BDT analysis is described in some detail, since it yields the highest sensitivity. The di-jet mass analysis is kept as a valuable cross-check of the BDT results.

2.4.1. *Main selection criteria*

In Table 9.1 the most important selection requirements for the ATLAS and CMS BDT analyses are presented. As already mentioned, events satisfying a mass window requirement of $0 < m_{b\bar{b}} < 250$ GeV are kept. In order to better exploit the boosted regime, described in Sec. 2.3, categories are defined by intervals of $p_T(V)$. In the CMS analysis the exact intervals have been optimized separately for each vector boson decay channel. In the ATLAS analysis the lowest $p_T(V)$ interval includes the region down to very low values of $p_T(V)$, except when experimentally inaccessible (in the $ZH \to \nu\bar{\nu}b\bar{b}$ channel). While the contribution to the signal sensitivity of the low $p_T(V)$ region is small, it helps in constraining the backgrounds. In the $ZH \to \ell\ell b\bar{b}$ channel, a relatively loose mass window requirement on the di-lepton invariant mass $(m_{\ell\ell})$ around the nominal Z boson mass is applied in both the ATLAS and CMS analyses. In the $ZH \to \nu\bar{\nu}b\bar{b}$ channel the requirement on E_T^{miss} naturally follows the requirement on $p_T(V)$. In the $WH \to \ell\nu b\bar{b}$ channel a cut on the minimum E_T^{miss} is useful to reject the multijet background, while in the $ZH \to \ell\ell b\bar{b}$ channel an upper cut can be useful to reject backgrounds with real E_T^{miss} in the final state, such as in $t\bar{t}$ events. The strategy differs slightly between ATLAS and CMS, with ATLAS choosing looser selection cuts.

The jets forming the Higgs boson candidate are selected in a slightly different way: in the CMS analysis the di-jet pair with the highest transverse momentum, computed from the vectorial sum of the jet momenta, is chosen; while in the ATLAS analysis the two leading jets in p_T are considered. In the ATLAS analysis a lower cut is imposed on the angular separation of the two jets $(\Delta R(jj) > 0.7)$, in order to avoid a region of phase space where the modeling of the $V+$ jet background is subject to large theory uncertainties. This requirement is not applied for $p_T(V) > 200$ GeV, where it would significantly impact the signal acceptance.

The jets from the Higgs boson candidate are then required to pass a basic b-tagging requirement (denoted as X_{ϵ_B} in the table), corresponding to an average b-tagging efficiency for real b-quark jets of ϵ_B. In ATLAS the same requirement with $\epsilon_B = 80\%$ is chosen everywhere. In CMS this requirement varies slightly from channel to channel, and in general differs

Table 9.1   Main selection requirements applied to the ATLAS and CMS multivariate analyses. When selection requirements depend on the $p_T(V)$ interval, they are encompassed by square brackets corresponding to the multiple intervals, otherwise a single inclusive requirement is quoted. Requirements on lepton isolation are not included in the table, but are described in the text.

Variable	$W(\ell\nu)H$	$Z(\ell\ell)H$	$Z(\nu\bar\nu)H$
		ATLAS	
$p_T(V)$ [GeV]	[0–120][> 120]	[0–120][> 120]	[100–120][> 120]
$m_{\ell\ell}$ [GeV]	—	[71,121]	
E_T^{miss} [GeV]	[-][> 20]	—	[0–120][> 120]
ΔR_{jj}		> 0.7 (for $p_T(V) < 200$)	
$p_T(j_1)$ [GeV]		> 45	
$p_T(j_2)$ [GeV]		> 20	
$p_T(jj)$ [GeV]	—	—	—
b-tag (tightest)		> $X_{80\%}$	
b-tag (loosest)		> $X_{80\%}$	
N_{aj}		0 or 1	
$N_{a\ell}$		=0	
H_T [GeV]	[> 180][-]	—	—
$\Delta\Phi(E_T^{miss}, \ell)$	—	—	—
$p_{T,miss}$ [GeV]	> 30	—	—
$\Delta\Phi(E_T^{miss}, jet)$	—	—	[-][> 1.5]
$\Delta\Phi(E_T^{miss}, p_{T,miss})$	—	—	< $\pi/2$
$E_T^{miss}/\sigma(E_T^{miss})$	—	—	—
$\sum_{jets=1}^{2(3)} p_T$ [GeV]	—	—	> 120 (150)
		CMS	
$p_T(V)$ [GeV]	[100–130][130–180][> 180]	[50–100][> 100]	[100–130][130–170][> 170]
$m_{\ell\ell}$ [GeV]	—	[75,105]	
E_T^{miss} [GeV]	> 45	< 60	[100–130][130–170][> 170]
ΔR_{jj}	—		
$p_T(j_1)$ [GeV]	> 30	> 30	> 60
$p_T(j_2)$ [GeV]	> 30	> 30	> 30
$p_T(jj)$ [GeV]	> 100	> 120	[> 100][> 130][> 130]
b-tag (tightest)	> $X_{80\%}$	[> $X_{75\%}$][> $X_{85\%}$]	> $X_{70\%}$
b-tag (loosest)	> $X_{80\%}$	> $X_{85\%}$	> $X_{85\%}$
N_{aj}	—	—	[< 2][-][-]
$N_{a\ell}$		=0	
H_T	—	—	—
$\Delta\Phi(E_T^{miss}, \ell)$	< $\pi/2$	—	—
$p_{T,miss}$ [GeV]	—	—	—
$\Delta\Phi(V, H)$	—	—	> 2.0
$\Delta\Phi(E_T^{miss}, jet)$	—	—	[> 0.7][> 0.7][> 0.5]
$\Delta\Phi(E_T^{miss}, p_{T,miss})$	—	—	< $\pi/2$
$E_T^{miss}/\sigma(E_T^{miss})$	—	—	[> 3][-][-]
$\sum_{jets=1}^{2(3)} p_T$ [GeV]	—	—	—

for the two jets: a slightly tighter requirement is applied in channels with a lower signal-to-background ratio. An asymmetric b-tagging requirement helps preserve signal efficiency while rejecting backgrounds without any b- or c-quark jet, as rejecting one of the two jets is sufficient to reject the event. This is the case especially for Z+jet production, which is the leading background in the $ZH \to \ell\ell b\bar{b}$ and $ZH \to \nu\bar{\nu}b\bar{b}$ channels. Beyond the two jets forming the Higgs boson candidate, additional jets can be present in the event. In ATLAS only events with either no or one additional jet, defined in the table as N_{aj}, are considered; while in CMS this limitation only applies to the lowest $p_T(V)$ region in the $ZH \to \nu\bar{\nu}b\bar{b}$ channel. A veto on extra leptons in the event, defined in the table as $N_{a\ell}$, is applied in all channels.

Further selection requirements are applied to reject the multijet background in the $WH \to \ell\nu b\bar{b}$ and $ZH \to \nu\bar{\nu}b\bar{b}$ channels. In the case of $WH \to \ell\nu b\bar{b}$ the main handle against the multijet background is lepton isolation, which has been introduced in Sec. 4.8 of Chapter 4, and helps to distinguish true leptons from jets misidentified as leptons. In CMS the ratio of the track-based isolation to the lepton p_T is required to be less than approximately 10%, with the exact value and the isolation cone depending on lepton p_T and η. In ATLAS the same ratio is required to be less than 4%, with the cone size being $\Delta R = 0.3$. In addition, in ATLAS a calorimeter-based definition is also used, where tracks are replaced by energy deposits in the calorimeter and, after subtracting the average contribution expected from pile-up, the total energy around the lepton is again required to be less than 4% of that of the lepton itself. To further reduce the multijet contribution in the lowest $p_T(V)$ region, where it is most significant, the ATLAS analysis applies a lower cut on H_T, the sum over the p_T of all objects in the event, as again reported in Table 9.1. In CMS, beyond the already mentioned E_T^{miss} cut, the azimuthal angle between E_T^{miss} and the lepton is required to be less than $\pi/2$, and the lepton isolation requirement is tightened further for the lowest $p_T(V)$ region (100–130 GeV).

In the $ZH \to \nu\bar{\nu}b\bar{b}$ channel the multijet background arising from high values of E_T^{miss} due to mismeasured jets is suppressed by several requirements. Firstly, the missing transverse energy vector, $\vec{E}_T^{\mathrm{miss}}$, and the track-based missing transverse momentum vector, $\vec{p}_{T,\mathrm{miss}}$, are required to point in the same direction (within $\pi/2$), to suppress the effect of bad energy measurements in the calorimeter which affect $\vec{E}_T^{\mathrm{miss}}$ but not the track-based $\vec{p}_{T,\mathrm{miss}}$. Secondly, a lower requirement is applied on the azimuthal angle between $\vec{E}_T^{\mathrm{miss}}$ and the closest jet, $\Delta\Phi(E_T^{\mathrm{miss}}, \mathrm{jet})$: this suppresses events with high E_T^{miss} due to jets whose energy has been partially lost, for example in uninstrumented regions of the calorimeter. The multijet background can be further suppressed by a lower cut on the $E_{\mathrm{T}}^{\mathrm{miss}}$ significance, $E_{\mathrm{T}}^{\mathrm{miss}}/\sigma(E_{\mathrm{T}}^{\mathrm{miss}})$, as in the CMS analysis, or by a lower cut on the sum of the transverse energy of all the jets, $\sum_{\mathrm{jets}=1}^{2(3)} p_T$, as was done in ATLAS.

The overall signal acceptance of the BDT analysis selection was estimated to be $\sim4\%$ ($\sim2\%$) for the $ZH \to \nu\bar{\nu}b\bar{b}$ channel, $\sim4\%$ ($\sim3.2\%$) for the $WH \to \ell\nu b\bar{b}$ channel[d] and $\sim13\%$ ($\sim10\%$) for the $ZH \to \ell\ell b\bar{b}$ channel[e] in the ATLAS (CMS) analysis. The higher signal acceptance of the ATLAS analysis is mainly due to the inclusion of the low $p_T(V)$ regions, which however do not add much to the overall signal sensitivity.

2.4.2. *Multivariate discriminant*

All the events passing the selection requirements are used as input to a BDT, which provides optimal discrimination between signal and background. The input to this BDT is a set of discriminating observables and the output is a single per-event value which orders the events by their expected S/B ratio. The optimization of the BDT, which is called *training*, is performed using large sets of simulated signal and background events. The input observables are listed in Table 9.2. Some of the most discriminating observables are the same as those presented in Sec. 2.2 in the context of the analysis selection.

The most discriminating observable is $m_{b\bar{b}}$. The b-tagging discriminants represent a measure of the likelihood of the jets forming the Higgs boson candidate to be originating from real b quarks. These help especially in channels which keep a significant residual non b-jet contribution even after the b-tagging requirements are applied in the selection, as detailed in Table 9.1. One important example is the $WH \to \ell\nu b\bar{b}$ channel in the boosted region, where $b+c$-jet combinations in $t\bar{t}$ events are misidentified as the Higgs boson candidate and constitute an important fraction of the background. An additional powerful handle against the backgrounds is obtained, especially in the boosted region, by combining $p_T(V)$, $\Delta\phi(V, bb)$, the p_T of the two b-jets from the Higgs boson candidate and their opening angle, given by $\Delta R(b_1, b_2)$. As discussed in Sec. 2.3, at high $p_T(V)$ the signal is expected to have the vector boson V recoiling back-to-back against the Higgs boson

---

[d]The acceptance quoted for the $WH \to \ell\nu b\bar{b}$ channel includes $W \to \tau\nu$ decays.
[e]The acceptance quoted for the $ZH \to \ell\ell b\bar{b}$ channel doesn't include $Z \to \tau\tau$ decays, since the expected acceptance for this decay mode is negligible.

Table 9.2   Observables used as input to the multivariate BDT discriminant. All observables are defined in the text.

Analysis Observable	ATLAS			CMS		
	$W(\ell\nu)H$	$Z(\ell\ell)H$	$Z(\nu\bar{\nu})H$	$W(\ell\nu)H$	$Z(\ell\ell)H$	$Z(\nu\bar{\nu})H$
$m_{b\bar{b}}$	×	×	×	×	×	×
$p_{T,V}$	×	×		×	×	
E_T^{miss}	×	×	×			×
b-tag discr. (jet 1)	×	×	×	×	×	×
b-tag discr. (jet 2)	×	×	×	×	×	×
$\Delta R(b_1,b_2)$	×	×	×	×	×	×
$p_T^{b_1}$	×	×	×	×	×	×
$p_T^{b_2}$	×	×	×	×	×	×
$|\Delta\eta(b_1,b_2)|$		×	×	×	×	×
$\Delta\phi(V,bb)$	×	×	×	×	×	×
$|\Delta\eta(V,bb)|$		×				
$\Delta\theta_{\text{pull}}$				×	×	×
H_T			×			
$m_T(W)$	×					
$m_{\ell\ell}$		×				
$\min[\Delta\phi(\ell,b)]$	×					
m_{VH}					×	
$\Delta\Phi(E_T^{\text{miss}},\text{jet})$						×
$\cos\left(\alpha_{Z,ZH}\right)$					×	
$\cos\left(\alpha_{\ell,\ell\ell}\right)$					×	
$\cos\left(\alpha_{b,H}\right)$					×	
N_{aj}				×	×	×
			Only in 3-jet events			
$\max(\text{b-tag})_{\text{aj}}$				×		×
$\min(\Delta R(H,\text{aj}))$				×		×
p_T^{aj}	×	×	×			
$m_{b\bar{b}j}$	×	×	×			

H in the transverse plane ($\Delta\phi(V,bb)$ close to $\pi/2$) and a decreasing opening angle between the two b-jets ($\Delta R(b_1,b_2)$), while a large fraction of the backgrounds fails to satisfy the same constraints.

Other less discriminating observables used are, in the ATLAS analysis, H_T (the sum over the p_T of all objects in the event), and, in the CMS analysis, the color pull angle $\Delta\theta_{\text{pull}}$, which is meant to separate a b-quark pair produced from a color singlet (the Higgs boson) from cases where the b quarks are color-connected with other quarks or gluons in the event.[8] A few observables based on the leptonic event signature, i.e. $m_T(W)$ and $\min[\Delta\phi(\ell,b)]$ for the $WH \rightarrow \ell\nu b\bar{b}$ channel and $m_{\ell\ell}$ for the $ZH \rightarrow \ell\ell b\bar{b}$ channel, are used in ATLAS, as a result of either removing or loosening the corresponding selection requirements. In the $ZH \to \ell\ell b\bar{b}$ channel, where the full final state can be reconstructed, the CMS analysis exploits the invariant mass of the ZH system, m_{VH}, and the helicity angle α of all the expected two-body decays (Z emitted from the virtual Z boson ($q\bar{q} \to Z^* \to ZH$), ℓ emitted from $Z \to \ell\ell$ and the b-jet emitted from $H \to b\bar{b}$). In many cases removing or adding a single input observable to the BDT does not change the sensitivity of the analysis significantly, since many of the observables are correlated and some of them can thus be obtained or approximated sufficiently well by exploiting the other observables.

Information about extra jets in the event is important to reject some of the main backgrounds, especially the $t\bar{t}$ background. Different strategies are adopted by ATLAS and CMS to include such information in the final statistical analysis of the data. In the CMS analysis an additional observable is added to the BDT (the number of additional jets N_{aj}) and a few extra input variables are filled only if at least one additional jet is present in the event. In ATLAS separate categories are defined for the case where either zero or one additional jet is present in the event, and therefore separate BDTs are trained in the two cases. The BDT for the case $N_{\mathrm{aj}} = 1$ contains a few extra inputs as well. The additional input observables used by ATLAS and CMS are slightly different, but exploit very similar information, as in the case of $\min(\Delta R(H, \mathrm{aj}))$, the angular separation between the Higgs boson candidate and the closest additional jet, and $m_{b\bar{b}j}$, the invariant mass of the system given by the Higgs boson candidate plus the extra jet. Both observables are meant to identify cases where an additional jet is radiated in the Higgs boson decay through final state radiation, resulting in either a small value of $\Delta R(H, \mathrm{aj})$ or in a value of $m_{b\bar{b}j}$ compatible with the nominal Higgs boson mass, which is less likely for the background processes. In the CMS analysis the b-jet discriminant for the additional jet in the event is also used as an input, while in ATLAS events with an additional b-tagged jet (with $\epsilon_B = 80\%$) are excluded already at the selection level. Additional b-jets appear most frequently in $t\bar{t}$ background events entering the $WH \to \ell\nu b\bar{b}$ channel selection, where a jet pair made of a b and a c-jet has been misidentified as the Higgs boson candidate and therefore an additional b-jet is present in the event.

2.5. *Signal extraction strategy*

Based on the events passing the selection requirements described in Sec. 2.4.1 and on the BDT discriminant defined in Sec. 2.4.2, a maximum likelihood fit is performed in multiple event categories and used to extract the Higgs boson signal rate in data. Although the basic procedure is very similar, the ATLAS and CMS analysis strategies differ in how the maximum likelihood fit and the BDT technique are used to provide constraints on the normalization of the main backgrounds, while simultaneously extracting the signal.

A multivariate analysis technique such as a BDT with a one-dimensional output provides close-to-optimal statistical separation between the signal and the combination of all backgrounds, by providing a discriminant which is monotonically increasing as a function of the expected signal-over-background (S/B) ratio. The BDT thus groups events in bins of expected S/B and the combination of these in a maximum likelihood fit provides the optimal statistical sensitivity to the signal, while simultaneously allowing the extraction of the overall background normalization.[f] This however ignores the effect of systematic uncertainties. In particular, since the BDT separates the signal only from the combined sum of all backgrounds, its sensitivity to the signal can be significantly impacted by the relative normalization uncertainty of the individual backgrounds, which enters as one of the most important shape uncertainties on the total background. An important effect is also played by systematic uncertainties distorting the BDT shape of each single background. A signal extraction technique based on a multivariate analysis addressing in an optimal way these two points is yet to be defined and would most likely require integrating systematic uncertainties into the BDT training. For the same reason, loosening completely the selection requirements on all variables which later enter as input to the BDT discriminant, while typically allowing to obtain a better statistical sensitivity due to the additional information provided, does not guarantee a higher final sensitivity to the signal once systematic uncertainties are considered, and the selection requirements therefore have been optimized taking systematic uncertainties into account.

To limit the effect of systematic uncertainties different strategies have been pursued. In the ATLAS analysis the main discriminant used in the maximum likelihood fit remains the BDT trained for separating the signal from the sum of all backgrounds, but several additional categories have been introduced to allow for a better normalization of the separate background contributions. The categories in which events are subdivided are

---

[f]The binning is chosen carefully, such that no significant loss of sensitivity is expected due to the limited number of bins.

Table 9.3 Event categories in the multivariate analysis. The $p_T(V)$ categories depend on channel: these are detailed in Table 9.1.

Categories	ATLAS	CMS
$p_T(V)$	2 categories	2 or 3 categories
Jet multiplicity	2 jets, 3 jets	—
b-tagging	2-tag loose, 2-tag medium, 2-tag tight	—

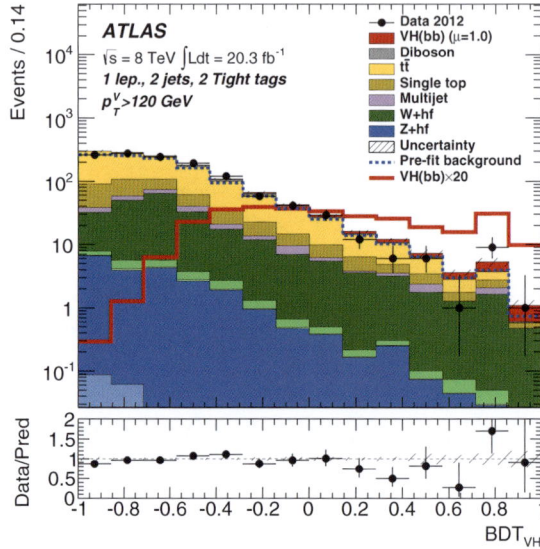

Fig. 9.6 The BDT discriminant distribution for the ATLAS multivariate analysis in the most sensitive event category of the $WH \rightarrow \ell\nu b\bar{b}$ channel, as observed in data (points with error bars) and expected from simulations (histograms). Taken from Ref. [4].

summarized in Table 9.3 for both ATLAS and CMS. The three-jet category helps mainly to control the $t\bar{t}$ background, while separating events in the loose, medium and tight two b-tag categories by using increasingly tighter b-jet tagging requirements helps control the uncertainty on the relative composition of the different flavour components of the V+jet background. This solution comes at the cost of significantly increasing the effective number of bins of the maximum likelihood fit and therefore its complexity. The BDT distribution for the $WH \rightarrow \ell\nu b\bar{b}$ channel in the most sensitive analysis category is shown in Fig. 9.6 for the ATLAS analysis.

The same challenge is addressed in the CMS analysis by using a different strategy. Additional BDTs are introduced, where the signal is trained separately against the main background components, given by the top, V+jet and resonant diboson background. For each BDT a cut value is defined. Events failing the top-background-related BDT cut are classified as top-like events, and binned in values of this BDT. The remaining events are further analysed. If they fail the V+jet-background-related BDT cut, they are binned in values of this BDT. The same happens for the diboson-background-related BDT cut and, finally, the events which survive are used as input to the BDT separating signal and all backgrounds. Through this procedure a single combined BDT distribution can be defined, with the four subsets of data, top-like events in the first quarter, V + jet-like in the second, diboson-like in the third and most signal-like in the last quarter. The resulting BDT distribution for the $WH \to \ell\nu b\bar{b}$ channel is shown in Fig. 9.7, for the most sensitive $p_T(V)$ region. This technique allows to separate out regions which are enhanced in different background components and therefore can significantly reduce the relative normalization uncertainty of the

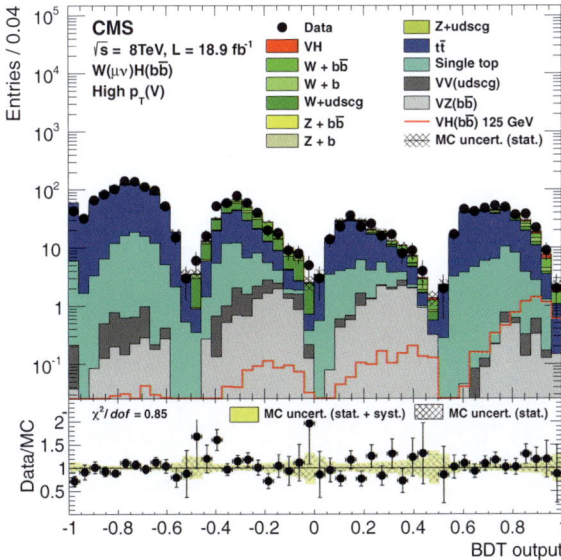

Fig. 9.7 The BDT discriminant distribution for the CMS multivariate analysis as observed in data (points with error bars) and expected from simulations (histograms) for the $WH \to \ell\nu b\bar{b}$ channel and the most sensitive $p_T(V)$ category of the $WH \to \ell\nu b\bar{b}$ channel. The BDT distribution is composed out of four different regions, the first enhanced in top background, the second in V+jet background, the third in diboson background and the last in Higgs boson signal, with different BDT discriminants evaluated in each, as described in the text. Taken from Ref. [3].

individual backgrounds. This method is not adopted for the $ZH \to \ell\ell b\bar{b}$ channel, where the uncertainty on the background composition plays a smaller role.

2.5.1. *Additional control regions*

Additional control regions are used to constrain the backgrounds further. A maximum likelihood fit to the b-tagging discriminant is applied to determine the flavour composition of the background and further improve on the relative normalization uncertainty of different background components. While the basic idea is again the same, the detailed strategy adopted by ATLAS and CMS is different. In ATLAS, such maximum likelihood fit is applied to the one b-tag regions, defined as the set of events in which only one of the two jets passes a loose b-tagging requirement corresponding to an average 80% b-jet efficiency. Since this sample is completely orthogonal to the other 2-tag regions used in the fit, this auxiliary fit can be integrated into the main signal extraction fit described in the previous section which relies on the BDT discriminant. In CMS several dedicated control regions are defined by adapting the event selection to enhance the purity of each individual background, and a simultaneous fit is performed in such regions, separately in each $p_T(V)$ region, determining MC to data normalization scale factors for the main backgrounds. These are then used in the main signal extraction fit, allowing them to further vary within their statistical and systematic uncertainties.

2.6. *Results*

The presence of a $H \to b\bar{b}$ signal is tested through a binned maximum likelihood fit. The main observable of the likelihood function is the value of the BDT discriminant, and the fit is subdivided into categories as described in the previous section. The signal and background components are parametrized as separate PDFs (probability density functions), and the parameters of the maximum likelihood fit are adapted to get the best description of the data. The shape of the PDFs is almost entirely estimated from simulation. The main free parameter of the fit is the signal strength, μ, which represents the ratio of the estimated signal yield in data divided by the prediction of the Standard Model. The normalization and the shape of the PDFs is allowed to vary within systematic uncertainties, of either theoretical or experimental nature, which are encoded in a set of nuisance parameters. These are constrained in the fit by auxiliary Gaussian or log-normal terms. In the case of the ATLAS analysis, some of the main background normalizations are also free parameters, since all control regions are directly integrated into the final fit; while for CMS they are constrained to the values and uncertainties found in the separate fits to the control regions.

The signal strength found by the ATLAS and CMS analyses (Refs. [4] and [3]) for a Higgs boson mass of ~ 125 GeV is reported in Table 9.4. Values of $\mu = 0.51^{+0.40}_{-0.37}$ (ATLAS) and $\mu = 0.89^{+0.47}_{-0.44}$ (CMS)[g] correspond to a signal above the background expectation at the level of 1.4 and 2.1 standard deviations, while the median expected sensitivity is 2.6 and 2.5 standard deviations, respectively.

Using the CLs method based on the profile likelihood ratio (for more details see Sec. A.2 of Appendix A), the levels of signal strengths were derived which are excluded at the 95% CL (confidence level). While in the absence of a signal the experiments are expected to exclude a $H \to b\bar{b}$ signal in a Higgs boson mass region approximately between 110 and 130 GeV, given the small observed excess of signal events only the mass region below 115–120 GeV is excluded or close to be excluded.

The well-known diboson $VZ \to Vb\bar{b}$ process is used to provide further validation of the analysis techniques used to extract the $VH \to Vb\bar{b}$ signal.

Table 9.4   Values of the signal strength μ extracted from data in the $VH \to Vb\bar{b}$ analysis, assuming $m_H \sim 125$ GeV.

Channel	ATLAS	CMS
$WH \to \ell\nu b\bar{b}$	$0.80^{+0.66}_{-0.60}$	$1.11^{+0.87}_{-0.83}$
$ZH \to \ell\ell b\bar{b}$	$0.94^{+0.88}_{-0.79}$	$0.70^{+0.79}_{-0.71}$
$ZH \to \nu\bar{\nu} b\bar{b}$	$-0.35^{+0.55}_{-0.52}$	$0.89^{+0.63}_{-0.61}$
Combination	$0.51^{+0.40}_{-0.37}$	$0.89^{+0.47}_{-0.44}$

---

[g]The CMS result published in Ref. [3] has been superseded by Ref. [9], where the combination of all channels is presented. In the combination, the differential contribution from the $gg \to ZH$ process, which is a NNLO correction to the $qq \to ZH$ process, has been considered as an additional correction, increasing the amount of expected signal events and therefore decreasing the observed value of μ in the $ZH \to \nu\bar{\nu} b\bar{b}$ and $ZH \to \ell\ell b\bar{b}$ channels by about 20%. More details can be found on the web page https://twiki.cern.ch/twiki/bin/view/CMSPublic/Hig14009PaperTwiki. The ATLAS analysis also includes this contribution.

The Standard Model prediction for the diboson process is well established and does not depend on the presence of the Higgs boson. A maximum likelihood fit is therefore applied where the parameter representing the Higgs boson signal is replaced by the analogous parameter for the diboson signal strength. In addition to this, in ATLAS the BDT trained to separate the Higgs boson signal from the backgrounds is replaced by an equivalent BDT trained to separate the diboson signal from the remaining backgrounds. In CMS, given the use of multiple BDTs, this procedure is not necessary. The level of diboson signal over the Standard Model expectation is found to be $0.74^{+0.17}_{-0.16}$ and $1.19^{+0.28}_{-0.23}$ by ATLAS and CMS, respectively, thus compatible with the Standard Model expectation. In both cases the background-only hypothesis can be excluded at the level of more than 5 standard deviations. The possible presence of a Higgs boson has only a small effect on the extraction of the diboson signal. While this is an important cross-check, the main limitation arises from the fact that the $p_T(H)$ spectrum for the Higgs boson signal is expected to be significantly harder than the $p_T(Z)$ spectrum in diboson events.

A clear visualization of the result is made difficult by the fact that in a BDT analysis the signal is not expected to emerge with a clear peak above the background level. In the absence of that, the best possible visualization can be obtained by combining the bins of the final BDT discriminant into bins of $\log(S/B)$, which displays both data and predictions from simulation in order of increasing signal purity. This is shown in Fig. 9.8.

A more direct visualization is given by the result of the di-jet mass analyses, which have been performed as a cross-check. These analyses have the disadvantage of having a lower sensitivity (by 25–30%) when compared to the corresponding BDT analyses. Nevertheless they allow to directly look for the Z and Higgs boson mass peaks, as shown in Fig. 9.9, where the $m_{b\bar{b}}$ distribution from data is shown after subtracting all expected background contributions, except for resonant diboson production, and where the contribution from the different fit categories is weighted by the expected signal purity. One can clearly see the diboson mass peak and a first possible hint of the $H \to b\bar{b}$ signal appearing in the results of both experiments.

2.7. *Systematic uncertainties and limiting factors*

Systematic uncertainties have a significant impact on the sensitivity of the $VH \to Vb\bar{b}$ analysis. They increase the uncertainty on the estimated signal strength μ with respect to the statistical-only component by about 25% for the ATLAS analysis and about 15% for the CMS analysis. The main reason

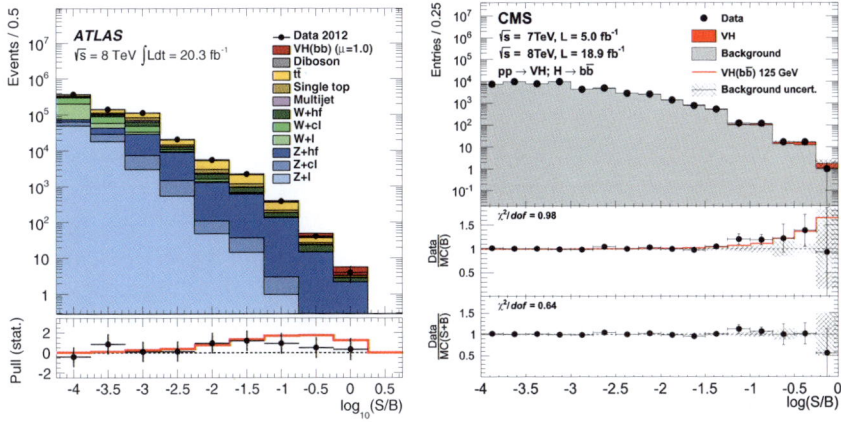

Fig. 9.8   Event yields as a function of $\log(S/B)$ for data, background and Higgs boson signal with $m_H = 125$ GeV, for the ATLAS analysis (left, 8 TeV data only) and the CMS analysis (right). Also shown are for ATLAS the pull of the data with respect to the background-only prediction, while for CMS the ratio of the data over the prediction from simulation, with and without the presence of a signal. Taken from Refs. [4] and [3], respectively.

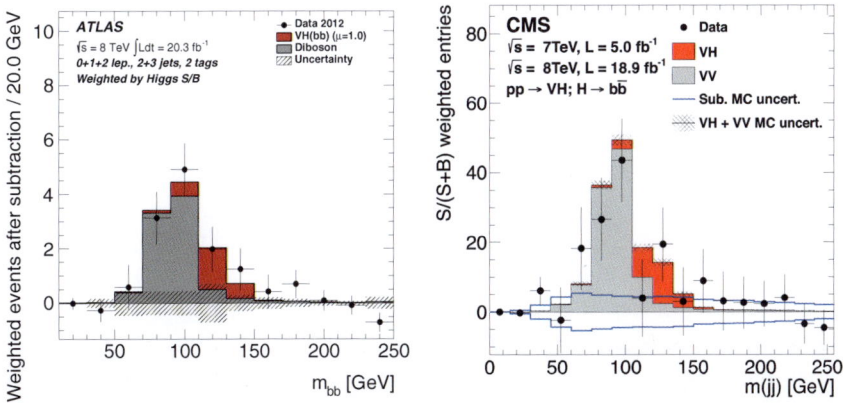

Fig. 9.9   The distribution of $m_{b\bar{b}}$ in data after subtraction of all backgrounds except for the diboson processes, in the case of the ATLAS (left) and CMS (right) di-jet mass analyses. In the case of ATLAS the result from the 8 TeV only dataset is shown. The contribution from all fit categories is summed up, weighting it by the expected signal purity. The Higgs boson signal expectation is also shown, corresponding to the SM expectation for $m_H = 125$ GeV. Taken from Refs. [4] and [3], respectively.

Table 9.5   Main systematic uncertainties affecting the $VH \rightarrow Vb\bar{b}$
analysis, expressed in terms of relative contribution to the total
uncertainty on the signal rate σ_μ.

Systematic uncertainty		ATLAS	CMS
Luminosity		7%	< 4%
b-tagging	b-jets	17%	
	c-jets	10%	20%
	light-jets	12%	
Jet energy uncertainty	scale	20%	10%
	resolution		12%
$E_{\mathrm{T}}^{\mathrm{miss}}$ uncertainty		7%	6%
Signal modelling	rate	17%	8%
	shape		8%
Diboson simulation		5%	10%
Single-top simulation		10%	10%
Simulation modelling	W+jets	27%	
	Z+jets	20%	15%
	$t\bar{t}$	12%	
Data normalization	W+jets	15%	
	Z+jets	7%	32%
	$t\bar{t}$	10%	
Simulation statistics		< 5%	27%

why the ATLAS analysis is impacted more by systematic uncertainties is
the inclusion of the lowest $p_T(V)$ categories, where the S/B level is lower
and therefore the relative impact of any uncertainty affecting the estimated
background rate higher.

Table 9.5 illustrates the main systematic uncertainties contributing to
the $VH \rightarrow Vb\bar{b}$ analysis. Since correlations between systematic uncertainties play an important role, the uncertainties in the table are estimated
by computing by how much the total error on μ is reduced by removing
a specific group of systematic uncertainties, and then by subtracting the
updated error in quadrature from the original error. Uncertainties are then
quoted relative to the total uncertainty on the signal rate. Only the leading
systematic uncertainties will be described in the following.

The leading systematic uncertainty is driven by how precisely the
yield ("data normalization") and shape ("simulation modelling") of the

backgrounds can be controlled by either data or simulation. The normalization is controlled in either the simultaneous fit to signal and control regions (ATLAS) or in the separate fits to the control regions (CMS). A particularly difficult background to normalize is $W + b\bar{b}$ production. It is difficult to find a phase space region where this background is dominant, given the simultaneous presence of the single-top and $t\bar{t}$ backgrounds. In addition the theory modelling for this background is relatively poor, resulting in large extrapolation uncertainties. The CMS analysis shows that the relative composition of the background in terms of the $W+1$ b-jet and $W+2$ b-jets components, which are extracted from data, is not reliably predicted by the MadGraph[10] generator used to simulate this background. A reliable theory modeling of this process is difficult, because of several issues: the possibility of having b quarks in the initial state, the matrix element to parton shower matching procedure which has to deal with gluon to $b\bar{b}$ splitting in the final state. In particular, the $W+1$ b-jet contribution, which is dominated by cases where the gluon splits to $b\bar{b}$, is predicted to be up to a factor of ~ 2 too low with respect to data.[3] Similarly, in ATLAS, where the Sherpa[11] generator is used for the description of the same process, the normalization of the $W + b\bar{b}$ process alone induces an uncertainty on the signal strength μ of $\sim 15\%$ with respect to the total uncertainty, and the relative $W+1$ b-jet vs $W+2$ b-jets composition induces an equivalent relative uncertainty of $\sim 15\%$ as well. Especially in CMS, there is also a significant impact on the result due to the uncertainty on the normalization of the Z+jet background as derived from the control regions. Another subleading but still important uncertainty is given by the modeling of the $t\bar{t}$ background. Despite the overall very good theory modelling of this process, thanks to adopting generators like POWHEG[12] and MCAtNLO[13] which match the NLO matrix elements with a parton shower to get simultaneously next-to-leading-order (NLO) and leading-log (LL) accuracy, a few distributions as for example p_T(top) are known not to be perfectly modelled. For example in ATLAS this distribution was re-weighted at truth level to the unfolded measurement performed in the data,[14] and half of the difference with respect to using the non-re-weighted distribution was taken as an estimate of the associated systematic uncertainty.

Experimental uncertainties have also an important effect on the measurement. While the impact of the lepton-based uncertainties (trigger turn-on, efficiency, energy scale and resolution) is very small, a visible effect is induced by the level of knowledge of the expected b-tagging efficiencies, the jet energy scale and the jet energy resolution, which are each between

10% and 20% with respect to the total uncertainty. Particularly important is the first of these, where the analyses rely on external measurements of the tagging efficiency for b-jets, c-jets and light-flavour jets. The b-jet efficiency is extracted in ATLAS using a high-purity sample of b-jets found in di-lepton $t\bar{t}$ events,[15,16] while in CMS the main method relies on jets from multijet events enriched in b-jets by requiring a muon inside the jet.[17] In the case of c-jets ATLAS relies on di-jet events where $D^{*+} \to D^0 (\to K^- \pi^+) \pi^+$ decays are selected and a lifetime fit is applied to disentangle c-jets from b- and light-flavour jets,[18] while CMS relies on extrapolation.[17] Finally, for the light-flavour jets, both ATLAS[18] and CMS[17] rely on the negative mistag method applied to di-jet events. In the central p_T region (100–150 GeV) the estimated precision of the calibration on a single tagged jet for a b-jet efficiency of 70% is $\sim 3\%$ for the b-jets, 6–8% for the c-jets and 15–17% for the light-flavour jets.

Finally, an important systematic uncertainty is induced by the limited amount of simulated events available to model the backgrounds in the corners of phase space which are most relevant for extracting the Higgs boson signal. In CMS this effect contributes to $\sim 27\%$ of the total error on μ. The effect is much smaller in ATLAS ($< 5\%$), mainly because dedicated simulation runs were produced using a faster simulation of the calorimeter, which relies on a parametrization of the shower shapes (FastCaloSim[19]), and enhancing the amount of events generated and simulated in the region of high $p_T(V)$ where the analysis has the highest sensitivity.

2.8. *Outlook*

Despite the fact that the expected sensitivities in both ATLAS and CMS VH searches are above 2σ and a combination of the two analyses yields an expected sensitivity above 3σ, no convincing evidence of a signal above the 3σ level has been observed yet. ATLAS measures a signal strength of $\mu = 0.51^{+0.40}_{-0.37}$, while CMS measures $\mu = 0.89^{+0.47}_{-0.44}$. The additional data from the LHC Run 2 at a center-of-mass energy of 13 TeV will therefore be crucial in order to finally establish the presence (or absence) of a $VH \to Vb\bar{b}$ signal, and to measure its rate more precisely. The ATLAS experiment should also profit from improved b-tagging performance due to the addition of the Insertable B-Layer (IBL), an additional pixel layer between the interaction region and the former first pixel layer.[20] While the Run 2 data represents a unique opportunity, at the same time it also presents several challenges. While the overall signal cross section is expected to increase by a factor of about 2, the $t\bar{t}$ background, which is one of the largest backgrounds in the

$VH \to Vb\bar{b}$ analysis, is expected to increase by a factor of approximately 3.5. This will make the rejection of the $t\bar{t}$ background even more important. Another challenge is that, while the statistical sensitivity is increased by the additional data and the increase of center-of-mass energy, the role of systematic uncertainties will become more important, and so, in order to benefit from the additional data, more effort will be required to better understand and reduce the modelling uncertainties related to the $(W/Z)+b$-jets, single-top and $t\bar{t}$ production processes as well as the uncertainties related to the b-tagging, the reconstruction of jets and the missing transverse energy.

3. VBF search

As mentioned in the introduction, the LHC cross section for Higgs boson production through the vector boson fusion process (VBF) is 1.6 pb at 8 TeV. This cross section is several times larger than for the Higgsstrahlung production processes. However, the VBF production process that includes two b quarks from the Higgs boson decay and two jets produced by the light quarks of the colliding protons results in an all-jet final state, making the study of this channel particularly challenging due to the copious multijet production from QCD processes. Still, the unique kinematical features of the VBF process allow for event selection criteria that help enhance the Higgs signal over the background and attain reasonable sensitivity in the search. So far such a search has been published only by the CMS experiment,[21] and in what follows the main elements of this challenging analysis are described.

3.1. *Analysis strategy*

As shown in Fig. 9.10, W or Z bosons are radiated off the colliding quarks and subsequently "fuse" in a vertex where the Higgs boson is produced. The momentum exchange is of order $Q^2 \sim m_Z^2, m_W^2$ such that the two hadronic jets resulting from the fragmentation of these light quarks are produced within the acceptance of the detectors, but mainly in the forward and backward directions and consequently with a large rapidity gap in between them. These light-quark jets are referred to as the "VBF-tagging jets" and later denoted as qq. In contrast, the two jets from the Higgs boson decay into b quarks populate the more central region of the detector.

Another feature of the VBF process is that no QCD color is exchanged in the Higgs boson production. The light valence quarks from which the

W or Z bosons are radiated are color-connected to the proton remnants in the positive and negative directions along the beam line, while the two b quarks are color-connected together. As a result, very little additional QCD radiation and hadronic activity is expected in the space outside these color-connected regions, and in particular in the rapidity gap between the two VBF-tagging jets, except for the b quarks from the Higgs boson decay which are color-connected to each other.

For signal events the p_T distributions of all four final state quarks are very similar regardless of the quark flavour, with an average $\langle p_T \rangle$ in the 60–70 GeV range, The b-quark jets are mostly contained within the tracker acceptance ($|\eta| < 2.5$) while the VBF-tagging jets have an average $|\langle \eta \rangle|$ of approximately 2.1, but are mostly contained within the forward calorimeter (HF) acceptance ($|\eta| < 5$).

By far the dominant background is the production of multijets. Other relevant backgrounds are: (i) hadronic decays of Z or W bosons in association with additional jets, (ii) hadronic or semi-leptonic decays of top quark pairs, and (iii) hadronic decays of events containing single top quarks. The contribution of the Higgs boson produced in the gluon-fusion (GF) process, with two or more associated jets, is considered as signal in this search.

The signal characteristics mentioned above are all useful in discriminating against the copious multijet background. In these processes jets originate mostly from gluons and light quarks, with only a few percent originating from heavy-flavoured quarks. Their p_T spectra is steeply falling and their di-jet invariant mass distributions do not form the peak expected from the Higgs boson decay into b quarks. In addition, for multijet events the separation in rapidity between jets is not as large as for signal events. Several other event-topology features also differ to some extent from those in signal events. Therefore, in order to enhance the separation of the signal from the backgrounds, part of the analysis strategy involves the use of multivariate discriminant algorithms that use as input a set of variables that are most useful for this purpose.

One of the main challenges for this search, also related to the very large cross section for multijet processes, is the capability of the experiment to trigger efficiently on signal events while not overwhelming the trigger bandwidth with background processes. To achieve this, compromises needed to be made in the way the triggers were defined.

An important component of the search for $H \to b\bar{b}$ in the VBF channel is the demonstration that with similar triggers and analysis techniques one can extract a well-known Standard Model signal in the same four-jet phase space in which the Higgs boson search is performed. This was achieved by measuring the expected event yield for the Z+jets process with $Z \to b\bar{b}$.

Some of the more salient elements of the search strategy are described below. The detailed analysis is documented in Ref. [21].

3.2. *Event selection*

3.2.1. *Triggers*

Two complementary sets of trigger requirements were used in this analysis; these are denoted A and B. The different requirements affect the properties of the events selected and therefore also have an effect on the subsequent event selection performed offline. As a result, the analysis is performed separately in two exclusive datasets, denoted as A and B. These datasets share many features and, for simplicity, only when necessary is the distinction between the two mentioned explicitly.

The only objects included in all trigger paths used in the analysis were jets. These paths were distinguished by the different jet p_T thresholds used, the b-tagging requirements applied, the different ways of pairing them, and the different kinematical requirements made on the event as a whole. As the LHC instantaneous luminosity increased during the run period it was necessary to raise the thresholds on kinematical variables in order to maintain a manageable rate for these triggers. In what follows, the ranges quoted for the various thresholds reflect this.

The trigger path A required the presence of at least four jets with p_T above the following thresholds: 75–82, 55–65, 35–48, and 20–35 GeV. At least one of these jets had to satisfy an online b-tagging requirement. Out of the four highest-p_T jets, the pair of candidate VBF-tagging jets was identified in two possible ways: (i) the pair with the two smallest values of the b-tagging discriminant used in the trigger or (ii) the pair with the maximum pseudorapidity (η) opening. Both pairs were required to exceed the minimum $|\Delta\eta_{jj}|$ and m_{jj} thresholds of 2.2–2.5 and 200–240 GeV, respectively.

For the trigger path B there was a requirement there be at least two jets with $p_T > 35$ GeV and that the scalar sum of the p_T of the event hadronic activity be greater than 175–200 GeV. Then out of all possible jet pairs in the event, with one jet lying at positive η (forward) and the other at negative η (backward), the pair with the highest invariant mass was selected as the most probable VBF-tagging pair. Their corresponding invariant mass, m_{jj}, was then required to be > 700 GeV and their pseudorapidity difference $|\Delta\eta_{jj}| > 3.5$.

3.2.2. Event reconstruction

All objects in the event were reconstructed offline and calibrated using the most advanced tools and techniques available at the CMS experiment. These tools include the particle-flow algorithm[22,23] that combines the information from all CMS sub-detectors to identify and reconstruct individual particles emerging from the proton–proton collisions: charged hadrons, neutral hadrons, photons, muons, and electrons. All jets in the event were reconstructed using this algorithm. Jets identified as originating from extra pile-up interactions were discarded, as were fake jets that result from possible detector noise. Energy corrections were applied to the final set of jets to account for different sources of miscalibration.[24]

The identification of jets that could originate from the hadronization of b quarks ("b-tagging") was performed using the CSV algorithm.[17] The algorithm combines the information about track impact parameters and secondary vertices within a jet and provides a continuous discriminator output that allows separation between jets originating from b quarks and jets originating from charm quarks, light quarks, or gluons.

All events were required to have at least four reconstructed jets. All jets were ordered in three different ways: according to their p_T, according to the value of the CSV b-tagging discriminant, and according to their pseudorapidity. The two b-jet candidates from the Higgs boson decay and the two VBF-tagging candidate jets were considered to be among the four highest-p_T jets. A BDT was then used to help decide if a jet was more likely to be of one type or the other by using as input variables its b-tag CSV value, its b-tagging ordering, its η value, and its η ordering. The BDT was trained on simulated signal events. The two jets with the highest discriminant score were identified as the b-jet candidates, while the other two were identified as the VBF-tagging jet candidates.

3.3. *Signal and background discrimination*

The data was split in two exclusive sets, A and B, and the event selection in each of these follows the different requirements for the trigger paths A and B. Set A was composed of events that passed the trigger path A together with the set of selection criteria listed in Table 9.6. Set B was composed of only events that do not belong to set A and that passed the trigger path B and the additional criteria listed in Table 9.6. The CSVL and CSVM are loose and medium working points of the CSV b-tagging algorithm. The offline criteria are designed to select events with characteristics of VBF

Table 9.6   Summary of selection requirements for the two VBF analysis datasets.

Offline selection	Set A	Set B
Trigger	Path A	Path B
Jet p_T	$p_{T,1,2,3,4} > 80, 70, 50, 40$ GeV	$p_{T,1,2,3,4} > 30$ GeV $p_{T,1} + p_{T,2} > 160$ GeV
Jet $\|\eta\|$	< 4.5	< 4.5
b-tag	at least 2 CSVL jets	at least 1 CSVM and 1 CSVL jets
$\Delta\phi_{bb}$	< 2.0	< 2.0
VBF topology	$m_{qq} > 250$ GeV $\|\Delta\eta_{qq}\| > 2.5$	$m_{qq}, m_{jj}^{\text{trig}} > 700$ GeV $\|\Delta\eta_{qq}\|, \|\Delta\eta_{jj}\| > 3.5$
Veto	none	events that belong to set A

production from the two sets and to enhance their b-quark content. This is shown in Fig. 9.10 where the distribution of pseudorapidity difference between the two candidate VBF-tagging jets for signal events is compared to background for events in set A.

To further improve the sensitivity of the search by enhancing the separation of signal from background, three other procedures were implemented in the analysis. First, the resolution of the invariant mass of the two b-jets was improved by applying the same multivariate regression technique used in the CMS *VH* search and mentioned in the introduction to this chapter. Second, a likelihood discriminant that helps separate jets that originate from light quarks from those that originate from gluons, QGL, was constructed from several jet composition and structure properties. Third, the soft QCD activity, outside of the jets, was quantified and used as a discriminating variable between the QCD processes with strong color flow, and the VBF signal, with no color flow. Figure 9.11 shows the nice separation that can be achieved between signal and background for these last two discriminants.

3.4. *Results*

3.4.1. *Z + jets signal extraction*

A multivariate discriminant was used to help distinguish this process from multijet production. The following variables were used: (i) the absolute

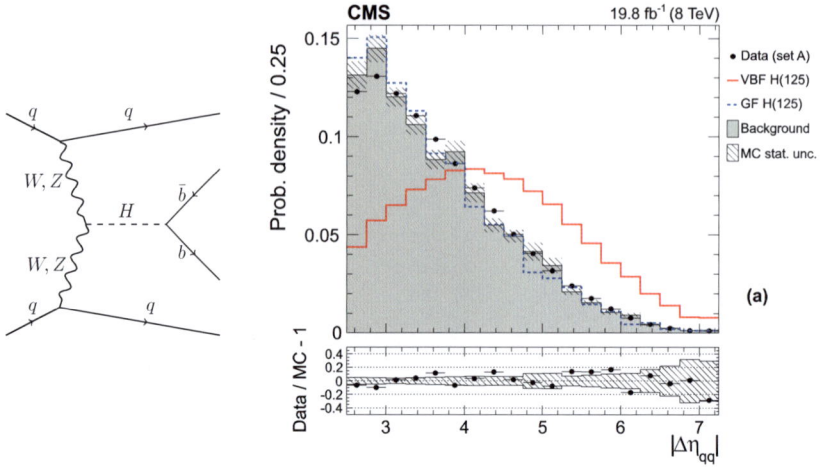

Fig. 9.10   (Left) SM Higgs boson VBF production, with Higgs bosons decaying to $b\bar{b}$. (Right) Normalized distribution of the absolute pseudorapidity difference between the two VBF jet candidates ($|\Delta\eta_{qq}|$). The selection corresponds to set A. The data are shown with filled circles, and background is shown with a filled histogram. The VBF Higgs boson signal is shown with a solid line, and the GF Higgs boson signal is shown with a dashed line. The panel at the bottom shows the fractional difference between data and background simulation, with the shaded band representing the statistical uncertainty from simulation.

η difference $|\Delta\eta_{qq}|$ of the VBF-tagging jets, (ii) the absolute η of the b-jet system $|\eta_{bb}|$, (iii) the CSV value of the most likely b-jet, (iv) the QGL values of the four leading jets. The value of the discriminant was used to define subsets of events that were examined for the presence of $Z \rightarrow b\bar{b}$ events. The invariant mass distribution of the two b-jet candidates, $m_{b\bar{b}}$, in these events was fit simultaneously to three template distributions: multijet events, top quark pair production and $Z \rightarrow b\bar{b}$. The $m_{b\bar{b}}$ template for top quark events and for $Z \rightarrow b\bar{b}$ events were obtained from simulations, while the multijet template was fit to a Bernstein polynomial that is general enough to allow for smooth, but non-resonant, features in the distribution. Figure 9.12 shows a visible $Z \rightarrow b\bar{b}$ signal in the $m_{b\bar{b}}$ distribution for the subset of events that is most signal-like according to the multivariate discriminant. All subsets were fit simultaneously, allowing for template shape and jet energy uncertainties, to result in a signal strength of $\mu_{Z+\text{jets}} = 1.10^{+0.44}_{-0.33}$, that corresponds to an observed (expected) significance of 3.6 (3.3) standard deviations. This is a nice confirmation of the analysis techniques used in the VBF Higgs boson search.

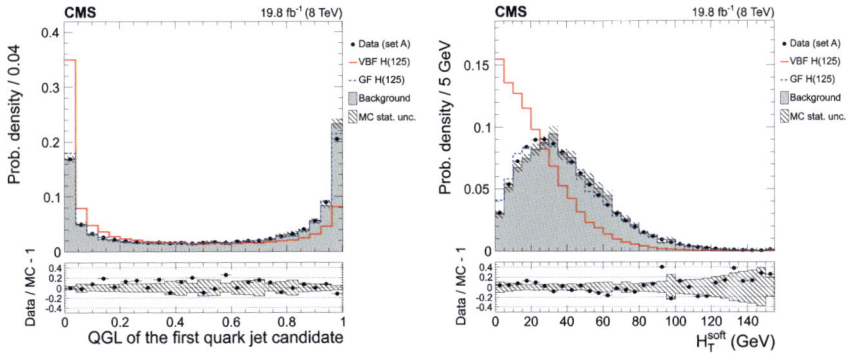

Fig. 9.11 (Left) Distribution of the quark–gluon likelihood discriminant (QGL) of the highest-p_T VBF-tagging jet candidate. Quark jets are expected to have low likelihood values (closer to 0), while gluon jets are expected to have higher ones (closer to 1). (Right) Distribution of the scalar p_T sum (H_T) of track jets that are associated to the soft QCD activity. The distributions are normalized to unity and the event selection corresponds to set A. The data are shown with filled circles, and background is shown with a filled histogram. The VBF Higgs boson signal is shown with a solid line, and the GF Higgs signal is shown with a dashed line. The panel at the bottom shows the fractional difference between data and background simulation, with the shaded band representing the statistical uncertainty in simulation.

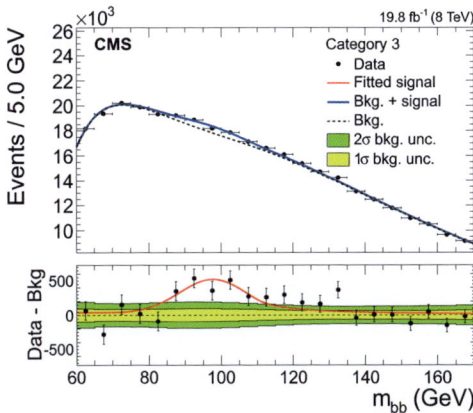

Fig. 9.12 Fit for the Z boson signal on the invariant mass distribution of the two b-jet candidates. The bottom panel shows the background-subtracted distribution of the invariant mass of the two b-jet candidates in the most signal-like event subset. The data are shown with filled circles. The solid line is the sum of the post-fit background and signal shapes and the dashed line is the background component alone, overlaid with the fitted signal, and with the background uncertainty bands corresponding to 1 and 2 standard deviations.

3.4.2. Extraction of the Higgs boson signal

The possible Higgs boson signal was extracted separately from events from set A and from events in set B, and the final results were combined. This was due to the different selection criteria used in selecting the events in each set, which affect their kinematical properties. Again a BDT discriminant was necessary to separate the overwhelmingly large multijet background from the VBF Higgs boson signal. Different event subsets were then selected using the value of the BDT discriminant. Then the $m_{b\bar{b}}$ distributions of these subsets were fit simultaneously to background (including Z+jets, top and QCD multijets) and Higgs boson signal templates, for different Higgs boson mass hypotheses.

All distinctive characteristics of the signal were used in order to optimize the signal-to-background separation. The variables used exploited five distinct features: (i) the dynamics of the VBF-tagging jets: $\Delta\eta_{qq}$, $\Delta\phi_{qq}$, and m_{qq}; (ii) the b-jet content of the event represented by the largest two CSV values among all b-tagged jets; (iii) the jet flavour of the event, represented by the QGL value of all four jets; (iv) the "soft" activity in the event, quantified by the number and scalar p_T sum of soft jets reconstructed from charged tracks alone; (v) the angular dynamics of the production mechanism, expressed by the cosine of the angle between the qq and bb vectors in the center-of-mass frame of the four leading jets $\cos\theta_{qq,bb}$. Figure 9.13 shows the output of the BDT for events in set A, together with the invariant mass distribution for the subset of events with largest signal-to-background content (with a BDT value > 0.84). The fit shown is for the Higgs boson mass hypothesis of 125 GeV. Similar distributions were obtained for four event subsets in set A and for three event subsets in set B.

All subsets were fit simultaneously to obtain limits on the signal strengh for VBF Higgs boson production. These limits are computed with the asymptotic CLs method (see Appendix A for details) and take into account several sources of uncertainty. The dominant sources are the shape and normalization of the multijet background, followed by the normalization of the other background processes. The uncertainty on many other effects was also included; such as those coming from the measurements of jet energy and resolution, b-tagging, quark–gluon classification, trigger efficiencies, and uncertainties related to the generation of signal processes such as scale, and underlying event, parton distribution and shower modelling. Figure 9.14 shows the limits obtained for VBF Higgs boson production in the $H \to b\bar{b}$ final state. For a Higgs boson mass of 125 GeV the expected 95% CL limit

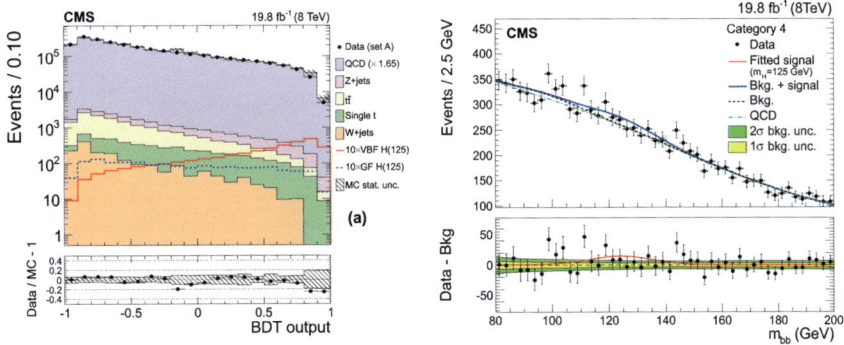

Fig. 9.13   (Left) Distribution of the BDT output for events in set A. The data are shown with markers, while the simulated backgrounds are stacked. The leading-order QCD multijet cross sections are scaled such that the total number of background events equals the number of events in data. The VBF and GF Higgs boson signal yields are scaled up by a factor of 10. The panels at the bottom show the fractional difference between data and background simulation yields, with the shaded band representing the MC statistical uncertainty. (Right) Fit for the Higgs boson signal ($m_H = 125$ GeV) on the invariant mass of the two b-jet candidates in the subset of events of set A with the highest signal-to-background content. The data are shown with markers. The solid line is the sum of the post-fit background and signal shapes, the dashed line is the background component, and the dashed-dotted line is the multijet component alone. The bottom panel shows the background-subtracted distribution, overlaid with the fitted signal, and with the background uncertainty bands corresponding to 1 and 2 standard deviations.

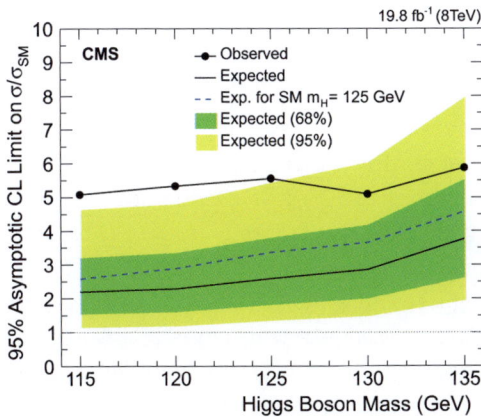

Fig. 9.14   Expected and observed 95% confidence level limits on the signal cross section in units of the SM expected cross section (μ), as a function of the Higgs boson mass, including all event categories. The limits expected in the presence of a SM Higgs boson with mass 125 GeV are indicated by the dashed curve.

on the rate for this process, relative to the standard model prediction, is $\mu < 2.6$ and the observed limit is $\mu < 5.5$. The expected signal significance is 0.8 standard deviations. The best-fit value is $\mu = 2.8^{+1.6}_{-1.4}$,[21] corresponding to an excess of events at $m_H = 125$ GeV, with an observed significance for signal of 2.2 standard deviations.

3.5. *Outlook*

For Run 2 at the LHC this search will continue to be very challenging. At 13 TeV, the gluon–gluon parton luminosities, out of which the QCD multijet backgrounds emerge, increase significantly with respect to the smaller increase in the VBF production cross section. Therefore the signal-to-background ratios will dilute further, making it more difficult to extract and measure both the $H \to b\bar{b}$ signal and the $Z \to b\bar{b}$ validation process. Moreover, due to the higher energy and larger pile-up, the trigger rates will be harder to contend with. It is therefore important for the experiments to develop specialized triggers that target very explicitly the topology of this process. It is also very important for the trigger implementations of the b-tagging algorithms to be very effective in selecting true b-jets and in rejecting lighter-flavour jets. With very large datasets it would be interesting to see how to revise the analysis technique to perhaps select regions of phase space that maximize the sensitivity. The search for $H \to b\bar{b}$ in this channel will certainly benefit from as large a dataset as possible.

Another important thing to keep in mind is that the same techniques used for trying to find a resonant $H \to b\bar{b}$ state can be applied to search for any other resonance in the $m_{b\bar{b}}$ spectrum after a 4-jet VBF selection is made. In that sense, this channel also presents a very good opportunity to search for new physics at 13 TeV.

4. *$t\bar{t}H$* searches

The process with the smallest cross section in which the $H \to b\bar{b}$ decay has been searched for at the LHC is $t\bar{t}H$ production. Figure 9.15 shows the tree-level Feynman diagrams that contribute to this process for which the inclusive production cross section is about 0.13 fb, approximately one tenth the size of the cross section for VH production. The $t\bar{t}H$ process with $H \to b\bar{b}$ not only contributes to the overall $H \to b\bar{b}$ search, but also tests directly the Higgs boson Yukawa coupling to the top quark. Processes in which the Higgs boson is produced through loops that include top quarks, such as

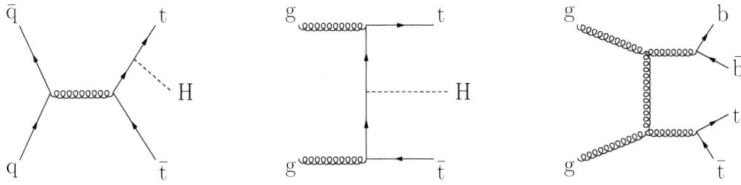

Fig. 9.15 Examples of tree-level Feynman diagrams that contribute to the $t\bar{t}H$ production process (left and middle). The diagram on the right is for the main background process $t\bar{t} + b\bar{b}$.

$H \to \gamma\gamma$ and $H \to ZZ$, also test this coupling. However particles from possible new physics processes could also contribute to these loops; making it difficult to extract the contribution of the pure top-quark Yukawa coupling. Therefore measuring the top-quark Yukawa coupling directly through the $t\bar{t}H$ production process is very important.

4.1. *Analysis strategy*

Since top quarks decay to Wb nearly 100% of the time, the different $t\bar{t}H$ final states, with $H \to b\bar{b}$, all contain four b quarks and are distinguished by the different decays of the W boson. The ATLAS and CMS experiments have studied final states in which at least one of the W bosons decays semi-leptonically, resulting in events containing either a single lepton or a lepton pair (dilepton). Events in these distinct final states are subsequently divided in exclusive categories that depend on the number of reconstructed jets and on the number of those jets that are b-tagged.

The main sources of background events, by far, come from processes in which $t\bar{t}$ pairs are produced in association with at least two extra jets. In particular the $t\bar{t} + b\bar{b}$ component of this background results in a final state that is identical to the signal. The contribution from components with charm jets, $t\bar{t}+c\bar{c}$, and with lighter-flavour jets, $t\bar{t} + \mathrm{lf}$, depend on the mistag rate of the b-tagging algorithms used. The expected background yield is very large compared to the signal and multivariate discrimination techniques are used, separately for the different subsets of events, to improve the sensitivity of the analyses. CMS performed two analyses, one that uses event weights computed with a BDT and another one that uses a probability derived from the Matrix Element (ME) of the processes studied. ATLAS uses Artificial Neural Networks (NN) in most categories, with ME variables used as input to the NN (ME + NN) in the two categories of highest sensitivity. For the

Table 9.7   Event categorization for ATLAS and CMS $t\bar{t}H$ analyses. See text for details.

Single-lepton	# jets	2 b-tags	3 b-tags	≥3 b-tags	4 b-tags	≥4 b-tags
CMS	4	–	BDT	–	BDT	–
	5	–	BDT	–	–	BDT
	≥5	–	–	–	–	ME
	≥6	BDT	BDT	–	–	BDT
ATLAS	4	H_T^{had}	H_T^{had}	–	–	NN
	5	H_T^{had}	NN	–	–	NN
	≥6	H_T^{had}	NN(+ME)	–	–	NN(+ME)

Dilepton	# jets	2 b-tags	3 b-tags	≥3 b-tags	4 b-tags	≥4 b-tags
CMS	3	BDT	–	–	–	–
	≥3	–	–	BDT	–	–
	≥4	–	–	–	–	ME
ATLAS	2	H_T	–	–	–	–
	3	H_T	NN	–	–	–
	≥4	H_T	NN	–	–	NN

least sensitive categories ATLAS uses a single discriminant variable: H_T^{had} (the scalar sum of the p_T of all jets in the event) for the single-lepton analyses and $H_T^{\ell,j}$ (the scalar sum of the p_T of the jets and leptons in the event) for the dilepton analyses. Table 9.7 lists the different event categories used in the ATLAS and CMS analyses.

Details on these three analyses can be found in the following publications: the ATLAS NN analysis in Ref. [25], the CMS BDT analysis in Ref. [26], and the CMS ME analysis in Ref. [27].

4.2. Event selection

The event selection for all analyses required a certain number of leptons, jets, and b-tagged jets as shown in Table 9.7. Each of these objects is required to be in the central fiducial region of the detectors with a p_T above a certain threshold, which depended on the final state under consideration. These thresholds were typically between 10–30 GeV for leptons and 20–40 GeV for jets and b-jets. Since leptons are present in all final states considered, the triggers used for these analyses all required the presence of at least one lepton.

Based only on the selection of these objects the signal-to-background ratio in the single-lepton (dilepton) channels varies from roughly about 1/40 to 1/400 (1/80 to 1/1200) in going from the most to the least sensitive final state categories. For example, for the CMS single-lepton channel with ≥ 6 jets and ≥ 4 b-tags, which is the most sensitive category in the BDT analysis, about 7 signal events were expected in a background of about 260 events.

4.3. $t\bar{t}H$ signal extraction

In order to improve the signal-to-background discrimination, the ATLAS NN and the CMS BDT analyses utilise sets of variables that characterize the distinct signal topologies of the final states under consideration. These algorithms were then trained using these variables as input from simulated samples of signal and backgrounds. This information was used to generate a single output discriminant that, in the final stage of the analysis, was tested (via a likelihood fit) for the presence of signal events in the data. The input variables to these algorithms included object kinematics, event shape variables, global event variables and object pair properties. Examples of the most discriminating variables include information about the mass and angular separation of b-tagged jets, the centrality of the event and the probability of the b-tagged jets to come from b hadrons. In addition, the ATLAS analysis uses two variables calculated using the ME method to improve further the signal-to-background separation in the $(\geq 6j, 3b)$ and $(\geq 6j, \geq 4b)$ categories. The first process probability was derived from the probabilities with which the different assignments of reconstructed objects to final state partons occur in a signal or a background event. The second one is based on the theoretical matrix elements of the $t\bar{t}H$ and the $t\bar{t} + b\bar{b}$ processes. This discriminant assigns density functions to each event according to how likely its kinematics and dynamics were consistent with those specific physics processes (at leading order).

The CMS ME analysis used two distinct discriminants to improve the separation between signal and background. A two-dimensional likelihood fit was made to extract the signal. The first variable helps discriminate between the $t\bar{t} + b\bar{b}$ and the $t\bar{t} + $ lf background processes, and is a likelihood ratio formed using the continuous output of the CSV b-tagging algorithm[17] applied to all relevant jets in an event under the hypothesis of it originating from $t\bar{t} + b\bar{b}$ or from $t\bar{t} + $ lf. The second discriminant, similar to ATLAS, is based on the theoretical matrix elements of the $t\bar{t}H$ and the $t\bar{t} + b\bar{b}$ processes and helps more directly with the signal extraction.

For any of these multivariate algorithms to perform reliably, it is critical that the simulation models describe well not only the distributions of the individual input variables but also the correlations between all of them. This

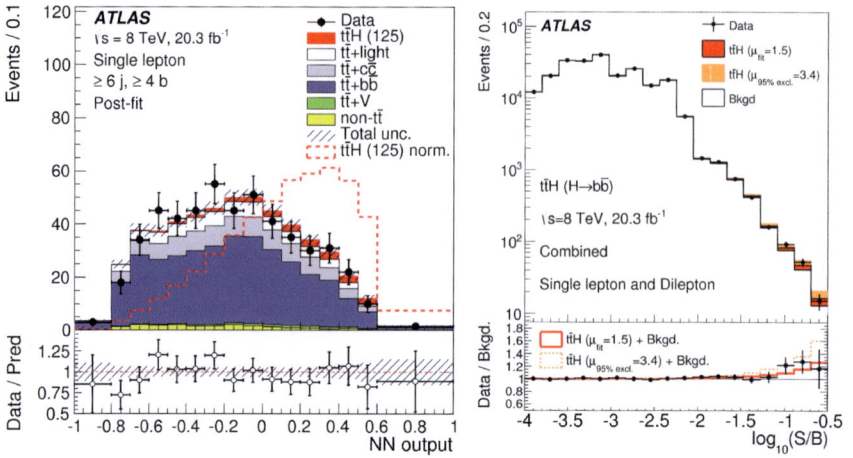

Fig. 9.16   ATLAS NN $t\bar{t}H$ analysis. (Left) Output of the NN distribution for the single-lepton category with ≥ 6 jets and ≥ 4 b-tags. (Right) Post-fit event yields per bin, ordered by $\log(S/B)$, for all bins used in the combined fit of the single-lepton and dilepton channels. The signal is normalized to the best-fit value of the signal strength μ. A signal strength of 3.4 times larger than predicted by the SM (excluded by the analysis at 95% CL) is also shown.

was tested in control regions that were selected in data with events that are close in kinematical and/or heavy-flavour properties to events in the signal region. Appropriate systematic uncertainties are assigned to account for possible discrepancies between the data and the simulated events.

The output distributions for the ATLAS NN, the CMS BDT and the CMS ME, for the most sensitive categories coming from the single-lepton channels with ≥ 6 jets and ≥ 4 b-tags can be seen on the left panels in Figs. 9.16 to 9.18. Signal-to-background ratios ranging approximately from 1/20 to 1/5 were achieved in the most sensitive bins of the output distributions of the multivariate analyses.

4.4. *Results*

Table 9.8 shows the results of the fits for all three analyses described above for an assumed Higgs boson mass of ~ 125 GeV. In each analysis the results were obtained from a simultaneous fit that combined all categories in each of the lepton+jets and dilepton channels. The categories with least sensitivity to signal where used to help further constrain the contributions from the various backgrounds.

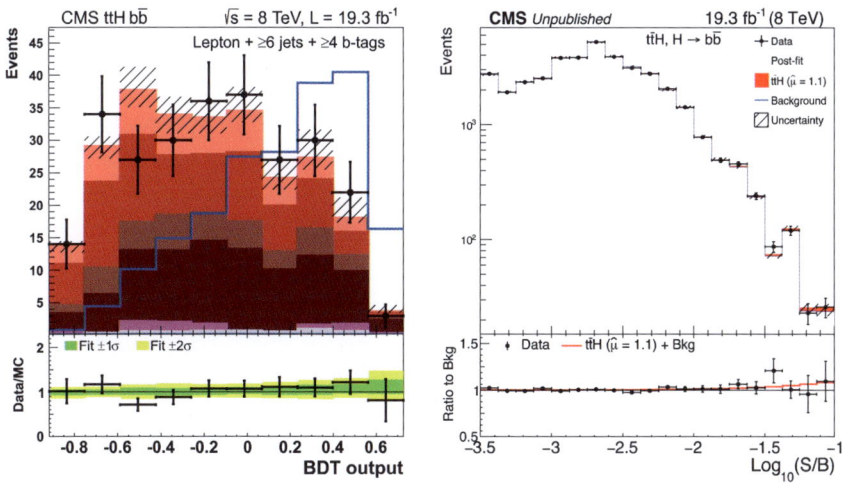

Fig. 9.17 CMS BDT $t\bar{t}H$ analysis, shown for 8 TeV data. (Left) Output of the BDT distribution for the single-lepton category with ≥ 6 jets and ≥ 4 b-tags. (Right) Post-fit event yields per bin, ordered by $\log(S/B)$, for all bins used in the combined fit of the single-lepton and dilepton channels. The signal is normalized to the best-fit signal strength.

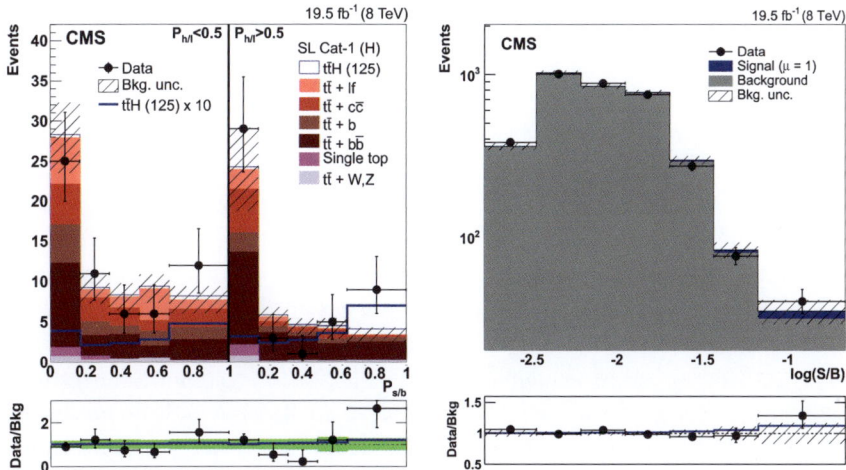

Fig. 9.18 CMS ME $t\bar{t}H$ analysis. (Left) Output of the BDT distribution for the single-lepton category with ≥ 6 jets and ≥ 4 b-tags. (Right) Post-fit event yields per bin, ordered by $\log(S/B)$, for all bins used in the combined fit of the single-lepton and dilepton channels. The distribution is obtained with the constraint that $\mu = 1$.

Table 9.8 $t\bar{t}H$ results for all $H \rightarrow b\bar{b}$ analyses. The CMS BDT analysis result includes 7 TeV data.

Analysis	Best-fit μ	Expected UL	Observed UL
ATLAS NN(+ME)[25]	1.1 ± 1.1	2.2	3.4
CMS BDT[26]	0.7 ± 1.9	3.5	4.1
CMS ME[27]	1.2 ± 1.6	3.3	4.2

The sensitivity is dominated by the lepton+jets channel which is roughly 1.6 times that of the dilepton channel. All three analyses observe a slight excess in the region of highest sensitivity that results in an observed 95% upper limit on the signal strength μ that is larger than the median value expected if a Higgs boson signal was not present in the data. The excess can be seen on the right panels of Figs. 9.16–9.18, where event yields, as obtained from the fit results ("post-fit") are displayed in bins that are ordered by $\log(S/B)$. The corresponding best-fit value for μ is therefore positive, albeit with rather large uncertainties.

4.5. *Oulook*

From the discussion above one can see that extracting a significant $H \rightarrow b\bar{b}$ signal in the $t\bar{t}H$ channel is very challenging. The final states contain multiple objects and differences in detector acceptance, event selection, classification of categories, and in the definition and usage of multivariate discriminants can result in significant differences in sensitivity. In Run 2 at the LHC the increased energy, together with the larger expected pile-up, will amplify these issues further as events will contain even more jets in the final state. The analyses will have to be even more creative and revisit the event selection and event classification to make sure that an optimum point that maximizes the sensitivity is found.

Another important challenge is the reduction of systematic uncertainties. The largest uncertainties are from the normalization and the modelling of the $t\bar{t} + b\bar{b}$ background and from the efficiency of the b-tagging algorithms. Both of these may be reduced when very large samples of data events are on hand. The large datasets will facilitate the selection of control regions in data in which the $t\bar{t} + b\bar{b}$ background can be better isolated and measured. Reducing the $t\bar{t} + b\bar{b}$ uncertainty will also require more accurate theoretical models and the tuning of the corresponding event generators (see the ATLAS $t\bar{t}H$ publication in Ref. [25] for a nice discussion on this).

5. Summary

The search for the $H \to b\bar{b}$ decay at the LHC has been a very successful enterprise in spite of the great difficulties expected due to the overwhelming background from many other Standard Model processes. In this chapter we discussed the strategies, and summarized the main features, of the analysis techniques used to successfully find regions of phase space in which a possible signal could be extracted from the data.

In all production modes studied, small excesses of events were found with measured signal strengths consistent with the production of the Standard Model Higgs boson with a mass of 125 GeV. However the uncertainties on the signal strengths are sizeable, and the estimated significances of these excesses are not at a level that establishes evidence for the $H \to b\bar{b}$ process.

The results from the individual channels and from the combined channels for each experiment are listed in Table 9.9. For $m_H = 125$ GeV the CMS combination yields a fitted $H \to b\bar{b}$ signal strength $\mu = 1.03^{+0.44}_{-0.42}$ with an overall observed significance of 2.6 standard deviations, while the median expected significance is 2.7 standard deviations. The ATLAS combination results in $\mu = 0.63^{+0.39}_{-0.37}$ with an overall observed significance of 1.8 standard deviations, while the median expected significance is 2.8 standard deviations.

Despite the fact that the expected sensitivities in both ATLAS and CMS *VH* searches are above 2 standard deviations, and such that a combination of the two analyses would yield an expected sensitivity close to or above 3 standard deviations, no unequivocal evidence of a signal has

Table 9.9 Summary of results for all $H \to b\bar{b}$ analyses (for $m_H = 125$ GeV): best–fit signal strength values μ, 95% CL expected and observed upper limits (UL), and expected and observed significances.

$H \to b\bar{b}$ channel	Ref.	Best-fit μ	Obs. UL	Exp. UL	Obs. Sig.	Exp. Sig.
VH (CMS)	[9]	$0.89^{+0.47}_{-0.44}$	1.68	0.85	2.08	2.52
$t\bar{t}H$ (CMS)	[26]	0.7 ± 1.8	4.1	3.5	0.37	0.58
VBF (CMS)	[21]	$2.8^{+1.6}_{-1.4}$	5.5	2.5	2.20	0.83
Combined (CMS)	[21]	$1.03^{+0.44}_{-0.42}$	1.77	0.78	2.56	2.70
VH (ATLAS)	[4]	$0.51^{+0.40}_{-0.37}$	1.21	0.76	1.37	2.61
$t\bar{t}H$ (ATLAS)	[25]	$1.5^{+1.1}_{-1.1}$	3.4	2.2	1.4	1.1
Combined (ATLAS)	[28]	$0.63^{+0.39}_{-0.37}$	-	-	1.8	2.8

been observed yet. ATLAS measures a signal strength of $\mu = 0.51^{+0.40}_{-0.37}$, while CMS measures $\mu = 0.89^{+0.47}_{-0.44}$, both expressed relative to the Standard Model signal expectation.

The collision data from Run 2 at the LHC, at a center-of-mass energy of 13 TeV, will prove critical in testing the $H \rightarrow b\bar{b}$ coupling and its consistency with the SM prediction. With roughly 30 fb^{-1} expected initially, and in combination with the Run 1 data, the overall sensitivity, dominated by the *VH* production mode, should reach well beyond the "evidence" level of 3 standard deviations per experiment. Reaching the canonical 5 standard deviations level required for "observation" will likely require significantly more data and possibly a combination of results from both ATLAS and CMS.

A very important measure of the ultimate success of this program at the LHC (with $\int Ldt \sim 300$ fb^{-1} of data expected by the end of 2022) will be the precision with which the $H \rightarrow b\bar{b}$ branching fraction is measured, as this further tests the consistency with the SM predictions. To improve the precision of the measurement it is paramount to reduce the systematic uncertainties. Some of these are of statistical nature and can be reduced further using larger datasets, as in the case of the normalization of the backgrounds in the very specific phase space region selected by the $H \rightarrow b\bar{b}$ analyses. In other cases, as for the experimental uncertainties related to the calibration of physics objects such as b-jets, the availability of larger datasets should allow to redefine the calibration strategies to profit from the larger statistics, for example by being more selective and by defining regions with higher purity. Significant theoretical work will also be needed to improve the Monte Carlo generators for key background processes, such as $V + b$, $V + b\bar{b}$ and $t\bar{t} + b\bar{b}$.

It is important to continue the study of all production mechanisms, beyond *VH*, not only because in combination they help the ultimate sensitivity but also because each has its own intrinsic interest: the $t\bar{t}H$ analysis allows a direct test of the Yukawa coupling of the Higgs boson to the top quark, while the VBF analysis could be sensitive to new physics distorting the $m_{b\bar{b}}$ spectrum.

As discussed in the outlook sections of this chapter, the challenges in Run 2 are significant: increased trigger rates and jet activity due to the increase in energy and more copious pile-up and reduced signal-to-background fractions due to the larger increase in the gluon–gluon partonic luminosities relative to quark–antiquark. At the same time, despite the challenges still ahead, there are excellent prospects to finally claim the observation of the $H \rightarrow b\bar{b}$

decay and measure the branching fraction with much improved precision, and maybe find a new physics process along the way.

References

1. C. Quigg, LHC physics potential versus energy (2009). arXiv:0908.3660.
2. ATLAS Collaboration, Search for the Standard Model Higgs boson decaying into $b\bar{b}$ produced in association with top quarks decaying hadronically in pp Collisions at $\sqrt{s} = 8\,\mathrm{TeV}$ with the ATLAS detector (2016). arXiv: 1604.03812.
3. CMS Collaboration, Search for the Standard Model Higgs boson produced in association with a W or a Z boson and decaying to bottom quarks, *Phys.Rev.* **D89**(1), 012003 (2014). doi: 10.1103/PhysRevD.89.012003.
4. ATLAS Collaboration, Search for the $b\bar{b}$ decay of the Standard Model Higgs boson in associated $(W/Z)H$ production with the ATLAS detector, *JHEP.* **1501**, 069 (2015). doi: 10.1007/JHEP01(2015)069.
5. CDF Collaboration, Improved b-jet energy correction for $H \rightarrow b\bar{b}$ searches at CDF (2011). arXiv:1107.3026.
6. J. M. Butterworth, A. R. Davison, M. Rubin, and G. P. Salam, Jet substructure as a new Higgs search channel at the LHC, *Phys. Rev. Lett.* **100**, 242001 (2008). doi: 10.1103/PhysRevLett.100.242001.
7. A. Hocker, J. Stelzer, F. Tegenfeldt, H. Voss, K. Voss, *et al.*, TMVA — Toolkit for multivariate data analysis, *PoS.* **ACAT**, 040 (2007).
8. J. Gallicchio and M. D. Schwartz, Seeing in color: Jet superstructure, *Phys. Rev. Lett.* **105**, 022001 (2010). doi: 10.1103/PhysRevLett.105.022001.
9. CMS Collaboration, Precise determination of the mass of the Higgs boson and tests of compatibility of its couplings with the Standard Model predictions using proton collisions at 7 and 8 TeV, *Eur. Phys. J. C* **75**(5), 212 (2015). doi: 10.1140/epjc/s10052-015-3351-7.
10. J. Alwall, M. Herquet, F. Maltoni, O. Mattelaer, and T. Stelzer, MadGraph 5: Going Beyond, *JHEP.* **1106**, 128 (2011). doi: 10.1007/JHEP06(2011)128.
11. T. Gleisberg, S. Hoeche, F. Krauss, M. Schonherr, S. Schumann, *et al.*, Event generation with SHERPA 1.1, *JHEP.* **0902**, 007 (2009). doi: 10.1088/1126-6708/2009/02/007.
12. S. Frixione, P. Nason, and C. Oleari, Matching NLO QCD computations with parton shower simulations: The POWHEG method, *JHEP.* **0711**, 070 (2007). doi: 10.1088/1126-6708/2007/11/070.
13. R. Frederix and S. Frixione, Merging meets matching in MC@NLO, *JHEP.* **1212**, 061 (2012). doi: 10.1007/JHEP12(2012)061.
14. ATLAS Collaboration, Measurements of normalized differential cross sections for $t\bar{t}$ production in pp collisions at $\sqrt{s} = 7$ TeV using the ATLAS detector, *Phys.Rev.* **D90**(7), 072004 (2014). doi: 10.1103/PhysRevD.90.072004.

15. ATLAS Collaboration, Performance of *b*-jet identification in the ATLAS experiment, *JINST.* **11**, P04008 (2016). doi: 10.1078/1748-0221/11/04/P04008.

16. ATLAS Collaboration, Calibration of *b*-tagging using dileptonic top pair events in a combinatorial likelihood approach with the ATLAS experiment. URL http://cds.cern.ch/record/1664335.

17. CMS Collaboration, Identification of *b*-quark jets with the CMS experiment, *JINST.* **8**, P04013 (2013). doi: 10.1088/1748-0221/8/04/P04013.

18. ATLAS Collaboration. Calibration of the performance of *b*-tagging for *c* and light-flavour jets in the 2012 ATLAS data. Technical Report ATLAS-CONF-2014-046, CERN URL http://cdsweb.cern.ch/record/1741020.

19. ATLAS Collaboration, The ATLAS simulation infrastructure, *Eur. Phys. J. C* **70**, 823 (2010).

20. ATLAS Collaboration, ATLAS Insertable B-Layer Technical Design Report. Technical Report CERN-LHCC-2010-013. ATLAS-TDR-19, CERN, Geneva (Sep, 2010). URL https://cds.cern.ch/record/1291633.

21. CMS Collaboration, Search for the Standard Model Higgs boson produced through vector boson fusion and decaying to $b\bar{b}$, *Phys. Rev. D* **92**(3), 032008 (2015). doi: 10.1103/PhysRevD.92.032008.

22. CMS Collaboration, Particle-flow event reconstruction in CMS and performance for jets, taus, and MET. (CMS-PAS-PFT-09-001) (Apr, 2009). URL http://cds.cern.ch/record/1194487.

23. CMS Collaboration, Commissioning of the particle-flow event reconstruction in minimum-bias and jet events from *pp* collisions at 7 TeV. CMS Physics Analysis Summary. CMS-PAS-PFT-10-002 (2010). URL http://cdsweb.cern.ch/record/1279341.

24. CMS Collaboration, Determination of jet energy calibration and transverse momentum resolution in CMS, *JINST.* **6**, P11002 (2011). doi: 10.1088/1748-0221/6/11/P11002.

25. ATLAS, Search for the Standard Model Higgs boson produced in association with top quarks and decaying into $b\bar{b}$ in *pp* collisions at $\sqrt{s} = 8$ TeV with the ATLAS detector, *Eur. Phys. J.* **C75**(7), 349 (2015). doi: 10.1140/epjc/s10052-015-3543-1. arXiv: 1604.3812.

26. CMS Collaboration, Search for the associated production of the Higgs boson with a top-quark pair, *JHEP.* **1409**, 087 (2014). doi: 10.1007/JHEP09(2014)087,10.1007/JHEP10(2014)106.

27. CMS Collaboration, Search for a Standard Model Higgs boson produced in association with a top-quark pair and decaying to bottom quarks using a matrix element method, *Eur. Phys. J.* **C75**(6), 251 (2015). doi: 10.1140/epjc/s10052-015-3454-1.

28. ATLAS Collaboration, Measurements of the Higgs boson production and decay rates and coupling strengths using *pp* collision data at $\sqrt{s} = 7$ and 8 TeV in the ATLAS experiment, *Eur. Phys. J.* **C76**(1), 6 (2016). doi: 10.1140/epjc/s10052-015-3769-y. arXiv:1507.04548.

Chapter 10

Higgs boson search in the $WW \rightarrow l\nu qq$ final state

Pietro Govoni* and Rikard Sandström[†]

*Milano-Bicocca University and INFN,
Pizza delle Scienze 3, 20126 Milano, Italy
[†]Max Planck Institut für Physik (Werner-Heisenberg-Institut)
Föhringer Ring 6, 80805 München, Germany

Among proton–proton collision events at the LHC where two W bosons are produced, the region defined by large masses of their system is sensitive to the presence of SM-like Higgs boson resonances. The final state with one W boson decaying leptonically and the other one hadronically features both the purity granted by the presence of a lepton, and the large cross section due to the hadronic branching ratio of the W boson. The ATLAS and CMS analyses of this final state performed with the LHC data acquired at 7 TeV and 8 TeV centre-of-mass energy are presented, in the frame of Standard Model (SM) Higgs boson searches: besides excluding high mass resonances beyond the one found at a mass of around 125 GeV, these studies pave the way for the investigation of the electroweak symmetry breaking that will be carried on in the forthcoming runs of the LHC.

1. Introduction

For masses below the production threshold of two on-shell W bosons, i.e. 160 GeV, the most favoured SM-like Higgs boson decays are dominated by light quark and fermion final states (see Chapter 1), while above this threshold the preferred final states are mostly pairs of W bosons (about 60% of the time), followed by pairs of Z bosons (about 30% of the cases). In turn, these vector bosons decay into leptons or quarks. A W boson decays roughly 30% of the time into lepton–neutrino pairs and 70% of the time into quark pairs. The first decay gives rise to the clean signature of a charged lepton. The drawback is the presence of a neutrino, which is not directly detectable and is reconstructed as missing transverse energy ($E_{\mathrm{T}}^{\mathrm{miss}}$), therefore with low energy resolution (of the order of 10%) and in the transverse plane only. The second decay generates quarks, that are not directly measured but produce jets, characterised by a lower energy resolution with respect to electrons and muons (of the order of 10% as well). The final state of the Higgs boson decay, therefore, can be characterised by the combination of leptonically- and hadronically-decaying W bosons, giving rise to four jets (fully hadronic configuration), two charged leptons and two neutrinos (fully leptonic), or two jets, one charged lepton and one neutrino (semi-leptonic).

The fully hadronic decay allows for the invariant mass reconstruction of the Higgs boson, but does not provide any leptons in the final state necessary for identifying the signal among the many multi-jet events produced in strong interactions. The fully leptonic decay, instead, is statistically limited and shows a poor resolution in the invariant mass reconstruction because of the presence of two neutrinos in the final state.

The semi-leptonic case features one lepton to trigger the events and, since only one neutrino is produced, the invariant mass of the WW pair system (four-body invariant mass) can be reconstructed by requiring the lepton–neutrino pair to have the W boson mass. The resolution is then dominated by the momentum resolution of the jets and of the missing energy. The Higgs boson width increases with its hypothetical pole mass, becoming of the order of 10% of its mass at $450\,\mathrm{GeV}$. As a consequence, the natural width dominates over the experimental mass resolution, which therefore does not constitute a limiting factor to the determination of the resonance line shape. All these considerations made the $WW \to l\nu qq$ decay channel one of the most promising in the high mass regime, for the SM-like Higgs boson search, which is presented in this chapter. Figure 10.1 shows the display of an event selected by the ATLAS analysis as signal candidate. The event contains a muon (the red track), well identified by the fact that it reaches the outer muon detectors, and two jets, visible as clusters of deposits in the electromagnetic (green) and hadron (yellow) calorimeters.

As for other final states, the analysis was performed separately in different categories of events. Typical variables used for the categorisation are the number of additional jets present in the events, the invariant mass range for the search, and the lepton flavour. For example, one category featured the typical signature of the vector boson fusion (VBF), characterised by the presence of two additional jets in the event with large separation

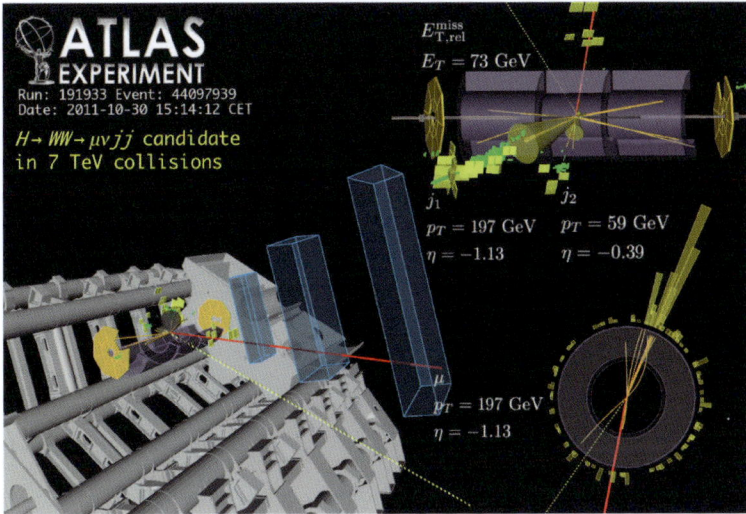

Fig. 10.1   The display of an event selected by the ATLAS analysis as signal candidate, with a muon in the final state. The event is shown from three different points of view, where only the main decay products originating in the hard scattering vertex and the relevant parts of the ATLAS detector are made visible. In the bottom-right part of the image the event is projected on a plane orthogonal to the beam axis. In this view, charged tracks exiting the vertex are drawn with light orange, while the muon is represented as a red line. The yellow towers show the amount of energy collected by the calorimeters. In the top-right corner, instead, a lateral view shows the beam direction and the ϑ angle, with a small tilt. The two cones show the two jets (j_1 and j_2) reconstructed in this event, while the yellow dashed line shows the $E_{\mathrm{T}}^{\mathrm{miss}}$ direction. In this system, the di-jet invariant mass is 87 GeV, the transverse mass of the muon–neutrino system is 47 GeV and the invariant mass of the Higgs boson candidate is 584 GeV. The large value of the invariant mass is correlated with the small angle between the two jets, which may originate from a W boson with a non-negligible boost.

in pseudorapidity. This channel has a smaller production cross section with respect to the total one, but exhibits a better signal-to-background ratio, and allows to test different couplings of the Higgs boson to the other elementary particles of the SM.

For very high Higgs boson masses (above about 600 GeV), the W bosons produced in the decay are highly boosted with respect to the laboratory reference frame and their decay products are very collimated. As a consequence, the two quarks produced by the hadronic decay of one W boson were most of the time reconstructed as a single energetic jet. The cases when one of the W bosons produced a τ lepton and a neutrino have not been considered in the analysis optimisations, since the subsequent decay of the τ lepton produced low momentum electrons and muons, typically below the analysis thresholds, or a jet, whose presence rendered the background contamination much higher with respect to the electron or muon cases. The contribution due to decays into τ leptons was anyhow included in the electron and muon channels respectively, through their leptonic decays: $\tau^+ \to e^+ \nu_e \bar{\nu}_\tau$ and $\tau^+ \to \mu^+ \nu_\mu \bar{\nu}_\tau$.

2. The signal signature and the backgrounds to the analysis

The events selected for the analysis were characterised by the presence of an electron or muon with large transverse momentum (p_T) produced by the decay of a W boson, missing energy due to an undetected neutrino, and the decay products of the hadronic W boson. Depending on the Higgs boson mass, the second vector boson would produce two jets, or a single one with large mass and energy. For the search in the VBF channel, two additional jets were required to be present in the events.

All processes that generate a charged lepton, missing transverse energy and a hadronic W boson candidate (one or two jets) were potential backgrounds to this analysis. The most important one was due to the production of W bosons in association with jets, where the W boson decays leptonically and the jets mimic the signal by showing an invariant mass compatible with the W boson one. The second largest background came from top quark pair production, where one W boson decayed leptonically and the other one hadronically, exactly like the Higgs boson signature. However, these top quark pair events also contained two b-quarks in the final state, which the Higgs boson events did not, and this difference was used to suppress their contributions. The non-resonant production of WW and WZ boson pairs in association with jets contaminated the signal region as well. Events with non-genuine missing transverse energy could also be confused for signal, for example in the Z boson production with jets. Finally, multi-jet production contributed to the background when a jet was misidentified as a lepton and the imbalance in the jet energy measurement generated missing transverse energy.

3. Reconstruction and identification of the single objects

The **electrons and muons** used in the analyses at the LHC (see Chapter 4) were required to be in the region covered by the triggers, in order to avoid biases in the trigger efficiency measurements. They were reconstructed from their tracks in the inner detectors with additional information from the calorimeters and muon spectrometers. Besides, they were required to be isolated by ensuring that there were no high energy particle tracks or large energy deposits in a cone around their direction of motion, and their distance with respect to the primary vertex of the event had to be small. This reduced backgrounds from leptons produced in heavy flavour particle decays and ensured that the lepton track was well reconstructed. The event was not allowed to contain any additional electrons or muons with $p_T > 20$ GeV in order to avoid contamination from $Z \to \ell\ell$ events and ensure that the same event was not appearing as a signal candidate in the analysis addressing final states with two final charged leptons.

The quarks produced by the hadronic W boson decay developed **jets of particles** in the detectors. The jets were reconstructed by algorithms combining the various energy deposits of these particles in the detectors. As for electrons and muons, jets were required to have a minimum transverse momentum of 25 GeV. Besides, quality criteria were applied to suppress jets that did not originate from the main interaction in the bunch crossing. The reconstructed jets were then categorised as originating from the decay of a b-quark on the basis of their internal structure. Furthermore, boosted W bosons produce very collimated quarks, that were reconstructed in a single jet in the detectors, characterised by a large energy and invariant mass. In this case, the internal structure of the jet in the $\eta \times \varphi$ plane featured a characteristic geometry induced by the hadronisation and showering of the two quarks, which was exploited to separate signal from background in the very high mass regimes (600 GeV < m_H < 1 TeV).[1]

The **neutrino transverse momentum** was estimated as the negative vector sum of the transverse momentum obtained from jets, photons and leptons in the event, also taking into account soft clustered energy in the calorimeters (the missing transverse energy).

4. Signal and background modelling

Physics processes happening during the LHC collisions were described theoretically, and their expected signature in the detector was simulated by means of Monte Carlo techniques, which generate virtual events mimicking the measurements. In the simulations, the particle interactions with the detectors were described using GEANT4,[2] and the events obtained were reconstructed using the same software used to reconstruct the collision data events. Contributions from additional pp interactions in the same bunch crossing, as well as contributions from nearby bunch crossings, were modelled by superimposing simulated minimum-bias events to the one containing the hard interaction. The simulation was validated by comparing it with collision data containing well known final states. In all the comparisons, the distributions of simulated events were normalised to the integrated luminosity corresponding to the analysed data sample.

Theoretical models describing the various processes typically rely on leading order predictions for the backgrounds and on next-to-leading order ones for signal. This analysis featured several generators, such as AcerMC,[3] Herwig,[4] Madgraph,[5] MC@NLO,[6] MCFM,[7] Powheg[8,9] and Pythia.[10] In most cases, the statistical uncertainty of the simulation exceeded the precision of cross sections used for the normalisation of the samples.

5. Analysis strategies

After the preliminary selections described so far, further steps of the analysis aimed at determining the amount of signal produced in the collisions. The amount of signal was compared to the yield foreseen by the SM, for any mass hypothesis in the range under investigation. Several independent strategies were exploited at the LHC, relying both on different topologies of events and on different analysis techniques. In fact, both the gluon–gluon fusion and vector boson fusion signatures were investigated in the ATLAS analysis,[11] relying on a combined fit of the four-body final state. The CMS analysis was based on a different strategy,[12] with no separation in the different production mechanisms. The analyses also featured a different choice in jet identification: in the ATLAS case, the candidates for the W boson decay products were chosen as the two that had the invariant mass closest to the nominal W boson mass, among all the possible combinations in each event; while in the CMS case the ones with the largest transverse momentum were chosen. In both cases, the main variables considered were the four-body mass, as it's the one where the Higgs boson resonance would have been observed, and the one of the hadronically decaying W boson (the two-body mass), since it's a powerful discriminant between the signal and the W+jets background. In the following, the details of the various approaches are presented. Since the ATLAS and CMS experiments followed independent strategies, the two analyses are described in separate sections.

5.1. The ATLAS gluon–gluon fusion analysis

The ATLAS analysis strategy was based on rather strict kinematic requirements designed to suppress the large W+jets background. These were optimised to improve the sensitivity for $m_H > 400$ GeV, at the expense of reduced signal efficiency for lower Higgs boson masses.

5.1.1. Selection strategy and motivation

In ATLAS the event was categorised by how many $p_T > 25$ GeV jets it contained. Events with less than two jets in addition to the jet pair from the W boson decay ($\ell\nu jj + 0/1j$) were characteristic for Higgs bosons produced by gluon–gluon fusion and events with two or more additional jets ($\ell\nu jj + 2j$) were considered as candidates of vector boson fusion.

In ATLAS $\ell\nu jj + 0/1j$ analysis, background contamination was reduced by requiring that offline electrons and muons must satisfy $p_T > 40$ GeV. Also $E_T^{\mathrm{miss}} > 40$ GeV was required, which reduced background contamination from physics processes without a real neutrino, where a small amount of missing transverse momentum was generated by poorly measured tracks and energy deposits. To reconstruct the Higgs mass the longitudinal momentum of the neutrino, p_z^ν, was needed. It was obtained by constraining $m_{\ell\nu} = m_W$ and solving its second degree equation. The solution closest to $p_z^\nu = 0$ was chosen as the neutrino longitudinal momentum since the final state particles from Higgs boson decays were usually produced with low pseudorapidity.

The two central $p_T > 40$ GeV jets closest to the W boson mass must pass the jet–jet invariant mass requirement 71 GeV $< m_{jj} < 91$ GeV to ensure that they were compatible with a $W \to q\bar{q}$ decay and thus reduced W+jets contamination. Demanding that one of them satisfied $p_T > 60$ GeV and that the two jets were close by (within 1.3 units in the (η, ϕ) coordinates) further improved the sensitivity for large Higgs boson masses since the W bosons from heavy Higgs boson decays were produced with large momentum and the two quarks were hence emitted at a small lab-frame angle.

Events that contained any jet tagged as a b-quark jet were vetoed to reduce background from $t\bar{t} \to WbWb \to \ell\nu qq + bb$.

5.1.2. Backgrounds

After this selection the accepted events were expected to be dominated by W+jets, where the W boson decays leptonically. Other important backgrounds were Z+jets, $t\bar{t}$ and di-boson production. These backgrounds were simulated using the previously described Monte Carlo methods, where Alpgen+Herwig was used for W+jets, MC@NLO and AcerMC for top. In addition, multi-jet contribution, where the missing transverse momentum did not correspond to a real neutrino or the charged lepton was produced in a QCD process, was evaluated using a control region in the real collision data composed of charged leptons inconsistent with $W \rightarrow \ell\nu$ decays.

A W+jets control region was defined as the two sidebands of the dijet invariant mass, 45 GeV $< m_{jj} <$ 60 GeV and 100 GeV $< m_{jj} <$ 115 GeV. This region was depleted of Higgs boson signal events and the collision data in this region matched the background estimated by the method described above. Due to a limited number of simulated background events a parametrized function was used to describe the background, motivated by the shape of the simulated mass distribution. When this background shape was fit to the W+jets control region and the signal region the χ^2 probability obtained was between 25% and 75% in all channels, supporting the choice of a single background model also for collision data. The invariant mass distribution and the parametrized background are shown in Fig. 10.2.

The difference between the fit result and the observed data in the signal region is shown in Fig. 10.3(a).

5.1.3. *Evaluation of systematic uncertainties*

In this section we describe the most important systematic uncertainties associated to the study of this channel.

Fitting the background model described above to the data resulted in uncertainties between 10% and 12% of the estimated background yield. In addition, two alternative functions were used to cross check this background model, and the differences between the nominal and the alternative fitted background yields were less than 5%, well within the uncertainties already associated to the likelihood fits.

The remaining uncertainties were related to the signal efficiency and the shape of the signal distributions. The uncertainty of the luminosity determination was 3.9%. The trigger efficiency, electron and muon efficiency, as well as uncertainties on the lepton momentum scale and resolution introduced uncertainties on the signal efficiency of less than one percent.

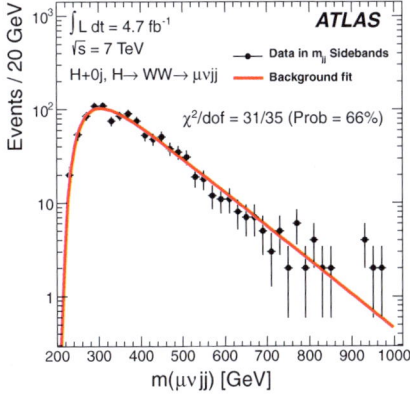

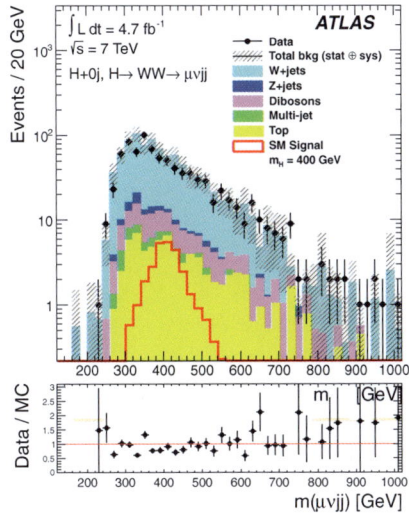

(a) Parametrised background   (b) Simulated background

Fig. 10.2 (a) Fits of the ATLAS background model to the reconstructed invariant mass $m_{\ell\nu jj}$ when m_{jj} was in the W boson sidebands for the $\ell\nu jj + 0j$ selection. The figure shows the muon channel distribution, taken with the full luminosity collected at 7 TeV. (b) Observed and simulated invariant mass distributions, for the same dataset. The bottom panel shows the ratio of observed data divided by expected background. The shaded areas illustrate the background uncertainties.

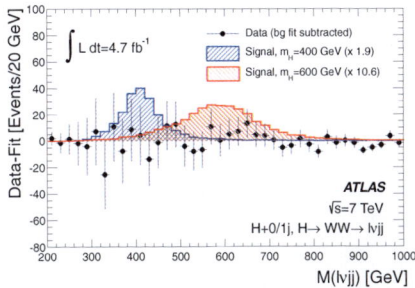

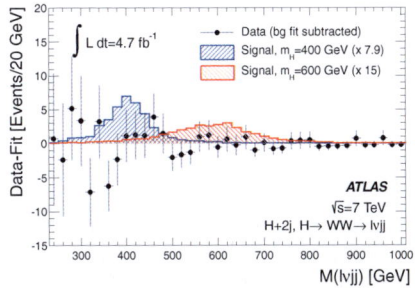

(a) $\ell\nu jj + 0/1j$ (b) $\ell\nu jj + 2j$

Fig. 10.3 The difference between data and the fitted background under a no-signal hypothesis for the $\ell\nu jj + 0/1j$ selection (left) and $\ell\nu jj + 2j$ (right) in ATLAS summed over lepton flavours. The expected contribution from SM Higgs boson decays is also shown for $m_H = 400$ GeV and $m_H = 600$ GeV, multiplied by a factor equal to the ratio of 95% CL limit on its production to the SM prediction.

Table 10.1   Overview of the uncertainties on the event yield considered in the ATLAS $\sqrt{s} = 7\,\mathrm{TeV}$ analysis.

Source	Impact
Signal cross section	19.4%
Interference	<30%
LHC luminosity	3.9%
Jet energy scale	<8% $(H + 0/1j)$, ∼11% $(H + 2j)$
Jet energy resolution	5 to 7% $(H + 0/1j)$, ∼16% $(H + 2j)$
b-tagging	<8%
Four-body mass fit	10 to 12%

The uncertainty of the jet energy scale gave up to 8% uncertainty on the signal efficiency for $m_H \geqslant 400$ GeV. The resolution of the energy of the jets gave 7% uncertainty on the reconstructed signal yield for the same mass range. These uncertainties included the secondary effects they had on the reconstructed missing transverse energy. The b-tagging uncertainty resulted in a maximum uncertainty of 8% on the signal yield. An overview of the important uncertainties is given in Table 10.1.

5.2. *The CMS gluon–gluon fusion analysis*

The analysis performed with the data collected by the CMS detector did not distinguish among the mechanisms that produce the Higgs boson. Only jets within the tracker acceptance were considered, which is where the W bosons were expected to fall for the signal hypothesis. Furthermore, since additional jets due to radiation might be present in the signal event, both events with two or three jets above the threshold of $p_T > 30$ GeV were considered as signal candidates.

5.2.1. *The signal phase space definition*

The SM Higgs boson decay kinematics is fully described, at leading order, by the two-body and four-body invariant masses, plus the angles shown in Fig. 10.4.[13,14] Here, θ^* is the polar angle between the beam axis z and the Higgs boson decay axis z' in the Higgs boson rest frame, Φ_1 is the angle between the zz' plane and the decay plane of the hadronic W boson. The angle Φ spans between the decay planes of the two W bosons in the Higgs boson rest frame, while θ_1 and θ_2 are the ones between the direction of the fermion f in the $W \to f\bar{f}$ decay and the direction opposite to the Higgs boson one in each W boson rest frame. In the case of the hadronically

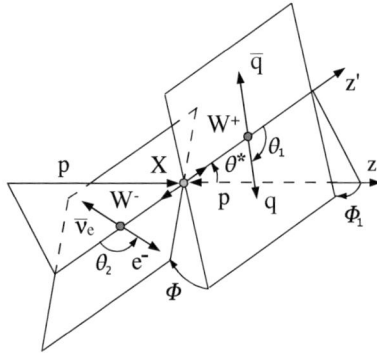

Fig. 10.4   Sketch describing the angles involved in the SM Higgs boson decay, used in the multi-variate discriminant for the $\ell\nu jj$ analysis.

decaying vector boson, the reference fermion is identified as the jet with the largest transverse momentum. The analysis strategy separated as much as possible the signal from the dominant W+jets background exploiting the correlations among these variables, which were combined into a single number through a multi-variate classifier[15] (a general description of multi-variate techniques can be found in Appendix B). The final selection was then chosen as the threshold on the classifier that maximised the expected sensitivity of the full analysis procedure. The two-body and the four-body invariant masses were used in the analysis procedure to evaluate the background normalisation and to search for the signal. Therefore, they were not included among the classifier input variables, while the transverse momentum and the pseudorapidity of the di-boson system were used, together with the lepton charge. This last variable is sensitive to the charge asymmetry in the production of the main W+jets background.

The Higgs boson mass hypothesis, the number of jets found in the event and the flavour of the charged lepton affect the behaviour of the classifier. Because of that, its definition was different in the several mass points chosen for the analysis optimization for the electron and muon cases, and for two or three jets. The full analysis procedure was performed for each of these operating points, and for each mass point the final results were combined.

5.2.2. *Expected signal after event selections*

The expected signal yield after the event selection was evaluated from the simulation, and the limited knowledge of the process translated into a systematic uncertainty, to be accounted for in the final fit used to obtain the analysis results. Therefore, it was of paramount importance to verify that the distribution of the classifier in data was well reproduced by the simulation, in a phase space similar to the signal region. To test this agreement, a sample of $t\bar{t}$ events was selected by requiring exactly four jets in the events, two of which were identified as being generated by b quarks, and two of them were not: this provided the best signal representation that could be isolated in data. A good agreement between data and simulation was observed for all the operating points. The relative difference in efficiency of the classifier selection between data and simulation was used to estimate the residual systematic uncertainty, which was conservatively quoted as the worst value obtained, namely 10%. Besides this, the limited knowledge of the LHC luminosity affected the simulated samples at the level of 4.4%, as well as the one due to the knowledge of the energy of each reconstructed particle which contributed at the level of 2%. The uncertainty on the knowledge of the pile-up conditions during the data taking was neglected, as its effect was less than a percent. All these effects influenced the expected number of selected events, without changing the four-body mass shape of the signal model.

As for the ATLAS analysis, the limited knowledge of the SM Higgs boson cross section was considered as a source of uncertainty of the analysis. Its impact, as calculated by the Higgs Cross Section Working Group,[16–18] was accounted for, with an effect at the order of 10% for the gluon–gluon fusion production and 5% for the VBF one. Selecting events with two or three jets only implied a further uncertainty ranging from 5% to 30%, depending on the operating point.

Eventually, the limited precision attained in the correction for the interference of the signal with the non-resonant WW boson pair background was accounted for by considering its effect on the four-body mass shape for the signal model.

5.2.3. *Background contamination in the signal region*

After the multi-variate discriminant selection, as for the ATLAS case, only events with the hadronically decaying W boson falling in a region around its mass (65 GeV $< m_{jj} <$ 95 GeV) were retained, and the Higgs boson was searched for in the four-body mass spectrum $m_{\ell\nu jj}$.

The expected number of events due to each background was measured with a maximum-likelihood fit on data of the m_{jj} spectrum after selections. The fit was performed only outside the signal region, in the ranges

55 GeV $< m_{jj} <$ 65 GeV and 95 GeV $< m_{jj} <$ 115 GeV. The m_{jj} shape of the sub-leading backgrounds given by di-bosons, top, and Drell–Yan was derived from the simulation and their cross section was left free to float around the theoretical prediction within a Gaussian constraint, proportional to the theoretical uncertainty on the cross section estimate. For the W+jets case, the simulation did not have enough events to provide a smooth m_{jj} distribution, therefore an analytical function was used to model the shape of the process, with an exponential decrease modulated by a turn-on function, where its parameters were free to float in the fits to the data. The W+jets normalisation was also left free to float in the fit. Figure 10.5 shows an example of the m_{jj} distribution for the case with a muon and two jets in the final state, where the various background contributions are visible, together with the measurement performed with 5.1 fb^{-1} of integrated luminosity collected at the energy of 8 TeV. Vertical lines indicate the limits of the signal region, not included in the fit.

The four-body mass shape description of the backgrounds was derived from the simulation as well for the case of the sub-leading backgrounds, while the W+jets one was measured in data. In fact, in the two regions 55 GeV $< m_{jj} <$ 65 GeV (lower sideband: lsb) and 95 GeV $< m_{jj} <$ 115 GeV (upper sideband: usb) only background events

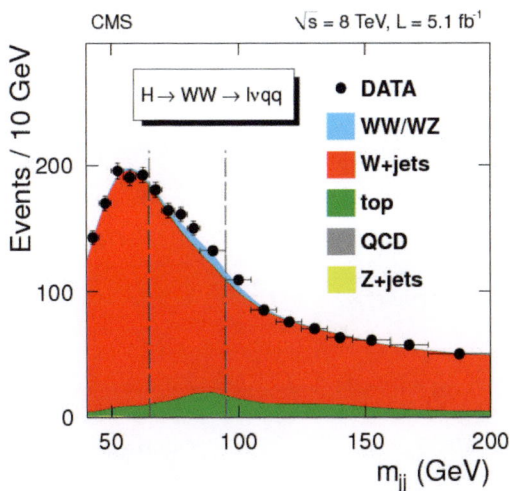

Fig. 10.5 The m_{jj} distribution for the Higgs boson mass hypothesis of 500 GeV, for events produced at the centre-of-mass energy of 8 TeV. Vertical lines indicate the limits of the signal region. The events correspond to the category with two jets and a muon found in the event.

were expected, therefore, after subtracting the sub-leading backgrounds (estimated with simulation), the W+jets shape could be measured there in bins of the four-body mass. The expected shape in the signal region 65 GeV $< m_{jj} <$ 95 GeV ($m_{WW}^{\mathrm{sig}}(i)$) was then assumed to be a linear combination of the two sidebands ($m_{WW}^{\mathrm{usb}}(i)$, $m_{WW}^{\mathrm{lsb}}(i)$), where the mixing factor α was determined with simulated events:

$$m_{WW}^{\mathrm{sig}}(i) = (1 - \alpha) \cdot m_{WW}^{\mathrm{usb}}(i) + \alpha \cdot m_{WW}^{\mathrm{lsb}}(i) . \qquad (10.1)$$

The index i runs over the bins in the four-body mass spectrum. The α parameter is the one that minimises the χ^2 function between the extrapolated shape and the expected one in the simulation. The uncertainty associated to this parameter was obtained by considering the variation due to a change of a unity in the χ^2 function when varying α itself. After the extrapolation, the points that compose $m_{WW}^{\mathrm{sig}}(i)$ were then fit with a parametric exponential shape, modulated with a turn-on when necessary, for $m_H < 250$ GeV. Figure 10.6 shows an example of the four-body mass spectrum, for the case with a muon and two jets in the final state, where the various background contributions are visible, together with the measurement performed with 5.1 fb^{-1} of integrated luminosity collected at the energy of 8 TeV.

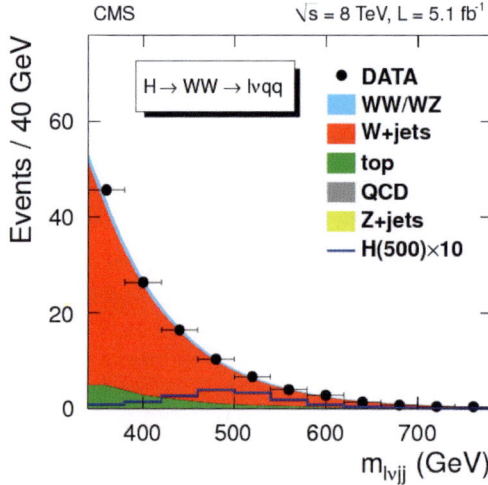

Fig. 10.6   The $m_{l\nu jj}$ distribution for the backgrounds in the signal region superimposed to the data, when searching for a Higgs boson with mass of 500 GeV, for events produced at the centre-of-mass energy of 8 TeV. The expected signal is shown, enhanced by a factor of ten, as a blue line. The events correspond to the category with two jets and a muon found in the event.

The background due to the multi-jet production was measured in data as well. The four-body mass shape for this sample was obtained from data by relaxing the lepton identification and isolation criteria, while its normalisation came from a simultaneous fit of the E_T^{miss} distribution, with the W+jets and multi-jet backgrounds free to float. For the muon final state the contamination was negligible, but for the electron case it was at the level of 10% of the W+jets events, before the multi-variate discriminant selection. A large uncertainty was assumed to account for discrepancies in the model for the W+jets used in the E_T^{miss} fit. The same procedure was applied after the multi-variate discriminant selection to produce the shapes used in the fit of the two-body and four-body mass distributions.

An overview of the systematic uncertainties considered in the analysis is presented in Table 10.2. As described in the text, the W+jets ones are dependent on the four-body mass, and are therefore indicated as "shape".

5.3. *The ATLAS vector boson fusion analysis*

The production of Higgs bosons from vector boson fusion generates two additional jets that are typically well separated in rapidity. In ATLAS, the $\ell\nu jj + 2j$ analysis searched for a Higgs boson produced by this process, so in addition to the selection criteria described in Sec. 5.1, two additional jets in opposite hemispheres and separated by $\Delta\eta_{jj} > 3$ were required. These two VBF-tag jets were required to have large invariant mass ($m_{jj} > 600\,\mathrm{GeV}$) with no additional jets in the range $|\eta| < 3.2$.

The $\ell\nu jj + 2j$ analysis used lepton p_T and E_T^{miss} cut of 30 GeV, lower than for $\ell\nu jj + 0/1j$, to maintain sensitivity for VBF production at lower Higgs boson masses.

Table 10.2 Overview of the systematic uncertainties considered in the CMS analysis.

Source	Impact
Signal theory model	~20%
Signal discriminant description	10%
LHC luminosity	4.4%
Final state objects reconstruction	~2% each
W+jets extrapolation	shape, from the $\Delta\chi^2(\alpha)$
W+jets four-body mass fit	shape, from the fitting procedure
Other backgrounds	negligible

The kinematic selection for $\ell\nu jj + 2j$ was looser than for the $\ell\nu jj + 0/1j$ analysis, since the tagging jet requirement already gives a large background rejection. As a result, the $m_{\ell\nu jj}$ shape for the background was different for $\ell\nu jj + 2j$, and was described by the sum of two exponential functions. The observed number of events after subtracting the background fit is shown in Fig. 10.3(b).

The looser selection criteria, and the larger jet multiplicity of the events accepted by the $\ell\nu jj + 2j$ analysis, generated larger jet-related uncertainties compared to the $\ell\nu jj + 0/1j$ analysis. For $m_H = 400$ GeV the uncertainty on the signal efficiency from the jet energy scale was 11%, while the uncertainty due to the jet energy resolution was 16%.

5.4. *The CMS very high mass analysis*

In the region of m_{WW} above 600 GeV, the hadronic decay of a W boson produced in the detectors a single jet with large invariant mass most of the times, therefore the analysis was performed by considering three objects in the final state: a lepton, the missing energy and a single, wide jet (J), selected as the one with the largest p_T, and identified as one of the W bosons of the final state. Because of the large m_{WW} invariant mass, tight selections on the event kinematics could be applied, to reduce to negligible levels the contamination of the QCD background. Therefore, the p_T of both the leptonically decaying W boson and of the large jet were required to be above 200 GeV, while the transverse missing energy was required to be above 50 GeV (70 GeV) for the muon (electron) channel. Also, the W bosons were required to be back-to-back. The amount of the top background was reduced by vetoing the presence of any b-tagged jets in the event. After these selections, W+jets events constituted the predominant background. Quality selections were applied on the internal structure of the large jet to further discriminate the ones coming from a hadronically decaying W boson from QCD jets originating from quarks and gluons. Since these variables rely on the precise shower development in the detector, which is hard to describe in the simulation, the selection efficiencies were measured with LHC data in a set rich of top events, and applied to the Monte Carlo samples. The signal region was eventually defined by events falling around the W boson mass in the wide-jet mass line-shape (65 GeV $< m_J < 105$ GeV). The same techniques for background estimates described in Sec. 5.2.3 were used, where the W+jets contribution was extrapolated from the lower sideband only.

5.5. *The ATLAS and CMS latest results*

During the preparation of the final release of this chapter, the CMS collaboration published a new paper[19] reporting on the combination of results of the search for high-mass SM-like Higgs bosons, analysing the full dataset collected at $\sqrt{s} = 8$ TeV. The $WW \to l\nu qq$ analysis was modified to best exploit the increased statistics available. In fact, in the intermediate mass interval (from 200 GeV to 600 GeV) W+jets background was described with a polynomial function, the parameters of which have been obtained in the same fit used to determine the exclusion limits on the signal hypotheses. In the high mass regime (above 600 GeV) events were categorised into those with zero or one jet besides the wide one produced by the boosted W decay, and those with two, satisfying VBF-like selections. For both cases, the analysis techniques described in Sec. 5.4 were applied. The results obtained confirm the ones presented in this book. The $\sqrt{s} = 8$ TeV $H \to WW \to l\nu qq$ BSM scan by varying the Higgs boson width.[20] No evidence of a high-mass Higgs boson was found.

5.6. *Results*

The Higgs boson search was conducted following the procedures described in Appendix A with maximum likelihood fits of the observed data in the four-body mass spectrum, with the data-driven background model and the Higgs boson signal simulation as input models. For each SM Higgs boson mass point, the gluon–gluon fusion and vector boson fusion ATLAS analysis were treated separately, as well as each working point of the CMS analyses.

No statistically significant excess over the background-only model was found for any Higgs boson mass hypothesis, hence upper limits on the Higgs boson production cross section have been set, using the profile likelihood ratio as a test statistic following the CL_s procedure. Figure 10.7 shows the expected (dashed line) and observed (solid line) 95% CL upper bound on the cross section times branching ratio for Higgs boson production through gluon–gluon fusion with respect to the SM prediction, as a function of the mass hypothesis, for the ATLAS (left) and CMS (right) experiment separately. Figure 10.8 shows the limits for Higgs bosons produced through vector boson production for the ATLAS experiment. In all the plots, the green and yellow bands indicate respectively the 1σ and 2σ fluctuations in the expected limit, due to statistical fluctuations of the expected number of events and to the systematic uncertainties in the background estimates and

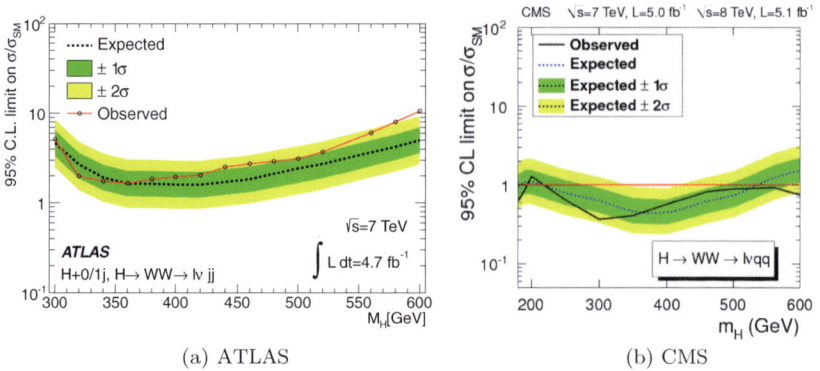

(a) ATLAS   (b) CMS

Fig. 10.7   The expected and observed 95% CL upper limits on the Higgs boson production cross section through gluon–gluon fusion divided by the SM prediction. The ATLAS results used $4.7\,\text{fb}^{-1}$ of 7 TeV data, while CMS results are based on $5.0\,\text{fb}^{-1}$ of 7 TeV plus $5.1\,\text{fb}^{-1}$ of 8 TeV data.

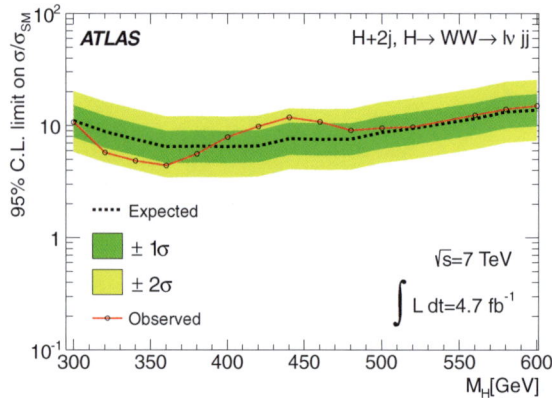

Fig. 10.8   The expected and observed 95% CL upper limits on the Higgs boson production cross section through vector boson fusion divided by the SM prediction, as measured by the ATLAS collaboration after $4.7\,\text{fb}^{-1}$ of 7 TeV data.

signal simulation. Figure 10.9 shows the latest CMS results, including all the categories described in this chapter. The limits are reported together with the other high-mass channels studied by the CMS collaboration, to appreciate the relative performance of the analyses. The total combination of the results is also reported.

The best sensitivity in the ATLAS analysis with 7 TeV data is at $m_H = 400$ GeV, where the observed 95% CL upper limit on the cross section is 2.2 pb for the $\ell\nu jj + 0/1j$ analysis, while the expected limit is 1.9 pb. This corresponds to 1.9 and 1.6 times the SM prediction, respectively. The

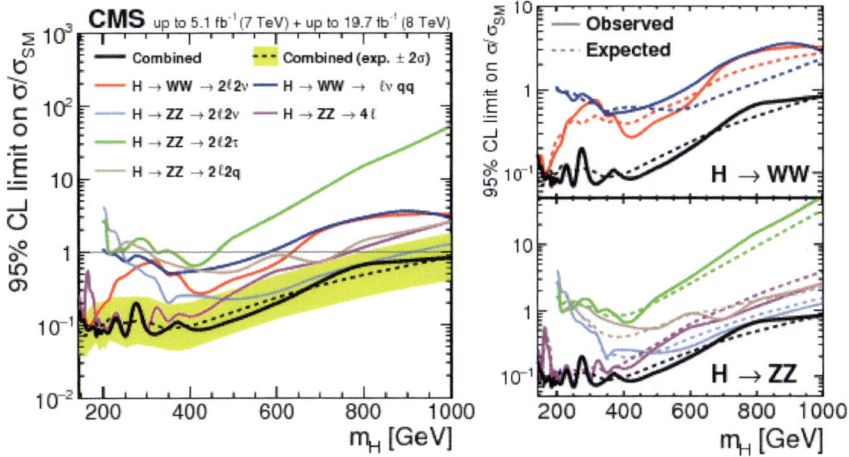

Fig. 10.9 The expected and observed 95% CL upper limits on the Higgs boson production cross section divided by the SM prediction, resulting from the analysis of the CMS data accumulated during the whole 7 and 8 TeV data taking. All the analysis categories described in this chapter are combined to determine the $H \to WW \to \ell\nu qq$ results. The limits are reported together with the other high-mass channels studied by the CMS collaboration, to appreciate the relative performance of the analyses. The total combination of the results is also reported.

CMS analysis was able to exclude at 95% CL most of the high mass range. The ATLAS observed upper limit for vector boson fusion production was 0.7 pb (0.6 pb expected) at $m_H = 400$ GeV in ATLAS, corresponding to 7.9 (6.5 expected) times the SM expectation.

As expected, the analyses showed their maximum exclusion power in the middle of the investigated range, while for low masses they were affected by the systematic uncertainties in the fit of the W+jets turn-on. At high masses, the worsening was due to the reduced production cross section for the Higgs boson, the increasing uncertainty due to the description of the interference between the signal and the non-resonant WW boson pair background, and also the large intrinsic width of the Higgs boson resonance rendered the separation between signal and background less pronounced with respect to low masses.

6. Conclusions and outlook

This study, besides contributing to the exclusion of the mass range up to about 600 GeV, laid the foundation to search for resonances beyond the SM, where the constraints to the theory are dictated by the mechanism of the electroweak symmetry breaking. In fact, the VBF signature gives access to the scattering spectrum of vector bosons as a function of the invariant mass of their system, where the unitarization of the WW bosons scattering will be confirmed, or more intriguingly significant deviations or even new resonances might be observed. The increase in the centre-of-mass energy to 13 TeV, put in place for the LHC running period that started in 2015, will allow to extend the search to higher masses, while for low ranges improved strategies for event triggering will be needed. Eventually, the search for a resonant WW boson pair production at high invariant mass will be fundamental to confirm the SM nature of the discovered Higgs boson, or it might indicate the existence of deviations from this theory. The semi-leptonic final state will play a major role in this investigation, as one of the processes with the largest branching ratio at the LHC.

References

1. M. Dasgupta, A. Fregoso, S. Marzani, and G. P. Salam, Towards an understanding of jet substructure, *JHEP.* **1309**, 029 (2013). doi: 10.1007/JHEP09(2013)029.
2. GEANT4 Collaboration, Geant4: A simulation toolkit, *Nucl.Instrum.Meth.* **A506**, 250–303 (2003). doi: 10.1016/S0168-9002(03)01368-8.
3. B. P. Kersevan and E. Richter-Was, The Monte Carlo event generator AcerMC versions 2.0 to 3.8 with interfaces to PYTHIA 6.4, HERWIG 6.5 and ARIADNE 4.1, *Comput.Phys.Commun.* **184**, 919–985 (2013). doi: 10.1016/j.cpc.2012.10.032.
4. G. Corcella, I. Knowles, G. Marchesini, S. Moretti, K. Odagiri, *et al.*, HERWIG 6: An event generator for hadron emission reactions with interfering gluons (including supersymmetric processes), *JHEP.* **0101**, 010 (2001). doi: 10.1088/1126-6708/2001/01/010.
5. J. Alwall, M. Herquet, F. Maltoni, O. Mattelaer, and T. Stelzer, MadGraph 5 : Going Beyond, *JHEP.* **1106**, 128 (2011). doi: 10.1007/JHEP06(2011)128.
6. S. Frixione and B. R. Webber, Matching NLO QCD computations and parton shower simulations, *JHEP.* **0206**, 029 (2002). doi: 10.1088/1126-6708/2002/06/029.
7. J. M. Campbell and R. Ellis, MCFM for the Tevatron and the LHC, *Nucl.Phys.Proc.Suppl.* **205–206**, 10–15 (2010). doi: 10.1016/j.nuclphysbps.2010.08.011.
8. P. Nason, A new method for combining NLO QCD with shower Monte Carlo algorithms, *JHEP.* **0411**, 040 (2004). doi: 10.1088/1126-6708/2004/11/040.
9. S. Alioli, P. Nason, C. Oleari, and E. Re, A general framework for implementing NLO calculations in shower Monte Carlo programs: the POWHEG BOX, *JHEP.* **1006**, 043 (2010). doi: 10.1007/JHEP06(2010)043.

10. T. Sjostrand, S. Mrenna, and P. Z. Skands, PYTHIA 6.4 Physics and Manual, *JHEP.* **0605**, 026 (2006). doi: 10.1088/1126-6708/2006/05/026.

11. ATLAS Collaboration, Search for the Higgs boson in the $H \to WW \to l\nu jj$ decay channel at $\sqrt{s} = 7$ TeV with the ATLAS detector, *Phys.Lett.* **B718**, 391–410 (2012). doi: 10.1016/j.physletb.2012.10.066.

12. CMS Collaboration, Search for a Standard-Model-like Higgs boson with a mass in the range 145 to 1000 GeV at the LHC, *Eur.Phys.J.* **C73**, 2469 (2013). doi: 10.1140/epjc/s10052-013-2469-8.

13. A. De Rujula, J. Lykken, M. Pierini, C. Rogan, and M. Spiropulu, Higgs look-alikes at the LHC, *Phys.Rev.* **D82**, 013003 (2010). doi: 10.1103/PhysRevD. 82.013003.

14. Y. Gao, A. V. Gritsan, Z. Guo, K. Melnikov, M. Schulze, *et al.*, Spin deter-mination of single-produced resonances at hadron colliders, *Phys.Rev.* **D81**, 075022 (2010). doi: 10.1103/PhysRevD.81.075022.

15. A. Hocker, J. Stelzer, F. Tegenfeldt, H. Voss, K. Voss, *et al.*, TMVA — Toolkit for Multivariate Data Analysis, *PoS.* **ACAT**, 040 (2007).

16. LHC Higgs Cross Section Working Group, Handbook of LHC Higgs Cross Sections: 1. Inclusive Observables (2011). doi: 10.5170/CERN-2011-002. arXiv: 1101.0593.

17. S. Dittmaier, S. Dittmaier, C. Mariotti, G. Passarino, R. Tanaka, *et al.*, Handbook of LHC Higgs Cross Sections: 2. Differential Distributions (2012). doi: 10.5170/CERN-2012-002. arXiv: 1201.3084.

18. LHC Higgs Cross Section Working Group, Handbook of LHC Higgs Cross Sections: 3. Higgs Properties (2013). doi: 10.5170/CERN-2013-004. arXiv: 1307.1347.

19. CMS Collaboration, Search for a Higgs boson in the mass range from 145 to 1000 GeV decaying to a pair of W or Z bosons (2015). arXiv: 1504.00936.

20. ATLAS Collaboration, Search for a high-mass Higgs boson decaying to a W boson pair in pp collisions at $\sqrt{s} = 8$ TeV with the ATLAS detector, *JHEP.* **01**, 032 (2016). doi: 10.1007/JHEP01(2016) 032.

Chapter 11

Higgs boson search in the $ZZ \rightarrow \ell\ell\nu\nu$ and $\ell\ell qq$ final states

Carl Gwilliam* and Adish Vartak[†]

*University of Liverpool, Liverpool L69 3BX, UK
[†]University of California San Diego, La Jolla,
California, 92093, USA

A Standard Model Higgs boson with a mass larger than 200 GeV decays almost exclusively to a pair of weak bosons. This chapter presents the search for such a high mass Higgs boson in the $H \rightarrow ZZ \rightarrow \ell\ell qq$ and $H \rightarrow ZZ \rightarrow \ell\ell\nu\nu$ final states performed by the ATLAS and CMS experiments. The search in these final states was instrumental in excluding a large range of Higgs boson masses beyond 200 GeV with the first few fb^{-1} of integrated luminosity delivered by the LHC. This helped to narrow down the window for the Standard Model Higgs boson to masses around 100 GeV.

1. Introduction

If the Standard Model (SM) Higgs boson has a large mass, it is expected to decay predominantly to a pair of gauge bosons; above twice the Z boson mass, it will decay to a pair of on-shell Z bosons, $H \rightarrow ZZ$, around 30% of the time. The resulting Z bosons will decay into a pair of quarks (about 70% of the time), a pair of neutrinos (about 20% of the time) or a pair of leptons (the remaining 10% of the time), leading to several possible final states for the Higgs boson decay. The latter decay mode, although less frequent, provides charged leptons, which in the case of electrons or muons, offer a clean experimental signature and a key handle to select the event online at the trigger level.

The "golden" $H \rightarrow ZZ^{(*)} \rightarrow llll$ channel is well known to give a very clean signal with excellent separation between the Higgs boson signal and background, as has been discussed in Chapter 5, but it has a low rate due to the relatively small branching fraction for both Z bosons to decay into leptons. The higher branching fraction channels offer larger signal rate but do so at the expense of significantly increased backgrounds. At high mass, however, kinematic requirements can be used to effectively suppress these backgrounds, and the inclusion of these channels can improve the sensitivity for a heavy Higgs boson. This is particularly true for the early data where signal rate is a limiting factor due to the limited luminosity, making these channels interesting early analyses.

The largest branching fraction comes from the fully-hadronic final state, $H \rightarrow ZZ \rightarrow qqqq$, but this channel suffers from an overwhelming background from QCD multijet production and with no leptons in the final state it is difficult to trigger on events of interest while keeping the background at a sustainable level. The $H \rightarrow ZZ \rightarrow qq\nu\nu$ channel has the second highest branching ratio. While it may be possible to trigger on high mass Higgs boson decays due to the large missing transverse momentum from the presence of two neutrinos in this final state, this channel is likely to still suffer from large multijet backgrounds and was not exploited by ATLAS or CMS thus far. The semi-leptonic decay modes, $H \rightarrow ZZ \rightarrow \ell\ell qq$ and $H \rightarrow ZZ \rightarrow \ell\ell\nu\nu$, on the other hand, retain the benefits of a charged lepton in the final state, while still offering a significantly increased rate of Higgs boson production compared to the $H \rightarrow ZZ^{(*)} \rightarrow llll$ channel.

The $H \rightarrow ZZ \rightarrow \ell\ell qq$ channel has a branching fraction which is about 20 times larger than the $H \rightarrow ZZ^{(*)} \rightarrow llll$ mode and the final state is fully reconstructed in the detector, allowing the invariant mass of the Higgs boson to be determined. The quarks hadronise in the detector to produce jets, which are reconstructed with a lower momentum resolution and hence dominate the resulting invariant mass resolution. However, the natural width of the Higgs boson increases exponentially with mass, becoming 20% of its mass at 600 GeV. Therefore, for a high mass Higgs boson the resolution of the individual reconstructed objects is less important than the width of the Higgs resonance itself. The flavour of the quarks, and hence the resulting jets, can also be used as a handle to improve the signal-to-background ratio in this channel.

The $H \rightarrow ZZ \rightarrow \ell\ell\nu\nu$ channel also offers a sizeable branching fraction, about 6 times as large as the $H \rightarrow ZZ^{(*)} \rightarrow llll$ mode, and has good separation from potential backgrounds for a high mass Higgs boson due to a boosted topology with large missing transverse momentum from the $Z \rightarrow \nu\nu$ decay balanced by a pair of leptons with high transverse momentum, p_T. However, the invariant mass of the Higgs boson cannot be fully reconstructed in this channel due to the escaping neutrinos. Nevertheless, this channel was found to provide the best sensitivity to a heavy Higgs boson.

Searches by ATLAS and CMS in these two channels[1-4] were able to exclude a large range of possible heavy Higgs boson masses with the early LHC data. Not only did this probe a region of Higgs boson masses, between 200 and 600 GeV, that was not accessible at previous experiments, but it also played an important role in quickly narrowing down the search range for the SM Higgs boson. Now that a candidate SM Higgs boson has been discovered, these channels can be utilised to search for new resonances, such as additional Higgs bosons predicted by many extensions to the SM, up to the highest masses (≥ 1 TeV).

2. $H \to ZZ \to \ell\ell qq$

$H \to ZZ \to \ell\ell qq$ events are characterised by a pair of high p_T leptons, either electrons or muons, and a pair of jets from the hadronisation of the quarks, both of which originate from the decay of a Z boson and are back-to-back in azimuthal angle. An example event from the ATLAS analysis is shown in Fig. 11.1. As mentioned above, the presence of the leptons not only allows the events to be selected online, using a combination of single and dilepton triggers, but also reduces significantly the background from QCD multijet processes. However, this channel still suffers from large electroweak backgrounds, dominated by the production of a leptonically-decaying Z boson in association with a pair of jets, $(Z \to \ell\ell)$+jets, which has the same final state as the signal. Reducing this background to a level that would allow a potential signal to be observed, and correctly modelling it, forms the main experimental challenge in this channel.

2.1. *Identifying potential signal events*

2.1.1. *Selection of $Z \to \ell\ell$ decay candidates*

The first step in identifying potential signal events is to select a lepton pair from a $Z \to ee$ or $Z \to \mu\mu$ decay. Leptons had to lie within the pseudo-rapidity coverage of the appropriate subdetector ($|\eta| < 2.4-2.5$) and be isolated in order to reduce backgrounds with fake leptons or leptons within jets (such as those produced from b-hadron decays). Both electrons and muons were required to satisfy $p_T > 20$ GeV for ATLAS, while for CMS the leading and sub-leading p_T leptons were required to have $p_T > 40$ GeV

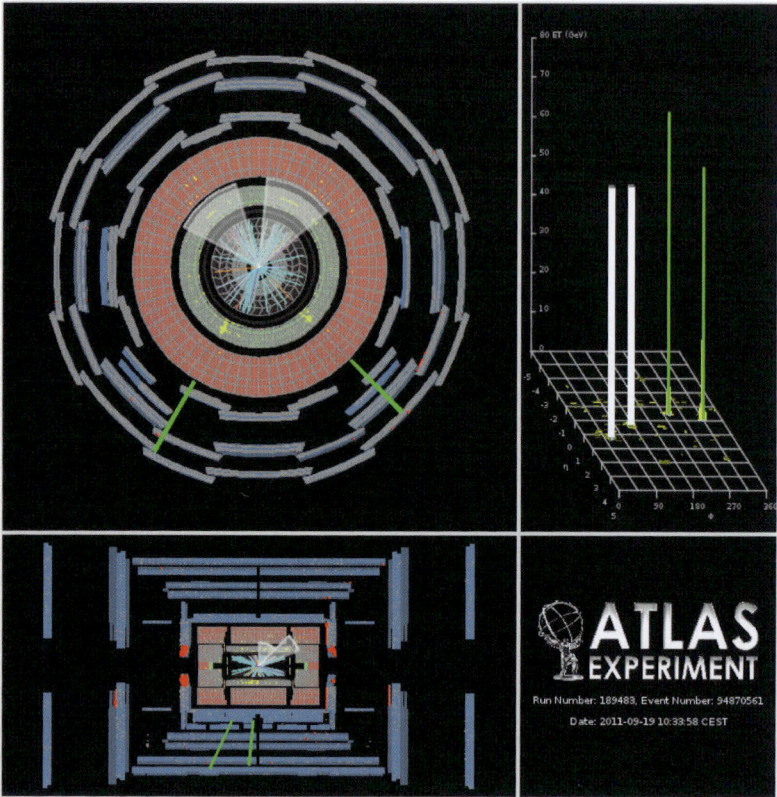

Fig. 11.1 A candidate ATLAS $H \to ZZ \to eebb$ event with $m_{eebb} = 326$ GeV. The event has two electrons with $p_T > 20$ GeV (indicated by the green lines) and two identified b-jets with $p_T > 25$ GeV (indicated by the grey cones). The left hand side shows the transverse (top) and longitudinal (bottom) views, while the top right plot shows the calorimeter energy versus η and ϕ.

and 20 GeV, respectively. These requirements were driven by the trigger thresholds of the two experiments. The two leptons in a pair were required to be oppositely charged (except for electrons in ATLAS where this requirement was dropped due to their higher charge misidentification rate relative to muons) and have an invariant mass, $m_{\ell\ell}$, consistent with the Z boson mass, m_Z. This significantly reduced the background from events without a real leptonically-decaying Z boson, such as top production, which are not produced resonantly and hence do not form a peak in the $m_{\ell\ell}$ spectrum. The mass requirement applied by ATLAS (83−99 GeV) was significantly tighter than that used by CMS (70−110 GeV), reducing the non-resonant background more strongly but at the expense of a slightly lower signal efficiency at this point. ATLAS required exactly one such pair, rejecting events with additional leptons to reduce the $WZ \to l\nu ll$ background, while CMS allowed multiple pairs, choosing the one closest to the nominal Z mass if more than one survived the full selection.

2.1.2. *Selection of* $Z \to q\bar{q}$ *decay candidates*

Events were then required to contain at least two jets with p_T above 25 GeV or 30 GeV for ATLAS or CMS, respectively, within the acceptance of the tracking detector ($|\eta| < 2.4$–2.5). Each pair of jets with invariant mass, m_{jj}, in the range 70–105 GeV for ATLAS or 70–110 GeV for CMS was considered as a $Z \to qq$ candidate; multiple candidates were allowed at this stage in the selection.

The type of the parton forming the jets provides a powerful handle for background rejection. For the signal, the jets originate from the hadronisation of quarks produced in $Z \to qq$ decays, where the flavour of the quarks is distributed almost equally amongst the five possible types (d, u, s, c, b). In contrast, the jets in the dominant $(Z \to \ell\ell)$+jets background come primarily from u- and d-quarks in the proton or gluon radiation. Consequently, the signal is differentiated by a large contribution of jets from heavy flavour quarks and an absence of gluon-produced jets, both of which can be exploited to improve the signal discrimination.

Jets originating from the hadronisation of b-quarks, known as b-jets, can be identified, or "tagged", in the detector by exploiting the relatively long lifetime of the b-quarks, as described in Chapter 9. The signal sensitivity was optimised by splitting the data into different categories based on the number of b-tags, and hence having different signal-to-background ratios, and analysing each separately. CMS split their events into 3 such categories: a 2 b-tag category where one jet is identified with medium ($\approx$65% efficiency) and the other with loose ($\approx$80% efficiency) b-tagging requirements; a 1 b-tag category containing those events not passing the 2 b-tag requirements but having at least one jet passing the loose b-tagging requirement; a 0 b-tag category containing the remainder of the events. ATLAS, on the other hand, used a single b-tagging working point ($\approx$70% efficiency) and separated their analysis into a "tagged" channel, containing events with exactly two identified b-jets, and an "untagged" channel, containing events with fewer than two b-jets. The small fraction of jets with more than two b-jets were rejected. Events with one identified b-jet were treated as

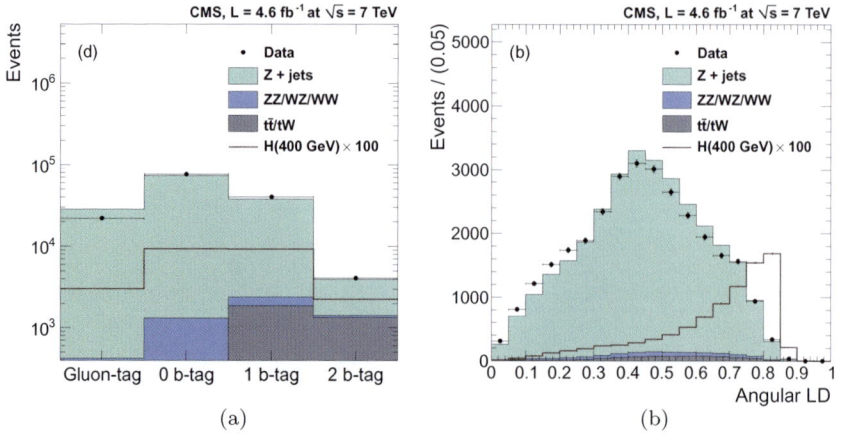

Fig. 11.2 Comparison between data (points) and simulated backgrounds (filled histograms) in the CMS analysis for (a) the different *b*-tag categories and the gluon-tagged category (which is not used in the analysis); (b) the distribution of the angular likelihood discriminant. The open histogram indicates the expected Higgs boson signal for a mass of 400 GeV, multiplied by a factor of 100.

untagged since splitting them into a separate category was not found to improve the expected sensitivity. The varying signal and background composition in the different categories in the case of the CMS analysis can be seen in Fig. 11.2(a).

As expected, events without *b*-jets are dominated by $(Z \to \ell\ell)$+jets background. In the case of CMS, this was reduced by constructing a quark–gluon likelihood discriminant, LD, which exploited the fact that jets produced by gluon hadronisation tend to have higher multiplicity than quark-initiated jets. Events where the two leading jets were consistent with being initiated by gluons (and hence unlikely to come from $Z \to q\bar{q}$ decays) were removed from the analysis. ATLAS did not discriminate quark and gluon jets, preferring to restrict themselves to variables felt to be more robust.

Events containing 2 *b*-jets have a substantial background from top-pair, $t\bar{t}$, production. Unlike signal events, these contain neutrinos from the leptonic decay of W bosons, and so can be suppressed by applying a cut on the missing transverse energy, E_T^{miss}. ATLAS required $E_T^{miss} < 50$ GeV for all events, both the tagged and the untagged (the latter can suffer from other background containing neutrinos, such as the Wt single-top process, or fake E_T^{miss}). In their 2 *b*-tag analysis, CMS applied a loose selection

$2\lambda(E_{\mathrm{T}}^{\mathrm{miss}}) < 10$, where λ is the ratio of the likelihood with the measured value of $E_{\mathrm{T}}^{\mathrm{miss}}$ to the likelihood with $E_{\mathrm{T}}^{\mathrm{miss}} = 0$.

Since the Higgs is expected to be a spin-0 boson, the angular distribution of its decay products does not depend on the production angle. As described in Chapter 10, this allows the kinematics of the $gg \to H \to ZZ \to \ell\ell qq$ process to be fully parametrised via five angles as defined in Ref. 5, which are analogous to those depicted in Fig. 10.4. These angles are weakly correlated with other kinematic variables of the H and Z bosons. CMS exploited this angular information by forming an angular LD based on the probability ratio of the signal and background hypothesis. For signal, the probability distribution was modelled by a correlated five-dimensional angular parameterisation; for background an empirical parameterisation taken directly from simulation was employed, relying on the Monte Carlo (MC) description of the background processes. The LD was parametrised as a function of the candidate Higgs boson mass, m_H, and a cut, varying linearly as a function of m_H, chosen to optimise the expected sensitivity in the different b-tag categories separately. The resulting LD, shown in Fig. 11.2(b), provided one of the main discriminators between signal and background in the CMS analysis. It can be seen, however, that the distribution of the LD in data is not perfectly modelled by the background simulation. ATLAS chose a different approach, preferring to perform a cut-based analysis rather than using multivariate techniques, which may be difficult to understand in early data.

At large m_H, the two Z bosons from the $H \to ZZ$ decay are produced with large momenta (due to the $m_H - 2m_Z$ mass difference), leading to a decrease in the opening angles between their decay products. ATLAS took advantage of this by splitting their analysis into low-m_H ($m_H < 300$ GeV) and high-m_H ($m_H \geq 300$ GeV) regions. In the high-m_H case, the p_{T} requirement on the jets was raised to 45 GeV and the azimuthal angles between the leptons or the jets in a pair are required to be less than $\pi/2$. CMS exploited the variation in the opening angles via the m_H dependence of their angular LD, but did not vary their p_{T} requirements. In both cases, this significantly suppressed the background at high mass where, unlike the signal, the leptons and jets have no significant boost.

Following the full event selection, events may contain multiple $Z \to qq$ candidates. At this point CMS retained those in the highest b-tag category and chose the one with m_{jj} closest to the Z boson mass. This has the side effect that backgrounds tend to be pulled towards m_Z, potentially biasing background determinations from data using sideband methods (see below).

In order to avoid this, ATLAS took a different approach. In their untagged channel, all candidates formed from the three highest p_T jets were retained, treating each with unit weight. In doing so, potential correlations between the candidates were neglected in the statistical analysis since the fraction of multiple candidates was relatively small ($< 5\%$ for the high-m_H region, which covers the majority of the mass range, and $\sim 10\%$ for the low-m_H region). In the tagged channel, the candidate with m_{jj} formed from the two b-jets was simply taken (additional b-jets having already been vetoed). In this case, the energies of the b-jets were corrected to account for semi-leptonic decays of the b-hadrons. Firstly, if a muon with $p_T > 4\,\text{GeV}$ was identified with $\Delta R < 0.4$ of a b-tagged jet its momentum is added back to the jet. Since it is difficult to identify electrons close to jets only muonic decays were corrected for. Secondly, the b-jet energy was scaled up by 5% to take into account the fact that the energy scale for heavy-quark jets tends to be lower than that of light-quark jets on average due, for example, to neutrinos from the semi-leptonic decays. This correction was derived by comparing the reconstructed $m_{\ell\ell}$ invariant mass peak between tagged and untagged signal events and allowed the same invariant mass window to be applied for both channels but did not affect the reconstructed Higgs mass due to the mass constraint described below.

2.1.3. *Selection of Higgs boson decay candidates*

The $H \to ZZ \to \ell\ell qq$ signal, should it exist, would appear as a peak in the invariant mass of the two Z boson candidates, i.e. the mass of the dilepton-dijet system, with m_{ZZ} ($\equiv m_{\ell\ell jj}$) around the candidate Higgs boson mass, m_H. In order to improve the Higgs mass resolution and hence the signal-to-background discrimination, the invariant mass of the dijet pair was constrained to the nominal Z boson mass in forming m_{ZZ}. This gave a significant improvement in resolution at low m_H (a factor of about 2.5 in the case of ATLAS), where the detector resolution dominates, but the improvement decreased with increasing m_H as the natural width of the Higgs boson becomes large. For ATLAS, the total efficiency for selecting potential signal events was around 10%.

The resulting m_{ZZ} invariant mass distributions in the different categories are shown in Fig. 11.3 for ATLAS (after the high-m_H selection) and Fig. 11.4 for CMS. It can be seen that the dominant background in this channel is ($Z \to \ell\ell$)+jets production, with a significant contribution from top production, mainly $t\bar{t}$ but also single top processes such as Wt, in events

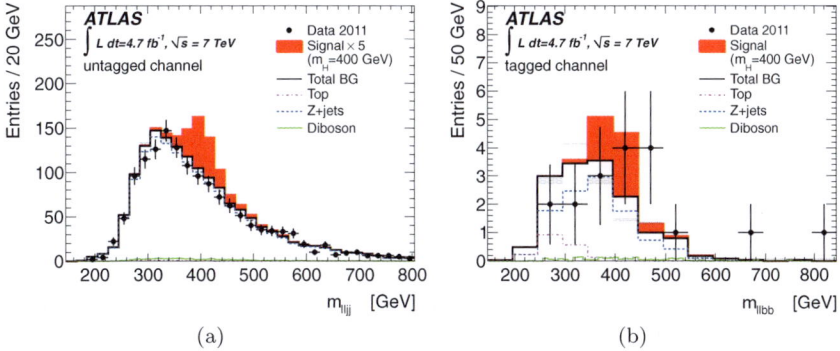

(a)   (b)

Fig. 11.3   The m_{ZZ} ($\equiv m_{\ell\ell jj}$) invariant mass distributions for the ATLAS analysis in the (a) untagged and (b) tagged channels for the full high-m_H selection. The data (points) are well described by the total background after the corrections described in Sec. 2.2 (solid histogram) within systematic errors (hatched band). The expected Higgs boson signal for a mass of 400 GeV (filled histogram) is also shown.

with 2 b-jets. There is also a small background from diboson $ZZ \to llqq$ or $WZ \to qqll$ production. The former is largely irreducible since the leptons and quarks are both produced resonantly from a Z boson (although the angular selection mentioned above offers some discrimination) but the rate is relatively low such that, in the end, it is not a significant background. In order to observe a potential signal, it is essential to correctly model these background processes and in both cases the background estimation was derived from data using the methods described in Sec. 2.2. No significant excess of data above the expected SM background was observed by either experiment and the shape of the resulting distributions were used to set upper limits on the SM Higgs boson cross section as described in Sec. 4.

2.2. *Background modelling*

The simulation cannot be relied on to perfectly model the distributions of the background processes, particularly in the ZZ threshold region around $2m_Z$ (see e.g. Fig. 11.4(a)), thus it is imperative to use data-driven methods to extract the background wherever possible. Of course, this cannot be achieved using the standard event selection since the results would be biased by the possible presence of signal. Instead, one or more orthogonal regions, known as control regions, must be defined which are similar to the signal region but do not suffer from signal contamination. Here the two experiment

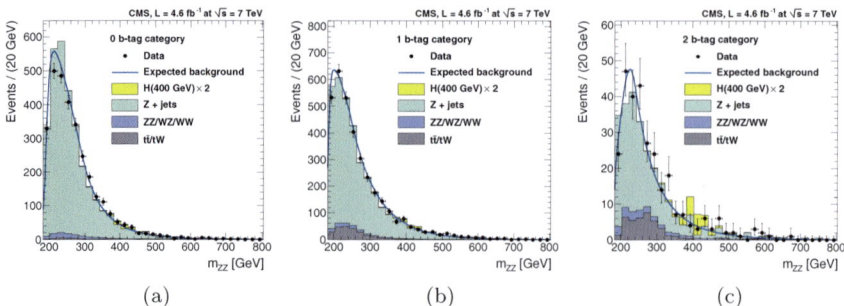

Fig. 11.4 The m_{ZZ} invariant mass distributions for that CMS analysis in the (a) 0 b-tag, (b) 1 b-tag and (c) 2 b-tag categories after the full selection. The data (points) are well described by the background prediction from the sideband estimation described in Sec. 2.2 (curved line). The solid histograms represent the background expectation from the uncorrected simulation for comparison only. The expected Higgs boson signal for a mass of 400 GeV (yellow histogram), multiplied by a factor of two, is also shown (although it is only visible in the highest-sensitivity 2 b-tag category).

followed somewhat different philosophies which are discussed below with particular attention to their relative strengths and weaknesses.

2.2.1. *The CMS approach*

CMS estimated the shape and normalisation of their total background distribution from data in the m_{jj} "sideband" regions. These are the regions on either side of the Z boson mass, which were defined as 60 GeV $< m_{jj} <$ 75 GeV and 105 GeV $< m_{jj} <$ 130 GeV, where the jets are inconsistent with being produced from a $Z \to qq$ decay and, therefore, have little or no contamination from the signal process. The composition and distribution of the dominant backgrounds (($Z \to \ell\ell$)+jets and top production) in this region were, however, verified in simulation to be similar to that in the signal region. The m_{ZZ} distribution in the sidebands was measured in data and parametrised with an empirical function to obtain both the shape and normalisation of the background. This was performed separately for each of the b-tag categories since they had different background compositions.

This background parameterisation cannot be directly applied to the signal region since it needs to be corrected for differences in the selection acceptance between the two regions and also for any differences in background composition. This was estimated from simulation by taking the ratio of the background in the signal region to that in the sideband and then parameterising it as a function of m_{ZZ} to obtain a correction factor $\alpha(m_{ZZ})$. The backgrounds in CMS were simulated using various MC generators: MADGRAPH[6] for $(Z \to \ell\ell)$+jets, PYTHIA[7] for diboson production and POWHEG[8–11] for top production.

The main advantage of this approach is that the majority of the systematic uncertainties on the background cancel out when forming the ratio $\alpha(m_{ZZ})$, although some will affect the two regions differently. The dominant uncertainty on the background normalisation estimate was due to statistical fluctuations in the number of events in the sideband. The choice of a functional form to parametrise the m_{ZZ} distribution was, however, not straight forward, particularly in the ZZ threshold region around $2m_Z$, where $\alpha(m_{ZZ})$ reached about 1.2, and can potentially bias the background estimate. Also, the extraction of a single background estimate, rather than determining the individual components separately, relies on the simulation to correctly model any difference in background composition between the sidebands and the signal region. While the dominant backgrounds had a similar composition in the two regions, this is clearly not true for the ZZ process since the jets are produced resonantly and hence it only populates the signal region. This background was relatively small, although it did reach $\approx 10\%$ in the 2 b-tag category, and was accounted for through $\alpha(m_{ZZ})$.

2.2.2. *The ATLAS approach*

ATLAS estimated the two main backgrounds (($Z \to \ell\ell$)+jets and top production) separately using two independent control regions. In each case the shape of the relevant kinematic distributions were taken from Monte Carlo (cross-checked in the control regions) while the normalisation was derived from data. The advantage of this approach is that it does not rely on arbitrary functional forms to parametrise the data but instead makes use of the full simulation, which also directly takes into account any differences in composition and distribution of background between the signal and control regions. However, uncertainties on the MC shape, due to both systematic effects and limited MC statistics, do not cancel and instead propagate through to the final result.

The ATLAS $(Z \to \ell\ell)$+jets background was simulated using two different leading order (LO) MC generators: ALPGEN[12] for Z+light-jets events and SHERPA[13] for Z+heavy-jets events. However, the simulation cannot be relied upon to correctly predict the flavour composition of the background, i.e. the amount $Z +$ light-jets, $Z + c$-jets and $Z + b$-jets. The relative fraction of these components was estimated from data by fitting the distribution of the b-tagging discriminant with exclusive MC templates for each flavour. The overall $(Z \to \ell\ell)$+jets normalisation was then determined using the data in the m_{jj} sidebands, which in the case of ATLAS were defined as 40 GeV $< m_{jj} <$ 70 GeV and 105 GeV $< m_{jj} <$ 150 GeV. The ratio of the number of events in the sideband, after subtracting the MC contribution from other background sources, was used to determined a scale factor that corrects the normalisation of the $(Z \to \ell\ell)$+jets simulation to that in data and which was subsequently applied to the signal region. The uncertainty on the shape of the simulated $(Z \to \ell\ell)$+jets background was estimated by reweighting various kinematic distributions to cover any data-to-MC differences in the sidebands. After applying the scale factors, the simulation describes the data well within uncertainties, as can be seen in Fig. 11.5(a).

The top background in ATLAS, both single top and $t\bar{t}$, was simulated using the next-to-leading order (NLO) MC@NLO[14] MC. Analogously to the $(Z \to \ell\ell)$+jets case, the normalisation was corrected to data using scale factors derived, in this case, from a control region defined by the $m_{\ell\ell}$ sidebands (60 GeV $< m_{\ell\ell} <$ 76 GeV and 106 GeV $< m_{\ell\ell} <$ 150 GeV) with the $E_{\mathrm{T}}^{\mathrm{miss}}$ selection reversed. This region is dominated by top production since the leptons from top decays (in contrast to the dominant $(Z \to \ell\ell)$+jets background) are not produced resonantly in $m_{\ell\ell}$. Again, the corrected simulation

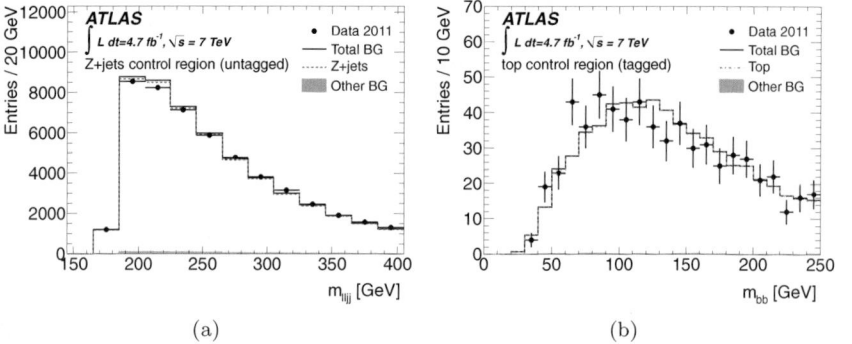

(a)   (b)

Fig. 11.5 Distributions of the background control regions, after correction, in the ATLAS analysis: (a) the $m_{\ell\ell jj}$ distribution in the m_{jj} sidebands for the low-m_H untagged channel and (b) the m_{jj} distribution in the $m_{\ell\ell}$ sidebands with $E_{\mathrm{T}}^{\mathrm{miss}} >$ 50 GeV for the low-m_H tagged channel. The points are the data and the sold line is the total simulated background.

describes the data well, as shown in Fig. 11.5(b). The extraction of the scale factors was performed simultaneously for the $(Z \to \ell\ell)$+jets and top background, and in both cases separately for the untagged and tagged channels. The uncertainty on the background normalisation procedure was dominated by statistical fluctuations in the number of simulated sideband events and amounted to 2–3% for the untagged channel and 5–6% for the tagged channel. Systematic uncertainties arising from detector effects, including the identification and calibration of leptons and jets and the b-tagging efficiency, were also (conservatively) applied for all backgrounds.

The small irreducible background from ZZ and WZ production was taken directly from LO HERWIG[15] MC simulation, with alternative PYTHIA and MC@NLO samples used to estimate systematic effects. The background due to QCD multijet events was estimated from data and found to be negligible.

3. $H \to ZZ \to \ell\ell\nu\nu$

The topology for a signal event in this channel consists of a pair of high p_T leptons with opposite charge $(e^+e^-$ or $\mu^+\mu^-)$ forming an invariant mass consistent with the Z boson peak. These leptons are accompanied by the presence of large missing transverse energy (E_T^{miss}) from the escaping neutrinos. An example event from the CMS analysis is shown in Fig. 11.6. Several physics processes presenting a similar event signature constitute the background in this channel. These processes include Z+jets events, $t\bar{t}$ and WW events with two same flavour leptons in the final state, WZ and ZZ events with the Z boson decaying leptonically and a small fraction of W+jets and QCD multijet events in which one or both of the leptons are actually misidentified jets.

3.1. *Identifying potential signal events*

3.1.1. *Basic selection*

Due to the presence of two high p_T leptons in the final state a combination of single and dilepton triggers was employed to select events for further analysis. In the offline analysis, two leptons were required, which must have the same flavour but opposite charge, and p_T greater than 20 GeV. This ensured a very high trigger efficiency (94–97% in the case of muons and nearly 100% in the case of electrons). In signal events, the two leptons

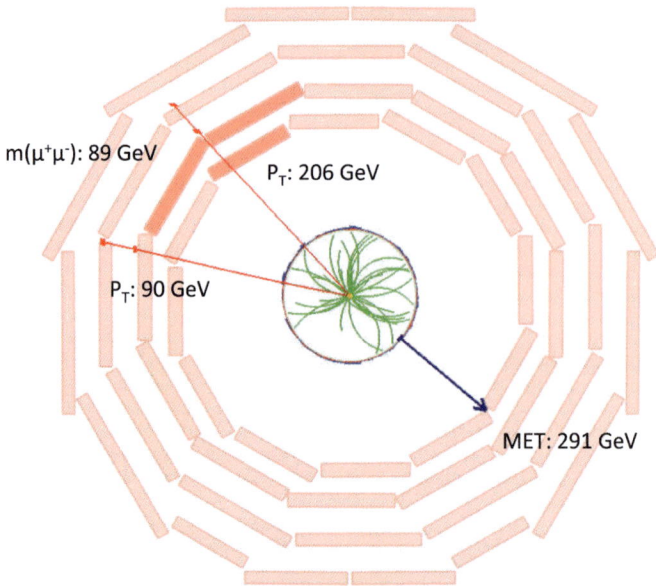

Fig. 11.6   A candidate $H \to ZZ \to \mu\mu\nu\nu$ event observed in the CMS detector. The red lines depict the reconstructed tracks of the two muon candidates in the event. These muon candidates have opposite charge and form an invariant mass of 89 GeV which is consistent with the mass of the Z boson. The two muon candidates have large transverse momenta (90 GeV and 206 GeV) and a small separation in ϕ, thus, indicating the decay of a boosted Z boson. There is very little energy seen recoiling against the dimuon system. Thus, a large $E_{\mathrm{T}}^{\mathrm{miss}}$ of 291 GeV is observed in the event.

are produced in the decay of a Z boson. Hence these leptons are expected to be isolated from hadronic activity. Furthermore, given the extremely small lifetime of the Z boson, these leptons are expected to be produced promptly in the pp collisions. Hence, certain isolation and identification requirements were imposed to remove events with jets misidentified as leptons. In addition, the lepton tracks were required to be consistent with a well-reconstructed primary vertex. This requirement helped to suppress background with leptons produced from b-decays.

3.1.2. Background categorisation and suppression

The various background processes in this channel can be categorised into two groups. One group consists of events in which the two leptons are produced in the decay of a Z boson. These can be referred to as "resonant" backgrounds since the dilepton mass distribution in these events forms the

Z boson peak. The other class of events can be termed as "non-resonant" backgrounds as these events do not have a Z boson peak in the dilepton mass spectrum. Since the signal events are themselves characterised by a Z mass peak, the non-resonant background can be reduced by requiring that the invariant mass of the dilepton pair be consistent with the Z boson peak. This was done by imposing a requirement that the leptons form an invariant mass within a 30 GeV window around the nominal Z boson mass.

While the search is designed to look for $H \rightarrow ZZ \rightarrow \ell\ell\nu\nu$ decays, signal events, if observed, would also consist of $H \rightarrow W^+W^{-(*)} \rightarrow \ell^+\nu\ell^-\nu$ decays in which the two leptons form an invariant mass consistent with the Z boson peak. The contribution from $H \rightarrow W^+W^{-(*)} \rightarrow \ell^+\nu\ell^-\nu$ events is substantial particularly at low Higgs boson masses. For a 200 GeV Higgs boson the contribution from $H \rightarrow W^+W^{-(*)} \rightarrow \ell^+\nu\ell^-\nu$ is expected to be as large as 70%, while for a 300 GeV Higgs boson it would contribute 13% of signal events. The ATLAS and CMS analyses adopted different approaches on how to treat the signal from $H \rightarrow W^+W^{-(*)} \rightarrow \ell^+\nu\ell^-\nu$ decays. In the ATLAS analysis, the $H \rightarrow W^+W^{-(*)} \rightarrow \ell^+\nu\ell^-\nu$ events were included as part of the signal expectation while in the CMS analysis, these events were included in the background prediction, and only $H \rightarrow ZZ \rightarrow \ell\ell\nu\nu$ events were treated as signal. The difference lies in how the results of the two analyses are to be interpreted. The ATLAS result should be considered as a combination of $H \rightarrow ZZ \rightarrow \ell\ell\nu\nu$ and $H \rightarrow W^+W^{-(*)} \rightarrow \ell^+\nu\ell^-\nu$ modes for an analysis designed to search for the $H \rightarrow ZZ \rightarrow \ell\ell\nu\nu$ channel. The CMS result, on the other hand pertains singularly to the $H \rightarrow ZZ \rightarrow \ell\ell\nu\nu$ mode.

Some of the background processes are characterised by the presence of a third lepton in the event. These include fully leptonic WZ and ZZ decays. In some cases, the third lepton may be produced in b-decays. These include events involving fully leptonic decay of a $t\bar{t}$ pair, and events in which a Z boson is produced in association with a $b\bar{b}$ pair. In order to suppress such backgrounds, events with a third, well identified, isolated lepton but with a lowered p_T threshold of 10 GeV were vetoed. In addition, the CMS analysis vetoed events with soft muons, i.e. muons with the p_T threshold lowered to 3 GeV, passing certain loose identification and isolation requirements to remove events involving leptonic b-decays. In both analyses, events were rejected if a b-tagged jet was identified to suppress the top background.

The core feature of signal events in this channel is the presence of E_T^{miss} as can be seen in Fig. 11.7. This is the most important handle

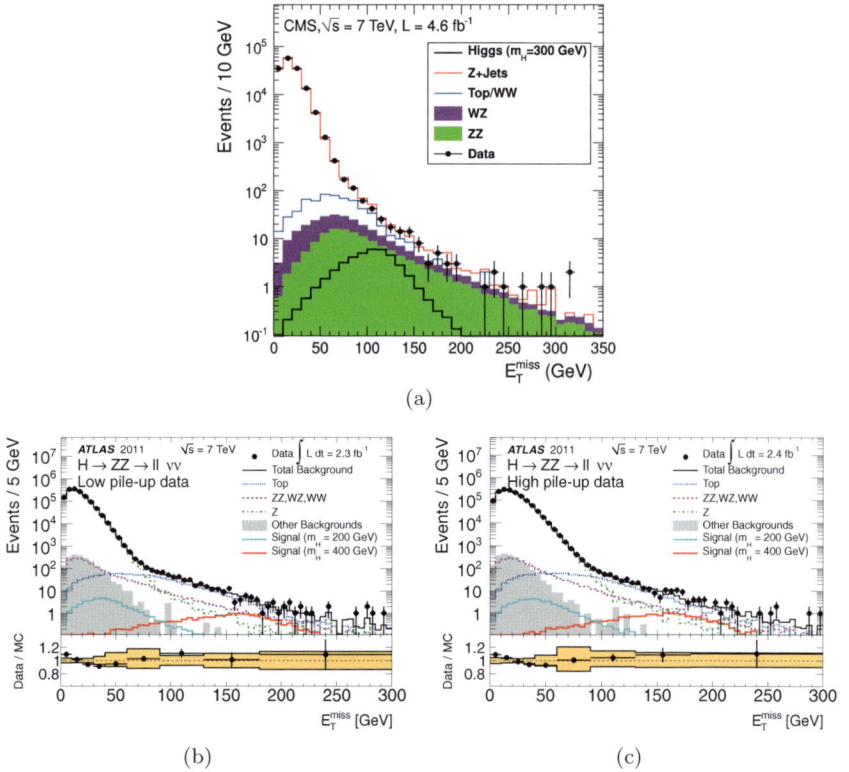

Fig. 11.7   The E_T^{miss} distributions for signal and background processes: (a) the E_T^{miss} distribution in the CMS analysis, (b) the E_T^{miss} distribution in the low pileup dataset of the ATLAS analysis and (c) the E_T^{miss} distribution in the high pileup dataset of the ATLAS analysis.

against the largest of all the backgrounds, namely Z+jets events. Without considering the E_T^{miss} in the event, the Z+jets background is almost 5 orders of magnitude larger in cross section than the signal. Since there are no high p_T neutrinos in a typical Z+jets event, the overall transverse momentum imbalance is small and arises mainly from the mismeasurement of the momentum of jets in the event. An additional contributing factor is pileup (described in Chapter 1). With the increase in the instantaneous luminosity delivered by the LHC, the average number of pileup interactions increased from six in the earlier periods of data-taking to about fifteen in the later periods of the 7 TeV run. Since E_T^{miss} is evaluated by taking into account the entire detector activity during an event, its resolution worsens as the amount of energy spilling into the detector from pileup events increases. This results in a significant degradation in the separation between signal and Z+jets background. Hence, to retain the best sensitivity the ATLAS analysis was split into two eras, with the earlier 2.3 fb^{-1} of integrated luminosity treated as low pileup data and the later 2.4 fb^{-1} of integrated luminosity treated as high pileup data. The analysis selection, however, was kept the same for both the periods.

3.1.3. *Kinematic selection*

The requirements on E_T^{miss} were varied depending on the Higgs boson mass hypothesis under investigation. Both analyses applied an optimised selection in two Higgs boson mass ranges. The requirement was looser for low Higgs boson mass region in which signal events typically have smaller E_T^{miss} and p_T of the dilepton system, due to smaller boost of the decaying Z bosons compared to the high Higgs boson mass region which is characterised by larger E_T^{miss} and p_T of the dilepton system. The ATLAS analysis defined the low mass region between Higgs boson mass of 200 to 280 GeV, and required E_T^{miss} to be larger than 66 GeV in this region. Higgs boson masses larger than 280 GeV were defined to constitute the high mass region, and the E_T^{miss} requirement was tightened to 82 GeV. In the case of the CMS analysis, E_T^{miss} was required to be larger than 70 GeV for Higgs boson masses between 250 to 300 GeV, while the requirement was tightened to 80 GeV for Higgs masses larger than 300 GeV. In spite of the large E_T^{miss} requirement, some Z+jets background is expected to survive comprising events with a large E_T^{miss} arising from significant mismeasurement of jet momenta. However, these events typically have the E_T^{miss} aligned close to the jet in the transverse direction. This feature was used in both analyses to suppress the background. In the CMS analysis, events were vetoed if the azimuthal separation between E_T^{miss} and a jet with p_T larger than 15 GeV was found to be smaller than 0.5 radians. In the ATLAS analysis, the requirement in the low mass region was to have the azimuthal separation between E_T^{miss} and a jet, with p_T larger than 25 GeV, greater than 1.5 radians, while in the high mass region it was required to be greater than 0.5 radians.

In addition to E_T^{miss}, the dilepton p_T is also a useful discriminant between signal and background, with signal events being characteristically higher in dilepton p_T compared to background. The p_T of the dilepton candidate is correlated to the azimuthal separation between the two leptons. As the p_T of the dilepton candidate increases, the azimuthal separation between the two leptons becomes smaller since the two leptons are produced from the decay of a single boosted particle i.e. the Z boson. ATLAS used this angular separation in the analysis selection. Upper bounds of 2.64 and 2.25 were imposed on the azimuthal separation in the low and high mass regions, respectively, while in the low mass region an additional lower bound was imposed requiring the angular separation to be larger than 1. This requirement helped to reduce the non-resonant background in which the two leptons are not produced from the decay of a single boosted particle and are hence uncorrelated. The CMS analysis, on the other hand, imposed a uniform requirement on the p_T of the dilepton candidate to be larger than 55 GeV.

To extract the signal from the events that pass the analysis requirements, an additional variable is constructed by combining the transverse momentum of the dilepton candidate with the E_T^{miss} in the event. This can be viewed as a mass-like variable constructed using the transverse momenta of the Z candidates instead of using the full momentum information. This "transverse mass" variable or m_T is defined as follows:

$$m_T^2 = \left(\sqrt{p_{T,\ell\ell}{}^2 + m_{\ell\ell}{}^2} + \sqrt{E_T^{miss}{}^2 + m_{\ell\ell}{}^2} \right)^2 - (\vec{p}_{T,\ell\ell} + \vec{E}_T^{miss})^2,$$

where $m_{\ell\ell}$ is the mass and $p_{T,\ell\ell}$ is the transverse momentum of the dilepton candidate. The shape of m_T is a useful discriminant between signal and background as can be seen in Fig. 11.8.

3.2. *Background modeling*

Having defined the event selection, the next key step in an analysis is to provide an estimation of the various background contributions and to qualify these estimates with the associated uncertainties. In the case of the CMS analysis, the backgrounds were grouped into four different types — $Z + \text{jets}$, non-resonant backgrounds, WZ and ZZ.

3.2.1. *Z+jets background estimation in CMS analysis*

The $Z + \text{jets}$ background was estimated from data using $\gamma + \text{jets}$ events as a control sample. These events are quite similar to $Z + \text{jets}$ events in their production mechanism, and the E_T^{miss} in these events can also be attributed to jet mismeasurement and detector effects. To make this control sample similar to the $Z+\text{jets}$ events in terms of kinematics and detector activity, it was reweighted in several ways prior to applying the E_T^{miss} selection. Firstly,

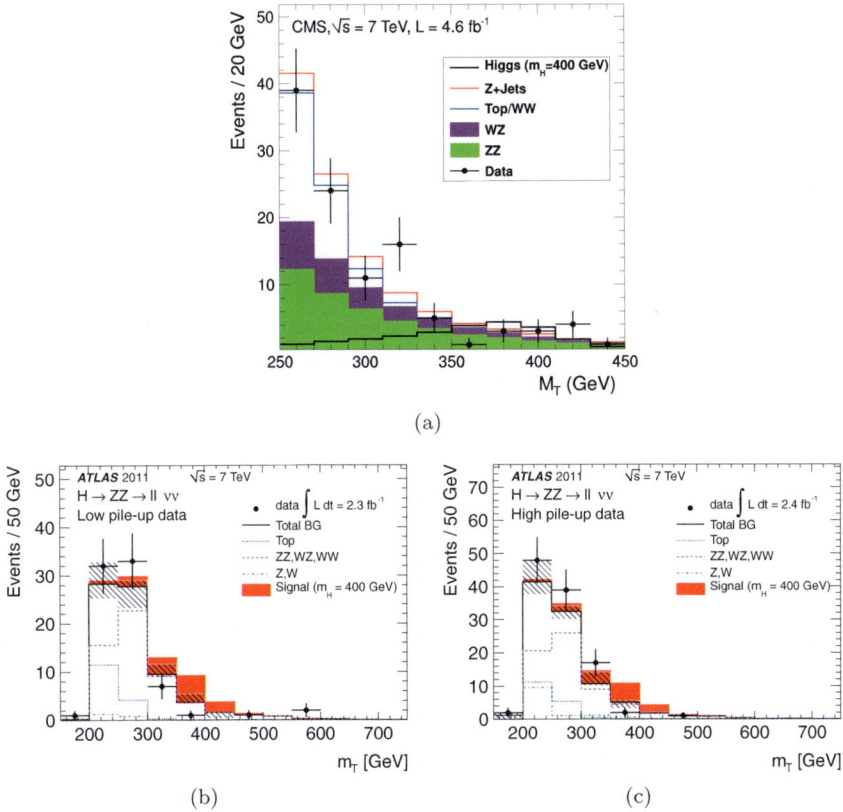

Fig. 11.8   The m_T distributions for 400 GeV SM Higgs boson signal and background processes: (a) the m_T distribution in the CMS analysis: (b) the m_T distribution in the low pileup dataset of the ATLAS analysis and (c) the m_T distribution in the high pileup dataset of the ATLAS analysis.

the photon p_T spectrum was reweighted to match that of events with a reconstructed Z boson candidate in data. Furthermore, the γ+jets events were reweighted to match the multiplicity of jets with $p_T > 30$ GeV in Z+jets events. The triggers used for collecting γ+jets events were increasingly prescaled at several stages of data-taking. Therefore, to match the pileup conditions between the γ+jets control sample and the Z+jets events, the γ+jets events were reweighted so that the distribution of the number of vertices matched with Z+jets events in data. Finally, the γ+jets sample was normalized to the total yield of Z boson events in data. The m_T variable for these events was constructed by using the p_T of the photon instead of a Z

boson candidate. The key advantage of this method is that it is completely data-driven and so it is not affected by the modelling of the Z+jets process in Monte Carlo generators and the simulation of the detector response. However, the main disadvantage is the fact that γ+jets events, particularly those with high E_T^{miss}, are contaminated with events containing a photon and genuine E_T^{miss}. These include Z+γ events in which the Z boson decays to neutrinos, or $W + \gamma$ events in which the W boson decays leptonically and the lepton falls outside the detector acceptance, or the case in which the lepton is a tau. This contamination could be nearly half of the Z+jets estimate for high Higgs boson masses. Thus, the estimate was treated as an upper bound on the Z+jets background with 100% uncertainty assigned to it.

3.2.2. *Non-resonant background estimation in CMS analysis*

The non-resonant backgrounds which include $t\bar{t}$, WW and a small proportion of W+jets events were all estimated together in the CMS analysis. In the case of all of these backgrounds, there are nearly as many events with same flavour leptons in the final state as there are events with opposite flavour leptons, i.e. an electron and a muon in the final state. This lends the $e^{\pm}\mu^{\mp}$ final state as a natural choice for a control region in data. These events were scaled to the expected yields in the e^+e^- or $\mu^+\mu^-$ final states by applying the corresponding scale factors α_μ and α_e computed from the sidebands (SB) of the Z boson peak (40 GeV $< m_{\ell\ell} <$ 70 GeV and 110 GeV $< m_{\ell\ell} <$ 200 GeV) by using the following relations:

$$\alpha_\mu = \frac{N_{\mu\mu}^{SB}}{N_{e\mu}^{SB}}, \quad \alpha_e = \frac{N_{ee}^{SB}}{N_{e\mu}^{SB}},$$

where N_{ee}^{SB}, $N_{\mu\mu}^{SB}$, and $N_{e\mu}^{SB}$ are the number of events in the Z sidebands in a top-enriched sample of e^+e^-, $\mu^+\mu^-$, and $e^{\pm}\mu^{\mp}$ final states, respectively. Such samples were selected by requiring $E_T^{\mathrm{miss}} >$ 70 GeV and a b-tagged jet in the events. The measured values of α with the corresponding statistical uncertainties were found to be $\alpha_\mu = 0.58 \pm 0.02$ and $\alpha_e = 0.42 \pm 0.02$. The difference between the values of α_μ and α_e can be attributed to the difference in the selection efficiencies of muons and electrons.

It should be noted that this control region includes, in addition to the aforementioned non-resonant backgrounds, also events from a possible Higgs boson signal in $H \to W^+W^{-(*)} \to \ell^+\nu\ell^-\nu$. Thus, the $H \to W^+W^{-(*)} \to \ell^+\nu\ell^-\nu$ part of signal events were counted as a part of the background in the CMS analysis, allowing only $H \rightarrow ZZ \rightarrow \ell\ell\nu\nu$ events to be treated as signal. The main uncertainty in this method arises from the limited number of $e^{\pm}\mu^{\mp}$ events passing the analysis selection. The backgrounds that remain are $WZ \rightarrow 3\ell\nu$ events and $ZZ \rightarrow \ell\ell\nu\nu$ events. These backgrounds were estimated from simulation and the uncertainties on these estimates, due to the parton distribution functions used to generate these events and the QCD scale, were found to vary between 10% to 20%.

3.2.3. *Background estimation in ATLAS analysis*

In the case of the ATLAS analysis, the background estimates were taken from simulation. This made it imperative to confirm that the simulation could accurately describe the backgrounds in data. The strategy adopted here was to verify the prediction for various backgrounds by comparing the simulation with certain control regions in data and to treat any differences as systematic uncertainties. For the WZ background, the normalization was verified in a control region with exactly three leptons with p_T greater than 10 GeV. The Z+jets background estimate was verified by comparing the simulation with events which would otherwise be rejected in the analysis, by inverting the cut on the azimuthal separation between E_T^{miss} and the nearest jet as can be seen in Fig. 11.9(a). In the case of the top-quark background the prediction was verified to agree with data in two separate control regions. In one case, events with a b-jet in the sidebands of the Z boson peak were used. The sidebands in this case included the dilepton

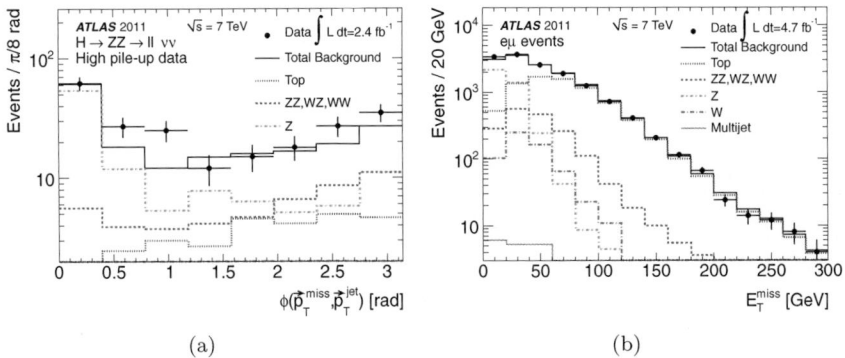

(a)   (b)

Fig. 11.9   Control regions used in the ATLAS analysis: (a) the distribution of the azimuthal separation between E_T^{miss} and the nearest jet in the high pileup dataset and (b) the E_T^{miss} distribution in events with an electron–muon pair.

mass region between 60 to 150 GeV excluding the 30 GeV window centered around the Z mass peak. Alternatively, a control region comprising events with $e^{\pm}\mu^{\mp}$ pairs was used as shown in Fig. 11.9(b). In the case of the W + jets background, the second lepton is a mismeasured jet, and so a certain fraction of these events have leptons with the same charge. Using a control region of such events in data, the normalisation of the W + jets background was checked. The background due to QCD multijet events was estimated from data and was found to be negligible.

The systematic uncertainties on these estimates could be attributed to several experimental sources. These include the selection and calibration of leptons and jets which also affect the $E_{\mathrm{T}}^{\mathrm{miss}}$ evaluation, and also the uncertainty on the efficiency of b-tagging selection used to veto b-jets in the analysis. Theoretical uncertainties on the signal and background yields were also taken into account. In the cases where the background normalisation uncertainty could be obtained by comparing simulation to certain control regions in data, additional systematic uncertainties related to detector effects were only taken into account if the behaviour in the control region was expected to deviate from the signal region.

4. Results

In both the $H \to ZZ \to \ell\ell qq$ and $H \to ZZ \to \ell\ell\nu\nu$ channels, the search for the SM Higgs boson was performed by exploiting a final discriminating variable whose shape is substantially different between signal and background. In the $H \to ZZ \to \ell\ell qq$ channel the natural choice was the m_{ZZ} variable, while in the $H \to ZZ \to \ell\ell\nu\nu$ channel the m_{T} variable was used. It is possible to use the m_{ZZ} and m_{T} variables in two possible ways.

One can apply selections on the m_{ZZ} and m_{T} variables to cut out a significant fraction of the background, and then count the number of events that are left to search for an excess due to the signal. However, such a "cut-and-count" approach has certain disadvantages. In the case of the $H \to ZZ \to \ell\ell\nu\nu$ channel, since m_{T} is not really the invariant mass of the system, and given that $E_{\mathrm{T}}^{\mathrm{miss}}$ has much poorer resolution compared to the leptons, the m_{T} distribution is typically quite broad. Thus, in order to cut away a significant fraction of the background the signal efficiency needs to be compromised. Another drawback of the cut-and-count approach is that the difference in the shape of the m_{T} or m_{ZZ} distributions between signal and background is not taken into account, and this leads to loss of performance. So a more suitable approach is to perform the search by fitting the shape of these distributions.

In the $H \rightarrow ZZ \rightarrow \ell\ell\nu\nu$ channel the binned shape of m_T was fitted to extract the signal. This can be viewed as performing a cut-and-count analysis for each of the m_T bins and then statistically combining the results of all the bins. Since very loose cuts were applied on m_T, more signal events could be retained in the analysis. Furthermore, the difference in m_T shapes is reflected in some bins having significantly higher signal-to-background ratio than the average. Thus, events in these bins contribute with a higher weight to the final result. The same approach was also used in the $H \rightarrow ZZ \rightarrow \ell\ell qq$ channel. The ATLAS analysis performed a binned fit of the m_{ZZ} variable while the CMS exeriment performed an unbinned fit using analytical functions to model the signal and background. The signal model in both analyses took into account next-to-next-to leading order and next-to-next-to leading log corrections by appropriately reweighting events generated using the next-to-leading order POWHEG Monte Carlo generator. This is relevant particularly for the $H \rightarrow ZZ \rightarrow \ell\ell\nu\nu$ analysis, since the m_T variable depends on the p_T of the Higgs boson. In the case of a high mass Higgs boson it is challenging to correctly describe the lineshape of the resonance.[16,17] As a result, a systematic uncertainty varying from 10% to 30% was applied to the signal yield for Higgs boson masses larger than 400 GeV.

No significant excess of events above the expected SM backgroud was observed in either of the two analyses. Hence, limits were set on the production cross section of the SM Higgs boson as a function of its mass, m_H, using the CL$_\text{s}$ modified frequentist formalism with the profile likelihood test statistic, as described in Appendix A.

4.1. *Results of the H → ZZ → ℓℓqq search with 7 TeV data*

The expected and observed upper limits at 95% CL for the CMS and ATLAS analyses in the $H \rightarrow ZZ \rightarrow \ell\ell qq$ channel are shown in Fig. 11.10. In this channel both analyses searched for the SM Higgs boson in the mass range of 200 GeV $\leq m_H \leq$ 600 GeV. The results of these analyses are tabulated in Table 11.1. The ATLAS analysis had an expected exclusion range of 351 GeV $\leq m_H \leq$ 404 GeV at 95% CL, while the observed exclusion was found to be in the range 300 GeV $\leq m_H \leq$ 322 GeV and 353 GeV $\leq m_H \leq$ 410 GeV at 95% CL. Here the contributions from the

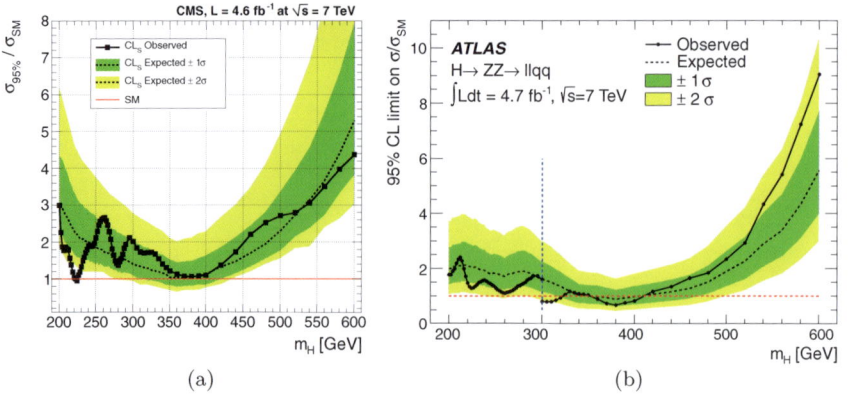

(a)   (b)

Fig. 11.10   The expected (dashed line) and observed (solid line) 95% CL upper limits
on the total cross section relative to the expected SM Higgs boson prediction for the
$H \rightarrow ZZ \rightarrow \ell\ell qq$ channel in (a) the CMS analysis and (b) the ATLAS analysis. The
expected exclusion is the range of Higgs boson masses for which the expected limit goes
below the value 1 (horizontal red line) which represents the SM expectation. Similarly,
the observed exclusion is the range of Higgs boson masses for which the observed limit
goes below the red line. The inner (green) and outer (yellow) bands indicate the one-
and two-sigma ranges in which the limit is expected to lie in the absence of a signal.
The discontinuity in the limit for (b), indicated by the vertical dotted line, is due to the
transition between the use of the low- and high-m_H selections.

Table 11.1   Expected and observed exclusion for the
SM Higgs boson in the $H \rightarrow ZZ \rightarrow \ell\ell qq$ search performed by ATLAS and CMS experiments.

	Expected exclusion (GeV)	Observed exclusion (GeV)
ATLAS	[351–404]	[300–322], [353–410]
CMS	—	—

tagged and untagged channels were on par with one another. The CMS
analysis was less sensitive in this channel and did not exclude the SM Higgs
boson at 95% CL for any mass hypothesis.

4.2. Results of the $H \rightarrow ZZ \rightarrow \ell\ell\nu\nu$ search with 7 TeV data

The $H \rightarrow ZZ \rightarrow \ell\ell\nu\nu$ channel was the most sensitive channel for Higgs
searches with 7 TeV data in the high mass range. The expected and observed
upper limits at 95% CL for the CMS and ATLAS analyses in this channel

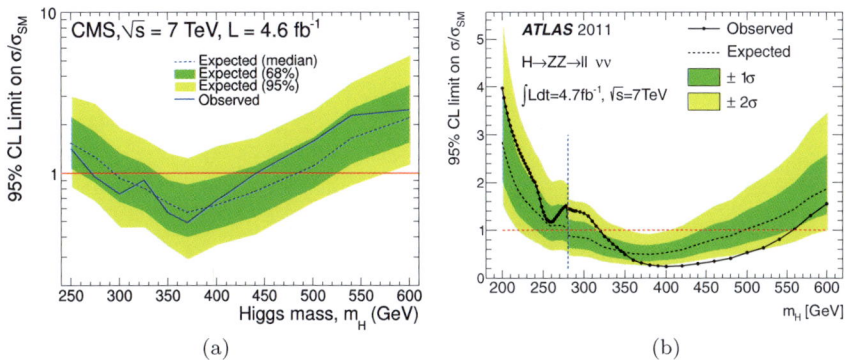

Fig. 11.11   The expected (dashed line) and observed (solid line) 95% CL upper limits on the total cross section relative to the expected SM Higgs boson prediction for the $H \to ZZ \to \ell\ell\nu\nu$ channel in (a) the CMS analysis and (b) the ATLAS analysis. The expected exclusion is the range of Higgs boson masses for which the expected limit goes below the value 1 (horizontal red line) which represents the SM expectation. Similarly, the observed exclusion is the range of Higgs boson masses for which the observed limit goes below the red line. The inner (green) and outer (yellow) bands indicate the one- and two-sigma ranges in which the limit is expected to lie in the absence of a signal. The discontinuity in the limit for (b), indicated by the vertical dotted line, is due to the transition between the use of the low- and high-m_H selections.

Table 11.2   Expected and observed exclusion for the SM Higgs boson in the $H \to ZZ \to \ell\ell qq$ search performed by ATLAS and CMS experiments.

	Expected exclusion (GeV)	Observed exclusion (GeV)
CMS	[290–490]	[270–440]
ATLAS	[280–497]	[319–518]

are shown in Fig. 11.11. The ATLAS search extended between m_H of 200 GeV to 600 GeV, while the CMS analysis restricted its search in the range of 250 GeV to 600 GeV. The results of these analyses are tabulated in Table 11.2. Both analyses had similar sensitivities in this mode. In the case of the ATLAS search the low pileup dataset was found to dominate the performance in the low mass region, while in the high mass region both the low pileup and high pileup datasets had similar sensitivity. The ATLAS search had an expected exclusion at 95% CL in the range 280 GeV $\leq m_H \leq$ 497 GeV, while the CMS experiment had an expected exclusion at 95% CL in the range 290 GeV $\leq m_H \leq$ 490 GeV. In the ATLAS analysis the observed exclusion was in the range 319 GeV $\leq m_H \leq$ 518 GeV at 95% CL, while in the CMS analysis the mass range of 270 GeV $\leq m_H \leq$ 440 GeV was excluded at 95% CL.

5. CMS updates with 8 TeV data

The analyses described so far were performed using a sample of 5 fb^{-1} of proton–proton collision data collected in 2011 at a centre-of-mass energy of 7 TeV. As of June 2015, the CMS experiment has updated these searches[18] by incorporating the 20 fb^{-1} of collision data collected in 2012 at a centre-of-mass energy of 8 TeV. This allowed the search range to be extended to Higgs boson masses up to 1 TeV. The analysis strategies were largely kept the same but some improvements were introduced. In the case of the $H \to ZZ \to \ell\ell qq$ search the result with 8 TeV data is presented while in the case of the $H \to ZZ \to \ell\ell\nu\nu$ search a combined result of 7 and 8 TeV data is presented.

In the $H \to ZZ \to \ell\ell\nu\nu$ search events were split into two categories. A vector-boson fusion (VBF) category was defined consisting of events with two or more jets with $p_T > 30$ GeV such that the two leading jets have a pseudorapidity gap larger than 4 and form an invariant mass greater than 500 GeV. Furthermore, the two leptons from the Z boson decay were required to lie between the two jets. In the VBF category, the m_T shape does not yield any significant improvement in performance owing to the low expected event yield. Therefore, the only selection imposed was $E_T^{\text{miss}} >$ 70 GeV. The remaining events were grouped into the gluon-fusion category which was further split into events with no jet activity and events with one or more jets. The m_T distributions in the gluon-fusion category and the E_T^{miss} distribution in the VBF category are shown in Fig. 11.12.

Similarly, a VBF category was also introduced in the $H \to ZZ \to \ell\ell qq$ search consisting of events with two additional jets with $p_T > 30$ GeV, having a pseudorapidity gap larger than 3.5, and forming an invariant dijet mass greater than 500 GeV. Furthermore, the analysis was modified for very high Higgs boson masses (beyond 600 GeV) to account for the merging of jets from the decays of Z bosons that are highly boosted. The $m_{\ell\ell jj}$ distributions for the three b-tag categories are shown in Fig. 11.13.

The exclusion limits with these updates at 95% CL are shown in Fig. 11.14. The results of the two searches are tabulated in Table 11.3. The $H \to ZZ \to \ell\ell qq$ search had an expected exclusion at 95% CL in the

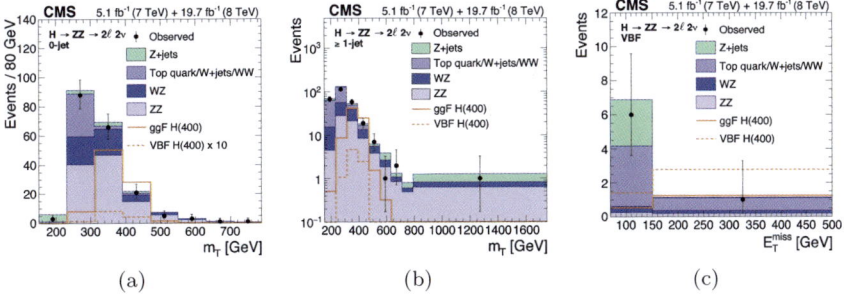

Fig. 11.12   The m_T and E_T^{miss} distributions for 400 GeV SM Higgs boson signal and background processes: (a) the m_T distribution in the gluon-fusion category for events with no jet activity: (b) the m_T distribution in the gluon-fusion category for events with one or more jets, and (c) the E_T^{miss} distribution in the VBF category. The distributions are shown by combining $5.1\,\mathrm{fb}^{-1}$ of 7 TeV collision data and $19.7\,\mathrm{fb}^{-1}$ of 8 TeV collision data.

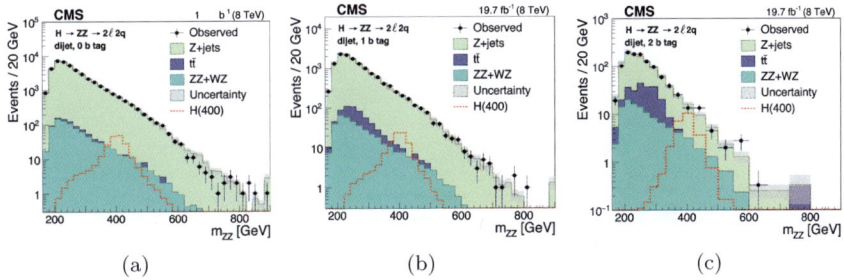

Fig. 11.13   The $m_{\ell\ell jj}$ distributions for 400 GeV SM Higgs boson signal and background processes in the (a) 0 b-tag, (b) 1 b-tag and (c) 2 b-tag categories. The distributions correspond to $19.7\,\mathrm{fb}^{-1}$ of 8 TeV collision data.

range $268\ \mathrm{GeV} \le m_H \le 660\ \mathrm{GeV}$, while the $H \to ZZ \to \ell\ell\nu\nu$ search had an expected exclusion at 95% CL in the range $254\ \mathrm{GeV} \le m_H \le 898\ \mathrm{GeV}$. The observed exclusion in the $H \to ZZ \to \ell\ell qq$ search was in the mass range of 305 GeV to 744 GeV at 95% CL, while in the $H \to ZZ \to \ell\ell\nu\nu$ search the mass range between 248 GeV to 930 GeV was excluded at 95% CL.

6. ATLAS updates with 8 TeV data

As this book was nearing completion, the ATLAS experiment also updated their searches in these channels using $20\ \mathrm{fb}^{-1}$ of 8 TeV data collected in

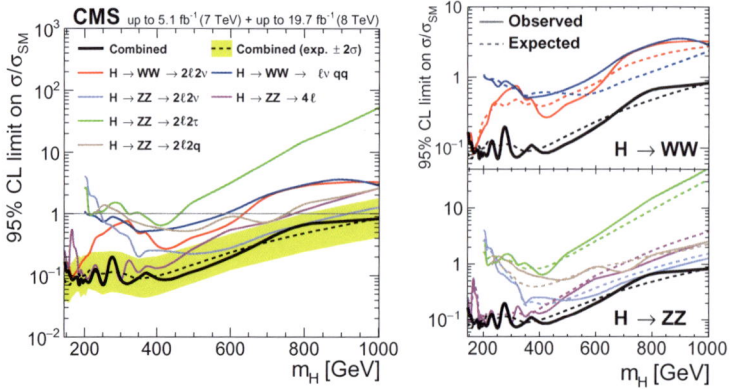

Fig. 11.14   The 95% CL upper limits obtained in the searches performed by the CMS experiment for the SM Higgs boson in the high mass region (145–1000 GeV) in several final states involving Higgs boson decays to a pair of Z or W bosons. The expected and observed limits for the $H \to ZZ \to \ell\ell qq$ search are denoted by the dotted and solid brown lines. Similarly, the expected and observed limits for the $H \to ZZ \to \ell\ell\nu\nu$ search are denoted by the dotted and solid grey lines.

Table 11.3   Expected and observed exclusion for the SM Higgs boson in the $H \to ZZ \to \ell\ell\nu\nu$ and $H \to ZZ \to \ell\ell qq$ search performed by the CMS experiment.

	Expected exclusion (GeV)	Observed exclusion (GeV)
$H \to ZZ \to \ell\ell qq$	[268–660]	[305–704]
$H \to ZZ \to \ell\ell\nu\nu$	[254–898]	[248–930]

2012.[19] Like in the CMS case, the analysis strategies largely remained the same but several improvements were introduced. The higher centre-of-mass energy allowed the search range to be extended up to Higgs boson masses of 1 TeV, targeting new additional Higgs bosons beyond the discovered SM Higgs boson. Due to this, the searches focused on Higgs boson having a narrow decay width, significantly smaller than the experimental resolution, since any additional Higgs boson will likely have a reduced width due to the presence of the observed low-mass Higgs boson.

Both channels were categorised into a gluon-fusion category, similar to that described above, and an additional vector-boson fusion category, which selected events with two forward jets having a high invariant mass and separated by a significant pseudorapidity gap. In addition, in the

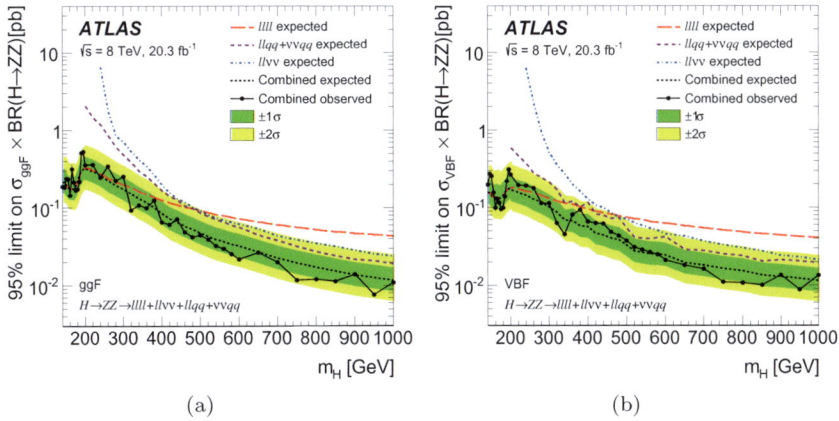

Fig. 11.15   The 95% CL upper limits on the cross section times branching ratio to a *Z* boson pair for a narrow-width Higgs boson with a mass in the range 140–1000 GeV from the ATLAS experiment. The expected (dashed black line) and observed (solid black line) results combine several *Z* boson decay channels and are shown for both the (a) gluon fusion and (b) vector-boson fusion categories. The expected contribution of the $H \to ZZ \to \ell\ell\nu\nu$ ($H \to ZZ \to \ell\ell qq$) channel is shown in the magenta dashed (blue dot-dashed) line. The $H \to ZZ \to \ell\ell qq$ result also includes a small additional contribution from the $H \to ZZ \to \nu\nu qq$ channel at the highest masses.

$H \to ZZ \to \ell\ell qq$ channel a further category was added to select very high m_H events, where the two jets from the highly-boosted Z boson decay merge together. The updated exclusion limits on the cross section times branching ratio to a Z boson pair at 95% CL for the two channels under the assumption of a narrow-width Higgs boson are shown in Fig. 11.15 for both the gluon-fusion and vector-boson fusion categories.

7. Summary and outlook

When the first run of the LHC started, the Higgs boson was undiscovered and its mass was unknown. Theoretical predictions allowed the mass of the SM Higgs boson up to the TeV scale. With just the first few fb^{-1} of integrated luminosity, the ATLAS and CMS experiments were able to exclude a heavy Higgs boson with mass larger than 200 GeV, thus paving the way for its eventual discovery at a mass of 125 GeV. The $H \to ZZ \to \ell\ell qq$ and $H \to ZZ \to \ell\ell\nu\nu$ channels played a crucial role in the exclusion of these masses. These channels still remain critical probes in searching for additional Higgs bosons which are predicted in theories beyond the Standard Model[20,21] in Run 2 of the LHC.

References

1. ATLAS Collaboration, Search for a Standard Model Higgs boson in the mass range 200–600-GeV in the $H \rightarrow ZZ \rightarrow \ell^+\ell^- q\bar{q}$ decay channel with the ATLAS detector, *Phys.Lett.* **B717**, 70–88 (2012). doi: 10.1016/j.physletb.2012.09.020.

2. ATLAS Collaboration, Search for a Standard Model Higgs boson in the $H \rightarrow ZZ \rightarrow \ell^+\ell^- \nu\bar{\nu}$ decay channel using 4.7 fb^{-1} of $\sqrt{s} = 7$ TeV data with the ATLAS detector, *Phys.Lett.* **B717**, 29–48 (2012). doi: 10.1016/j.physletb.2012.09.016.

3. CMS Collaboration, Search for a Higgs boson in the decay channel $H \rightarrow ZZ(*) \rightarrow q\bar{q}\ell^-\ell^+$ in pp collisions at $\sqrt{s} = 7$ TeV, *JHEP.* **1204**, 036 (2012). doi: 10.1007/JHEP04(2012)036.

4. CMS Collaboration, Search for the Standard Model Higgs boson in the $H \rightarrow ZZ \rightarrow 2\ell 2\nu$ channel in pp collisions at $\sqrt{s} = 7$ TeV, *JHEP.* **1203**, 040 (2012). doi: 10.1007/JHEP03(2012)040.

5. Y. Gao, A. V. Gritsan, Z. Guo, K. Melnikov, M. Schulze, *et al.*, Spin determination of single-produced resonances at hadron colliders, *Phys.Rev.* **D81**, 075022 (2010). doi: 10.1103/PhysRevD.81.075022.

6. J. Alwall, M. Herquet, F. Maltoni, O. Mattelaer, and T. Stelzer, MadGraph 5 : Going Beyond, *JHEP.* **1106**, 128 (2011). doi: 10.1007/JHEP06(2011)128.

7. T. Sjostrand, S. Mrenna, and P. Z. Skands, PYTHIA 6.4 Physics and Manual, *JHEP.* **0605**, 026 (2006). doi: 10.1088/1126-6708/2006/05/026.

8. S. Frixione, P. Nason, and C. Oleari, Matching NLO QCD computations with parton shower simulations: The POWHEG method, *JHEP.* **0711**, 070 (2007). doi: 10.1088/1126-6708/2007/11/070.

9. S. Alioli, P. Nason, C. Oleari, and E. Re, A general framework for implementing NLO calculations in shower Monte Carlo programs: The POWHEG BOX, *JHEP.* **1006**, 043 (2010). doi: 10.1007/JHEP06(2010)043.

10. S. Alioli, P. Nason, C. Oleari, and E. Re, NLO Higgs boson production via gluon fusion matched with shower in POWHEG, *JHEP.* **0904**, 002 (2009). doi: 10.1088/1126-6708/2009/04/002.

11. P. Nason and C. Oleari, NLO Higgs boson production via vector-boson fusion matched with shower in POWHEG, *JHEP.* **1002**, 037 (2010). doi: 10.1007/JHEP02(2010)037.

12. M. L. Mangano, M. Moretti, F. Piccinini, R. Pittau, and A. D. Polosa, ALPGEN, a generator for hard multiparton processes in hadronic collisions, *JHEP.* **0307**, 001 (2003). doi: 10.1088/1126-6708/2003/07/001.

13. T. Gleisberg, S. Hoeche, F. Krauss, M. Schonherr, S. Schumann, *et al.*, Event generation with SHERPA 1.1, *JHEP.* **0902**, 007 (2009). doi: 10.1088/1126-6708/2009/02/007.

14. R. Frederix and S. Frixione, Merging meets matching in MC@NLO, *JHEP.* **1212**, 061 (2012). doi: 10.1007/JHEP12(2012)061.

15. G. Corcella, I. Knowles, G. Marchesini, S. Moretti, K. Odagiri, *et al.*, HERWIG 6: An event generator for hadron emission reactions with interfering gluons (including supersymmetric processes), *JHEP.* **0101**, 010 (2001). doi: 10.1088/1126-6708/2001/01/010.

16. G. Passarino, C. Sturm, and S. Uccirati, Higgs pseudo-observables, second Riemann sheet and all that, *Nucl.Phys.* **B834**, 77–15 (2010). doi: 10.1016/j. nuclphysb.2010.03.013.

17. S. Goria, G. Passarino, and D. Rosco, The Higgs boson lineshape, *Nucl.Phys.* **B864**, 530–579 (2012). doi: 10.1016/j.nuclphysb.2012.07.006.

18. CMS Collaboration, Search for a Higgs boson in the mass range from 145 to 1000 GeV decaying to a pair of W or Z bosons (2015). arXiv:1504.00936.

19. ATLAS Collaboration, Search for an additional, heavy Higgs boson in the $H \to ZZ$ decay channel at $\sqrt{s} = 8$ TeV in pp collision data with the ATLAS detector (2015). arXiv:1507.05930.

20. G. Branco, P. Ferreira, L. Lavoura, M. Rebelo, M. Sher, *et al.*, Theory and phenomenology of two-Higgs-doublet models, *Phys.Rept.* **516**, 1–102 (2012). doi: 10.1016/j.physrep.2012.02.002.

21. A. Hill and J. van der Bij, Strongly interacting singlet–doublet Higgs model, *Phys.Rev.* **D36**, 3463–3473 (1987). doi: 10.1103/PhysRevD.36.3463.

Chapter 12

Higgs combination and properties of the Higgs boson

Michael Duehrssen and Giovanni Petrucciani

CERN

CH-1211 Geneva 23, Switzerland

The combination of different Higgs boson searches was instrumental for the discovery of the Higgs boson in summer 2012. After the discovery the ATLAS and CMS collaborations have used the LHC Run 1 data to determine the properties of the Higgs boson. The mass of the Higgs boson is $125.09 \pm 0.21(\text{stat.}) \pm 0.11(\text{syst.})$ GeV. Measurements of kinematic properties of Higgs boson production and decay as well as measurements of signal yields and the coupling strengths to other particles show very good consistency with the predictions of the Standard Model.

1. Introduction

The previous chapters describe the searches for the Higgs boson and measurements of its properties in the various Higgs boson decay modes accessible at the LHC. These searches and measurements are in most cases already combinations of several exclusive final states. The first combinations of different Higgs boson decay modes were performed in order to reach the best sensitivity for the discovery of the Higgs boson itself.[1,2] The main goal of the post-discovery combinations was to compile a coherent picture of the properties of the Higgs boson from all individual measurements and compare them with the predictions for the Standard Model (SM) Higgs boson. After the discovery, the most fundamental measurement was the determination of the mass of the Higgs boson, as this was the last unknown parameter in the SM. Once the mass is measured, the SM is completely predictive and all other measurements can be used to evaluate the consistency of observations in the Higgs sector with the SM (see Chapter 1, Sec. 2). Within the context of Higgs boson measurements, the consistency of the observations with the SM were tested by the ATLAS and CMS experiments in three almost independent sectors:

- The kinematic properties of Higgs boson production.
- The kinematic properties of Higgs boson decays in the $H \to \gamma\gamma$, $H \to WW^{(*)}$ and $H \to ZZ^{(*)}$ channels, which are used to establish the spin and CP nature of the Higgs boson.
- Yields of signal events from different Higgs boson production and decay modes, which are then used to infer the Higgs boson coupling strength to the bosons W, Z, γ, and g, and the fermions t, b, and τ.

2. Overview of analyses included in the Higgs combination

The combination of Higgs boson searches and measurements is based on the Higgs boson analyses described in Chapters 5–9. Depending on the particulars of a given combination, different channels and analyses are used. Table 12.1 gives a schematic overview of which Higgs boson production and decay modes enter in the various combinations.

The Higgs boson searches that contributed most to the discovery of the Higgs boson[1,2] were dominated by the gluon fusion production mode in the $H \to \gamma\gamma$, $H \to ZZ^{(*)} \to llll$ and $H \to WW^{(*)} \to l\nu l\nu$ channels. Dedicated searches for other production and decay modes, as listed in Table 12.1, were included as well and helped to improve the overall search sensitivity of the observations. The discovery of the Higgs boson was based on a limited Run 1 dataset corresponding to integrated luminosities of about 5 fb^{-1} collected at $\sqrt{s} = 7$ TeV and about 5 fb^{-1} collected at $\sqrt{s} = 8$ TeV for each of the two experiments. The $H \to ZZ$ and $H \to WW$ decay modes, including the final states described in Chapters 10 and 11, were the only ones used to exclude a high-mass Higgs boson.

Following the discovery, the final Run 1 measurements of both experiments are based on the full Run 1 dataset of about 5 fb^{-1} collected at $\sqrt{s} = 7$ TeV and about 20 fb^{-1} collected at $\sqrt{s} = 8$ TeV for each of the two experiments (some measurements are based only on the dominant $\sqrt{s} = 8$ TeV dataset). The combined measurement of the Higgs boson mass is based on the two channels with a narrow peak in the mass distributions: $H \to \gamma\gamma$ and $H \to ZZ^{(*)} \to llll$. Fiducial differential cross section measurements, which provide information on the kinematic production properties, were performed in the $H \to \gamma\gamma$ and $H \to ZZ^{(*)} \to llll$ channels. The spin-parity hypothesis tests make use of kinematic decay properties in the

Table 12.1 Overview of which different Higgs boson production and decay modes enter into the different Higgs combinations. The red triangles labeled with "5σ" indicate the channels for which dedicated searches contributed to the discovery of the Higgs boson. The yellow triangles labeled with "m_H" indicate which channels contribute to the mass measurement. The green triangles labeled with "J" and "J^{CP}" indicate which channels contribute to the spin and CP hypothesis tests using kinematic decay properties. The blue triangles labeled with "$\frac{d\sigma}{dx}$" indicate which channels contribute to the differential measurements of kinematic production properties. The black tick marks indicate which channels enter into signal and coupling strength measurements.

	$H \to ZZ^{(*)}$ Chapter 5	$H \to \gamma\gamma$ Chapter 6	$H \to WW^{(*)}$ Chapter 7	$H \to \tau\tau$ Chapter 8	$H \to b\bar{b}$ Chapter 9
$gg \to H$	5σ / m_H / J^{CP} / $\frac{d\sigma}{dx}$ ✓	5σ / m_H / J / $\frac{d\sigma}{dx}$ ✓	5σ / J^{CP} ✓	5σ ✓	
VBF	m_H / $\frac{d\sigma}{dx}$ ✓	5σ / m_H / $\frac{d\sigma}{dx}$ ✓	5σ ✓	5σ ✓	
$WH+ZH$	✓	✓	✓	✓ (only CMS)	5σ ✓
$t\bar{t}H$		✓	✓ (see text)	✓ (see text)	✓

$H \to \gamma\gamma$, $H \to ZZ^{(*)} \to llll$ and $H \to WW^{(*)} \to l\nu l\nu$ channels. Finally, the combined measurements of signal and coupling strengths make use of all analyses of the 125 GeV Higgs boson described in the previous chapters, which cover all combinations of Higgs boson production and decay modes with some sensitivity for a SM Higgs boson in Run 1.

One analysis not discussed in the previous chapters is the search for $t\bar{t}H$ production in leptonic final states. In both experiments the analysis selects events with either 2 leptons of the same charge or 3 leptons or 4 leptons, in all cases associated with several jets of which at least one is required to be b-tagged. ATLAS also searched for final states with hadronic decays of the τ lepton. The expected Higgs boson signal is dominated by $H \to WW^{(*)}$ and $H \to \tau\tau$ decays with subsequent leptonic W and τ decays. In principle $H \to ZZ^{(*)}$ decays contribute as well, but their contribution is close to negligible with the LHC Run 1 data. The main background is $t\bar{t}$ and $t\bar{t}V(V)$ production, where V is either W, Z or γ. The details of the analysis can be found in Refs. [3,4]. The observed signal strength μ, defined as $\mu = (\sigma \cdot \mathcal{B})/(\sigma_{\mathrm{SM}} \cdot \mathcal{B}_{\mathrm{SM}})$, is $\mu = 2.1^{+1.4}_{-1.2}$ for ATLAS and $\mu = 3.9^{+1.7}_{-1.4}$ for CMS.

A number of searches for rare or invisible Higgs boson decay modes, which are not discussed in this book, are helpful to complete the picture of the properties of the Higgs boson. The 95% CL upper limits on the signal strength μ in the rare Higgs boson decay modes $H \to \mu\mu$,[5,6] $H \to ee$[6] and $H \to Z\gamma$[7,8] are listed in Table 12.2. The fact that no signal is seen in the clean $H \to \mu\mu$ and $H \to ee$ final states, while a signal is seen in $H \to \tau\tau$ (see Chapter 8), shows that the Higgs boson couplings to fermions from different generations are not universal as is the case for all gauge interactions mediated by photons, gluons, W or Z bosons. For example, should the Higgs boson couple to muons and electrons with the same strength as to taus, one would expect to observe a signal strength of $\mu(H \to \mu\mu) \sim 280$ and $\mu(H \to ee) \gg 10^6$, respectively, which are both strongly excluded by the measurements.

For the searches for Higgs boson decays to invisible particles the limits on $\sigma \cdot \mathcal{B}(H \to \mathrm{inv})/\sigma_{\mathrm{SM}}$ are shown in Table 12.2 for the combination of the ZH and VBF production modes, in ATLAS[9] and CMS.[10] No indication of invisible decays were found with the LHC Run 1 data. Within the SM the branching ratio to invisible particles, through the $H \to ZZ^{(*)} \to 4\nu$ decay, is only $\sim 0.1\%$ and hence no sensitivity to the SM is expected.

3. Systematic uncertainties and correlations

When combining the different searches or measurements into a single analysis, it is important to correctly account for the correlations of the systematical uncertainties across the different analyses.

In order to account for these correlations, systematic uncertainties on signals and backgrounds are factorized into individual physics sources, which can then be taken to be mutually independent. Then, the impact of each source of uncertainty on all signals and backgrounds can be evaluated, and accounted for assuming that the effect is correlated across all the channels where relevant. It should be noted that the correlation does not

Table 12.2   Observed and expected 95% CL limits on the signal strength $\mu = (\sigma \cdot \mathcal{B})/(\sigma_{\mathrm{SM}} \cdot \mathcal{B}_{\mathrm{SM}})$ for the Higgs boson decay modes $H \to \mu\mu$, $H \to ee$ and $H \to Z\gamma$ and on $\sigma \cdot \mathcal{B}(H \to \mathrm{inv})/\sigma_{\mathrm{SM}}$ for a decay of the Higgs boson into an invisible final state. In all cases, σ represents the production cross section and $\mathcal{B}$ the branching ratio into the respective final state. All results are given for an approximate Higgs boson mass of $m_H \sim 125$ GeV.

	ATLAS		CMS	
	Obs.	Exp.	Obs.	Exp.
$H \to \mu\mu$	7.0	7.2	7.4	6.5
$H \to ee$	—	—	$\sim 3.7 \cdot 10^5$	$\sim 4.7 \cdot 10^5$
$H \to Z\gamma$	11	9	9.5	10
$H \to \mathrm{inv}$	0.25	0.27	0.58	0.44

imply that the size of the corresponding uncertainty in different channels is the same.

Experimental uncertainties correlated across multiple final states include the knowledge of the integrated luminosity, affecting the predictions for the yields of the signal and of any background estimated from simulations, and systematical uncertainties on the reconstruction efficiencies, energy scales and resolutions for the different physics objects. As the uncertainties are defined at the level of the individual objects, their correlated effects across all analyses can be evaluated directly, and accounted for in the combination.

Theoretical uncertainties from unknown higher orders in the perturbative expansion of the cross section for each individual physical process are assumed to be correlated across the different decay modes, since the uncertainties mainly affect the overall normalization of the cross section and not the acceptance for each analysis. For Higgs boson production via gluon fusion, theoretical uncertainties on the predicted multiplicity of additional hadronic jets (see Chapter 1, Sec. 7) also have an important role, as they affect the amount of gluon fusion contamination in the analyses aimed at measuring VBF production. Such uncertainties are assumed to be fully correlated across the different decay modes, neglecting the small differences in the jet selections used in each analysis. No correlation is assumed between different Higgs boson production processes.

The correlation of the uncertainties from the knowledge of the parton density functions (PDF) across different physics processes is in principle fully known, but technically challenging to implement, especially for the acceptance-related uncertainties. The full covariance matrix of the inclusive

PDF uncertainties between different physics processes has been used[11] to check the relevance of these correlations and no sizeable impact was found, as only the PDF uncertainty on the $gg \rightarrow H$ production gives a substantial contribution to current uncertainties in Higgs measurements, while other PDF uncertainties are small with respect to statistical and systematic uncertainties. Hence, for practical considerations, these correlations are either neglected (ATLAS) or modelled more simply assuming processes dominated by the same partonic initial state to be either totally correlated or anti-correlated, and the others to be independent (CMS). A more consistent treatment of these correlations may be necessary for future measurements, as for example the PDF uncertainties for the $gg \rightarrow H$ production cross section, which probes the gluon density mainly at small x values, is mostly anti-correlated with the VBF production cross section, which probes the quark densities at higher x.

4. Combined searches for a SM Higgs boson

In order to achieve the best sensitivity for the discovery or exclusion of a SM Higgs boson, the results of the searches in the different final states are combined into a single analysis. In this context, the theoretical predictions for the SM Higgs boson are used to relate the expected signal event yields in the various final states. The combination of multiple final states also allows one to benefit from their complementary strengths, e.g the good mass resolution from $H \rightarrow \gamma\gamma$ and $H \rightarrow ZZ^{(*)} \rightarrow llll$ and the good sensitivity to the signal yield from $H \rightarrow WW^{(*)} \rightarrow l\nu l\nu$.

The results of the searches for a SM Higgs boson are presented in two complementary ways: exclusion limits and significances of any observed excess above the background-only expectations. While the searches target a SM Higgs boson, the results are interpreted and presented in a slightly more general model, allowing the expected signal event yield to deviate from the SM predictions. An overall signal strength modifier $\mu = \sigma/\sigma_{SM}$ is introduced. It rescales linearly the expected signal yields in all final states, so that $\mu = 1$ corresponds to the predictions for a SM Higgs boson, and $\mu = 0$ corresponds to the absence of a Higgs boson signal. Exclusion limits are then presented as upper limits on μ, and significances are computed by testing and comparing the two hypotheses $\mu = 0$ and $\mu > 0$. In either case, the search is performed by testing each hypothesized value of the Higgs boson mass separately on a search grid finer than the observable width of a

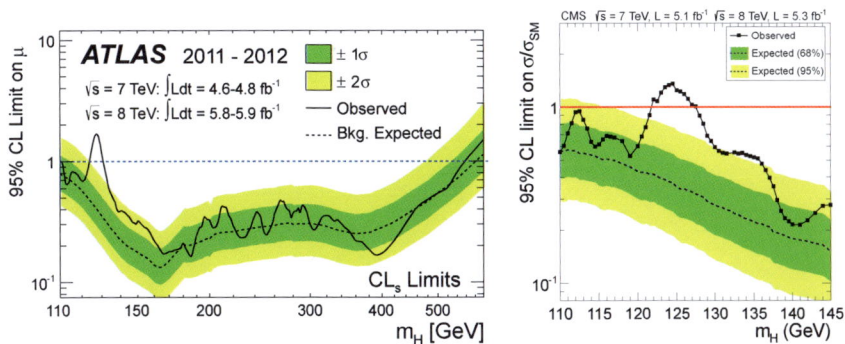

Fig. 12.1 95% CL exclusion limits on $\mu = \sigma/\sigma_{SM}$ from ATLAS (left) and CMS (right), with the dataset corresponding to the Higgs boson discovery. A SM Higgs boson is excluded at all masses m_H for which the observed upper limit on μ is below unity.

possible signal. The statistical methodology used in the search is described in Appendix A.

Results of the SM Higgs boson searches presented in this section are mainly based on the dataset available at the time of the Higgs boson discovery, corresponding to integrated luminosities of about 5 fb^{-1} collected at $\sqrt{s} = 7$ TeV and about 5 fb^{-1} collected at $\sqrt{s} = 8$ TeV for each of the two experiments.

The 95% confidence level upper limits on the signal strength modifier μ for the ATLAS and CMS searches on the discovery dataset[1,2] are shown in Fig. 12.1. The limits are computed using the CL$_s$ prescription. Both experiments excluded a SM Higgs boson ($\mu = 1$) in a wide mass range, 111–559 GeV for ATLAS and 110–710 GeV for CMS, with the exception of a narrow window at a low Higgs boson mass where an excess of events compared to the background-only expectations was observed.

The statistical significance of the observed excess in the low mass range is quantified in terms of p-values for the background-only hypothesis, as defined in Appendix A, Sec. A.3. Figure 12.2 shows the p-values for the ATLAS and CMS combinations and for the individual analyses contributing to them. In both experiments, a strong excess is observed near a mass of 125 GeV, consistently across the three most sensitive channels $H \to \gamma\gamma$, $H \to ZZ^{(*)} \to llll$, $H \to WW^{(*)} \to l\nu l\nu$. Moreover the excess appears consistently in data collected at $\sqrt{s} = 7$ TeV and at $\sqrt{s} = 8$ TeV.

The observed statistical significance of the excess was $5.9\,\sigma$ for ATLAS and $5.0\,\sigma$ for CMS, compatible with the expectations for a SM Higgs boson

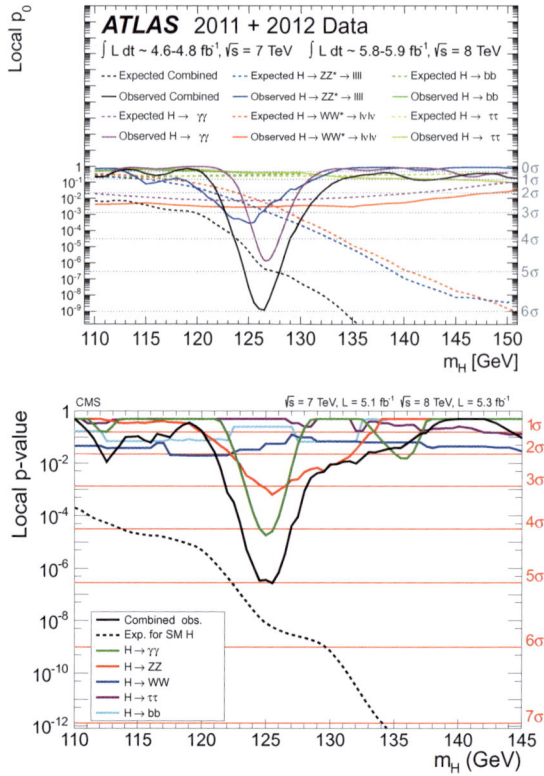

Fig. 12.2 Local p-values observed by ATLAS (top) and CMS (bottom), in the low mass range, with the discovery dataset. The observed results are displayed as solid lines, both for the combination of all channels and for the contributions from individual decay modes. Dotted lines indicate at each m_H value the expected p_0 for a SM Higgs boson of that mass. Narrow dips are observed, as expected, in the individual results for the two decay modes with excellent mass resolution, $\gamma\gamma$ and ZZ. The searches in the WW decay mode also exhibit 2–3σ excesses, but the corresponding p-value is approximately flat, as the analyses do not have a good discrimination between different m_H hypotheses due to the relatively poor Higgs boson mass resolution.

with a mass of about 125 GeV, 4.9 σ and 5.8 σ for ATLAS and CMS respectively. This observation allowed both experiments to claim the discovery of a new heavy state, of mass near 125 GeV.

5. Higgs boson mass measurement

In the $H \to \gamma\gamma$ and $H \to ZZ^{(*)} \to llll$ channels, the momenta of all final state decay products is reconstructed with excellent experimental accuracy by ATLAS and CMS. Under the assumption that the natural width of the new particle is negligible compared to the experimental resolution, true both for the SM Higgs boson and most alternative models proposed in the literature, the mass of the boson can then be directly measured without theoretical ambiguities.

Measurements of the Higgs boson mass have been performed by each experiment separately,[12,13] and then combined.[14] In all cases, the measurements use the full LHC Run 1 dataset.

5.1. *Compatibility of the mass measurement from individual channels*

In order to make the measurement independent of any assumption on the production cross section and decay branching ratio of the boson, the mass measurement is first performed considering each decay mode independently and without constraining the overall signal yield to the SM Higgs boson prediction.

In the analysis of the $H \rightarrow \gamma\gamma$ final state, some dependency of the result on the Higgs boson production mechanism could arise from the event categorization, which relies on the diphoton p_T and, in the CMS analysis, also experimental signatures of the different production modes: events contribute to the mass measurement with weights dependent on the expected purities of the category they belong to, and thus the combined result depends on how the signal yield is assumed to be distributed across the different categories. To reduce this potential model dependency, the analysis is performed leaving unconstrained the signal strengths for the production modes related to fermions (gluon fusion, $t\bar{t}H$) and those related to vector bosons (VBF, VH).

The measurements in the $H \rightarrow ZZ^{(*)} \rightarrow llll$ final state rely on dedicated analyses that are by design less sensitive to the production mechanism, and thus in the analyses it is sufficient to have a single unconstrained signal strength for all Higgs boson production modes.

To illustrate the compatibility of the different measurements, confidence regions in the (m_H, μ) plane are constructed for each final state separately, as shown in Fig. 12.3. For the $H \rightarrow \gamma\gamma$ final state, the analysis would naturally define a confidence region in the space of the three parameters used (m_H, $\mu_{\mathrm{ggH},t\bar{t}H}$, $\mu_{\mathrm{VBF,VH}}$), but for the purpose of this illustration and to allow a comparison with the $H \rightarrow ZZ^{(*)} \rightarrow llll$ analysis the data is re-analysed assuming a single signal strength across all production modes in the $H \rightarrow \gamma\gamma$ analysis ($\mu = \mu_{\mathrm{ggH},t\bar{t}H} = \mu_{\mathrm{VBF,VH}}$).

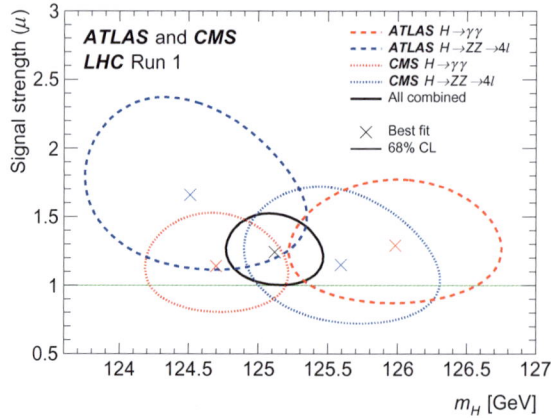

Fig. 12.3   Confidence regions in the (m_H, μ) plane for $\gamma\gamma$ and 4ℓ final states in the two experiments; the solid and dashed lines show the boundaries of the 68% and 95% CL regions, respectively.

Differences are visible between the measurements in the two final states for both the experiments. The statistical compatibility of the results in the two final states has been evaluated quantitatively by performing a fit to the data allowing for a mass difference $\Delta m_H = m_{\gamma\gamma} - m_{4\ell}$, while allowing different signal strengths for each final state: the fitted Δm_H values are found to be compatible with zero at the $1.6\,\sigma$ level for CMS and $2.0\,\sigma$ level for ATLAS.

5.2. *Combined results*

ATLAS and CMS results are combined assuming each of the three signal strengths to be the same in the two experiments, since they are normalized to the same hypothesis of a SM Higgs boson, and thus depend only on the physics of the Higgs boson production, and not on experimental aspects like event selection efficiencies. The results for the individual mass measurements and their combinations are summarized in Fig. 12.4. The combined result is $125.09 \pm 0.21(\text{stat.}) \pm 0.11(\text{syst.})$ GeV, where the systematic uncertainty is dominated by the experimental uncertainties on the energy and momentum scale and resolution, uncorrelated between the two experiments. The impact of the correlated systematic uncertainties of theoretical origin is quite small, about 0.01 GeV.

A very good agreement is observed between the combined results from ATLAS and CMS, and similarly for the LHC combinations of the two final

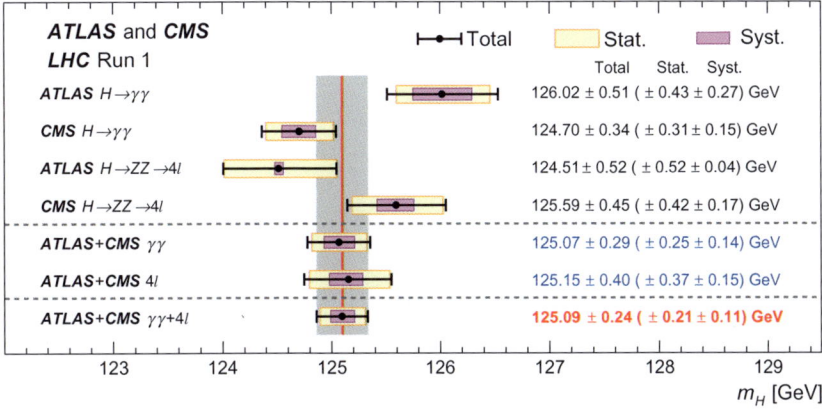

Fig. 12.4   Summary of individual and combined mass measurement results from ATLAS and CMS.

states separately. This is assessed quantitatively by performing alternative fits allowing a mass shift between the two final states ($\Delta m_H = m_{\gamma\gamma} - m_{4\ell}$) or between the two experiments ($\Delta m_H^{\text{exp}} = m_H^{\text{ATLAS}} - m_H^{\text{CMS}}$). Both results are compatible with zero within uncertainties: $\Delta m_H = -0.1 \pm 0.5$ GeV and $\Delta m_H^{\text{exp}} = 0.4 \pm 0.5$ GeV.

As discussed in Chapter 1, Sec. 2, the SM predicts all properties of Higgs boson production and decay with the last unknown parameter of the SM determined by experimental measurements: the mass m_H of the Higgs boson itself. In the following sections the comparison of the experimental measurements of the Higgs boson production and decay properties with those from the SM predictions are discussed.

6.   Differential production cross section measurements

The SM predictions for the kinematic properties of Higgs boson production are discussed in Chapter 1, Sec. 7. Corresponding differential measurements of the production properties are obtained from the $H \to \gamma\gamma$[15,16] and $H \to ZZ^{(*)} \to llll$[17,18] channels, as discussed in Chapter 5, Sec. 10 and Chapter 6, Sec. 6.5, respectively. Both channels allow a precise reconstruction of the 4-momentum of the Higgs boson candidate in each event and hence measurements of the differential distribution of the kinematic observables in Higgs boson production.

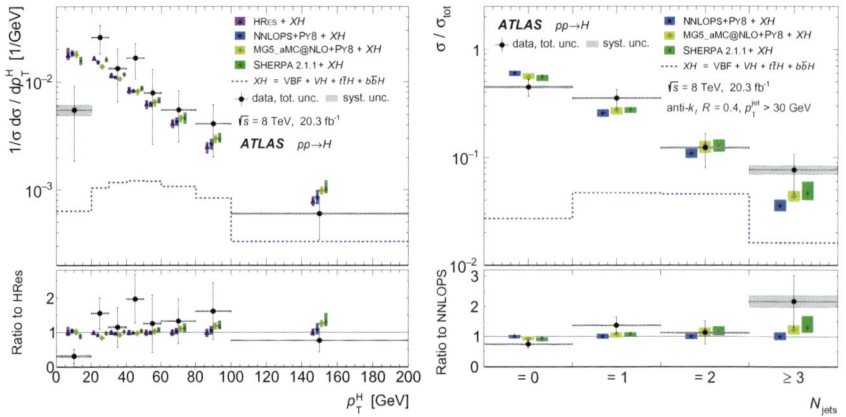

Fig. 12.5 Comparison of the predictions of several SM theory calculations (see Chapter 1, Secs. 7 and 8) to the measured normalized distributions of the Higgs boson transverse-momentum p_T^H (left) and the number of jets N_{jets} associated with the Higgs boson production (right). HRES[20] and NNLOPS[21] currently provide the best SM prediction for p_T^H and N_{jets}, respectively.

The combination of both channels[19] performed by the ATLAS experiment for several Higgs boson production-related quantities significantly improves the sensitivity compared to the individual measurements, as the two channels have very similar statistical power (the corresponding CMS combination results were not available at the time of writing of this book). For example, the combined measurements of the Higgs boson transverse momentum $p_T(H)$ and the number of jets N_{jets} associated with the Higgs boson production are shown in Fig. 12.5 together with predictions from the best theory calculations available (see Chapter 1, Secs. 7 and 8). The $H \rightarrow \gamma\gamma$ and $H \rightarrow ZZ^{(*)} \rightarrow llll$ measurements are each normalized to the observed total Higgs boson cross section to make them independent of the inclusive signal yield and independent of assumptions on the ratio between the branching fractions to $H \rightarrow \gamma\gamma$ and $H \rightarrow ZZ^{(*)}$. This ensures that conclusions drawn from these measurements are valid even if deviations from the SM in the Higgs boson signal and coupling strength are found.

Within the SM, the $gg \rightarrow H$ production mode dominates and these measurements are especially sensitive to the intricate QCD aspects of this process. In addition, the VBF and VH production modes are expected to give a sizeable contribution to the high p_T and high jet multiplicity part of the phase space, where dedicated measurements for VBF-related observables were done in the $H \rightarrow \gamma\gamma$ analysis.[15]

The overall agreement of the observed distributions with the SM predictions is within the current uncertainties from both the measurements and the theory. The most noticable difference with respect to the SM prediction is in the lowest bin of the p_T^H distribution. Besides a possibility of a statistical fluctuation, unaccounted-for QCD effects in the theory calculations could contribute to this deviation. Modifications caused by beyond SM (BSM) physics contributing to Higgs boson production are expected to appear foremost in the high p_T and high jet multiplicity part of the phase space, where, however, the agreement with the SM predictions is good.

7. Spin-parity properties

Deviations from the SM predictions in kinematic decay distributions are analyzed with an effective Lagrangian approach introduced in Chapter 1, Sec. 9 and Chapter 5, Sec. 11, which augments the SM contributions with leading new physics contributions generated by heavy unknown particles. Analyses are performed in the $H \to ZZ^{(*)} \to llll$ and $H \to WW^{(*)} \to l\nu l\nu$ channels only as not all Higgs boson decay modes are suitable for measurements of kinematic distributions in the decay. For a scalar particle, like the Higgs boson, no kinematic correlation exists between production and decay. Furthermore, the particles Y in a two-body Higgs boson decay $H \to YY$ are back-to-back in the rest frame of the Higgs boson and hence no information on the properties of the Higgs boson can be obtained from angular correlations between these two direct decay products. On the other hand, kinematic correlations between secondary decay products can be used to measure Higgs boson properties, which singles out the $H \to ZZ^{(*)} \to llll$ and $H \to WW^{(*)} \to l\nu l\nu$ channels. Invariant masses and angular correlations between the final state leptons are used by ATLAS and CMS in both channels to obtain information on possible deviations from the SM behaviour from the LHC Run 1 dataset as discussed in Chapter 5, Sec. 11 and Chapter 7, Sec. 6.3.

Similar analysis techniques can also be used to obtain information on the spin of the new boson in these two channels. In the case of a Higgs boson imposter with spin 1 or higher, angular correlations between the production and the decay of the new boson can also be exploited, as in the $H \to \gamma\gamma$ channel discussed in Chapter 6, Sec. 6.4 or the Tevatron analysis of the $VH, H \to b\bar{b}$ channel discussed in Chapter 3, Sec. 9.2.2. As a spin 1 or higher spin particle cannot take the role of the Higgs boson in the theory of electroweak symmetry breaking, it will be referred to as a "new boson" in the following.

With regards to kinematic properties in decays the overall analysis goal of ATLAS and CMS discussed in the next subsections is to exclude several non-SM hypotheses of spin J and parity P and then measure, for a spin zero ($J = 0$) Higgs boson, deviations from the SM behaviour in $H \to VV$ decays within an effective Lagrangian approach.

7.1. *Constraints on the spin of the new boson*

The SM predicts a spin zero Higgs boson with quantum numbers $J^{CP} = 0^{++}$. Kinematic decay properties in the $H \to \gamma\gamma$, $H \to ZZ^{(*)} \to llll$ and $H \to WW^{(*)} \to l\nu l\nu$ channels can be used to set constraints on both the spin J as well as the parity P. This also determines J^{CP} if either C or CP is conserved (see Chapter 1, Sec. 9).

In order to be less dependent on the implementations of concrete BSM models in the exclusion of J^P states, the overall observed signal yields are not used for the J^P hypothesis tests. For example, spin 2 models with universal couplings predict branching ratios that are not compatible with the observed signal yields in different final states. This approach also ensures that conclusions drawn from these exclusions are valid when deviations from the SM in the observed signal and coupling strength are considered.

Hypothesis tests for the $J^P = 0^-$, 1^+ and 1^- states are performed in the $H \to ZZ^{(*)} \to llll$ and $H \to WW^{(*)} \to l\nu l\nu$ channels. The $H \to \gamma\gamma$ decay has no discriminating sensitivity between various spin-zero Lagrangian terms (Eq. (1.22)) as discussed above and the observation of the $H \to \gamma\gamma$ final state already excludes a single spin 1 boson due to the Landau–Yang theorem[22,23] (see Chapter 1, Sec. 9 for a detailed explanation). Hypothesis tests for a new boson with spin 2 are performed for a large number of spin 2 models in the $H \to ZZ^{(*)} \to llll$, $H \to WW^{(*)} \to l\nu l\nu$ and $H \to \gamma\gamma$ channels. For the combined spin hypothesis tests,[24–26] a well motivated spin 2 model, corresponding to a massive graviton-like particle (2^+_m) as introduced in Chapter 5, Sec. 11.1.3, is discussed here as an example. States with a spin higher than two were not considered by either experiment, and are even more challenging than spin 2 for a fundamental particle from a theoretical view point. Table 12.3 shows that the J^P states discussed here are excluded at >99% CL. This also applies to all other models considered in Refs. [25,26]. A model-independent test of the spin 2 nature may eventually be possible with more data.

Table 12.3 Observed CL_s values from ATLAS[24,25] and CMS[26] for the exclusion of different J^P hypotheses tested against the SM Higgs boson hypothesis of $J^P = 0^+$. The largest value of the CL_s and hence the more conservative limit is given in cases where different production modes were tested.

J^P	ATLAS	CMS
0^-	<0.026%	<0.01%
1^+	0.03%	0.004%
1^-	0.27%	<0.001%
2^+_m	0.011%	0.13%

7.2. *Measurements of the decay properties for a spin zero Higgs boson*

Following the exclusion of many spin 1 and spin 2 states, both experiments focused on the detailed measurement of decay properties for a spin zero Higgs boson in the $H \to ZZ^{(*)} \to llll$ and $H \to WW^{(*)} \to l\nu l\nu$ channels. This also covers the case of CP-mixing for a spin zero Higgs boson, which has not been discussed so far.

The experimental analyses in the $H \to ZZ^{(*)} \to llll$ and $H \to WW^{(*)} \to l\nu l\nu$ channels are performed in the context of the generic Lagrangian given by Eq. (1.22) as discussed in Chapter 1, Sec. 9. Separating the HWW and HZZ interactions and putting the SM value of 2 for κ_W explicitly into the Lagrangian (such that $\kappa_W = 1$ corresponds to the SM hypothesis hereafter) yields:

$$
\begin{aligned}
\mathcal{L}_{HVV} \sim \, & \kappa_Z \frac{m_Z^2}{v} H Z^\mu Z_\mu + \kappa_W \frac{2m_W^2}{v} H W^\mu W_\mu \\
& + \frac{\alpha_Z}{v} H Z^\mu \Box Z_\mu + \frac{\alpha_W}{v} \left(H W^{+\mu} \Box W_\mu^- + H W^{-\mu} \Box W_\mu^+ \right) \\
& + \frac{\beta_Z}{v} H Z^{\mu\nu} Z_{\mu\nu} + \frac{2\beta_W}{v} H W^{\mu\nu} W_{\mu\nu} \\
& + \frac{\gamma_Z}{v} H Z^{\mu\nu} \tilde{Z}_{\mu\nu} + \frac{2\gamma_W}{v} H W^{\mu\nu} \tilde{W}_{\mu\nu}.
\end{aligned}
\tag{1.22'}
$$

This Lagrangian allows for both CP-even and CP-odd tensor structures in addition to the SM gauge boson interaction. Within the effective Lagrangian calculation, a non-vanishing value of any of the parameters α_V, β_V and γ_V results, in general, in a change of the kinematic distributions as well as the overall signal yields with respect to the SM. Only κ_V, which is associated with the SM term, modifies only the signal yield and corresponds to the κ_W

and κ_Z coupling scaling factors defined for Higgs couplings measurements in Chapter 1, Sec. 9 and used in Sec. 9 of this chapter for Higgs boson coupling measurements.

As in the measurement of kinematic production properties and the J^P hypothesis tests, the overall observed signal yields are not used in the analyses, as this not only considerably simplifies the analyses but also allows for a more general interpretation of the results. This implies that only ratios of Lagrangian parameters are determined, which are chosen as α_V/κ_V, β_V/κ_V and γ_V/κ_V. Without further assumptions, these ratios of parameters are unrelated between the HZZ and HWW interaction vertices. The results from the independent measurements are presented in Chapter 5, Sec. 11 and Chapter 7, Sec. 6.3. For the combination of the measurements from the $H \rightarrow ZZ^{(*)} \rightarrow llll$ and $H \rightarrow WW^{(*)} \rightarrow l\nu l\nu$ channels, the same values of these ratios are assumed in the HZZ and HWW interactions: $\alpha^{ZZ}/\kappa^{ZZ} = \alpha^{WW}/\kappa^{WW}$, $\beta^{ZZ}/\kappa^{ZZ} = \beta^{WW}/\kappa^{WW}$ and $\gamma^{ZZ}/\kappa^{ZZ} = \gamma^{WW}/\kappa^{WW}$. While this assumption might seem natural, there is no strong theoretical argument for it and unrelated values for the HZZ and HWW interaction parameters are equally motivated. The allowed 95% CL intervals for these parameter ratios are listed in Table 12.4.

Unfortunately the Lagrangian parameters do not give an intuitive idea of the order of magnitude of the allowed deviations from the SM. References [27,28] introduce an alternative parametrization for the parameters in the effective Lagrangian calculation based on anomalous couplings translated into fractions of effective cross sections. While the definition of fractions of effective cross sections has some process dependence and ignores interference effects, it provides an intuitive upper limit of approximately 20% to 40% on the fraction of anomalous $H \rightarrow VV$ decays.

8. Signal strength measurements

Having established the existence of a new boson and measured its mass, both experiments rely on the observed signal yields in the different final states to assess the compatibility of the data with the SM Higgs boson hypothesis. For the calculations of the signal yields, the measurements assume SM-like kinematic properties in both the Higgs boson production and in Higgs boson decays, as experimentally tested in the previous sections.

Table 12.4 The expected and observed 95% CL intervals of the allowed amounts of alternative, not SM-like, spin-zero interactions admixed to the primarily SM-Higgs-boson-like gauge boson interaction for the combination of the $H \to ZZ^{(*)} \to llll$ and $H \to WW^{(*)} \to l\nu l\nu$ channels. The results obtained by ATLAS and CMS in Refs. [25,26] are translated into the notation introduced in Chapter 1, Sec. 9. Identical values for the α/κ, β/κ and γ/κ ratios are assumed for the HZZ and HWW interactions. Small non-vanishing values for these ratios are also expected in the SM from higher order loop contributions,[29,30] but the sensitivity is not sufficient to observe these.

		α/κ	β/κ	γ/κ
ATLAS	observed	—	$[-0.63, 0.73]$	$[-0.83, 2.18]$
	expected	—	$[-4.80, 0.55]$	$[-2.30, 2.33]$
CMS	observed	$[-1.66, 1.57]$	$[-0.76, 0.58]$	$[-1.57, 1.54]$
	expected	$[-\infty, 1.24] \cup [9.0, +\infty]$	$[-1.67, 0.45]$	$[-2.65, 2.65]$
Expected SM loop contributions		$<\mathcal{O}(10^{-2})$	$<\mathcal{O}(10^{-2})$	$<\mathcal{O}(10^{-11})$

Results in this section and the following one are based on the combinations of all ATLAS searches[11] and of all CMS searches,[13] with the LHC Run 1 dataset.

In the simplest test, an overall signal strength modifier μ is fit from the combination of all channels: the results from ATLAS and CMS are $\mu = 1.18 \pm 0.10(\text{stat.})^{+0.08}_{-0.07}(\text{theo.}) \pm 0.07(\text{syst.})$ and $1.00 \pm 0.09(\text{stat.})^{+0.08}_{-0.07}(\text{theo.}) \pm 0.07(\text{syst.})$, respectively. In reporting these results, the theoretical part of the uncertainty includes all contributions to the signal cross sections and branching ratios from missing higher orders in the perturbative expansion, parton distribution functions, signal acceptance, underlying event description, and uncertainties on the input parameters of the theoretical predictions (e.g. quark masses). In the CMS case, it also includes theoretical uncertainties affecting the background estimations, while in the ATLAS case those are included among the other systematic uncertainties. The statistical compatibility of the ATLAS results with the SM expectation of $\mu = 1$ is 18%.

Another consistency test is performed allowing independent signal strength modifiers in each analysed decay mode (Fig. 12.6, left). In this test, the relative contributions of different Higgs boson production modes in all final states are assumed to be as in the Standard Model. The overall level of consistency with the SM is good.

An alternative consistency test is made by allowing for independent signal strengths for each Higgs boson production mode, and assuming the SM Higgs boson decay branching fractions (Fig. 12.6, right).

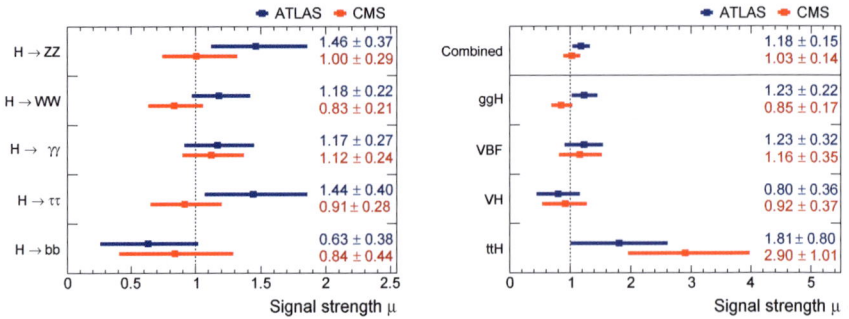

Fig. 12.6 Signal strength modifiers for each decay mode (left), and for each production mode (right); in the right hand side plot, the overall signal strengths measured by ATLAS and CMS are also shown. For both plots, the uncertainties on the individual signal strength measurements are partially correlated. Numerical values of the signal strengths are also given in the plot (with uncertainties symmetrized for simplicity of presentation).

Different decay modes can normally be separated well experimentally, so the first of the two consistency tests described above could be performed to a good approximation by measuring the signal strength for each decay mode using only the analyses targeting that decay mode, as is the case for the CMS result. Conversely, signals from different production modes cannot often be separated with high purity, and thus experimental final states often receive contributions from multiple production modes. This is especially true for channels targeting VBF production, which often have a significant contamination from gluon fusion production with additional jets from initial state radiation. Because of this, signal strengths by production mode can only be extracted from a simultaneous four-parameter fit to the combination of all channels. In the ATLAS case, also the signal strengths by decay mode are extracted from a similar simultaneous fit to the combination of all channels.

Both the signal strength by the production mode and by the decay mode show an overall agreement with the SM prediction, except for an excess in the signal strength for the $t\bar{t}H$ production mode, which is at the level of about 1σ in ATLAS and 2σ in CMS.

New physics beyond the SM can simultaneously affect the production and the decay of a Higgs boson, while the two previous tests assumed SM rates either for production modes or for decay modes, making it hard to interpret possible deviations, if they were observed. By introducing independent signal strengths that scale the different production modes for each decay mode separately, the previous tests can be generalized in a simple way that preserves a direct, linear connection between the model parameters

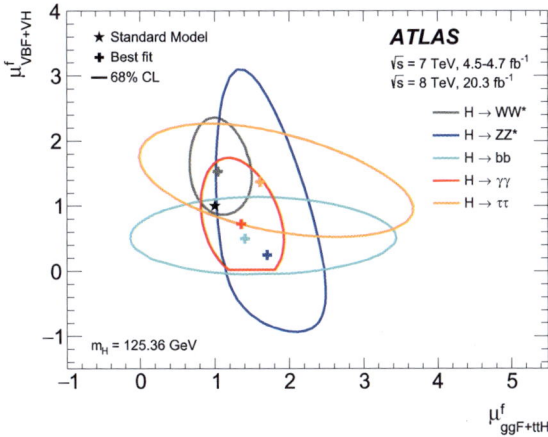

Fig. 12.7 The 68% CL regions for signal strength in the production mechanisms associated with couplings to quarks ($\mu_{\mathrm{ggH},t\bar{t}H}$) and to electroweak bosons ($\mu_{\mathrm{VBF,VH}}$), in each of the analyzed decay modes. Results from ATLAS and CMS are similar, so only the ATLAS result is shown here. Channels with good sensitivity to both production modes, such as $H \to WW^{(*)}$ and $H \to \gamma\gamma$, yield a region that is of comparable size on both axes, while channels with limited sensitivity to one kind of production modes yield more elongated regions, as visible for $H \to b\bar{b}$ and $H \to ZZ$. The tilt in the contours, visible e.g. for $H \to \tau\tau$ and $H \to ZZ$, derives from event categories receiving substantial signal contributions from both kind of production modes, and hence yield a measurement of a linear combination of the two signal strength modifiers.

and the expected signal yields. A convenient parameterization is done by assigning one common signal strength to all production modes involving the interaction between the Higgs boson and the electroweak gauge bosons (VBF, WH, ZH) and a second signal strength modifier to the production modes associated with the coupling between the Higgs boson and quarks (gluon fusion and $t\bar{t}H$). Bi-dimensional confidence regions in the plane of the signal strength modifiers for gluon fusion and $t\bar{t}H$ ($\mu_{\mathrm{ggF},t\bar{t}H}$) and VBF and VH production mechanisms ($\mu_{\mathrm{VBF},VH}$) are shown in Fig. 12.7 for each Higgs boson decay mode. Since decay modes are not combined, the signal strengths can reflect differences from SM predictions either in the measured cross sections or in the branching ratios.

9. Higgs boson coupling strength

Within the SM the Higgs sector is completely determined and the coupling strength between the Higgs boson and all other fermions and gauge bosons are set by their known masses (see Eq. (1.18)). Potentially existing new

heavy particles or deviations from the SM electroweak symmetry breaking
mechanism are expected to lead to changes in the Higgs boson couplings.
Hence, a search for deviations from the SM in the Higgs boson coupling
strength is a sensitive probe of new physics.

The measurements of the signal strength μ or cross sections in topologies
targeting different production and decay modes are already sensitive to new
physics effects. The use of Higgs boson couplings can combine the information from production and decay modes and hence reach higher sensitivity
to new physics in the Higgs sector.

The searches for deviations from the SM in the Higgs boson coupling
strength are implemented using the leading-order framework introduced in
Chapter 1, Sec. 9, where predictions for the cross sections in the different
Higgs boson channels are computed in terms of coupling scale factors or
coupling modifiers κ_i. These searches rely on the combination of all available
Higgs channels, as each channel is sensitive to only a few couplings.

With the data available from LHC Run 1, the sensitivity of a generic
measurement of all LHC accessible coupling scale factors κ_W, κ_Z, κ_t, κ_b,
κ_τ, κ_γ and κ_g is limited by the large uncertainties in the more challenging production and decay modes (e.g. $t\bar{t}H$ or $H \to b\bar{b}$) or rare processes
(e.g. $H \to \mu\mu$). On the other hand, probing potential deviations in specific aspects of the SM Higgs sector with a minimal number of coupling
scale factors could reveal deviations which are not significant in the measurements of the signal strength μ of individual channels. Hence a series of
searches for deviations are performed first in minimal benchmark models,
where different aspects of the SM Higgs sector are probed individually, and
then in more generic models.

The coupling scale factors are defined such that a value of 1 for all factors
κ_i corresponds to the SM hypothesis. In order to search for deviations from
the SM, it would, hence, be natural to try to directly extract the coupling
scale factors κ_i from a combined fit. Unfortunately this is not possible
without further assumptions, as can be illustrated with the general cross
section expression for the Higgs boson production via fusion of particles i
and decay into a pair of final state particles f:

$$\sigma \cdot \mathrm{BR}(ii \to H \to ff) = \frac{\sigma_i \cdot \Gamma_f}{\Gamma_H} = \frac{\kappa_i^2 \cdot \kappa_f^2}{\kappa_H^2} \cdot \frac{\sigma_i^{\mathrm{SM}} \cdot \Gamma_f^{\mathrm{SM}}}{\Gamma_H^{\mathrm{SM}}}. \tag{12.1}$$

All measurements of signal cross sections and hence signal event yields
depend on the total width Γ_H of the Higgs boson, or its scale factor κ_H^2,
as it appears in the denominator of all cross section expressions. However,

the total width is neither directly nor indirectly measureable with sufficient precision for a Higgs boson with a mass of ~ 125 GeV, as discussed in Chapter 1, Sec. 9 and Chapter 5, Sec. 12. The consequence of this limitation is that an increase in Γ_H by a factor α caused by some new invisible or at the LHC undetectable Higgs boson decay mode cannot be disentangled from a change of all coupling scale factors κ_i by a common factor $1/\sqrt[4]{\alpha}$, as the expected signal event yields in all production and decay modes are identical for these two scenarios.

Hence, direct measurements of the scale factors κ_i can only be made under some assumption which breaks this degeneracy between κ_i and the total width. The most widely used assumption, which is used predominantly in this section, is to postulate a vanishing branching ratio $\mathrm{BR}_{\mathrm{BSM}} = 0$ into new invisible or at the LHC undetectable Higgs boson decay modes mediated by BSM particles. Invisible decay modes are characterized by a missing transverse momentum signature as exploited in the direct searches for invisible decays summarized in Sec. 2, while at the LHC undetectable decays are for example characterized by many jets in the final state which are inseparable from QCD backgrounds. Alternative assumptions, explored in the measurements in Refs. [11,13], require either the SM coupling strength as upper limit on the gauge couplings[27,31–33] $\kappa_W, \kappa_Z \leq 1$ (this assumption is valid in a wide class of BSM models including models with an arbitrary number of Higgs singlets and doublets), or use the indirect width measurement via off-shell Higgs boson production as discussed in Chapter 1, Sec. 9 and Chapter 5, Sec. 12 , which assumes that BSM physics does not affect the Higgs boson off-shell coupling strength differently than the on-shell coupling strength ($\kappa_{\mathrm{on\text{-}shell}} = \kappa_{\mathrm{off\text{-}shell}}$). On the other hand, measurements of ratios of coupling scale factors $\lambda_{XY} = \kappa_X/\kappa_Y$ do not have this limitation, as the total width cancels out in such ratios, and are hence more model independent. However, the interpretation of any deviation from the SM value of $\lambda_{XY} = 1$ is less intuitive.

The loop-induced $H \to \gamma\gamma$ and $gg \to H$ processes as well as the already discussed total width are expected to be especially sensitive to direct contributions from unkown BSM particles. Therefore effective coupling scale factors κ_γ, κ_g and κ_H are used for these components in some measurements in order to distinguish the effects of direct BSM contributions from changes to the more fundamental coupling scale factors κ_W, κ_Z, κ_t, κ_b and κ_τ. In other measurement no such direct BSM contributions are assumed and hence κ_γ, κ_g or κ_H are expressed as a function of κ_W, κ_Z, κ_t, κ_b and κ_τ (only the dominant fermion contributions are indicated here for simplicity.

Both experiments use the full expressions[27] involving all second and third generation fermions). In all cases the VBF production is expressed as a weighted sum of the W and Z scale factors. The relevant relationships for a ~ 125 GeV Higgs boson are approximately:

$$\kappa_\gamma^2(\kappa_t, \kappa_W) \sim 1.59 \cdot \kappa_W^2 - 0.66 \cdot \kappa_W \kappa_t + 0.07 \cdot \kappa_t^2 \tag{12.2}$$

$$\kappa_g^2(\kappa_b, \kappa_t) \sim 1.06 \cdot \kappa_t^2 - 0.07 \cdot \kappa_t \kappa_b + 0.01 \cdot \kappa_b^2 \tag{12.3}$$

$$\kappa_{\mathrm{VBF}}^2(\kappa_W, \kappa_Z) \sim 0.74 \cdot \kappa_W^2 + 0.26 \cdot \kappa_Z^2 \quad . \tag{12.4}$$

A negative interference in $H \to \gamma\gamma$ appears in Eq. (12.2) as the Feynman rules for fermions and bosons have opposite signs in loop calculations. The negative interference in $gg \to H$ (Eq. (12.3)) is not easy to explain, but is essentially caused by the large mass hierarchy between the top- and the b-quark mass. Interference between W- and Z-fusion is kinematically strongly suppressed in VBF (Eq. (12.4)) and hence neglected.

The total width is the sum of the SM partial decay widths scaled by the appropriate coupling scale factors κ_i:

$$\Gamma_H = \sum_i \kappa_i^2 \cdot \Gamma_i^{\mathrm{SM}} \tag{12.5}$$

which then yields for $\kappa_H^2 = \Gamma_H / \Gamma_H^{\mathrm{SM}}$:

$$\kappa_H^2(\kappa_i) \sim 0.57 \cdot \kappa_b^2 + 0.22 \cdot \kappa_W^2 + 0.09 \cdot \kappa_g^2$$
$$+ 0.06 \cdot \kappa_\tau^2 + 0.03 \cdot \kappa_Z^2 + 0.03 \cdot \kappa_c^2 \quad . \tag{12.6}$$

In all cases the constants are obtained from SM calculations.[27] The third generation fermion coupling scale factors are assumed to apply also to the first and second generation fermions. Only κ_μ is treated independently in some cases.

9.1. *Tests of couplings to fermion and vector bosons*

One of the most fundamental aspects of the SM Higgs sector is the difference between the Yukawa couplings of the Higgs boson to fermions and the gauge couplings to the W and Z bosons. In order to experimentally test this aspect, all fermion couplings are assumed to be modified by the same scale factor $\kappa_f \equiv \kappa_t = \kappa_b = \kappa_\tau = \ldots$ and all gauge couplings by an independent scale factor $\kappa_V \equiv \kappa_W = \kappa_Z$. For this test, the loop-induced processes $H \to \gamma\gamma$ and $gg \to H$ are assumed to proceed only via the known SM particles and the total width is computed using contributions from only

SM particles. Hence, Eqs. (12.3), (12.4) and (12.6) are simplified to $\kappa_g = \kappa_f$, $\kappa_{\rm VBF} = \kappa_{\rm V}$ and $\kappa_H^2 \sim 0.75 \cdot \kappa_f^2 + 0.25 \cdot \kappa_{\rm V}^2$. Only in Eq. (12.2) a non-trivial mixing of κ_f and $\kappa_{\rm V}$ remains. However, this mixing through the interference of the W and t quark amplitudes in the $H \to \gamma\gamma$ decay (see Fig. 1.4) allows the fit to become sensitive to the relative sign between the Yukawa and gauge couplings. Since only the relative sign of the two contributions can be assessed, it is conventionally assumed that $\kappa_{\rm V}$ is positive, and a difference in the relative sign is attributed to κ_f.

Results from individual decay modes and their combination are shown for CMS in Fig. 12.8 (results for ATLAS are similar). From the figure, it is apparent that analyses in all decay modes except $H \to \gamma\gamma$ yield very similar constraints in the two planes $\kappa_f > 0$ and $\kappa_f < 0$, since predictions in those modes depend predominantly on κ_i^2. In addition, for the other decay modes small differences between the two planes can be caused by the low rate tH[34] and $gg \to ZH$[35,36] production modes, as they contain interference effects between top-coupling and W- or Z-coupling mediated Higgs production. However, these differences are very small for the LHC Run 1 data, and were explicitly explored in Ref. [11].

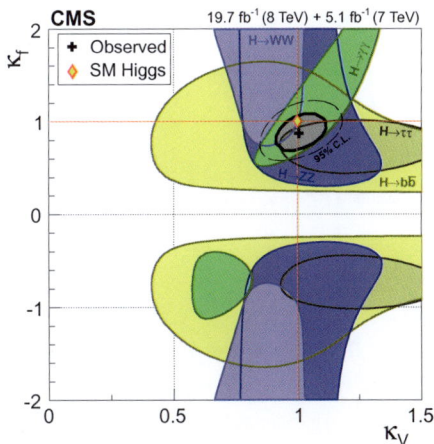

Process	Approx. scaling
$gg \to H \to VV$ $VV \to H \to ff$	$\dfrac{\kappa_f^2 \cdot \kappa_{\rm V}^2}{\frac{3}{4}\kappa_f^2 + \frac{1}{4}\kappa_{\rm V}^2} \sim \kappa_{\rm V}^2$
$gg \to H \to \gamma\gamma$	$\dfrac{\kappa_f^2 \cdot \left\vert \frac{5}{4}\kappa_{\rm V} - \frac{1}{4}\kappa_f \right\vert^2}{\frac{3}{4}\kappa_f^2 + \frac{1}{4}\kappa_{\rm V}^2} \sim \kappa_{\rm V}^2$
$gg \to H \to ff$	$\dfrac{\kappa_f^2 \cdot \kappa_f^2}{\frac{3}{4}\kappa_f^2 + \frac{1}{4}\kappa_{\rm V}^2} \sim \kappa_f^2$
$VV \to H \to VV$	$\dfrac{\kappa_{\rm V}^2 \cdot \kappa_{\rm V}^2}{\frac{3}{4}\kappa_f^2 + \frac{1}{4}\kappa_{\rm V}^2} \sim \dfrac{\kappa_{\rm V}^4}{\kappa_f^2}$

Fig. 12.8 Left: 2D confidence regions in the $(\kappa_{\rm V}, \kappa_f)$ plane for individual CMS measurements and their combination. Right: A table of approximate scalings of the signal strength as function of the coupling for some different classes of processes. $VV \to H$ stands for either VBF or VH production, while $t\bar{t}H$ production has the same coupling dependence as gluon fusion. The numerical coefficients used $(\frac{3}{4}, \frac{1}{4}, \frac{5}{4})$ are approximate numbers for a Higgs boson of mass around 125 GeV.

A fermiophobic Higgs ($\kappa_f \sim 0$) is excluded, both because it predicts no signal in the fermionic final states and because it would require all signal in bosonic final states to be produced via VBF or *VH* production, in disagreement with what is observed.

The remaining features can be understood by observing that for a light Higgs boson with couplings not too different from the SM ones, the total width is driven by decays to fermions or mediated by fermions (e.g. $H \rightarrow gg$), and thus scales approximately as κ_f^2, while the $H \rightarrow \gamma\gamma$ partial decay width is mainly driven by the vector boson coupling, κ_V^2. Most search channels with good sensitivity rely either on gluon fusion production but decays to vector bosons (e.g. $gg \rightarrow H \rightarrow WW$), or vector-boson related production and decays to fermions (e.g. $VH \ H \rightarrow b\bar{b}$), and thus probe mainly κ_V since the κ_f dependency in the numerator is approximately cancelled by that of the total width in the denominator. Sensitivity to κ_f is achieved only from purely fermionic processes (e.g. $gg \rightarrow H \rightarrow \tau\tau$), or indirectly from purely bosonic processes (e.g. VBF $H \rightarrow WW$) when combined with other processes that constrain κ_V. Because of this, κ_V is probed with a significantly better accuracy than κ_f. The results for a simulatenous fit of the two couplings from the combination of all measurements within each experiment are: $\kappa_V = 1.09 \pm 0.07$, $\kappa_f = 1.11 \pm 0.16$ (ATLAS), and $\kappa_V = 1.01 \pm 0.07$, $\kappa_f = 0.87 \pm 0.13$ (CMS). Such a coupling model was also explored, although with limited sensitivity, in the Higgs boson studies at the Tevatron, as described in Chapter 3, Sec. 9.2.2 and Fig. 3.10 therein.

9.2. *Tests of custodial symmetry and fermion universality*

In many extensions of the SM containing a light Higgs-like boson, the ratio between the couplings of such a boson to the W and Z fields is expected to be the same as the SM because of custodial symmetry. Measurements of the ratio of coupling modifiers $\lambda_{WZ} = \kappa_W/\kappa_Z$ have been done at the LHC in two ways: (a) using only the WW and ZZ final states and (b) including all final states, with several different assumptions on the other couplings. In the former case, especially when restricting the measurement to Higgs boson decay final states with at most one accompanying hadronic jet, the signal yield is expected to be dominated by gluon fusion production, and λ_{WZ} is accessed from the ratio of the Higgs boson decay branching ratios in WW and ZZ. When considering instead the combination of all final states, additional sensitivity to deviations in the W and Z couplings comes from the WH and ZH cross sections, from the branching ratio in the diphoton final state (sensitive to κ_W and not to κ_Z), and from the VBF production cross section.

The result from ATLAS and CMS are $\lambda_{WZ} = 1.00^{+0.15}_{-0.11}$ and $\lambda_{WZ} = 0.92^{+0.14}_{-0.12}$, respectively. Both results are derived in a minimal benchmark model that has the same assumptions as the $\kappa_V - \kappa_f$ model described earlier, except for allowing different couplings to the W and Z bosons through the parameter λ_{WZ}, while the other two model parameters are profiled (the assumption on only SM particles contributing to the total width does not influence the ratios λ considered in this section).

These observations are in good agreement with the SM predictions. This is also true for all other measurements of λ_{WZ} performed by both experiments under various assumptions.

While in the SM a single Higgs field is used to provide masses to both the up-type and down-type fermions, this is not true in other scenarios. The measurement of the ratio of coupling modifiers $\lambda_{du} = \kappa_d/\kappa_u$ is therefore an interesting probe for new physics. Similar to λ_{WZ}, this ratio is probed in a minimal extension of the $\kappa_V - \kappa_f$ model, in which only the constraint $\kappa_d = \kappa_u$ is relaxed. In the λ_{du} measurement, the sensitivity to the coupling to up-type fermions is from the gluon fusion and $t\bar{t}H$ production cross sections, all driven by the coupling to top quarks. The coupling to down-type fermions is mainly probed via the decays into tau leptons and bottom quarks. The results from ATLAS and CMS are $\lambda_{du} = 0.90^{+0.14}_{-0.15}$ and $\lambda_{du} = 0.99^{+0.19}_{-0.18}$, respectively, both in agreement with SM predictions.

A very similar test for fermion non-universality has been done assuming a different scaling for the Yukawa couplings to quarks and to leptons, i.e introducing a parameter $\lambda_{\ell q} = \kappa_\ell/\kappa_q$. For this test, results from the two experiments are close to the SM predictions: $\lambda_{\ell q} = 1.12^{+0.22}_{-0.18}$ (ATLAS) and $\lambda_{\ell q} = 1.03^{+0.23}_{-0.21}$ (CMS).

The results of all three tests are summarised graphically in Fig. 12.9.

9.3. *Tests of contributions from BSM particles*

While the previous benchmarks assumed that the loop-induced processes $H \to \gamma\gamma$ and $gg \to H$ proceed only via the known SM particles, tests for BSM particle contributions were performed by reversing these assumptions. The most simple benchmark model has only two parameters, the loop-induced coupling scale factors κ_γ and κ_g for the processes $H \to \gamma\gamma$ and

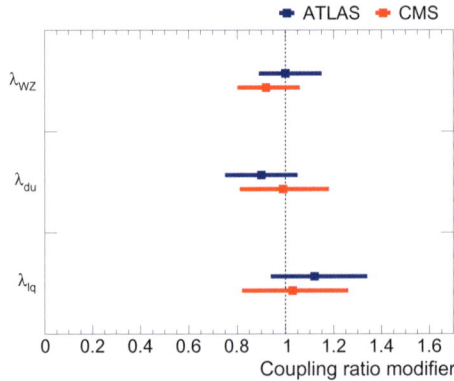

Fig. 12.9   Constraints on the coupling ratios from the three minimal benchmark models that extend the $\kappa_V - \kappa_f$ model allowing for deviations in the couplings to W and Z bosons, or to up-type and down-type fermions, or leptons and quarks.

$gg \to H$. All tree-level couplings are as in the SM: $\kappa_W = \kappa_Z = \kappa_t = \kappa_b = \kappa_\tau = 1$. This fit assumes no invisible or undetectable contributions to the Higgs boson total width as discussed in Sec. 9.1. The results on the coupling scale factors κ_γ and κ_g are $\kappa_\gamma = 1.00 \pm 0.12$ and $\kappa_g = 1.12 \pm 0.12$ for ATLAS and $\kappa_\gamma = 1.14^{+0.12}_{-0.13}$ and $\kappa_g = 0.89^{+0.11}_{-0.10}$ for CMS. No indications of BSM contributions to the loop-induced $H \to \gamma\gamma$ and $gg \to H$ processes were found.

The strong assumption of $\kappa_i = 1$ on the tree-level couplings allows the assumption on the total width to be relaxed and to indirectly test a possible non-zero branching ratio $\mathrm{BR}_{\mathrm{BSM}}$ to final states not predicted by the SM and thus not considered in this combination. To a good approximation the total width scales like $\kappa_H^2 = (0.086\kappa_g^2 + 0.0023\kappa_\gamma^2 + 0.91)/(1 - \mathrm{BR}_{\mathrm{BSM}})$ in this model. The obtained 95% CL upper limits on $\mathrm{BR}_{\mathrm{BSM}}$ are 27% for ATLAS and 32% for CMS, respectively.

9.4. *General tests of couplings*

In the minimal benchmark models, constraints were imposed on the couplings in sectors not directly tested by the model. However, at the price of larger uncertainties, it is possible to fit simultaneously all coupling modifiers related to final states accessible at the LHC.

Three results of this kind will be presented here, in increasing level of generality.

9.4.1. *Tree-level couplings only*

The first result assumes the SM particle content, and only allows modifications to the tree-level couplings of the Higgs boson to the SM fields. Modifications to loop-induced processes, and to the total width of the Higgs boson, are computed in terms of κ_W, κ_Z, κ_t, κ_b, κ_τ, κ_μ, i.e. assuming no BSM contributions to loops and no BSM decays. Results for this model are shown in Fig. 12.10 (left), and are in excellent agreement with SM predictions, except for a small deficit in κ_b. Within this model the coupling to the top quark is probed with good accuracy via the gluon fusion production cross section, that is approximately proportional to κ_t^2, as the top quark loop gives the dominant contribution to this process. The coupling to the bottom quark is probed both directly from the $H \to b\bar{b}$ channel and indirectly via its contribution to κ_H; the slight deficit in κ_b is a result of the smaller-than-SM signal strength observed in $H \to b\bar{b}$ and the slight excess observed in other modes, since a suppression of the $H \to b\bar{b}$ decay increases the branching ratios of all other decay modes.

At the moment, both experiments have only limited sensitivity to the Higgs boson coupling to muons, through the $H \to \mu\mu$ analyses mentioned in Sec. 2. However, the current results already set an upper limit on κ_μ at approximately the SM value, which supports the hypothesis that the Higgs boson couplings to the fermions from different generations are not

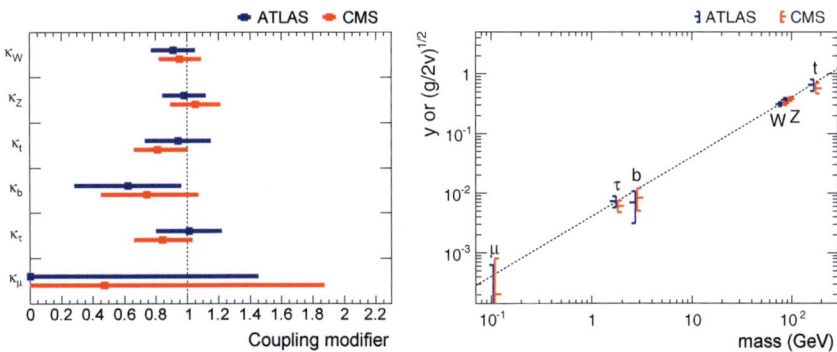

Fig. 12.10 Left: Results for the simultaneous fit of all tree-level κ_i, from ATLAS and CMS. Right: Results of the same fit, but expressed in absolute scale and plotted as function of the particle mass (as defined in the text); the dashed diagonal line represents the proportionality between coupling and mass predicted for the SM Higgs boson.

universal. This is better illustrated in Fig. 12.10 (right), where the couplings are presented in absolute scale instead of being normalized to the SM value. For fermions, the measured Yukawa couplings are given by $y_i = \kappa_i \cdot (m_i/v)$, where v is the SM vacuum expectation value of the Higgs boson field (≈ 246 GeV), and m_i is the fermion mass evaluated at the energy scale relevant for the Higgs boson decay (the latter is important only for the bottom quark, for which $m_b(q^2 = m_H^2) \approx 2.76$ GeV differs significantly from the value evaluated at the scale of heavy flavour processes $m_b(q^2 = m_b^2) \approx 4.2$ GeV). For gauge bosons, the measured couplings are $g_i = \kappa_i \cdot 2\, m_i^2/v$. If one rescales them as $\sqrt{g_i/2v} = \sqrt{\kappa_i} \cdot (m_i/v)$, they are expected to be on the same line defined by the fermionic Yukawa couplings.

9.4.2. Tree-level and loop-induced couplings

This second fit extends the previous one by introducing two additional free parameters that scale the effective coupling of the Higgs boson to photons and gluons, thus allowing for possible contributions of beyond SM particles to the loops. However, it is still assumed that there are no beyond SM decay modes of the Higgs boson, e.g. because the beyond SM particles are all heavier than the Higgs boson. The results are shown in the left panel of Fig. 12.11. The main difference compared to the first model is in the modifier to coupling to the top quark, which now is probed only through the cross section of $t\bar{t}H$ associated production, a tree-level process proportional to κ_t^2, and no longer via gluon fusion. The slight excess observed in the signal strengths for the $t\bar{t}H$ searches is thus reflected in the measured $\kappa_t \sim 1.5$.

Instead of assuming no BSM decay modes of the Higgs boson ($\text{BR}_{\text{BSM}} = 0$) in this second fit, it is also possible to make the alternative assumption of $\kappa_W, \kappa_Z < 1$ and then determine an upper limit on the branching ratio BR_{BSM} in a similar way as discussed in Sec. 9.3. The 95% CL upper limits on BR_{BSM} are 49% for ATLAS and 57% for CMS, respectively. Under the assumption of no BSM Higgs boson decays to SM detectable particles, a combination with the direct searches for invisible Higgs decays (see Sec. 2) improves the CMS limit to 49%.

As a third alternative to the assumptions on BR_{BSM} or κ_W, κ_Z, ATLAS combines the on-shell measurements with the off-shell measurements described in Chapter 5, Sec. 12 to find a 95% CL upper limit of 68% under the strong assumption of identical on-shell and off-shell Higgs boson couplings.

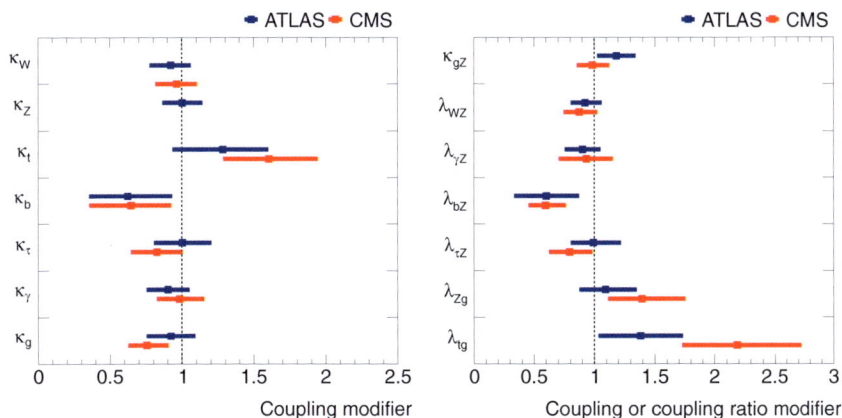

Fig. 12.11 Left: Results for the simultaneous fit of all κ_i, from ATLAS and CMS. In the CMS case, custodial symmetry is imposed ($\kappa_W = \kappa_Z$). Right: Results for the simultaneous fits of κ_{gZ} and all the coupling ratios λ_{XY}, from ATLAS and CMS.

9.4.3. *Coupling ratios*

This third and most general fit is performed without the assumptions of no beyond SM Higgs boson decays, so that κ_H is no longer a function of the κ_i. To remove the unknown κ_H, the fit is parameterized in terms of coupling ratios $\lambda_{XY} = \kappa_X/\kappa_Y$ plus a single parameter $\kappa_{gZ} = (\kappa_g \cdot \kappa_Z)/\kappa_H$ that absorbs both the overall scale of the couplings and any deviation in the total width from possible BSM decays. Results from this last fit are shown in the right hand panel of Fig. 12.11. In this model, the ratio between top and gluon couplings λ_{tg} tests directly the consistency between the top quark coupling probed at tree-level via the $t\bar{t}H$ cross section and the SM-like top quark contribution to the gluon fusion production; the observed $\lambda_{tg} > 1$ thus reflects the larger signal strengths observed in $t\bar{t}H$ compared to gluon fusion. Similarly, the low value on the ratio λ_{bZ} reflects the relatively low signal strength in $H \to b\bar{b}$ decays. All other ratios show good consistency with the SM hypothesis, including the ratio $\lambda_{\gamma Z}$ which can be seen as a probe for new physics contributions to the $H \to \gamma\gamma$ decay.

10. Summary

The combination of different Higgs boson searches resulted in the discovery of the Higgs boson in the summer of 2012. After the discovery the ATLAS and CMS collaboration have used the LHC Run 1 data to determine the properties of the Higgs boson. The mass of the Higgs boson is measured to be $125.09 \pm 0.21(\text{stat.}) \pm 0.11(\text{syst.})$ GeV. Measurements of the kinematic properties in Higgs boson production are consistent with the state-of-the-art predictions from the SM. The measured kinematic properties in the observed Higgs boson decays exclude all tested spin 1 and 2 hypotheses as well as a pure CP-odd Higgs boson and limits are provided on non-SM CP-even and CP-odd admixtures in $H \rightarrow VV$ decays.

Measurements of Higgs boson signal yields reach precisions of ~20% and results are in good agreement with the SM predictions. Only the measurements of the signal strength in $t\bar{t}H$ production show some excess with respect to the SM prediction. No significant discrepancy is seen in the Higgs boson coupling strength in all analysed benchmark models that probe different types of deviations from the SM. The best measurements reach a precision of approximately 15%. Only quantities related to the top-quark coupling show some excess and will be a primary target for Higgs boson measurements in LHC Run 2.

11. Towards future measurements

While Higgs boson measurements of both signal yields and kinematic distributions are already sensitive enough to exclude some concrete BSM models, most not yet excluded models predict deviations from the SM that are substantially smaller than the current experimental sensitivity. Future measurements at the LHC, using substantially more data, will help to reduce the uncertainty on Higgs boson signal yields and couplings to the level of 2–5% in the best cases,[37,38] where these types of models can also be tested.

More data will also allow new types of measurements which were not possible with the LHC Run 1 datasets. An example is the search for CP-mixing in the Higgs sector, where measurements of CP-odd admixtures to the fermion couplings of the Higgs boson in $t\bar{t}H$ production or $H \rightarrow \tau\tau$ decays, which are not possible in Run 1, are expected to be sensitive to the predictions of many BSM models. Also searches for rare Higgs boson production and decay modes will likely require the large foreseen integrated luminosities of LHC and its successor HL-LHC in order to observe a signal.[37,38]

Finally, with more data, the overall strategy of Higgs boson property determination will evolve. Most BSM models predict deviations with respect to the SM in both the signal yields and kinematic distributions. Currently these two aspects are analyzed separately, while a combined approach could yield a more consistent picture of the Higgs sector and substantially better sensitivity.

References

1. ATLAS Collaboration, Observation of a new particle in the search for the Standard Model Higgs boson with the ATLAS detector at the LHC, *Phys.Lett.* **B716**, 1–29 (2012). doi: 10.1016/j.physletb.2012.08.020.
2. CMS Collaboration, Observation of a new boson at a mass of 125 GeV with the CMS experiment at the LHC, *Phys.Lett.* **B716**, 30–61 (2012). doi: 10.1016/j.physletb.2012.08.021.
3. ATLAS Collaboration, Search for the associated production of the Higgs boson with a top quark pair in multilepton final states with the ATLAS detector, *Phys.Lett.* **B749**, 519–541 (2015). doi: 10.1016/j.physletb.2015.07.079.
4. CMS Collaboration, Search for the associated production of the Higgs boson with a top-quark pair, *JHEP.* **1409**, 087 (2014). doi: 10.1007/JHEP09(2014)087,10.1007/JHEP10(2014)106.
5. ATLAS Collaboration, Search for the Standard Model Higgs boson decay to $\mu^+\mu^-$ with the ATLAS detector, *Phys.Lett.* **B738**, 68–86 (2014). doi: 10.1016/j.physletb.2014.09.008.
6. CMS Collaboration, Search for a Standard Model-like Higgs boson in the $\mu^+\mu^-$ and e^+e^- decay channels at the LHC, *Phys.Lett.* **B744**, 184–207 (2015). doi: 10.1016/j.physletb.2015.03.048.
7. ATLAS Collaboration, Search for Higgs boson decays to a photon and a Z boson in pp collisions at $\sqrt{s} = 7$ and 8 TeV with the ATLAS detector, *Phys.Lett.* **B732**, 8–27 (2014). doi: 10.1016/j.physletb.2014.03.015.
8. CMS Collaboration, Search for a Higgs boson decaying into a Z and a photon in pp collisions at $\sqrt{s} = 7$ and 8 TeV, *Phys.Lett.* **B726**, 587–609 (2013). doi: 10.1016/j.physletb.2013.09.057.
9. ATLAS Collaboration, Constraints on new phenomena via Higgs boson couplings and invisible decays with the ATLAS detector, *JHEP.* **11**, 206 (2015). doi: 10.1007/JHEP11(2015)206. arXiv:1509.00672.
10. CMS Collaboration, Search for invisible decays of Higgs bosons in the vector boson fusion and associated ZH production modes, *Eur.Phys.J.* **C74**, 2980 (2014). doi: 10.1140/epjc/s10052-014-2980-6.
11. ATLAS Collaboration, Measurements of the Higgs boson production and decay rates and coupling strengths using pp collision data at $\sqrt{s} = 7$ and 8 TeV in the ATLAS experiment, *Eur. Phys. J.* **C76**(1), 6 (2016). doi: 10.1140/epjc/s10052-013-3769-y. arXiv:1507.04548.

12. ATLAS Collaboration, Measurement of the Higgs boson mass from the $H \to \gamma\gamma$ and $H \to ZZ^* \to 4\ell$ channels with the ATLAS detector using 25 fb^{-1} of pp collision data, *Phys.Rev.* **D90**, 052004 (2014). doi: 10.1103/PhysRevD. 90.052004.

13. CMS Collaboration, Precise determination of the mass of the Higgs boson and tests of compatibility of its couplings with the standard model predictions using proton collisions at 7 and 8 TeV, *Eur.Phys.J.* **C75**(5), 212 (2015). doi: 10.1140/epjc/s10052-015-3351-7.

14. ATLAS and CMS Collaborations, Combined measurement of the Higgs boson mass in pp collisions at $\sqrt{s} = 7$ and 8 TeV with the ATLAS and CMS experiments, *Phys.Rev.Lett.* **114**, 191803 (2015). doi: 10.1103/PhysRevLett. 114.191803.

15. ATLAS Collaboration, Measurements of fiducial and differential cross sections for Higgs boson production in the diphoton decay channel at $\sqrt{s} = 8$ TeV with ATLAS, *JHEP.* **1409**, 112 (2014). doi: 10.1007/JHEP09(2014) 112.

16. CMS Collaboration, Measurement of differential cross sections for Higgs boson production in the diphoton decay channel in pp collisions at $\sqrt{s} = 8$ TeV, *Eur. Phys. J.* **C76**(1), 13 (2016). doi:10.1140/epjc/510052-015-3853-3. arXiv:1508.07819.

17. ATLAS Collaboration, Fiducial and differential cross sections of Higgs boson production measured in the four-lepton decay channel in pp collisions at $\sqrt{s} = 8$ TeV with the ATLAS detector, *Phys.Lett.* **B738**, 234–253 (2014). doi: 10.1016/j.physletb.2014.09.054.

18. CMS Collaboration, Measurement of inclusive and differential fiducial cross sections for Higgs boson production in the $H \to 4\ell$ decay channel in pp collisions at $\sqrt{s} = 7$ and 8 TeV. Technical Report CMS-PAS-HIG-14-028, *JHEP.* **04**, 005 (2016). doi: 10.1007/JHEP04(2016)05. arXiv:1512008377.

19. ATLAS Collaboration, Measurements of the total and differential Higgs boson production cross sections combining the $H \to \gamma\gamma$ and $H \to ZZ^* \to 4\ell$ decay channels at $\sqrt{s} = 8$ TeV with the ATLAS detector, *Phys.Rev.Lett.* **115**(9), 091801 (2015). doi: 10.1103/PhysRevLett.115.091801.

20. D. de Florian, G. Ferrera, M. Grazzini, and D. Tommasini, Higgs boson production at the LHC: Transverse momentum resummation effects in the $H \to 2\gamma$, $H \to WW \to l\nu l\nu$ and $H \to ZZ \to 4l$ decay modes, *JHEP.* **1206**, 132 (2012). doi: 10.1007/JHEP06(2012)132.

21. K. Hamilton, P. Nason, E. Re, and G. Zanderighi, NNLOPS simulation of Higgs boson production, *JHEP.* **1310**, 222 (2013). doi: 10.1007/ JHEP10(2013)222.

22. L. Landau, On the angular momentum of a two-photon system, *Dokl.Akad.Nauk Ser.Fiz.* **60**, 207–209 (1948).

23. C.-N. Yang, Selection rules for the dematerialization of a particle into two photons, *Phys.Rev.* **77**, 242–245 (1950). doi: 10.1103/PhysRev.77.242.

24. ATLAS Collaboration, Evidence for the spin-0 nature of the Higgs boson using ATLAS data, *Phys.Lett.* **B726**, 120–144 (2013). doi: 10.1016/j. physletb.2013.08.026.

25. ATLAS Collaboration, Study of the spin and parity of the Higgs boson in diboson decays with the ATLAS detector, *Eur.Phys.J.* **C75**(10), 476 (2015). doi: 10.1140/epjc/s10052-015-3685-1.

26. CMS Collaboration, Constraints on the spin-parity and anomalous *HVV* couplings of the Higgs boson in proton collisions at 7 and 8 TeV, *Phys.Rev.* **D92**(1), 012004 (2015). doi: 10.1103/PhysRevD.92.012004.

27. LHC Higgs Cross Section Working Group, Handbook of LHC Higgs Cross Sections: 3. Higgs Properties (2013). doi: 10.5170/CERN-2013-004. arXiv:1307.1347.

28. I. Anderson, S. Bolognesi, F. Caola, Y. Gao, A. V. Gritsan, *et al.*, Constraining anomalous *HVV* interactions at proton and lepton colliders, *Phys.Rev.* **D89**(3), 035007 (2014). doi: 10.1103/PhysRevD.89.035007.

29. A. Soni and R. M. Xu, Probing CP violation via Higgs decays to four leptons, *Phys.Rev.* **D48**, 5259–5263 (1993). doi: 10.1103/PhysRevD.48.5259.

30. Y. Gao, A. V. Gritsan, Z. Guo, K. Melnikov, M. Schulze, *et al.*, Spin determination of single-produced resonances at hadron colliders, *Phys.Rev.* **D81**, 075022 (2010). doi: 10.1103/PhysRevD.81.075022.

31. D. Zeppenfeld, R. Kinnunen, A. Nikitenko, and E. Richter-Was, Measuring Higgs boson couplings at the CERN LHC, *Phys.Rev.* **D62**, 013009 (2000). doi: 10.1103/PhysRevD.62.013009.

32. M. Duhrssen, S. Heinemeyer, H. Logan, D. Rainwater, G. Weiglein, *et al.*, Extracting Higgs boson couplings from CERN LHC data, *Phys.Rev.* **D70**, 113009 (2004). doi: 10.1103/PhysRevD.70.113009.

33. B. A. Dobrescu and J. D. Lykken, Coupling spans of the Higgs-like boson, *JHEP.* **1302**, 073 (2013). doi: 10.1007/JHEP02(2013)073.

34. M. Farina, C. Grojean, F. Maltoni, E. Salvioni, and A. Thamm, Lifting degeneracies in Higgs couplings using single top production in association with a Higgs boson, *JHEP.* **1305**, 022 (2013). doi: 10.1007/JHEP05(2013)022.

35. O. Brein, A. Djouadi, and R. Harlander, NNLO QCD corrections to the Higgs-strahlung processes at hadron colliders, *Phys.Lett.* **B579**, 149–156 (2004). doi: 10.1016/j.physletb.2003.10.112.

36. C. Englert, M. McCullough, and M. Spannowsky, Gluon-initiated associated production boosts Higgs physics, *Phys.Rev.* **D89**(1), 013013 (2014). doi: 10.1103/PhysRevD.89.013013.

37. ATLAS Collaboration, Projections for measurements of Higgs boson signal strengths and coupling parameters with the ATLAS detector at a HL-LHC. (ATL-PHYS-PUB-2014-016) (2014). URL https://cds.cern.ch/record/1956710.

38. CMS Collaboration, Projected performance of an upgraded CMS detector at the LHC and HL-LHC: Contribution to the Snowmass Process (2013). arXiv:1307.7135.

Chapter 13

Summary and outlook

About fifty years after the conjecture by Englert & Brout, Higgs, and Guralnik, Hagen & Kibble, we now have good experimental proof for the existence
of the mechanism that breaks the electroweak symmetry and generates the
masses of the known elementary particles. Since the initial observation[1,2]
of a resonance with an observed mass[3] near 125 GeV, a series of detailed
measurements investigating the nature of this particle have been performed
by the ATLAS and CMS collaborations. The measurements in Ref. [4] combine the individual results from ATLAS and CMS as discussed in Chapter
12 and provide a more statistically precise profile of this particle.

Here is what we have learnt from the experimental measurements and
investigations so far:

- **Search and discovery**: The LEP, the Tevatron and the LHC experiments have searched for a Standard Model (SM) Higgs boson and
  excluded its presence below 1000 GeV except for a statistically significant resonance with a mass of around 125 GeV, observed at the LHC in
  several decay final states, the H_{125} boson. The observed yield of the H_{125}
  boson, averaged over various production mechanisms and decay modes,
  is consistent with that for the Standard Model Higgs boson to within a
  10% measurement uncertainty.
- **Mass and natural width** : The measured mass of the H_{125} particle is
  125.09 ± 0.21(stat.) ± 0.11(syst.) GeV. The upper limit from the direct
  measurement of the natural width of the H_{125} is 3.4 GeV at 95% confidence level (CL), well above the predicted intrinsic width for the SM
  Higgs boson of 4.1 MeV. Model-dependent measurements based on on-
  and off-shell Higgs boson production cross sections place a 95% CL upper
  limit of 22 MeV.
- **Spin-parity**: The quantum numbers of the H_{125} boson have been probed
  and are consistent with $J^{PC} = 0^{++}$, as expected for the Standard Model

411

Higgs boson. All its properties appear to be consistent with that of an elementary scalar particle, the first of its kind observed in nature.

- **Production rate in pp collisions**: The gluon fusion mechanism is found to be the dominant H_{125} production process; its rate measured in units of Standard Model expectation is $\mu_{ggF} = 1.03^{+0.17}_{-0.15}$. While there is limited evidence for production via the vector boson fusion process from both ATLAS and CMS, the significance of the observation exceeds 5 standard deviations only after the two sets of measurements are combined. There is limited evidence for H_{125} production accompanied by a massive vector boson or $t\bar{t}$. Nevertheless, the overall pattern of the measured cross sections in various production processes is consistent, within uncertainties, with the calculations based on the Standard Model.

- **Couplings to massive vector bosons**: The observed decay rates of the H_{125} particle to W^+W^- and Z^0Z^0 and their ratio are in excellent agreement with SM Higgs boson couplings that are proportional to the squares of the vector boson masses.

- **Couplings to fermions**: In the SM, the Higgs boson coupling is proportional to the fermion mass. So far the experiments have only observed direct evidence for decay to third generation fermions: the τ lepton and the bottom quark. Within large experimental uncertainties, these measurements of couplings to the third generation fermions are consistent with the SM expectations. While the observation of H_{125} decay to second generation quarks at hadron colliders is unrealistic due to the lack of a characteristic signal signature, searches for the decay of H_{125} to the $\mu^+\mu^-$ final state have led to a 95% CL upper limit that is seven times the SM expectation. This limit is sufficient to conclude that, as in the SM, the fermionic couplings of the H_{125} boson are not generation-universal.

- **Exotic decays**: A variety of characteristic decays of the H_{125} predicted in beyond-Standard Model scenarios have been searched for but with sensitivities limited by the size of the LHC Run 1 data sample. For example, ATLAS has placed 95% confidence level upper limits on branching fractions $\mathcal{B}(H_{125} \rightarrow \text{invisible}) < 25\%$ and $\mathcal{B}(H_{125} \rightarrow \mu\tau) < 1.85\%$ from direct searches for these processes.

- **Search for partners in electroweak symmetry breaking**: Straightforward extensions of the Standard Model such as the Two-Higgs Doublet model introduce four additional scalars: H^0, A^0 and $H^\pm$. These particles have been searched for in a variety of final states but none have been found so far.

While the properties of the H_{125} boson observed so far appear to be consistent with those of the Standard Model Higgs boson, it will require a broader and more precise set of measurements to determine whether the H_{125} boson is the simplest and the only manifestation of the Higgs mechanism that occurs in nature or if it is part of a bigger picture. In the next decades the LHC will be the only existing facility for such investigations. As we write this, the LHC has already completed its commissioning run at $\sqrt{s} = 13$ TeV and plans to deliver more than 100 fb^{-1} by 2018, and about 300 fb^{-1} at $\sqrt{s} = 14$ TeV by 2023. In the high luminosity phase (HL-LHC) about 3000 fb^{-1} at $\sqrt{s} = 14$ TeV is expected to be recorded by the upgraded ATLAS and CMS experiments. Such rapidly evolving datasets will form the basis for a precise exploration of the H_{125} boson properties, searches for additional states belonging to the electroweak symmetry breaking sector, and use of the H_{125} boson as a portal to new physics.

With a 300 fb^{-1} data sample the emphasis will be on establishing the $H_{125} \to b\bar{b}$ and $H_{125} \to \tau^+\tau^-$ decay modes and VH and VBF production channels at greater than 5σ level of significance in each experiment. The expected precision[5,6] on all observables should be 3–5 times better than from the 8 TeV data. The coupling precision will improve further by a factor of $\sim$3 at the HL-LHC. Unique observations at the HL-LHC should include rare decays such as $H \to Z\gamma$ and the coveted decay $H \to \mu^+\mu^-$, which probes the Higgs boson coupling to fermions of the second generation. Higgs boson production via $pp \to t\bar{t}H$, which allows a direct determination of the top quark Yukawa coupling and probes CP properties of the Higgs boson, should be precisely measurable at the HL-LHC. With this dataset the H_{125} boson mass should be measurable with a precision of $\sim$ 50 MeV and several Higgs couplings with a precision of 3–7%.

Measurement of the Higgs boson self-coupling is the holy grail of the Higgs physics program since it allows full determination of the shape of the potential responsible for electroweak symmetry breaking and probes non-SM physics it may contain. Higgs trilinear self-coupling λ_{HHH} can be probed via detection of Higgs boson pair production at the HL-LHC but the rate is tiny. By combining several di-Higgs final states from ATLAS and CMS, a sensitivity at best of 2–3 σ is projected, but translating such observations into a significant measurement of λ_{HHH} will be challenging due to background pollution.

A more precise exploration of the Higgs boson sector, including its total width, trilinear and quartic self-couplings, will require new colliders. A summary and comparison of the Higgs physics reach of some of the proposed machines can be found in the Snowmass Higgs group report,[7] in the INFN *What Next?*[8] document and in the *Physics Briefing Book* for the *European Strategy for Particle Physics*.[9] In addition, impressive gains in precision Higgs physics at a future 100 TeV pp collider are described in Refs. [10,11].

References

1. ATLAS Collaboration, Observation of a new particle in the search for the Standard Model Higgs boson with the ATLAS detector at the LHC, *Phys.Lett.* **B716**, 1–29 (2012). doi: 10.1016/j.physletb.2012.08.020.
2. CMS Collaboration, Observation of a new boson at a mass of 125 GeV with the CMS experiment at the LHC, *Phys.Lett.* **B716**, 30–61 (2012). doi: 10.1016/j.physletb.2012.08.021.
3. ATLAS and CMS Collaborations, Combined measurement of the Higgs boson mass in pp collisions at $\sqrt{s} = 7$ and 8 TeV with the ATLAS and CMS experiments, *Phys.Rev.Lett.* **114**, 191803 (2015). doi: 10.1103/PhysRevLett.114.191803.
4. ATLAS and CMS Collaborations, Measurements of the Higgs boson production and decay rates and constraints on its couplings from a combined ATLAS and CMS analysis of the LHC pp collision data at $\sqrt{s} = 7$ and 8 TeV, ATLAS-HIGG-2015-07, CMS-HIG-15-002 (2016). Submitted to *JHEP*. URL http://cds.cern.ch/record/2158863. arXiv: 1606.02266.
5. ATLAS Collaboration, Projections for measurements of Higgs boson signal strengths and coupling parameters with the ATLAS detector at a HL-LHC, ATL-PHYS-PUB-2014-016 (2014). URL https://cds.cern.ch/record/1956710.
6. CMS Collaboration, Projected performance of an upgraded CMS detector at the LHC and HL-LHC: Contribution to the Snowmass Process (2013). arXiv:1307.7135.
7. S. Dawson *et al.*, Working Group Report: Higgs Boson. In *Community Summer Study 2013: Snowmass on the Mississippi (CSS2013)* Minneapolis, MN, USA, July 29–August 6, 2013 (2013). arXiv:1310.8361.
8. A. Andreazza *et al.*, What Next: White Paper of the INFN-CSN1, *Frascati Phys. Ser.* **60**, 1–302 (2015). URL https://inspirehep.net/record/1374543/files/Volume60.pdf.
9. R. Aleksan *et al.*, Physics Briefing Book: Input for the Strategy Group to draft the update of the European Strategy for Particle Physics. Technical Report CERN-ESG-005, Geneva (2013). URL https://cds.cern.ch/record/1628377. Open Symposium held in Cracow from 10th to 12th of September 2012.
10. J. Baglio, A. Djouadi, and J. Quevillon, Prospects for Higgs physics at energies up to 100 TeV (2015). arXiv:1511.07853.
11. N. Arkani-Hamed, T. Han, M. Mangano, and L.-T. Wang, Physics opportunities of a 100 TeV proton–proton collider (2015). arXiv:1511.06495.

Appendix A

Statistical methods

Thomas R. Junk*, Andrey Korytov†, and Alexander L. Read‡

*Fermilab, Batavia, Illinois, 60510, USA
† University of Florida, Gainesville, FL 32611, USA
‡ University of Oslo, Postboks 1048 Blindern, 0316 Oslo, Norway

A.1. Introduction

This appendix summarises the statistical methods which are used to set limits on possible signals (Sec. A.2), to compute the significances of observed excesses of events (Sec. A.3), and to measure signal model parameters (Sec. A.4). While these techniques are by now standard, having been put to use in a broad variety of scientific and other applications over the last century, there is always more than one valid technique which can be used to interpret experimental data in order to produce the results. Often, even within a single publication, a variety of techniques may be chosen in order to optimise the sensitivity or to best incorporate the effects of systematic uncertainty on the model predictions. For a comprehensive overview, see, for example, Refs. [1–5].

Henceforth, the expected event yields for the nominal signal (the Standard Model Higgs boson in this book) will be generically denoted as s, and the background predictions as b. Depending on the context, these will stand for event counts in one or multiple bins (e.g., s_i) or for unbinned probability density functions of some observables, whichever approach is used in an analysis. The notations b and $s+b$ will be also used to represent symbolically the background-only and the signal+background hypotheses, respectively; $\mu s + b$ will stand for a signal+background hypothesis, in which all nominal signal event yields are scaled by μ. Predictions for the signal and the background yields are subject to multiple uncertainties that are handled by introducing a set of nuisance parameters $\boldsymbol{\theta} = (\theta_1, \ldots, \theta_n)$, so that signal and background expectations become functions of the nuisance parameters: $s(\boldsymbol{\theta})$ and $b(\boldsymbol{\theta})$. Nuisance parameters may affect either the overall event rates, the shapes of distributions, or both. A set of observed events collected by the experimental apparatus, again in a binned or unbinned format, is referred to as the "data" or the "observation". The term "pseudo-data" refers to simulated experimental outcomes, or pseudo-observations.

A.2. Limits

The Bayesian method and the classical frequentist method, the latter with a number of modifications, are two statistical approaches commonly used in high energy physics for characterising the results of a search for a possible signal. In the absence of an observed signal, the upper limit on the signal strength is the primary result of the search.

The frequentist methods set limits that are characterised by a confidence level (CL). A statement that a signal of strength μ is excluded at the $1 - \beta$ CL (where β is some small number, usually set at 5%) is expected to imply that if a signal is truly present at the quoted signal strength, then in repetitions of the experiment, a fraction of at most β of them will falsely exclude it. A limit-setting procedure that satisfies this requirement for the error rate is said to have proper coverage. A limit-setting procedure with a larger error rate than stated is said to undercover, and a test with a smaller error rate is said to overcover. The false exclusion of a signal in its presence is known as a Type II error.[a]

Bayesian results are characterised by a credibility level, which is also abbreviated CL. At a CL of $1 - \beta$, the integral of the "belief" probability density function of the signal event rate over values greater than the limit is β. No claim is made regarding the coverage of Bayesian methods, although in practice they tend to overcover when flat priors on signal strength are used (see Sec. A.2.1).

Limits can and should be set even in the case that an excess of events is observed. Doing so is a condition for proper coverage and eliminates the flip-flop hazard of quoting limits only when no signal is observed. Limits quoted when an excess is observed also have important physical interpretations as they exclude signal strengths stronger than observed, which

---

[a] A Type I error, to be discussed in Sec. A.3, refers to a claim of a signal that is not actually present.

may test interesting models which predict anomalously large cross sections, branching ratios, or may have kinematic properties that enhance the signal acceptance.

In addition to reporting the exclusion confidence level for a fully specified model (e.g., the Standard Model (SM) Higgs boson of a given m_H), null results of a search targeting a specific signal production mechanism and a particular decay mode can be reported as approximately model-independent limits on the signal cross section times the branching ratio ($\sigma \times$ BR) for the decay mode targeted by the analysis. Less model dependence is induced by setting limits on the cross section times the branching ratio times the experimental acceptance ($\sigma \times$ BR $\times \mathcal{A}$). However, neither is perfect. The former explicitly depends on assumptions made on the fraction of a signal cross section in the phase space not covered by an experiment. The latter does not introduce such dependencies, but, in order to allow theorists to calculate the signal cross section within experimental acceptance $\mathcal{A}$, one has to provide a model of that acceptance, the exact definition of which may be too complicated for a practical use. If an analysis is based on the shape of the distribution of a discriminating observable then, in any case, its results (limits) can be interpreted only in models that yield the same distribution of the discriminating observable as the signal for which they were derived. Therefore, extrapolating results of an analysis to make a statement about a signal that has different kinematic properties from the one assumed in a given analysis is not trivial and, in general, requires additional dedicated studies.

In a combination of multiple analyses sensitive to different signal production mechanisms and different decay modes, presenting results in a form of limits on $\sigma \times$ BR or $\sigma \times$ BR $\times \mathcal{A}$ is impossible. The customary alternative is to set limits on a common signal strength modifier μ that is taken to change event yields in each (production)$\times$(decay) mode by exactly the same scale. The Standard Model Higgs boson is said to be excluded at, say, 95% CL, when the 95% CL limit on μ becomes smaller than one. In the next sub-sections, we will follow this convention and discuss limits on the common signal strength modifier μ.

A.2.1. *Bayesian approach*

In the Bayesian approach, a degree of belief is assigned to each value of μ, as a probability density function for μ. Bayes' theorem is then invoked to calculate the impact of the experimental data to update the prior probability density $\pi(\mu)$ to obtain the posterior probability density $L(\mu)$:

$$L(\mu) = \frac{1}{C} \int_{\boldsymbol{\theta}} p(\text{data}\,|\,\mu\,s(\boldsymbol{\theta}) + b(\boldsymbol{\theta}))\,\rho(\boldsymbol{\theta})\,\pi(\mu)\,d\boldsymbol{\theta}, \qquad (\text{A.1})$$

where $p(\text{data}\,|\,\mu\,s + b)$ is the probability to observe the data as seen in an experiment assuming the $\mu\,s + b$ hypothesis. The function $\rho(\boldsymbol{\theta})$ is the density function describing our prior belief in the values of the nuisance parameters which affect the predicted signal and background event yields and distributions, and is typically a product of prior densities for each of the θ_i. Popular functional choices for individual nuisance parameter prior densities $\rho(\theta_i)$ are: Gaussian (often truncated so that all signal and background predictions are non-negative), log-normal, Gamma, or flat (either constrained, a so-called box distribution, or not). The function $\pi(\mu)$ is the prior probability density for the signal strength, and is commonly taken to be uniform for $\mu \geq 0$ and zero for $\mu < 0$. Other priors are possible, but have hardly ever been used in high energy physics. The constant C is set by requiring that $\int L(\mu)d\mu = 1$.

The probability $p(\text{data}\,|\,\mu\,s + b)$ can be expressed as the product of Poisson probabilities for the number of observed (or simulated) events (n_k) in each bin (k) given the expected event rates per bin $\mu\,s_k + b_k$:

$$p(\text{data}\,|\,\mu\,s + b) \;=\; \prod_k \frac{(\mu\,s_k + b_k)^{n_k}}{n_k!}\,e^{-(\mu\,s_k + b_k)} \qquad (\text{A.2})$$

$$=\; e^{-(\mu\,S+B)}\prod_k \frac{(\mu\,s_k + b_k)^{n_k}}{n_k!}, \qquad (\text{A.3})$$

where $\mu\,S + B = \mu\sum_k s_k + \sum_k b_k$ is the total expected event rate. An unbinned approach to data can be thought of as a binned analysis in the limit of infinitely narrow bins in some observable x, which in general can be multi-dimensional. In this case, the function $p(\text{data}\,|\,\mu\,s + b)$, up to an irrelevant constant factor, becomes:

$$p(\text{data}\,|\,\mu\,s + b) \;\sim\; e^{-(\mu\,S+B)}\prod_i \mathcal{P}(\,x_i\,|\,\mu\,s + b\,), \qquad (\text{A.4})$$

where index i runs over all events, and $\mathcal{P}(\,x_i\,|\,\mu\,s + b\,)$ is an event density function of x such that the expected event rate in the vicinity of a given value of x is predicted as $\mathcal{P}(\,x\,|\,\mu\,s + b\,)dx$.

Integration over nuisance parameters in Eq. (A.1) is known as marginalisation. Marginalising the nuisance parameters sums the credibility of the parameter of interest over all possible values of the nuisance parameters, including the effects of systematic uncertainty, even far from the central predictions $\tilde{\boldsymbol{\theta}}$. This integration step also serves to constrain the values of the nuisance parameters *in situ* because the kernel of the integral is large for nuisance parameter values that fit the data well and is vanishingly small for nuisance parameter values that fit the data poorly. The inclusion in the combination of data sets that constrain nuisance parameters helps improve the sensitivity of the Bayesian limits in much the same way that fits to nuisance parameter values improve the sensitivity of $\mathrm{CL_s}$ limits as discussed in Sec. A.2.2.2. The integrals over the space of nuisance parameters are often performed using Markov Chain Monte Carlo methods, such as the Metropolis–Hastings algorithm.[6] A benefit of this procedure is that the posterior credibility density distributions for the nuisance parameters can be calculated alongside that for the signal, and inspecting these is an important validation step of the analysis. If one or more nuisance parameters are pulled multiple sigmas from their central values, or if the posterior uncertainties are unusually small for one or more nuisance parameters, this behavior ought to be investigated and explained. It is also useful to know if a nuisance parameter is driven against a boundary in its prior distribution.

Figure A.1 gives examples of Bayesian posterior probability densities $L(\mu)$ for experimental situations without or with an event excess. The distinction is whether the maximum of the posterior probability density is reached at zero signal strength or at a positive value.

The Bayesian one-sided 95% CL upper limit on μ is extracted using the following equation:

$$\int_0^{\mu_{95\%\mathrm{CL}}} L(\mu) \, d\mu = 0.95. \tag{A.5}$$

This equation implies that one decides *a priori* to set limits only on high values of signal strength μ, even in a situation when a large excess of events is observed and small values of μ become just as unlikely as high values. Defining Bayesian credibility regions with an upper and a lower bound is performed with the same posterior probability density function; the procedure is described in Sec. A.4.

The experimental sensitivity is characterised by the limits expected to be set in the absence of a signal. These are computed by simulating many repeated runs of the experiment under the assumption of the background-only hypothesis, and computing an observed limit for each set of pseudo-data. Pseudo-data are simulated according to the Poisson distribution,

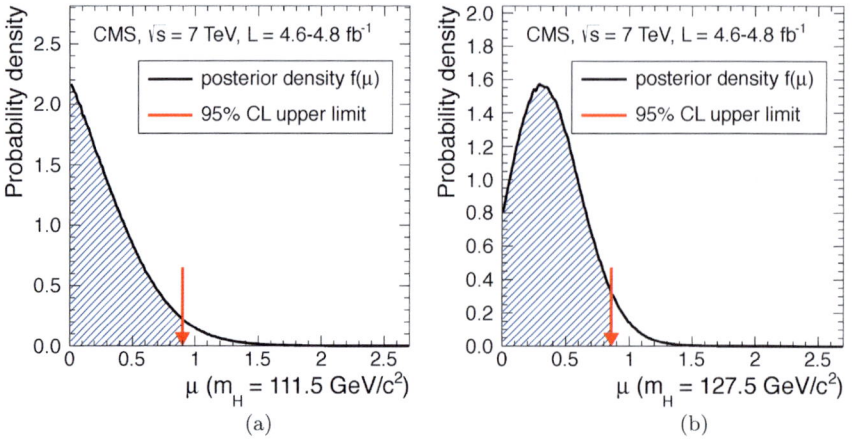

Fig. A.1   Examples of the Bayesian posterior probability density $L(\mu)$ for cases (a) without and (b) with an event excess observed. The 95% CL limit $\mu_{95\%\mathrm{CL}}$ is defined such that the integral of the shaded area for $0 \leq \mu \leq \mu_{95\%\mathrm{CL}}$ equals 0.95. Note that $\mu = 1$ is excluded in both cases. The plots are taken from Ref. [7].

assuming event rates $b(\boldsymbol{\theta})$. The Bayesian approach is to fluctuate the values of the nuisance parameters $\boldsymbol{\theta}$ for each pseudo-experiment according to their priors $\rho(\boldsymbol{\theta})$, so that the expected limit distribution is summed over all values the nuisance parameters can take, according to how much credibility they have. This is an important step when computing the sensitivity of an experiment that has not yet run and only highly uncertain *a priori* predictions are available for the signal and background yields. After the experiment has collected data, more is known about the expected backgrounds and the signal efficiencies, and the sensitivity may be updated. Alternatively, one may be tempted to take a seemingly conservative approach by setting the nuisance parameter values to the values that correspond to the least predicted sensitivity. One must be careful however, as nuisance parameter values set to increase the background prediction weaken the sensitivity, but strengthen the observed limit (for a given observed event count, the larger the assumed background is, the less room is left for a possible signal).

Since the distribution of expected limits is typically asymmetrical, the sensitivity is quoted as the median expected limit. Typically the distribution of expected limits is indicated by showing intervals containing 68% and 95% of the integral of the distribution, centered on the median. The quantiles are thus 0.025, 0.16, 0.5, 0.84, and 0.975. When these intervals are indicated on

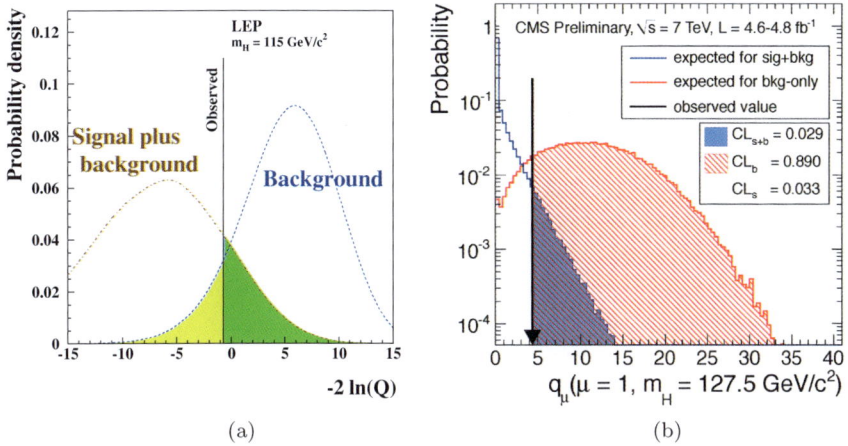

Fig. A.2 Examples of distributions of the test statistics $-2\ln Q$ and q_μ defined respectively by (a) Eq. (A.6) and (b) Eq. (A.7) for the signal+background and background-only hypotheses. Plots (a) and (b) are taken from Refs. [10] and [7], respectively.

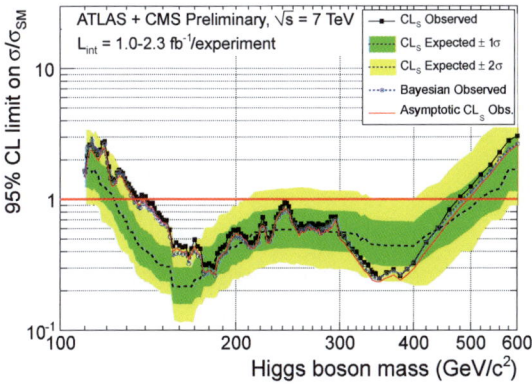

Fig. A.3 An example of limits on signal strength μ as obtained by generating pseudo-observations (exact) and via asymptotic formula (approximate). Also shown are limits obtained with the Bayesian technique. The results of the three calculations are very similar in the full Higgs boson mass range. The plot is taken from Ref. [16].

a plot of observed and expected limits versus m_H, they are usually shown with colored bands (e.g., see Fig. A.3).

The Bayesian method, in addition to the frequentist methods described in the next section, was frequently used to quantify results of Higgs boson searches at the Tevatron. Its usage at LEP and the LHC was less prominent.

A.2.2. *Frequentist approach and its modifications*

A.2.2.1. *Classical frequentist*

The classical frequentist approach is formulated for the case of no systematic uncertainties and begins by defining a test statistic q_μ designed to discriminate signal-like from background-like events. The test statistic summarises all signal-vs-background discriminating information in one number. By the Neyman–Pearson lemma,[8] the ratio of likelihoods Q is the most powerful discriminator. For a number of practical reasons, the actual quantity used is a logarithm of the ratio, or more accurately, $-2 \ln Q$:

$$q_\mu = -2 \ln \frac{p(\text{data} \mid \mu\, s + b)}{p(\text{data} \mid b)}. \tag{A.6}$$

Modulo the modifications associated with handling systematic uncertainties, this is the test statistic used in quantifying null Higgs boson search results at LEP and the Tevatron in the frequentist paradigm context. In LEP papers, this test statistic was refered to as $-2 \ln Q$, and in Tevatron papers, it was denoted LLR. There is another definition of the test statistic that has taken a prominent role at the LHC:

$$q_\mu = -2 \ln \frac{p(\text{data} \mid \mu\, s + b)}{p(\text{data} \mid \hat{\mu}\, s + b)}, \quad \text{with a constraint: } 0 \le \hat{\mu} \le \mu, \tag{A.7}$$

where $\hat{\mu}$ maximises the likelihood $p(\text{data} \mid \mu\, s + b)$. The advantage of this test statistic is that its distribution can be approximated by asymptotic formulae based on the theorems of Wilks and Wald, as derived in Ref. [9]. The upper bound on $\hat{\mu}$ ($\hat{\mu} \le \mu$) is needed when one desires to set one-sided limits only on high values of signal strength μ, even if an excess of events is observed.

Having chosen the test statistic q_μ, its distributions are constructed under the signal+background and background-only hypotheses by means of generating toy pseudo-observations according to the very same Poisson probabilities $p(\text{data} \mid \text{rate})$. Figure A.2 shows examples of distributions of the test statistic q_μ defined by Eqs. (A.6) and (A.7) for the hypotheses of signal+background ($\mu = 1$) and background-only ($\mu = 0$).

For the test statistic defined by Eq. (A.6), experimental outcomes with $q_\mu > 0$ are more likely to appear under the background-only hypothesis than under the background+signal assumption. Assuming the signal+background hypothesis, the smaller number of observed events, the larger value of the test statistic is. For the test statistic defined by Eq. (A.7), the test statistic is always positive definite; the smaller number of observed events, the larger value of the test statistic is.

Using these distributions, one can then evaluate the probability for the observed value q_μ^{data} to be as or less compatible with the background+signal hypothesis. Such a probability, $P(q_\mu \geq q_\mu^{\text{data}} \mid \mu\, s + b)$, is denoted as $\text{CL}_{\text{s+b}}$. These probabilities correspond to the green and blue areas in Fig. A.2 (a) and Fig. A.2 (b), respectively. In the classical frequentist approach, one says that the signal is excluded at, say, 95% CL, if $\text{CL}_{\text{s+b}} = 0.05$.

However, such a definition has a pitfall: by taking the signal strength equal to zero, one expects, by construction, that $\text{CL}_{\text{s+b}} \leq 0.05$ with a 5% chance in the background-only hypothesis — hence, 5% of all searches will end up excluding a signal of zero strength. In these cases, what has actually been observed is a downward fluctuation of the background. The exclusion of a zero-strength signal is certainly a questionable physics-wise result, even though proper mathematical coverage is guaranteed by the method. The problem with excluding a signal of zero strength is that an experiment cannot possibly test for the presence or absence of such a signal, and thus should not make a statement about it. To prevent, at least partially, the inference of a signal in presence of such downward fluctuations, a number of solutions have been suggested.

A.2.2.2. *Modifications of the classical frequentist method*

A method of constructing unified (i.e. one/two-sided) confidence intervals was suggested in Ref. [11] by Feldman and Cousins (FC). In this method, confidence intervals are constructed using ranking of experimental outcomes based on the value of the likelihood-ratio test statistic:

$$q_\mu \;=\; -2 \ln \frac{p(\text{data} \mid \mu\, s + b)}{p(\text{data} \mid \hat{\mu}\, s + b)}, \quad \text{with a constraint: } 0 \leq \hat{\mu}, \qquad (A.8)$$

where $\hat{\mu}$ maximises the likelihood $p(\text{data} \mid \mu\, s + b)$. Such construction automatically protects the limits on signal strength from the undesired effects of downward fluctuations of background, preserves coverage, and does not suffer from undercoverage due to having to make flip-flop decisions between reporting one-sided upper limits (no excess) and two-sided intervals when a considerable excess of events is observed. One, however, faces a conundrum: the FC method starts giving a lower limit on signal strength μ for excesses not yet significant enough for claiming a discovery (see Sec. A.3). To avoid the "inconvenience" of giving a statistical interpretation of reporting a lower limit on signal strength, while not claiming an observation of a signal, one can choose to report upper limits only — the price is overcoverage for the cases in which an excess of events is observed.

At the time of LEP, the so-called modified frequentist approach was introduced with the same goal to "protect" against too-strong statements made about vanishingly weak signals when downward fluctuations occur in the observed data.[12–14] In this method, in addition to probability $CL_{s+b} = P(q_\mu \geq q_\mu^{data} | \mu s + b)$, one also calculates $CL_b = P(q_\mu \geq q_\mu^{data} | b)$, by simulating pseudo-data for assuming the background-only hypothesis, and then calculates the quantity CL_s as the ratio of these two probabilities:

$$CL_s = \frac{CL_{s+b}}{CL_b}. \tag{A.9}$$

The method does not prescribe the test statistic to be used. In the modified frequentist approach, it is this value, CL_s, that is required to be less than or equal to 0.05 in order to declare the 95% CL exclusion. By construction, the CL_s-based limits are one-sided. For $\mu = 0$, $CL_s \equiv 1$; hence, $\mu = 0$ cannot be excluded, regardless of how low the observed event count is. The price of the protection from background downward fluctuations is a gradual increase in the overcoverage as one observes fewer and fewer events. For an observation right on the top of the background-only expectation ($CL_b \sim 0.5$), CL_s is twice as large as CL_{s+b}.

Between the two modifications, Feldman–Cousins and CL_s, the latter was most frequently used at LEP, the Tevatron, and the LHC. However, there were distinct variations of the CL_s method, stemming from the differences in the choice of the test statistic and in the methods used to incorporate systematic uncertainties.

A.2.2.3. *Introducing systematic uncertainties*

Systematic uncertainties on the predicted signal and background rates, $s(\boldsymbol{\theta})$ and $b(\boldsymbol{\theta})$, are introduced via modifications to the test statistic itself and/or the way pseudo-data are generated. In the following, the prior densities for the nuisance $\boldsymbol{\theta}$ will be written as $\rho(\boldsymbol{\theta}|\tilde{\boldsymbol{\theta}})$, where $\tilde{\boldsymbol{\theta}}$ is the "nominal" best-guess value of the nuisance parameter.

At LEP, the test statistic given by Eq. (A.6) was used; it was always evaluated at the nominal values of the signal and background rates, i.e. at $s(\tilde{\boldsymbol{\theta}})$ and $b(\tilde{\boldsymbol{\theta}})$. The effect of systematic uncertainties was then introduced via modifying $s(\boldsymbol{\theta})$ and $b(\boldsymbol{\theta})$ before each pseudo-data set was generated by drawing random numbers from the $\rho(\boldsymbol{\theta}|\tilde{\boldsymbol{\theta}})$ distributions. This method was first introduced to the field by Cousins and Highland[15] and is now known as a hybrid Bayesian–frequentist method, since the treatment of nuisance parameters in this case is explicitly Bayesian.

At the Tevatron, the hybrid Bayesian–frequentist approach to generating the pseudo-data remained the same as at LEP, but the test statistic given by Eq. (A.6) was redefined in order to improve the sensitivity in the face of large systematic uncertainties. The Poisson-like likelihoods were extended to include the nuisance parameter densities $\rho(\boldsymbol{\theta}|\tilde{\boldsymbol{\theta}})$:

$$\mathcal{L}(\text{data} \,|\, \mu, \boldsymbol{\theta}) = p(\,\text{data} \,|\, \mu \cdot s(\boldsymbol{\theta}) + b(\boldsymbol{\theta})\,) \cdot \rho(\boldsymbol{\theta}|\tilde{\boldsymbol{\theta}}). \qquad (\text{A.10})$$

Before taking the ratio, both the numerator and the denominator likelihoods were maximised with respect to nuisance parameters. The test statistic then takes the following form:

$$q_\mu \;=\; -2 \ln \frac{\mathcal{L}(\text{data} \,|\, \mu, \hat{\boldsymbol{\theta}}_\mu)}{\mathcal{L}(\text{data} \,|\, 0, \hat{\boldsymbol{\theta}}_0)}, \qquad (\text{A.11})$$

where $\hat{\boldsymbol{\theta}}_\mu$ and $\hat{\boldsymbol{\theta}}_0$ are maximum likelihood estimators for the signal+background hypothesis (with the signal strength factor μ) and for the background-only hypothesis ($\mu = 0$).

At the LHC, the ATLAS and CMS experiments started to use the profile likelihood test statistic given by Eq. (A.7), which was further modified to incorporate systematic uncertainties in the definition of likelihoods, as described below. The overall treatment of systematic uncertainties was conceptually different with respect to that used at LEP and the Tevatron; it was brought to be closer to the frequentist treatment of data fluctuations. First, following the Bayesian paradigm, systematic uncertainty densities $\rho(\boldsymbol{\theta}|\tilde{\boldsymbol{\theta}})$ were reinterpreted as posteriors of some measurements of $\tilde{\boldsymbol{\theta}}$, either real (e.g., measurements in control regions) or imaginary (e.g., uncertainties on theoretical cross sections):

$$\rho(\boldsymbol{\theta}|\tilde{\boldsymbol{\theta}}) \sim p(\tilde{\boldsymbol{\theta}}|\boldsymbol{\theta}) \cdot \pi_\theta(\boldsymbol{\theta}), \qquad (\text{A.12})$$

where priors $\pi_\theta(\boldsymbol{\theta})$ were assumed to be flat. Second, these initial best-guess values $\tilde{\boldsymbol{\theta}}$ were treated on par with data in construction of the likelihoods and in generation of pseudo-data. The likelihood was as follows:

$$\mathcal{L}(\text{data}, \tilde{\boldsymbol{\theta}} \,|\, \mu, \boldsymbol{\theta}) = p(\text{data} \,|\, \mu \cdot s(\boldsymbol{\theta}) + b(\boldsymbol{\theta})) \cdot p(\tilde{\boldsymbol{\theta}}|\boldsymbol{\theta}), \qquad (\text{A.13})$$

which formally coincided with that used at the Tevatron. Then, as in the case of the Tevatron, to take advantage of data constraining *a priori* uncertainties, the test statistic was defined with the numerator and denominator likelihoods maximised:

$$q_\mu \;=\; -2 \ln \frac{\mathcal{L}(\text{data}, \tilde{\boldsymbol{\theta}} \,|\, \mu, \hat{\boldsymbol{\theta}}_\mu)}{\mathcal{L}(\text{data}, \tilde{\boldsymbol{\theta}} \,|\, \hat{\mu}, \hat{\boldsymbol{\theta}})}, \qquad 0 \le \hat{\mu} \le \mu, \qquad (\text{A.14})$$

where the pair of parameters $\hat{\mu}$ and $\hat{\boldsymbol{\theta}}$ gives the global maximum of the likelihood. Finally, the treatment of nuisance parameters in generation of pseudo-observations is where the approach taken by LHC was very different from that used at LEP and the Tevatron. Using the best values of the nuisance parameters for the background-only and for the signal+backhround hypotheses ($\boldsymbol{\theta_0}$ and $\boldsymbol{\theta_\mu}$, respectively), the pseudo-data and pseudo-measurements of $\tilde{\boldsymbol{\theta}}$ were generated. In other words, instead of using a Bayesian–frequentist hybrid method, the nuisance parameters were treated in a nearly pure frequentist way.

A.2.2.4. Summary of the frequentist approaches used in the Higgs boson search at LEP, the Tevatron, and the LHC

For comparison purposes, the differences in the CL_s method, as used at LEP, the Tevatron, and the LHC, are summarised in Table A.1 below. The LEP prescription does not allow one to take full advantage of the constraints imposed on the nuisance parameters by the data used in the statistical analysis. The Tevatron and LHC versions of CL_s, though constructed differently, in practice give nearly identical results. The benefit of the LHC-type CL_s is that it uses a test statistic with useful asymptotic properties, as shown in Fig. A.3. Also, the sampling distributions of the test statistic can be built following the pure frequentist language.

In all cases, the sensitivity of the experiment is given by the median limit expected to be set in the absence of a signal, and in nearly all cases, the intervals containing 68% and 95% of the limits expected in pseudo-data are shown, centered on the median.

Table A.1   Comparison of CL_s definitions as used at LEP, Tevatron, and LHC.

Collider	Test statistic	Profiled?	Test statistic sampling
LEP	$q_\mu = -2 \ln \dfrac{\mathcal{L}(\text{data}\|\mu, \tilde{\theta})}{\mathcal{L}(\text{data}\|0, \tilde{\theta})}$	no	Bayesian–frequentist hybrid
Tevatron	$q_\mu = -2 \ln \dfrac{\mathcal{L}(\text{data}\|\mu, \hat{\theta}_\mu)}{\mathcal{L}(\text{data}\|0, \hat{\theta}_0)}$	yes	Bayesian–frequentist hybrid
LHC	$q_\mu = -2 \ln \dfrac{\mathcal{L}(\text{data}\|\mu, \hat{\theta}_\mu)}{\mathcal{L}(\text{data}\|\hat{\mu}, \hat{\theta})}$ $(0 \le \hat{\mu} \le \mu)$	yes	frequentist

A.3. Significance of an excess of events

A.3.1. *Quantifying an excess of events for a given model*

In the case of observing an excess of events, characterisation of it begins with evaluating the p-value, i.e. the probability for background alone to yield an outcome as signal-like as observed. This can be done by generating background-only pseudo-data and building up the corresponding probability distribution for the test statistic of choice.

The four test statistics given in Eqs. (A.6), (A.11), (A.8), and (A.14) can be used. The first two compare the models $\mu = 0$ with $\mu = 1$, while the profile likelihood ratio used by the LHC is constructed for $\mu = 0$ and $\hat{\mu}$, where $\hat{\mu}$ is either unconstrained or constrained to be positive, which makes no difference to the tail of the distribution:

$$q_0 = -2 \ln \frac{\mathcal{L}(\text{data} \,|\, 0, \hat{\boldsymbol{\theta}}_0)}{\mathcal{L}(\text{data} \,|\, \hat{\mu}, \hat{\boldsymbol{\theta}})}. \tag{A.15}$$

For the first two test statistics, Eqs. (A.6) and (A.11), observations with a large excess of events would form a left-hand tail (see Fig. A.2), while the profile likelihood test statistic would stretch to the right as shown in Fig. A.4.

The p-value, i.e. the probability of getting an observation as or less compatible as seen in data for the background-only hypothesis, is then defined as $P(q_1 \leq q_1^{\text{data}})$ for the test statistics given by Eqs. (A.6) and (A.11), and $P(q_0 \geq q_0^{\text{data}})$ for the profile likelihood test statistic given by Eq. (A.15).

In addition to the p-value, the significance Z, commonly described as the number of standard deviations, is reported. A significance Z of three standard deviations is the customary criterion for "evidence", while a significance of five standard deviations is the commonly accepted criterion for "observation" of a new particle or process. Two conventions have been used to compute Z from the p-value (one-sided or two-sided normal distribution tail probability):

$$p = \int_Z^\infty \frac{1}{\sqrt{2\pi}} \exp(-x^2/2)\, dx, \tag{A.16}$$

$$p = \int_{-\infty}^{-Z} \frac{1}{\sqrt{2\pi}} \exp(-x^2/2)\, dx + \int_Z^\infty \frac{1}{\sqrt{2\pi}} \exp(-x^2/2)\, dx. \tag{A.17}$$

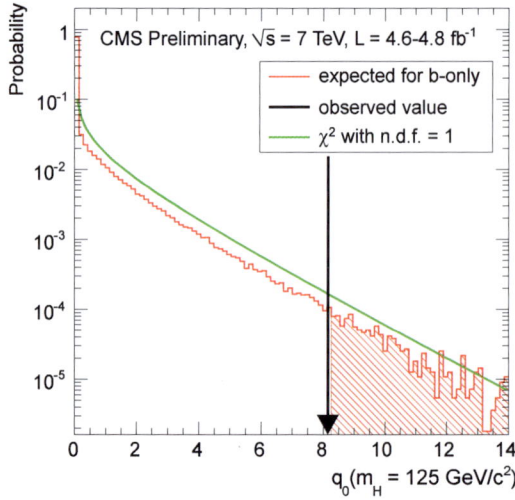

Fig. A.4   Example of a distribution of the profile likelihood test statistic q_0 (Eq. (A.15)). The shaded area represents the p-value, or the probability $P(q_0 \geq q_0^{\text{data}})$. The solid green curve shows the asymptotic χ^2-distribution for one degree of freedom. The plot is taken from Ref. [7].

The two-sided convention of Eq. (A.17) was used by the four LEP collaborations, while the one-sided convention of Eq. (A.16) was used by the two Tevatron collaborations and the two LHC collaborations. In the one-sided convention, the Type-I error rate, that is, the probability that the null hypothesis will generate an experimental outcome that elicits a false claim of evidence ($Z > 3$), is approximately 0.00135, and the Type-I error rate for observation ($Z > 5$) is approximately 2.87×10^{-7}. For the two-sided convention, the Type-I error rates are twice as large.

In the asymptotic regime the profile likelihood test statistic (Eq. (A.15)) has the very attractive property of being distributed as a half χ^2 for one degree of freedom, which allows one to approximately estimate the significance, Z, as defined by Eq. (A.16) from the following formula:

$$Z \approx \sqrt{q_0^{\text{data}}}. \tag{A.18}$$

The asymptotic approximation gives very satisfactory results for significance estimations even when one is far from the asymptotic regime (e.g., if one observes a few events, while the expected background rate is less than one event).

A.3.2. Look-elsewhere effect

In the Higgs boson search, the experimental collaborations scan over Higgs boson mass hypotheses and look for the one giving the minimum local p-value $p_{\text{local}}^{\min}$ (corresponding to local significance Z_{local}), which describes the probability of a background fluctuation for that particular Higgs boson mass hypothesis, as shown in Fig. A.5 (b). The probability to find a fluctuation with a local p-value lower or equal to the observed $p_{\text{local}}^{\min}$ anywhere in the explored mass range is referred to as the global p-value, p_{global}:

$$p_{\text{global}} = \text{P}(p_0 \leq p_{\text{local}}^{\min} \,|\, b). \tag{A.19}$$

The fact that the global p-value can be significantly larger than $p_{\text{local}}^{\min}$ is often referred to as the look-elsewhere effect (LEE). The global significance (and global p-value) of the observed excess can be evaluated in this case by generating pseudo-datasets, which, however, becomes too CPU-intensive and not practical for very small p-values. Therefore, the method suggested in Ref. [18] was used. The relationship between the global and local p-values is given by:

$$p_{\text{global}} = p_{\text{local}}^{\min} + C \cdot e^{-Z_{\text{local}}^2/2}. \tag{A.20}$$

Assuming one can simulate correlations in data selected for different Higgs boson mass hypotheses, the constant C can be found by generating a relatively small set of pseudo-data and then use it to evaluate the global p-value corresponding to the value $p_{\text{local}}^{\min}$ observed in the experiment.

For a very wide mass range, the constant C can be evaluated directly from the data[19] by counting the number N_{up} of times that $\hat{\mu}(m_H)$ crosses the line $\mu = 0$ in the upwards direction, as shown in Fig. A.5 (c), and setting $C = N_{\text{up}}$.

A.3.3. Discovery sensitivity

In analogy to the procedure to compute the sensitivity of the experiment using the median expected limit, the discovery sensitivity is quantified using the median expected p-value assuming the presence of a signal. This is often quoted as the median expected Z value. Sensitivities are often shown without correction for the LEE, as curves on plots of median expected p-values as functions of the Higgs boson mass (e.g, see Fig. A.5 (b)).

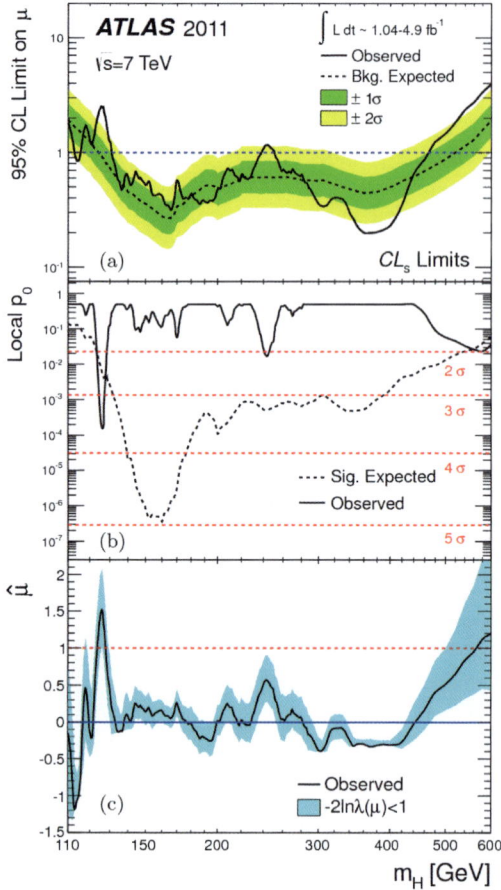

Fig. A.5 (a) Observed and expected limits on signal strength vs hypothesized Higgs boson mass. (b) Observed p-value vs hypothesized Higgs boson mass. (c) Best-fit μ as a function of a hypothesized Higgs boson mass. The number of times μ crosses 0 from negative to the positive value is called the number of upcrossings. The number of upcrossings can be used to evaluate the look-elsewhere effect directly from data as described in the text. The plot is taken from Ref. [17].

A.4. Extracting signal model parameters

Signal model parameters $\boldsymbol{a}$ (the signal strength modifier μ can be one of them) are evaluated from a scan of the profile likelihood ratio $q(\boldsymbol{a})$:

$$q(\boldsymbol{a}) = -2 \ln \frac{\mathcal{L}(\mathrm{obs} \,|\, s(\boldsymbol{a}) + b, \, \hat{\boldsymbol{\theta}}_a)}{\mathcal{L}(\mathrm{obs} \,|\, s(\hat{\boldsymbol{a}}) + b, \, \hat{\boldsymbol{\theta}})}. \qquad (A.21)$$

The parameter values $\hat{a}$ and $\hat{\theta}$ that maximise the likelihood, $\mathcal{L}(\text{obs} \mid s(\hat{a}) + b, \hat{\theta}) = \mathcal{L}_{\max}$, are called the best-fit set. The 68% (95%) CL on a single parameter of interest a is evaluated from $q(a) = 1$ (3.84) with all other unconstrained model parameters treated in the same way as the nuisance parameters. The 2D 68% (95%) CL contours for pairs of parameters are derived from $q(a_1, a_2) = 2.3$ (6.0), as shown in Fig. A.6 (a) for a pair of parameters of interest (m_H, μ). One should keep in mind that boundaries of 2D confidence regions projected on either parameter axis are not identical to the 1D confidence interval for that parameter.

Alternatively, model parameters can be extracted using the Bayesian technique. For example, the posterior probability density $L(\boldsymbol{a})$ is computed by marginalising over the nuisance parameters, usually using a uniform prior density for the parameters of interest $\boldsymbol{a}$. The best-fit values $\hat{\boldsymbol{a}}$ are those which maximise $L(\boldsymbol{a})$, and the 68% (95%) CL region is the smallest-area region that contains 68% (95%) of the integral of the posterior density. Figure A.6 (b) shows the Bayesian posterior density $L(\boldsymbol{a})$ and 68% (95%) CL contours for the same datasets used for the profile likelihood scan presented in Fig. A.6 (a). The CL contours on both plots are remarkably similar. As is the case with limits, the marginalisation of the nuisance parameters explores the behavior of the likelihood function for all values of the nuisance parameters and not just those near the maximum.

The measurement sensitivity is quantified by computing the distribution of expected measurement uncertainties in pseudo-data drawn from

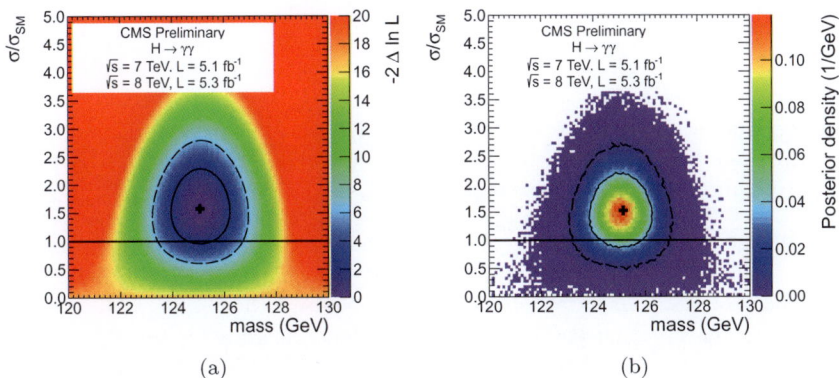

(a)

(b)

Fig. A.6 Examples of (a) 2D profile likelihood scan and (b) Bayesian posterior likelihood function. Solid (dashed) lines indicate 68% (95%) CL intervals. The plots are taken from Ref. [7].

models with known values of the parameters of interest. Usually the uncertainty obtained in the data fit is compared with the distribution of expected uncertainties in the process of validating a result in order to determine if the experiment is much luckier or unluckier than expected, and the median expected uncertainty is the figure of merit used to optimise the analysis. An observed uncertainty that is very different from what is expected requires investigation and explanation. The expected measurement uncertainty is also of value when averaging parameters, as the observed uncertainty is often correlated with the value of the parameter being measured, and this can bias weighted averages. Optimising simultaneous measurements of two or more parameters leaves more choices for figures of merit to use.

References

1. L. Lyons, *Statistics for Nuclear and Particle Physicists.* Cambridge University Press (1986).
2. B. P. Roe, *Probability and Statistics in Experimental Physics.* Springer (1992).
3. R. J. Barlow, *Statistics: A Guide to the Use of Statistical Methods in the Physical Sciences.* John Wiley & Sons (1993).
4. G. Cowan, *Statistical Data Analysis.* Oxford University Press (1998).
5. F. James, *Statistical Methods in Experimental Physics.* World Scientific Publishing Company, Inc. (2006).
6. N. Metropolis, A. W. Rosenbluth, M. N. Rosenbluth, A. H. Teller, and E. Teller, Equation of state calculations by fast computing machines, *J.Chem.Phys.* **21**, 1087 (1953). doi: 10.1063/1.1699114.
7. G. Petrucciani, *The Search for the Higgs Boson at CMS.* Springer (2013).
8. J. Neyman and E. Pearson, On the problem of the most efficient tests of statistical hypotheses, *Phil. Trans. of the Royal Soc. of London A.* **31**, 289 (1933).
9. G. Cowan, K. Cranmer, E. Gross, and O. Vitells, Asymptotic formulae for likelihood-based tests of new physics, *Eur.Phys.J.* **C71**, 1554 (2011). doi: 10.1140/epjc/s10052-011-1554-0.
10. LEP Working Group for Higgs boson searches, ALEPH Collaboration, DELPHI Collaboration, L3 Collaboration, OPAL Collaboration, Search for the Standard Model Higgs boson at LEP, *Phys.Lett.* **B565**, 61–75 (2003). doi: 10.1016/S0370-2693(03)00614-2.
11. G. J. Feldman and R. D. Cousins, A unified approach to the classical statistical analysis of small signals, *Phys.Rev.* **D57**, 3873–3889 (1998). doi: 10.1103/PhysRevD.57.3873.
12. T. Junk, Confidence level computation for combining searches with small statistics, *Nucl.Instrum.Meth.* **A434**, 435–443 (1999). doi: 10.1016/S0168-9002(99)00498-2.

13. A. Read, Modified frequentist analysis of search results (The CL_s method). In eds. F. James, Y. Perrin, and L. Lyons, *Workshop on Confidence Limits*, CERN, Geneva, Switzerland, 17–18 Jan 2000 (2000). URL https://cds.cern.ch/record/451614.

14. A. L. Read, Presentation of search results: The CL_s technique, *J. Phys.* **G28**, 2693–2704 (2002). doi: 10.1088/0954-3899/28/10/313.

15. R. D. Cousins and V. L. Highland, Incorporating systematic uncertainties into an upper limit, *Nucl. Instrum. Meth.* **A320**, 331–335 (1992). doi: 10.1016/0168-9002(92)90794-5. Revised version.

16. ATLAS and CMS Collaborations, Combined Standard Model Higgs boson searches with up to 2.3 inverse femtobarns of pp collision data at $\sqrt{s} = 7$ TeV at the LHC, ATLAS-CONF-2011-157, CMS-PAS-HIG-11-023 (2011). URL http://cds.cern.ch/record/1399599, https://cds.cern.ch/record/1399607.

17. ATLAS Collaboration, Combined search for the Standard Model Higgs boson using up to 4.9 fb^{-1} of pp collision data at $\sqrt{s} = 7$ TeV with the ATLAS detector at the LHC, *Phys. Lett.* **B710**, 49–66 (2012). doi: 10.1016/j.physletb.2012.02.044.

18. E. Gross and O. Vitells, Trial factors for the look elsewhere effect in high energy physics, *Eur. Phys. J.* **C70**, 525–530 (2010). doi: 10.1140/epjc/s10052-010-1470-8.

19. ATLAS and CMS Collaborations, Procedure for the LHC Higgs boson search combination in summer 2011 (2011). URL https://cds.cern.ch/record/1379837.

Appendix B

Multivariate analysis techniques

Josh Bendavid*, Wade C. Fisher[†], and Thomas R. Junk[‡]

*CERN
CH-1211 Geneva 23, Switzerland
[†]Department of Physics and Astronomy
Michigan State University
East Lansing, Michigan, 48824, USA
[‡]Fermi National Accelerator Laboratory
Batavia, Illinois, 60510, USA

B.1. Introduction

The end products of experimental data analysis are designed to be simple and easy to understand: hypothesis tests and measurements of parameters. The experimental data themselves however are voluminous and complex. In modern collider experiments, many petabytes of data must be processed in search of rare new processes which occur together with much more copious background processes that are of less interest to the task at hand. The systematic uncertainties on the background may be larger than the expected signal in many cases. The statistical power of an analysis and its sensitivity to systematic uncertainty can therefore usually both be improved by separating signal events from background events with higher efficiency and purity.

Events are customarily selected using definite requirements on reconstructed quantities. Events that fail the selection requirements can be used to constrain background rates and kinematic properties. Since no selection requirements can separate the samples into pure signal and background samples, event samples will be mixtures. One may devise selection requirements that purify samples of signal and examine the rates of events on either side of each cut. A more optimal handling of the data is to form the distribution of a discriminant variable that has a characteristic distribution for the signal and a very different distribution for the sum of the backgrounds.

A typical variable of this type is the reconstructed invariant mass of a combination of final-state particles. This choice has been very effective in the discovery and measurement of new resonances. Not only is the signal present in a very intuitive way in a histogram of invariant mass with a sharp peak in it, but the sidebands can be used to estimate the background under the peak, reducing the uncertainty even if the background rate is not known *a priori*. Variables other than invariant masses, such as jet counts and measured angles, have been used in similar ways.

Higgs boson analyses frequently require the measurement of many-particle final states. Associated production of Higgs bosons and vector bosons requires selection of leptons, jets, missing transverse momentum, and frequently b-tags. Leptonic decays of $H \to W^+W^-$ and $H \to ZZ \to \ell^+\ell^-\ell^+\ell^-$ involve multiple leptons, and there are frequently jets and missing transverse energy accompanying the leptons in both the signal and the background processes. Particles may be missing or misidentified, and some objects measured by the detectors are more clearly identifiable than others. For example, tags for b-flavored jets are usually uncertain. Jets with clearly displaced vertices and/or soft lepton candidates from b decay are more cleanly tagged than other jets. Optimal usage of all of this information for each event is desired.

A further consideration is the optimization of an analysis for discovery, exclusion, or measurement. The criteria for these three tasks can be in conflict with each other if events are simply selected and counted. An example of this is the low-background, low-signal case. Using either the Bayesian or the CL_s methods, a signal with a strength of less than $-\ln(0.05) \approx 3$ events cannot be excluded at the 95% C.L., even with zero events observed in the data and zero expected background. But a discovery can be made with just one event as long as the background is small enough. An analysis would be more optimal for discovery if it can purify a small sample of events with very little background, while for exclusion or measurement, it may be more optimal to allow more signal and more background in the selection. Frequently, no set of cuts will simultaneously optimize both. To solve this dilemma, and also improve the sensitivity, analyzers should retain all events passing a loose selection and classify them according to how signal-like they are, giving each one its appropriate contribution to the statistical interpretation.

Multivariate analysis (MVA) techniques are natural tools to achieve these goals. They are methods that define mathematical functions of multiple input variables, such as an invariant mass or a lepton identification variable that by themselves may help to separate the signal from the background by varying amounts. These functions produce output variables, or discriminants, that optimally separate the signal from the background. They thus are tools for reducing the dimensionality of the input dataset in a way that attempts to minimize the loss of discriminating information. These output variables may be thought of as "scores," or rankings, of events as signal-like or background-like. These discriminant variables often have smooth distributions, but they are not required to. Their output distributions for the observed data and the predictions are easily input to the statistical framework described in Appendix A; indeed the statistical tools used in modern Higgs boson analyses are designed to make maximum use of the distributions measured for any variable and produce optimal results in simple forms. These methods however require effort and care in order to optimize them, to validate their performance, and to reliably estimate the effects of systematic uncertainties. This appendix discusses these topics in some detail. The historical development of MVA algorithms is discussed first, and then brief descriptions of common MVA methods used in the Higgs boson analyses in this book are given. Following that is a discussion of best practices and suggestions for validation, optimization, and evaluation of systematic uncertainty.

B.2. Historical development

The field of multivariate analysis techniques has been very active over the course of the last decades in many scientific fields, not just high-energy physics, and many techniques used in Higgs boson analyses were developed in other fields. Notably, however, decision trees and matrix-element techniques got their start in high-energy physics analyses.

Multivariate analysis methods found early application in particle physics via the use of decision trees.[1] These and other multivariate analysis methods became more widely used, especially in Higgs boson analyses, at LEP.[2-5] Typically events were selected using preselection requirements to identify events with the desired topology. Likelihood functions and neural networks were used to separate the signals from the remaining backgrounds. In many of the $ZH \to Zb\bar{b}$ searches, two-dimensional discriminant distributions were formed, such as m_{jj} vs. a neural-network variable that included the quality of the b-tags in the event as well as kinematic variables independent of m_{jj}. Two-dimensional discriminants allowed for easily-understood m_{jj} distributions to be examined with varying neural-network requirements. Nonetheless, it is difficult to fill a two-dimensional histogram with sufficient simulated background events to provide reliable predictions. With the goal of the preselection and the MVA to minimize the contributions of background processes in bins with prominent signals present, bins often became depleted in Monte Carlo (MC) simulated events. Smoothing techniques were applied, and systematic effects of these smoothing techniques came under intense scrutiny. At LEP, MVA techniques extended to the more basic tools, such as b-tagging of jets.[6]

Nearly all of the Tevatron searches for Higgs bosons used MVAs in some way, with widespread use in the basic tools such as b-tagging and lepton identification as well. It was found that neural networks could also be trained to output estimators for reconstructed variables, and not simply separate signals on one side of a histogram from backgrounds on another. Analyses commonly used cascades of separate MVAs, trained to separate signals from one background at a time.[7,8] This cascading technique gave a significant boost to the statistical power of the searches in which they were used because the variables that best separate a signal from one background may be different from those that separate it from another, and because a portion of the signal that is buried under a particler copious background may be useless in the training of a network to separate it from another background. Cascading MVAs also gave an additional benefit — separate samples enriched in each background were purified, so that the backgrounds could be separately fit and constrained with the data, instead of having just aggregate background yield with uncertain contributions from several component processes. More details on the use of MVA techniques in Higgs boson searches at the Tevatron can be found in Chapter 3, including differential MVA distributions for various final states.

The LHC Higgs searches and measurements also implemented MVA methods, often extending the applications beyond what was done at LEP and the Tevatron. These methods are described in the relevant chapters of this book and the reader is encouraged to look there for a more complete overview.

B.3. Common MVA methods

The following sections describe the details of methods that are commonly put to use in collider-physics data analysis. As the field of machine learning is rapidly evolving, more methods are expected to be developed in the future, either with improved power to separate signals from backgrounds in difficult cases, to be able to handle input variables with unusual distributions, or simply to be more convenient and automatic to apply, optimize, and validate. The literature contains more thorough reviews of MVA methods; see for example Refs. [9] and [10].

In the absence of systematic uncertainties, the optimal separation between two classes of events given a set of input variables is given by the likelihood ratio,[11] constructed from the ratio of probability densities in the multidimensional space, i.e. the local s/b ratio. The practical issue in high-energy physics is that the input variables are usually the result of a complex detector response, and therefore the local s/b is not known analytically or numerically *a priori*. Often a representative but finite set of events is available for both signal and background through MC simulation or data control regions. A common theme in multivariate classifier techniques is therefore to approximate the local s/b for a given set of input variables starting from finite training samples for signal and background. This estimate for the local s/b over the input variable space can then be used to construct an estimate of the optimal likelihood ratio test statistic for any given dataset, effectively reducing the dimensionality of the input variable space to a single output variable.

The procedure of optimization varies among MVA methods, but is typically linked to a learning algorithm or a procedure for improving the MVA's approximation of the multidimensional likelihood ratio. This is generally performed by preparing an independent sample of randomly simulated instances of the feature space (events) for a given pattern class. These *training samples* are used by the MVA with the known pattern class to optimally choose values for the degrees of freedom within the MVA. This procedure can be repeated for each pattern class included in the MVA's construction. To avoid the pitfall of the MVA learning features unique to the training samples, the performance of the MVA itself is then evaluated using a second, independent sample of simulated events referred to as the *testing sample*. Issues related to this training and testing procedure are discussed in more detail in Sec. B.7.

B.3.1. *Fisher's discriminant*

Fisher's linear discriminant (FLD) is a dimensional reduction technique that is designed to map a d-dimensional feature space to a scalar discriminant that is represented as a weighted, linear sum of the random variables in the feature space of two or more distinct classes. There are many such mappings for a linear model, which can be represented as:

$$f(\mathbf{x}) = \sum_i w_i x_i = \mathbf{w}^{\mathrm{T}} \mathbf{x} \tag{B.1}$$

where $\mathbf{w}$ is a vector of coefficients and the x_i represent the random variables in the d-dimensional feature space. In a model with two classes (for example, signal and background) the FLD is defined by choosing the coefficients that maximize the separation between the probability distribution functions (PDFs) of the linear discriminant for the signal and background classes. The method of maximizing the separation proposed by Fisher[12] is to choose coefficients that maximize:

$$S(\mathbf{w}) = \frac{(\mu_s - \mu_b)^2}{\sigma_s^2 + \sigma_b^2} \tag{B.2}$$

where μ_s and μ_b (σ_s^2 and σ_b^2) are the mean values (variances) of $f(\mathbf{x})$ PDF for the signal and background classes, respectively. The coefficients that maximize the optimization criterion ($\mathbf{w}^{\mathrm{max}}$) are generally estimated using simulated events to populate the feature space of the signal and background classes.

As can be inferred from the optimization criterion $S(\mathbf{w})$ and the linear discriminant $f(\mathbf{x})$, the FLD is most powerful in cases where the PDFs of the input feature variables are uncorrelated and normally distributed.

B.3.2. *Likelihood discriminants*

Likelihood discriminants (LDs) can be understood as a generalization of the Law of Total Probability ($p(s) = \sum_i p(s|x_i)p(x_i)$) and Bayes' theorem ($p(x|s) = p(s|x)p(x)/p(s)$), for a given class s and data x.[a] These two

---

[a] Here $p(A)$ represents the prior probability of outcome A and $p(A|x)$ is the conditional probability of A given x. In the context of theory A and data x, $p(x|A)$ is commonly referred to as the likelihood of A and $p(A|x)$ is the posterior probability of A.

concepts can be combined to describe the posterior probability for a given class, for example the signal class:

$$p(s|x) = \frac{p(x|s)p(s)}{\sum_i p(x|c_i)p(c_i)} \tag{B.3}$$

where the sum in the denominator runs over the total number of candidate classes (c_i), including the signal class. In practice, we typically have more than one feature and we do not have priors for the various classes (or do not wish to use them). To address this we replace the likelihood for a single feature $(p(x|s))$ with the joint likelihood over the full feature space $P(\mathbf{x}|s) = \prod_i p(x_i|s)$ and define the LD without priors. Here we assume only two event classes, signal (s) and background (b):

$$D(x) = \frac{P(\mathbf{x}|s)}{P(\mathbf{x}|s) + P(\mathbf{x}|b)} \tag{B.4}$$

Thus, the LD can be seen to vary between 0 and 1 depending on general agreement with the predicted signal feature space. It should be noted that this discriminant does not provide meaningful information about the degree of agreement with the background feature space. As with the Fisher discriminant, the LD is most powerful when the feature space PDFs are uncorrelated.

B.3.3. *Neural networks*

Artificial neural networks, or just neural networks (NNs), are examples of nonlinear discriminants that attempt to overcome the inherent deficiencies of linear discriminants, such as Fisher's discriminant. The NN is designed to emulate the structure of an interconnected network of neurons that communicate with one another. The general structure of the NN is layered groups of nodes that generate an output response based on the values of the inputs to the node. In a feed-forward NN, each layer receives inputs from an upstream layer and its output generates the input to the downstream layer. The two exceptions to this are the *input layer* that receives the raw information provided to the network and the *output layer* that represents the final NN response. The layers intermediate to the input and output layers are referred to as hidden layers. A single-layer NN discriminant that receives n inputs (x_i) to the output node can be viewed mathematically as:

$$D(\mathbf{x}) = A\left(w_0 + \sum_{i=1}^{n} w_i x_i\right) \tag{B.5}$$

where the w_i are weights that adjust the relative importance of the inputs and the activation function $A(\cdot)$ is a linear function of the weighted inputs, including the offset w_0 that typically sets the threshold for the activation function. The activation function is often taken as a logistic sigmoid, though other functions are common. So long as the activation function is monotonic, the single-layer NN is highly similar to a linear discriminant.

This single-layer NN formalism can be easily expanded to describe NN with hidden layers (often referred to as multi-layer perceptrons, or MLPs). For example, for a two-layer NN with one hidden layer containing m nodes:

$$D(\mathbf{x}) = A_{\text{out}} \left(w_0 + \sum_{i=1}^{m} w_i H_i(\mathbf{x}) \right) \tag{B.6}$$

$$H_i(\mathbf{x}) = A_i \left(v_{i0} + \sum_{j=1}^{m} v_{ij} x_j \right) \tag{B.7}$$

where the hidden layer weights v_i are independent of the input layer. This description of the NN discriminant can be easily generalized to a larger number of hidden layers and nodes.

When properly optimized, the functional form of the NN discriminant is intended to approximate the intrinsic likelihood ratio between the signal and background classes. The NN optimization is typically achieved by minimizing an error function that describes the expectation value of the squared separation of classes. In the case of two classes, signal and background, this minimization criterion can be written as:

$$f = E_s[(D - t_s)^2] + E_b[(D - t_b^2)] \tag{B.8}$$

where the t parameters are the target NN values and the E represent the expectation value, with subscripts s and b for the signal and background classes, respectively.

B.3.4. *Matrix-element techniques*

In cases where the input variables consist entirely of final state particle kinematics, the multidimensional probability density ratio may be directly known at fixed perturbative order through the analytic or numerical calculation of the corresponding matrix elements. Remaining effects due to higher perturbative orders, hadronization, and detector response can then be either neglected, or taken into account through analytic or numerical convolution. This technique therefore tends to work best when the correspondence between detector level and matrix element kinematics is good, i.e. where the detector resolution is negligible compared to the relevant features of the distributions. This tends to be the case in particular for the kinematics of charged leptons, and less so for missing transverse energy or jets.

B.3.5. *Boosted decision trees*

The use of boosted decision trees for classification is a method for estimating the local s/b through a series of decision trees. A decision tree is a simple method in which events are successively partitioned into two subsets using cuts on the input variables, and with each partition depending on the result of the previous partition. After events have been subdivided in this way, each subset of events, corresponding to a "terminal node" of the tree, is assigned a binary classification, or a real-valued response. In the context of multivariate classification, the terminal nodes would typically contain a real-valued response which estimates the local s/b based on the composition of training events in each node, with the structure of the tree defined from the training samples in order to optimize this estimate. Since the phase space regions which correspond to the terminal nodes are constructed from a set of simple cuts on the input variables, this corresponds to a piecewise-continuous estimate of the local s/b over a serious of hypercube-shaped phase space regions in the input variable space. For a single input variable, this would correspond to a histogram with variable bin sizes.

In practice, a single decision tree classifier constructed in this way is very sensitive to statistical fluctuations in the training datasets, and is not able to properly map the shape of the local s/b over the input phase space without significant overtraining. "Boosting" represents a practical solution to this, where the final likelihood ratio estimate comes from a weighted sum over a large number of decision trees. This tends to smooth out the effect of statistical fluctuations in the training samples on the structure of the individual trees. There are two commonly used and closely related boosting procedures. One procedure is ADABoost,[13] in which the trees are constructed by training each tree on a set of training events which has been reweighted such that events which are poorly classified by the previous trees are given higher weight. The second boosting method is gradient boost,[14,15] in which, for the classification case, the response F as a function of the input variable space $\bar{x}$ is optimized to directly minimize the negative log-likelihood ratio

$$L = -\sum_s \ln \frac{1}{1 + e^{-2F(\bar{x})}} - \sum_b \ln \frac{1}{1 + e^{2F(\bar{x})}}$$

where $F(\bar{x})$ is determined by summing the response over all of the trees and is a direct proxy for the probability density ratio in the input variable space such that

$$e^{2F} \equiv \frac{\mathcal{L}_s(\bar{x})}{\mathcal{L}_b(\bar{x})}.$$

B.4. Multivariate regression

While the description of multivariate techniques in Sec. B.3 is focussed mainly on the classification use case, there is another class of problems which frequently arises in high-energy physics which requires instead the use of multivariate regression, in which the goal is to build a prediction for the value of an unknown dependent variable as a function of the input variables. A common realization of this uses BDTs with gradient boosting[14] in which the decision trees and responses are optimized to minimize the loss function over the training sample

$$L = \sum \begin{cases} \dfrac{1}{2}(F(\bar{x}) - y)^2 & |F(\bar{x}) - y| \le \delta \\[2mm] \delta\,(|F(\bar{x}) - y| - \delta/2) & |F(\bar{x}) - y| > \delta \end{cases}$$

where $F(\bar{x})$ is the predicted value and y is the true value of the dependent variable for each event. This minimizes the square deviation in the core of the distribution, and the absolute deviation in the tails, in order to reduce the impact of outlier events. The parameter δ which controls the transition is typically set to correspond to about 1 standard deviation of the training sample.

A generalization of this procedure can be used to regress unknown parameters appearing in arbitrary likelihood functions, for example using BDTs to predict the response of $\mu(\bar{x})$ and $\sigma(\bar{x})$ in order to minimize the negative log-likelihood sample

$$L = -\sum \ln p(y|\mu(\bar{x}), \sigma(\bar{x}))$$

$$= -\sum \ln \frac{1}{\sigma(\bar{x})\sqrt{2\pi}} e^{-\frac{(y-\mu(\bar{x}))^2}{2\sigma(\bar{x})^2}}$$

in which the dependent variable y is assumed to follow a Gaussian distribution where the mean $\mu(\bar{x})$ and width $\sigma(\bar{x})$ are *a priori* unknown functions of the input variable space $\bar{x}$ which are determined in the course of the regression. This procedure has been developed in the context of the CMS Higgs$\rightarrow \gamma\gamma$ measurement,[16] in which a variation (using a two-sided Crystal Ball distribution in place of the simple Gaussian example) has been used for the photon energy regression.

B.5. Systematic uncertainty

Mismodeling of the signal or background rates and distributions can easily be the cause of measurement error or a hypothesis test that arrives at the wrong conclusion. It is the purpose of the systematic uncertainty estimation to ensure that the quoted uncertainties cover the expected deviations of the measured value from its true value and that the hypothesis tests do not yield incorrect results more frequently than the stated error rates. For example, a signal that is truly present should not be excluded at the 95% confidence level in more than 5% of repetitions of an experiment.

Handling of systematic uncertainty typically proceeds in the same way in an MVA-based analysis as it does for a non-MVA based analysis. The sensitivity of the predictions to sources of uncertainty must be evaluated and included in the statistical interpretation. Nuisance parameters are identified, one for each source of systematic uncertainty, such as the integrated luminosity, the trigger efficiency, lepton ID efficiency, jet energy scale, and so forth. Each of these parameters is varied one at a time and the impacts on the predictions are propagated to the rates and shapes of the MVA distributions. A nuisance parameter may affect both the rate (due to selection requirements) and the shape of the same distribution, and it may affect both the signal and the background, or affect multiple channels.

It is typically the job of the statistical tools package to interpolate and extrapolate the impact of systematic uncertainty beyond the cases explored by the analyzer. This process has its pitfalls. Incorporating the distortions from multiple sources of shape uncertainty is typically an easier operation when the distortions are small and are linear functions of the nuisance parameters. It is good practice for analyzers to consider the effects of non-linearity in multiple-sigma variations of nuisance parameters from their central values and to consider the correlated effects of multiple simultaneous shape distortions. If these are found to be significantly different from the linear approximations typically used, then custom fitting software may need to be developed. This concern also holds for non-MVA analyses.

B.6. Validation

It is possible that the set of systematic uncertainties initially enumerated by the analysis is incomplete. One or more additional effects may cause the predictions to diverge from the data. Sometimes the effects cannot be identified with the available data from the detector: two or more alternate hypotheses may explain the differences and it may not be possible to distinguish between them, or if a mixture is at work. Alternatively, the signal and background predictions may be correct within the statistical and systematic uncertainties and no deviations are present. Even in this case, a similar amount of work must be done in order to demonstrate that there is nothing mismodeled in the inputs or outputs of the MVA that could be reason to believe that the quoted systematic uncertainties may be incorrectly estimated.

An important validation method is to check the distributions of all of the input variables to see if the data are well predicted by the models, be they simulations, data-driven background estimates, or, more commonly, mixtures thereof. Deviations between the observed distributions and the corresponding predictions after systematic uncertainties are accounted for are reason to believe that additional uncertainties, and sometimes corrections, must be applied. This step must be done carefully, as the presence or absence of a signal, or its strength, can affect the outcome of the procedure. One does not want to absorb a signal into the definition of an uncertain background just to ensure that all distributions match with the background prediction.

This validation study therefore is most useful in cases in which the signal is either absent, or diluted so much as not to distort the rates or shapes of the validation samples. In the case of the SM Higgs boson searches, the signals are so small compared to the background that any variable except for the final discriminant variable may be checked for possible distortion due to mismodeling with only a tiny difference arising from the signal. Nonetheless, it is good practice to include possible signals in the validation step.

If a mismodeling is seen in the distribution of an input variable, it is best to understand the root cause of the mismodeling and correct it.

Mismodeling can be investigated by looking at different classes of events with similar objects in them to see if the detector response is well simulated by the MC. For example, if the jets in lepton+jets events have a deviation in some distribution from the prediction, a sample of inclusive jets or photon+jets events may provide some independent checks of the modeling of the detector response. It may also be true that the detector response is well modeled, but that a calibration obtained from one type of data was extrapolated to another type of data without adequate systematic uncertainty included due to the extrapolation procedure.

A last resort can be to correct one or more predictions, usually by reweighting the simulated events, in order to match the input variable distribution, though this is usually done using data in a control sample depleted in signal. If done improperly, not only can this procedure wash away a true signal, but if it is applied to more than one variable, then it can fix the same problem twice. One should check to see if the correction thus applied makes other variables' validation plots look worse or better.

If no satisfactory explanation for the mismodeling can be found, and if correcting the predictions to improve the modeling results in concerns of invalidating the analysis, then a prudent way forwards is to drop the variable from the list of input variables.

Correlations between pairs of input variables are also customary to check, although different kinds of correlation are possible. Multivariate analysis software packages often compute matrices of covariances between all pairs of input variables and these can be compared between data and the background simulation, or the signal+background simulation. As before, care must be taken as the presence of a signal may affect the covariances. Because absence of covariance does not imply absence of correlation, comparing two-dimensional distributions of one input variable against another, or comparing functions of two variables that reduce the dimensionality of the comparison may be useful. For example, it may be useful to check the modeling of the sum and the difference of two highly-correlated input variables, or even pairs of variables that are not expected to be highly correlated.

The distribution of the MVA method's output variable must also be validated. Typically this is done in one or more control samples. Events that fail the preselection requirements for one or more reasons can form a sample of events that can be used to compare predictions against observations using the same MVA function as used in the signal region. Typical cases include using non-b-tagged events in a b-tagged analysis, or by looking at neighboring regions in missing transverse energy, or by requiring a different number of leptons or leptons with a different charge combination from what the signal selection requires. Sources of mismodeling in these samples ought to be investigated and understood, and only as a last resort, corrected via reweighting. Uncertainties should be applied if necessary, though variations within already-considered uncertainties should not result in additional uncertainty.

Comparing output distributions in control samples, while ideally checking the MVA method using similar data and MC events as are used in the signal region, is also accompanied by challenges. The background composition in the control regions is usually rather different from that in the signal region. A good strategy is to form as many control regions as there are different sources of background, and to find a control region that constrains each background, one at a time. Control samples with different linear combinations of the background contributions are also acceptable and are usual, but they require extra work to find the right fit of the predictions to all of the control regions and the background-rich portions of the signal-region discriminant distributions.

The MVA function may not even be defined properly in the control regions. For example, events in a control region may lack a reconstructed object that is present in the signal region and for which measured parameters are input variables to the MVA. Two options are possible for handling this case — "typical" fixed values of the missing inputs may be supplied, or a new MVA may be trained without the missing variables. If the first approach is tried, different values for the "typical" values should be explored in order to check whether something is missed by picking a special value.

The distributions of the input variables also should be checked in the control samples as well. An additional validation of the MVA function is to check its distributions in bins of each of the input variables.

B.7. Optimization

One of the attractive features of MVA techniques is that the optimization is automated, given the choice of MVA method and the list of input variables. Simulated events from MC models of each source of signal and background are presented to the MVA method in ratios similar to those expected in the data for the backgrounds, though typically a larger fraction of signal events is supplied. It has been seen that many MVA methods, when trained with a different ratio of signal and background events, produce MVAs that are simply transformations of one another. Since an invertible function of an MVA discriminant is as optimal as the original discriminant, the net effect is not to change the optimality, though it does amount to rebinning the output variable, which can have an impact on the optimality if the bins are not also adjusted.

It may be useful to remove signal-like events from the background training samples in order to improve the sensitivity. It may also be useful to train an MVA with a series of input events representing different systematic variations of the nuisance parameters.

B.7.1. *Choice of input variables*

A typical analysis proceeds by computing many reconstructible variables — energies, invariant masses, angles between reconstructed final-state objects, missing energy, etc. Some of these variables may exhibit more separation of the signal from the background than others. Variables are usually constructed to take advantage of known properties of the signal and the background. It may be possible that the most powerful variables exploit characteristic features of the background more than those of the signal. Indeed, since the SM Higgs boson is a scalar particle, the backgrounds frequently have more complex kinematic structures than the signal.

Boosted decision trees have the advantage of being rather insensitive to the introduction of many variables that contribute little to the output discriminant, as they do not get included in the cuts in the tree. Other methods, such as likelihood discriminants and some NN implementations, may degrade in performance if many low-information variables are included. They essentially add random noise without providing discriminating power.

Many MVA software packages rank variables by their importance to the output. Unfortunately, there is no single algorithm for doing so. One may compute the coefficient of linear correlation between the input variable and the output discriminant. This strategy may over-rank many highly-correlated input variables which contain redundant information. Or they may be under-ranked, if the difference between two highly correlated variables in fact is the most discriminating combination.

Alternatively, one may remove one variable at a time from the list of input variables and train new discriminants and see how optimal each one is, compared with the full discriminant function. This too may misrepresent the importances of variables that perform well only in conjunction with other variables.

B.7.2. *Choice of MVA method*

The automated optimization procedures in popular MVA methods maximize the values of figures of merit that do not correspond to the best median expected limit in the absence of a signal, the best median expected discovery in the presence of a signal, or the smallest expected uncertainty on the measurement of a parameter. Instead, the figures of merit are chosen for the ease of the training, either because they can be differentiated analytically with respect to the free parameters inside the MVA functions, or because they are convenient to calculate based on the outputs. Because of this difference between what an MVA optimizes and what analysers desire, it is good practice to try several MVA techniques and choose the one that has the best sensitivity in the analysis at hand. Incorporation of systematic uncertainties may change the choice of the optimal method. While this approach may require significant computing effort and labor, packages like TMVA[17] allow automation of this process.

B.7.3. *Overtraining*

One possible source of poor performance of an MVA is inadequate numbers of simulated events presented to it for training. The MVA therefore does not have the opportunity to "learn" all of the features of possible events in all the combinations in which the features appear. On the other hand, if the MVA function has too much internal freedom — weights in an NN, or nodes in a decision tree, then it may "learn" the features of the specific events in the training sample. It may therefore perform better separation on the training sample events than a random set of simulated events not used to train the MVA. Such an MVA is said to be "overtrained."

One mistake to avoid is to use the same events to train an MVA and to compare against the data when performing validation or statistical interpretation. Typically a fraction of the available simulated sample is used to train the MVA and is called the "training" sample, and the rest, called the "testing" sample, is used to model the data. Packages like TMVA also produce comparisons of the distributions of the MVA output functions for the training and the testing samples. Deviations between the two distributions, especially in the signal-like regions which may be depleted in simulated background events, are indications of overtraining.

B.7.4. *Binning*

Each bin of a binned MVA output distribution is a cut-and-count search that can be combined with the others using the statistical interpretation methods of Appendix A. As such, for a reliable interpretation of the data in that bin, sufficient numbers of simulated signal and background events must populate it. A bin devoid of data events is not a problem for an analysis, but a bin that has insufficient background Monte Carlo events in it can yield incorrect results. A bin that has no signal prediction has less of an impact, though such a bin, if its background prediction also suffers from inadequate numbers of simulated events for the background prediction to be reliable, may cause the background rates to float in a fit improperly if the data do not agree well with the background prediction. This can occur with as few as one data event if the background is small enough.

In the limit that an MVA discriminant is binned extremely finely, then each simulated signal event and each simulated background event has its own bin, and most bins are empty. This case is very similar to the overtraining case of an MVA described above. The bins have "learned" the features of the individual events in the testing sample in this case.

Bins should be chosen so that the structure of the underlying distribution is visible, with enough simulated events in the high signal-to-background regions to adequately predict the data. One may attempt to optimize the binning of an MVA histogram by assigning uncertainties to predictions with small numbers of simulated events, though it is not clear what uncertainty needs to be assigned to a bin with zero events from a particular background source. Since contributions to the background may come from many sources, and one source may be poorly predicted while others are better predicted, it is best to bin them so that the contribution from high-weight MC events, or control-sample events in the data, does not have empty bins in the high signal-to-background region unless it is understood why this must be the case.

Unbinned MVA analyses are fraught with difficulties, as there is usually not an *a priori* functional form for the PDFs of the signal and background distributions. Smoothing Monte Carlo distributions to form the necessary continuous functions needed to perform the unbinned analysis is subject to problems of undersmoothing and oversmoothing, which can yield incorrect results. Multivariate discriminants have been used however to select events which are then used in an unbinned analysis using a more traditional discriminant, such as an invariant mass of a reconstructed system of particles.

B.7.5. *Optimizing Beyond the MVA*

It is often useful to give an MVA extra assistance in separating signals from backgrounds, especially if features of the data are known *a priori*, which can assist in the classification. A strategically-placed event categorization cut at the preselection stage can help the training process, and one should not expect the MVA to discover it automatically. Such categories may correspond to natural classes of events already inherent in an analysis — the number and types of reconstructed lepton candidates, the number of reconstructed jets, and the number and quality of *b*-tags, for example, are often used to divide events into categories for which separate MVAs are trained. The signal and background contributions in the different categories may be very different, even within an analysis, and it may give the MVA extra help to concentrate on only one category of events at a time.

After an analysis has matured, and all of the MVA optimization and validation techniques applied, it may be useful to revisit the event selection requirements and find ways to loosen them and create new categories of more loosely-selected events in order to increase the sensitivity. If, on the other hand, a signal has been well established, such as the SM Higgs boson signal, it may be more optimal to divide the high-quality samples into subsamples of even higher quality in order to de-weight events that have poorer resolutions or that reside in samples with larger systematic uncertainties, in order to reduce the expected uncertainties on measured parameters.

References

1. J. H. Friedman, A recursive partitioning decision rule for nonparametric classification, *IEEE Trans.Comput.* **26**, 404 (1977). doi: 10.1109/TC.1977. 1674849.
2. ALEPH Collaboration, Observation of an excess in the search for the Standard Model Higgs boson at ALEPH, *Phys.Lett.* **B495**, 1–17 (2000). doi: 10.1016/S0370-2693(00)01269-7.
3. DELPHI Collaboration, Search for the Standard Model Higgs boson at LEP in the year 2000, *Phys.Lett.* **B499**, 23–37 (2001). doi: 10.1016/S0370-2693(01) 00069-7.
4. L3 Collaboration, Standard Model Higgs boson with the L3 experiment at LEP, *Phys.Lett.* **B517**, 319–331 (2001). doi: 10.1016/S0370-2693(01)01010-3.

5. OPAL Collaboration, Search for neutral Higgs bosons in e^+e^- collisions at $\sqrt{s} \approx 189$ GeV, *Eur.Phys.J.* **C12**, 567–586 (2000). doi: 10.1007/s100520000286.

6. OPAL Collaboration, A measurement of $R(b)$ using a double tagging method, *Eur.Phys.J.* **C8**, 217–239 (1999). doi: 10.1007/s100529901087.

7. CDF Collaboration, Search for the Standard Model Higgs boson decaying to a bb pair in events with two oppositely-charged leptons using the full CDF data set, *Phys.Rev.Lett.* **109**, 111803 (2012). doi: 10.1103/PhysRevLett.109.111803.

8. CDF Collaboration, Updated search for the Standard Model Higgs boson in events with jets and missing transverse energy using the full CDF data set, *Phys.Rev.* **D87**, 052008 (2013). doi: 10.1103/PhysRevD.87.052008.

9. P. C. Bhat, Multivariate analysis methods in particle physics, *Ann.Rev. Nucl.Part.Sci.* **61**, 281–309 (2011). doi: 10.1146/annurev.nucl.012809.104427.

10. I. Narsky and F. C. Porter, *Statistical Analysis Techniques in Particle Physics*. Wiley-VCH, Weinheim, Germany (2014). ISBN 9783527410866, 9783527677313.

11. J. Neyman and E. Pearson, On the problem of the most efficient tests of statistical hypotheses, *Phil. Trans. of the Royal Soc. of London A.* **31**, 289 (1933).

12. R. A. Fisher, The use of multiple measurements in taxonomic problems, *Annals of Eugenics.* **7**(2), 179–188 (1936). doi: 10.1111/j.1469-1809.1936.tb02137.x.

13. Y. Freund and R. E. Schapire, Experiments with a new boosting algorithm. In *International Conference on Machine Learning*, pp. 148–156 (1996). URL http://citeseerx.ist.psu.edu/viewdoc/summary?doi=10.1.1.51.6252.

14. J. H. Friedman, Greedy function approximation: A gradient boosting machine, *Annals of Statistics* **29**, 1189–1232 (2000).

15. J. Friedman, T. Hastie, and R. Tibshirani, Additive logistic regression: A statistical view of boosting, *Annals of Statistics* **28**, 2000 (1998).

16. CMS Collaboration, Observation of the diphoton decay of the Higgs boson and measurement of its properties, *Eur.Phys.J.* **C74**(10), 3076 (2014). doi: 10.1140/epjc/s10052-014-3076-z.

17. A. Hocker, J. Stelzer, F. Tegenfeldt, H. Voss, K. Voss, *et al.*, TMVA — Toolkit for Multivariate Data Analysis, *PoS.* **ACAT**, 040 (2007).

Acknowledgements

The discovery and characterization of the Higgs boson described in this book is the result of decades of world-wide collaborative effort by a large number of scientists and engineers. While we, the authors, had the fortune to participate in this discovery at a critical phase, the science described in this book would not be possible without the dedicated efforts of the ALEPH, DELPHI, L3 and OPAL collaborations at LEP, CDF and D0 collaborations at the Tevatron, and ATLAS and CMS collaborations at the LHC. This is truly a collaborative discovery!

We commend the LEP, Tevatron and LHC machine teams for designing, commissioning and operating some of the most advanced tools for science. We thank Fermilab and CERN for hosting these machines as well as the experiments and the scientists associated with them.

We acknowledge with pleasure the crucial role played by the theory community in all aspects of the Higgs boson discovery. Not only did they conjecture this mass-generating mechanism, they also proposed various Higgs boson production mechanisms and have refined the Standard Model Higgs phenomenology to a remarkable level of precision. Their synergistic effort with the experimentalists has been crucial in extracting the most precise profile possible of the observed boson from the available data.

The role of computing in the event reconstruction and data analysis has grown with the volume and complexity of the recorded collider data. We thank the Worldwide LHC Computing Grid project for providing the reliable computing infrastructure that allowed analysis of petabytes of data, sometimes just a few weeks after its acquisition!

We most gratefully acknowledge the generous and enthusiastic financial support for this research from funding agencies, state, federal and private, worldwide. We are also thankful to millions of citizens of the world who cheered us on during this breathtaking chase. Their goodwill fueled the adrenalin which powered this hunt.

Finally, we dedicate this book to the memory of our colleagues who did not live to see the full impact and significance of their valuable contributions to the Higgs boson search and discovery.

About the Authors

Vivek Sharma is a Distinguished Professor of Physics at the University of California, San Diego, and a member of the CMS collaboration. He was the co-leader of the Higgs search group of the CMS collaboration in 2010–2011. Since 2012, he has served on the CMS Editorial Board overseeing all publication on the Higgs boson studies. He is a fellow of the American Physical Society.

Aleandro Nisati is a Senior Research Physicist at the Istituto Nazionale di Fisica Nucleare (INFN), Rome, Italy. From early 1990s to today his main scientific activity focused first on the scientific case for the Large Hadron Collider (LHC) at CERN, with particular emphasis on the search for the Standard Model Higgs boson; and then on the design and construction of the ATLAS experiment. With the start of LHC operations in 2009, Nisati served as ATLAS Physics Coordinator till the end of 2011. Since 2012, after the discovery of Higgs boson, he serves as the Coordinator of the physics studies in ATLAS for the luminosity upgrade of the LHC.

Printed in Great Britain
by Amazon

25802470R00272